Constructing Competence in Failure Analysis:

A Technical and Human Factors Guide

Debbie Aliya

President, Aliya Analytical

Fellow, ASM International

The Materials Information Society™

Koho Pono, LLC

Constructing Competence in Failure Analysis:
A Technical and Human Factors Guide

Published by Koho Pono, LLC
Clackamas, Oregon USA
https://KohoPono.com

First Print Edition 28March2024

ISBN: 978-1-938282-33-1

Library of Congress Control Number: 2023920564

"It seems extraordinary to me that [in Constructing Your Competence in Failure Analysis] you have combined human factors such as reasoning and thought that I have not seen addressed in any other book with technical-analytical content. I applaud this. You are an extraordinary engineer and… are incomparable.

- Jose Medina, Engineer and Seminar Attendee

"I enjoyed reading the book and getting deeper in your mindset for failure analysis. In addition to the mindset and technical parts, there is one huge contribution of this book for us, non-native English speakers… and that is engineering vocabulary. You have explained all the main terms widely enough to understand the background of the word. We all know that terminology between languages is not possible to exchange one-to-one with a direct translation. In my role working part-time at the local university as a researcher and teacher, I will always highly recommend this book to my students, to gain the proper language skills and mindset for their future in materials engineering.

Finally, let me congratulate you for capturing your lifetime of knowledge for this book. Thank you very much for including me as part of the review process."

- Libor Trško, PhD, Research Centre of the University of Žilina & Materials Engineer at Danfoss Power Solutions

"My compliments for the seventh chapter: it covers many crucial topics in real world Failure Analysis, through a series of Case Histories. I should mention that the style of the presentation transmits a vivid and very spontaneous spirit which is found in classical novels and literature. It is certainly not usual in technical books and that makes this contribution truly original and innovative."

- Dr. George Pantazopoulos, Technology and Quality Manager, ELKEME, Hellenic Research Centre for Metals S.A.

"Chapter 1 is great! I just read it in one go. Thank you for sending it to me. There are a lot of people that would benefit from reading that. I've already got some friends in mind that will be gifted the book when it's ready and others that will get a very warm recommendation to buy it for themselves on their company's money.

"It's exactly the kind of things young engineers need to read.

"We tended to be too strong on the math/technical side of things and not enough on the 'think, please' side of things for quite a long time, but this new generation is ready to go back to thinking for themselves (at least those I teach)."

- Véronique Vitry, Ph.D., Department of Metallurgy, Universite de Mons, Belgium

I dedicate this book to Dr. William T. Becker

They say, "Those who can, do'; and those who can't, teach." However, the original phrase by Aristotle was, "Those who can, do; but those who understand, teach." Dr. William (Bill) Becker could do, understand, and teach.

After meeting Dr. Becker in the late 1990s, through ASMI failure analysis programming committee meetings, and getting the opportunity to co-teach short courses with him, my practical learning really took off. Bill developed a unique method for performing fractography during his thirty-five years teaching at the University of Tennessee-Knoxville. This breakthrough allows interested practitioners to imagine the stresses causing the fracture *based on the shape of the fragments*. Bill's method integrates metallurgical and mechanical engineering concepts. Certainly, he was not the first to notice the basic facts of this type of macro-fractography, but from what I have been able to gather, he systematized these facts in a new way. He generously trained professionals, too, which helped so many of us failure analysts investigate more meticulously, thoughtfully, thoroughly and reliably. Bill's macro-fractography system facilitates interdisciplinary understanding between metallurgists and mechanical/design engineers. What I learned from Bill strengthens my confidence even when I face a PhD mechanical engineer who had created a computer model of the stress state and I need to tell them they were wrong about what forces drove a crack.

Bill died in 2006,[1] not long after publication of the 2002 Edition of the Metals Handbook Volume 11, which he had edited. He never got to write his own book on failure analysis so I dedicate this book to him.

---

[1] Go to the following URL and scroll down to find the entry: *https://mse.utk.edu/in-memoriam/*

Acknowledgments

To all of my colleagues from the ASM International Failure Analysis Society, I owe the breadth and depth of my knowledge base.

Working with the whole team that produced Volume 11-2002 was a huge learning experience for me. And since the Metals Handbook series is considered an authoritative source, being a section editor and author of four articles meant others thought that I was competent. This was very important to me, as I have always struggled with self-confidence.

The field of failure analysis investigations was fundamentally changed with breakthroughs by Dr. William Becker, ASM International Fellow. His work is documented in the ASM Handbook Volume 11, 2002 Edition.

Some time after Dr. Becker (Bill) passed away, I convinced my colleague Larry Hanke, of Materials Evaluation and Engineering in Plymouth, Minnesota, to teach with me. Larry grew up on a farm and his knowledge of the mechanical aspects of engineering underpin his skills in Materials Engineering. Larry came up with the question, "What is it about that ABC (screwdriver handle, cylindrical object) that makes you say XYZ (it's made of plastic, a forging)?" The visual inspection method that is the subject of Chapter 3 in this book would not have been so comprehensive and useful without Larry's insight and collaboration.

This brings us to Dr. David Levy, of Pepperdine University. His 1996 book, *Tools for Critical Thinking: Metathoughts for Psychology,* has revolutionized how I see the world over the 25 years that I have been working with it; this includes my failure analysis work. One day in 2022, I was looking for a link to his book to give to a colleague and saw that Dr. Levy had listed contact info. I reached out to him, labeling my email "From a Long-Time Admirer". He replied and set an appointment to talk. He told me that he thought that someone, someday, would use his ideas to write a book for another field, but it never occurred to him that it would be an engineer. Over the decades of using Dr. Levy's book, I have learned to see the specialty engineering aspects of my work more and more clearly. Perhaps more importantly for an independent consultant, there has also been a big improvement in my ability to explain aspects of HOW I KNOW what I know.

Without Bill, Larry, and Dr. Levy, nothing like this book could have existed. Even were he still with us, I could never express my thanks to Dr. Becker. Likewise, words are inadequate to thank Dr. Levy and my friend and colleague, Larry Hanke.

Also, I can never adequately thank my colleagues in India. How we came to meet began with a strange series of events centering around Jackie Earle (Thank you, Jackie! At the time Jackie was ASM Trustee and had invited me to give my failure analysis talk to the Peoria ASMI Chapter, her home chapter). After the talk, and on the way home, I stopped for lunch in Chenoa, Illinois. That's where I received an invitation to present the talk I had just given in Peoria to a group in Chennai, Tamil Nadu, India. I had already planned to attend a retreat in a rural town nearby and my presentation was scheduled to follow after that. Shankar Subburathinam, from the Chennai ASMI chapter, has been very encouraging to me. He arranged opportunities to test my teaching concepts in a culture very different from that of the USA. Shankar was the one who articulated what I knew, but not how to say it, "I am spiritually attached to India." Having the words means a lot to me; so, thank you much.

I am also grateful to Mr. Sampath Kumar and the other Chennai Chapter board members. A woman who attended my first Chennai seminar has become a colleague and lets me believe that I might be doing something helpful for my "sister metallurgists" in India. Thank you for keeping in touch, Valarmathi!

The Pune, Maharashtra ASMI Chapter was the first to invite me to speak, way back in 2008, and I thank Mr. Ravindra T. Kulkarni for listening to an early version of my Thinking Skill Optimization in Failure Analysis talk at the Pittsburgh 2007 MS&T conference and stopping to ask for my slides. My connections to the India ASMI Chapters all started there. I soon met Mr. Subhash Kulkarni, which eventually forced me to understand that there was a reason that Mr. Kulkarni in 2018 looked so different from my memory of 2007 (although the name Kulkarni sounds exotic to me, it is a common name in Maharashtra).

Ms. Ruta Bharve and Mr. Udayan Pathak, who serve on the ASMI Pune board, have also helped further my failure analysis teaching ideas. The original India Chapter, currently chaired by Sudhakar Bonde, has also supported my teaching development by hosting my seminar based on this book.

Finally, I would like to express my gratitude to my Indian business representative, Gautam Sharma, who has worked tirelessly to introduce my training material to a broader audience in India. He managed to arrange for me to go on a book promotion tour in January of 2023. I taught two-day workshops at Cummins and Bharat Forge in Pune, Boeing in Bengaluru, Karnattaka, and Mumbai, for the ASM India Chapter. Special thanks to Visesh Challa and his family, for hospitality and tour guide services while I was in Bengaluru in January of 2023 to promote the book. To invite someone who you have only ever met through LinkedIn and Meetup to stay at your house for 10 days is truly brave and kind. I am not sure what adjective I should use for my own acceptance of such generosity, but I feel spiritually and culturally enriched by the experience.

I can't leave the subject of my Indian colleagues without extra thanks to Shankar Subburathinam for introducing me to Dr. Sathya Prasad Mangalaramanan. Shankar arranged for me to present to Sathya-ji's group at Ashok-Leyland, in 2017. At the end of the two-hour presentation, Dr. Prasad joined me at the speaker's platform and informed his group with astonishment, "Madam has just given you a recipe for intuition!" Wow. I had never thought of it in those words, and I would use the word "methodology" rather than "recipe", but YES! That's certainly a big part of what I was trying to do. Hearing it articulated so thoughtfully gave me a great boost in confidence.

A thank you is also due to my former boss at the Cummins Corporate Technology Center, where I did a short stint (2011) in Columbus, IN through a contract firm. Dr. YC Chen was the first one to allot me regular time to hone my "Thinking Skill Optimization" training material for technical learners.

Other colleagues who deserve special thanks include Aaron Tanzer, who revolutionized my understanding of damage and failure as well as Steve McDanels, who shared Section 4 editing duties for Volume 11-2002 with me and Aaron. Having two other people who insisted that failure analysis can't be done by recipe was crucial to my professional development at that time. Aaron also helped me improve Chapter 5, on damage categories, for which I am very grateful. Aaron was also the mover and shaker who arranged a teaching gig for me with Bill Becker, at Westinghouse, where Aaron worked at the time. Bill and I presented two versions of our material: one geared to technicians and one geared to engineers. Over the years, Aaron and I have supported each other's missions, each encouraging the use of philosophy in failure analysis. At the same time, we argue over many aspects of the work, which, hopefully, leaves each of us in a position to think more clearly about the best ways to look at the world of engineering.

Marc Pepi is a member of ASMI, a board member of the Machinery Failure Prevention Technology Society, and an employee of the Army Research Lab in Aberdeen, MD. He encouraged me to refine and present my ideas on the human aspects of FA work, and was an early skillful editor of some of my writings.

He also introduced me to MFPT, which is where I developed the ideas for Chapter 7 (on Epistemologically Negotiated Project Scopes). I am very grateful, Marc.

Zayna Conner, who was on the ASMI Technical Books Committee, had the idea that I should write a Root Cause book. Apparently, several people on that committee thought I would be a good candidate to write that book. For various reasons, it has not worked out, but that was an impetus for me to get going on the larger book I wanted to write.

Reed Miller, of Industrial Heating Magazine (BNP), has been supportive since 2008, when we met at a heat treating equipment show. Soon thereafter, he facilitated a new failure analysis blog on their website – written by me. I don't think I would have had the courage to tackle writing Chapter 8 if I hadn't organized the content mentally over the years through this blog. Thank you, Reed. Bill Mayer, who manages the monthly magazine aspect of the business, arranged for me to get my first regular PAID technical writing gig. THANK YOU, Bill!

The Metallurgy and Materials Engineering curriculum at Carnegie-Mellon in 1979 included half of a semester on fracture analysis, and I definitely benefitted from Dr. Anthony Thompson's teachings on the theoretical aspects of why materials and components get fragmented. Dr. James Williams' classes on physical metallurgy were also extremely interesting to me. I must not overlook the fact that my international outlook on life, and devotion to the work required for inter-cultural appreciation, would not be nearly as well developed had Dr. Chuck Bauer not inaugurated an exchange with the Swiss Federal Polytechnic (or École Polytechnique Fédérale de Lausanne) where I traveled to be part of the second class of the program in 1978. I would be remiss to leave out Dr. David Laughlin, whose metallurgical thermodynamics class eventually laid the foundation for many of my social theory ideas and provided opportunities to practice applying concepts across the boundaries of disciplines. To me, this is a powerful and still sadly lacking problem solving tool. Dr. Mel Bernstein, who was my freshman advisor, encouraged me to keep going with my studies when it all seemed too much.

It has become my passion to avoid giving admonishments to people about how to do, or not do, failure analysis work. Rather, I have sought ways to teach people *how to* avoid the snares that await the newbie or inattentive failure analyst. Dr. Chuck White, retired from Kettering University (also an ASMI colleague), helped me articulate the idea of making technical seminars a coaching process that imparts skills, rather than a chunk of time that lets the student go home with "a sore butt and a certificate". Thanks, Chuck!

The next group of people deserves appreciation for reviewing ALL 9 chapters of this book. Dr. George Pantazopoulos was the first person who read and truly understood the multi-level purposes of the book. I can't say how important this felt when I had gotten several reviewers in a row who just could not understand why I would talk about my mother buying a step-ladder, or baking a cake in an engineering book. Hello! Commercial bakeries as well as the companies who make commercial equipment employ engineers. Thank you, George, from the bottom of my heart. Danny Brinkley responded to my announcement looking for reviewers at MS&T 2021 and gave me feedback on all nine chapters. Thank you, Danny.

Jake Auliff, another ASMI colleague, recruited several of his group members to read the book, and one actually finished it and gave me helpful feedback. Thanks to you Jake, and to your team, and particularly Dr. Libor Trsko.

Veronique Vitry, Bill Rossey, Jose Medina, KC Santosh, Visesh Challa, Pierre DuPont, Ravi Putrevu, Chanchal Vyas, Zanele Nxumalo all read one or two chapters, and provided feedback. I must thank Brian Horstman for noting what would have been an embarrassing error in the translation of a common acronym. Each of these people lifted me up at a critical point.

Dr. Al Segall, of Penn State University, was a late reviewer of Chapter 1. He liked the approach so much he volunteered to help with the second edition. We shall see!

Dr. Kip Findley, of the Colorado School of Mines, kindly reviewed Chapter 8, and motivated me to do a major clarification of what is now the New Three-Level Fractographic Analysis Method. The revision forced me to improve my argument about why all the materials and mechanical engineering faculty members who teach failure analysis, statics, mechanical behavior and other related subjects should check it out. Thank you, Kip!

I cannot leave out the members of my two writing groups. I joined the first one around 2014. Charlott Perry, who helped me to clarify how to explain loading geometries to non-engineers, deserves a nod. My eternal thanks to the founder of Mid-Michigan Word Gatherers, Karen Bota, whose devotion to helping others express themselves is unsurpassed. Cathy LaPointe-Blundy, another group member, and her husband, Paul (who worked as a heat treating furnace operator for many years), read most of the book, and assured me that they could follow it. Cathy, Paul and Karen are my closest friends in beautiful Ionia County, Michigan. In 2018, attorney Christine Yared started a Grand Rapids, Michigan non-fiction writing group through Meet-Up. My special thanks to Chris and fellow members Phil VanHuffel, Sonya Vann DeLoach, "Mighty Joe" Stankowski, Skyler Lambert, Cedar Leverett, and Sterling Hawkins.

There were numerous additional people who were not interested in reading a draft, but told me "I want to buy a copy when it's done!" One who must not be overlooked is Dr. Roch Shipley, who co-edited the 2002 version of the *Metals Handbook Volume 11*, and with whom I co-chaired what was then the ASM International Failure Analysis Committee. Another acquaintance, Mel Weaver from my creative writing group told me: "I'll take two!" I replied, "But Mel, they're going for like $\$175^{\underline{00}}$ each." He repeated, even more enthusiastically, "I'll take two!" It's heartwarming to have someone express such support!

Steve Lampman, technical publication staff at ASM who shepherded Volume 11-2002, helped me build an understanding of the patience, persistence, and perseverance required to get a reference-quality technical publication into print.

Scott Burr and Dayna Hubenthal of Koho Pono have created a truly unique publishing company. Even after being published in multiple peer-reviewed and professionally edited journals, conference proceedings, and handbooks, I have been completely surprised by their editing process. I selected this press for multiple reasons, but their assurance that they would keep my style, along with important business considerations, were paramount. Scott took somewhere around 15 hours to read the 50 page, 21,000 word draft of Chapter 4 ALOUD to me, so we could improve the flow and clarify the content. At the end of that particular week, I felt like a different person as I looked over the submissions of my fellow non-fiction writing group members for our upcoming meeting.

Peter Ried introduced me to Scott and Dayna of Koho Pono. He also helped me develop the Systems Thinking aspect of my Thinking Skill Optimization methodology. Peter has been an especially supportive colleague over the last 12 years.

My parents, Rita and Gerry Rosenkrantz, raised me, gave me a set of values and established standards for me. They supported my engineering and humanities education. My mother passed the CPA exam in West Virginia in 1956, the youngest person ever to do so at the time. She got her law degree and was sworn in to practice law in Washington, DC and Maryland in 1966. In doing so, she set an example for women breaking professional barriers. My dad kindled my early interest in science. I got a microscope for my 5th or 6th birthday.

I would also like to acknowledge my ex-spouse, Daniel Aliya, who discovered the already-mentioned Levy book as well as the Michael Gelb and Donella Meadows books, which will be introduced later in the body of this work. These books are truly the shoulders of the giants that I stood on to be able to write this opus. Back to the ex: Having a failed marriage clearly also provides, as Ram Dass used to say, "grist for the mill" of future understanding and compassion. This is especially true for the philosophically-minded failure analyst. As the great Rumi is said to have written (paraphrased): Let the darkness be your candle.

I have tried to make the book as clear as possible. Even as I use metaphor to communicate ideas, when writing technical content, I always make every effort to avoid both mistakes and language that could readily be misinterpreted. Any oversights and mistakes are my responsibility. For better and worse, that's what it means to be "the author".

Table of Contents

PREFACE

During 38-plus years in the business of analyzing failures, I have learned that methodologies, concepts, and critical and creative thinking are more powerful in satisfying clients and courts than following a recipe for a procedure. With practice, this book will help you learn to see the situational reality in a failure analysis project more clearly.

While this book has wide-ranging application as far as the concepts go, most of my case studies are from medium technology industries. The materials in question are those commonly used for structural components. I don't confine myself to large civil structures, such as bridges or buildings. I include any component whose primary function is to either transmit a load or "sit there" and protect an assembly or device (such as a housing). Many of my examples will be from machinery, either mobility related or equipment that is used in a manufacturing process.

My work has also included field returns for warranty complaints and durability test failures. Sometimes I have been asked to evaluate the characteristics of durability test successes. Maybe the component or assembly was tested well beyond the minimum durability limit and the company wanted to know the details of high-performing material characteristics that went the extra mile. In every case we must interpret the data. Fracture is inherently a process that may have a high degree of natural variation. Understanding fracture and resistance to fracture in an integrated way includes some basic understanding of probability and statistics. In Chapter 7, we will explore the issue of "Why THIS part broke at THIS moment".

Some of my projects were part of litigation issues. Most of these projects offered excellent opportunities for me to sharpen my technical skills. As costs or consequences of a failure increase, a larger budget is generally provided to determine what happened in greater detail than usual. Failure analysis work that is supervised by the courts is different from routine work in that it is often required to assign blame. In contrast, for routine commercial work, often the important objective is to be able to move forward with a lower chance of repeat problems.

While the examples presented in this book belong to medium-tech industries, they do not limit the power of the broadly applicable concepts presented in all of the chapters. The concepts inherently liberate you to apply them in wholly unique ways to a variety of novel situations in any level of technology, but also in your life. With diligence, you will learn to question the context of the situations you face. You will learn to think more clearly. Keep this frame of reference in mind as you read the book. Discover how constructing your competence is a conscious act which involves both growing your technical competence and growing as a human.

INTRODUCTION

This is not a standard engineering textbook or technical guide. That is because great Failure Analysts are competent beyond technical knowledge. They have important capabilities for clear thinking, for provocative questioning, for making effective decisions, and for skilled communications in addition to the failure analysis body of knowledge. The information in this book is intended to strengthen the reader in all of these competencies. That is what makes this book distinctly different from engineering texts which focus exclusively on technical details. In your embrace of these competencies, you will be transformed and the book will become your true companion, whether you are still in school or a lifelong learner with decades of experience.

The competencies of thinking, questioning, deciding and communicating remain distinctly human, especially in a failure analysis. Some of my colleagues believe strongly in Artificial Intelligence (AI) and automation but to me, being a human is a powerful advantage.[2] We are flexible. We question context and motives. We have a sense of justice and fairness. We have a sense that we are mortal and we understand that our actions, and those of others, have consequences that could be detrimental if we are careless. We were all humans prior to choosing a profession. Even if our parents forced us into a profession, and knew they would bend us towards that even before we were born, it is meaningless to practice engineering if you are not a human. The meaning of a competent human's work can be sensed and understood by another human beyond a failure analyst's report. That nuance can be very valuable in understanding failures, which helps others assign responsibility in the right measure.

Maybe the abuses of unfettered capitalism have rendered toxic the idea that engineers are humans first. Maybe some people now prefer to come in, do their jobs, and go home. It is a bit baffling to me that anyone would have this reaction. It is valuable to be curious, especially regarding what others may consider mundane details. I just heard about a movie where company employees were offered the chance to get a chip implanted in their brains so that they never thought about home when at work, and vice versa. If this is your ideal, you are not going to be very successful as a failure analyst – not by any meaningful definition.

One reviewer said, "You have included what is often missing from our educational system that has come to focus on tasks and knowledge. We need to learn to look at the bigger picture."

Being well-rounded or acquiring a "liberal education" used to be the goal of a college education. Constructing your competence in failure analysis requires more than a technical education, because failure includes human error and may include deception, hidden information, or fear of consequences. All of this and more requires the analyst to stand in reality and have correct and defensible results. Seeing the big picture is critical to being able to spot inconsistencies, and thus, to experiencing success.

Another difference between this and standard technical books is the presentation of stories. In an engineering class, good instructors provide the stories that deepen our understanding of concepts. The stories in this book are ones I wish I had known much earlier in my career. In this way, this work is both a textbook and the instructor. Thus, this book offers special value for use in engineering failure analysis classes with actual live instructors, because it's always helpful to have more than one mentor when studying interdisciplinary fields.

[2] NIST.org; There's More to AI Bias Than Biased Data, NIST Report Highlights: Rooting out bias in artificial intelligence will require addressing human and systemic biases as well. March, 16, 2022; *https://www.nist.gov/news-events/news/2022/03/theres-more-ai-bias-biased-data-nist-report-highlights*

When constructing your competence, you gradually acquire a thirst for comprehensive knowledge of factors (physical, mental and organizational) that lead to failures. These empower you to understand and communicate the means for averting those preventable factors.

Stories have been used for tens of thousands of years as a way of warning, educating, and entertaining, often all at once. Humans are considered to be storytelling animals as much as tool making or rational animals. When I tell a story, since this is a technical book, you can rest assured that there are multiple extractable lessons available to the careful, attentive reader. Sometimes you will have to think about the story for the deeper lessons to reveal themselves in a useful way. The famous story of the blind men and the elephants, which shows up in Chapter 7, has continued to reveal deeper layers of wisdom to me for several decades.

We were all humans before we became engineers and that will be your advantage. The purpose of this book is to improve your critical, rational and creative thinking skills, to improve your ability to question, to improve your decision-making and to improve your communications, no matter your current expertise level. This will help you on your path to becoming a great failure analyst.

WHO IS THIS BOOK FOR?

The content of this book will be of use to those

- doing failure analysis work;
- specifying or negotiating a failure analysis, but not directly performing the physical evaluation of the project;
- reviewing a failure analysis report.

Others may also benefit from this book:

- Engineers with a background in materials or mechanical engineering will find the book useful and empowering both technically and for the so-called soft skills.
- Materials lab technicians will find it useful when communicating with engineers.
- Maintenance engineers (these engineers usually have a broad and deep interdisciplinary background) especially when negotiating failure analysis or doing some themselves!
- Products liability attorneys will find this information helpful when communicating with expert witnesses.
- University engineering professors will find it useful to teach failure analysis classes.
- Those working in a manufacturing environment who have failures in their equipment or products; including quality engineers and technician colleagues.
- Those working as a design engineer, who must address testing failures.
- Forensic investigators and civil engineers can benefit significantly from the book.
- Even purchasing agents for components and materials can find information here to communicate more effectively with both internal engineers and suppliers.

In fact, this book explains how to figure out why some suppliers reliably produce quality products, while others reliably produce what is effectively junk. Because companies are choosing new suppliers to strengthen supply chains, this is important. Often, a longstanding supplier makes necessary adjustments over the years to meet requirements, even though these features are not specified on any print. If a new supplier works to print, they may produce a product that is found wanting, but not through any fault of their own.[3]

With such a broad range of backgrounds served by this book, some content may seem superfluous to some and incomprehensibly difficult to others. But remember, start where you are! You will soon gain skills and understanding despite your initial level. The book will be useful to anyone who is curious and is willing to embrace their vulnerabilities as a beginner. Why did it break, get deformed, corrode, or wear out? Learn to ask better questions, in order to get better answers.

> *Here, I have a confession. I received very poor grades in my first two semesters in engineering school. I almost flunked out. Over the years, I have strengthened my understanding into a useful working knowledge of the topics that were problematic for me. I now find that these are the concepts that are most useful when explaining ideas to attorneys, technicians from other fields, and mechanical or electrical engineers. My poor grades – my vulnerability – gave me the determination to solidify a deep understanding and help others who may not have been the stars of their class on these topics.*[4]

Conceptual understanding is the key to doing competent failure analysis work. So, slow down. Read. Think. Come back to this book over the years. Through practice, your thinking will be clarified and your ability to evaluate creative new ideas for merit will improve. I believe that those who read and reread will gain new insights each time but especially over a long career.

The best way to master thinking skills and failure analysis techniques is to practice, practice, practice, and then teach, teach, teach! Teach your colleagues. You don't need to become a professor to benefit from the practice of having to clarify your thoughts enough to be able to explain them to others.

Please relax as you read the book. Some things are repeated, or expressed in a conversational and perambulating (wordy!) manner, to allow time for the ideas to sink into the reader's pre-existing mental structure. Some of my sentences seem long and full of "interruptions". Pay extra attention to these passages. This degree of precise communication (even to ourselves) is typical during failure investigations. Our goal is to integrate what we're learning with what we already know. Siloed knowledge is much less valuable than integrated knowledge.

Exploring Key Branches of Knowledge will Compound your Competence in FA

To gain high level competence as a failure analyst, you will need to explore many branches of the tree of knowledge. Cultivating your curiosity in a broad range of technical and human factors fields will facilitate both the confidence and competence required to excel in your failure analysis work. What are the most important relevant technical background subject areas? We will cover these later in much more detail.

1) A Basic Theory of Knowledge: How to develop the ability to question what we think we know.
2) Mechanics of Materials: How loads and forces affect components and assemblies.
3) Chemistry and Thermodynamics: How a component's environment and composition affect its behavior.
4) Testing and Analysis: How to develop the ability to determine and reveal chemical, mechanical and physical characteristics as well

[3] Many companies now use Source Control Documents as the place to document materials and process information for purchasing, engineers and vendors to capture knowledge so it is not lost. That doesn't help when the manufacturer is holding proprietary knowledge.

[4] Editor's note: The author, Debbie Aliya, is a Fellow of the American Society of Materials International, acknowledging her distinguished contributions to the fields of materials science and engineering, especially for teaching the use of critical thinking in failure analysis. | https://www.asminternational.org/2019-fellows

as the macro, meso, micro, and crystal structures that influence them.

5) Basic Materials Engineering: How to determine the relationship between the processes used to manufacture a Subject Component, the nested layers of crystal, micro, meso and macro structures created by the process, and the resultant associated behavioral characteristics.
6) Human Factors: How to process visually, to communicate, to perform failure investigation work in an ethical manner, and to leverage general critical and creative thinking skills.

Take time to think while facing uncertainty. When you are exploring new ideas, give the mind time to see reality in a new light. Acknowledging the value of this fuzzy thinking that author Michael Gelb called "Sfumato[5]" or smokiness, is a key aspect of creative thinking. Learning to tolerate uncertainty is an important skill leading to developing the benefits of Sfumato and all the other thinking skills presented here.

It's important to understand that there will never be an end point where you know all there is to know about these branches of study. Following up on your continually cultivated curiosity will reveal new opportunities to find interesting work.

Reiteration: The Path to Comprehensive Knowledge – Practice and Engage

Some of the mental tune-up lessons have broad applicability in life; but most of the specific content is geared to participating in failure investigations. Over time, practice of the techniques presented here will allow you to acquire far more comprehensive knowledge of all the physical, mental, and organizational factors that lead to failures. Participating in this profession will strengthen your ability to provide useful prevention advice. That is part of what "Constructing Competence in Failure Analysis" is about.

There are many resources available on the subject of failure analysis. Even if we confine ourselves to failure analysis of structural components using the body of knowledge acquired by mechanical and materials engineers, there is an overwhelming amount of published content. See my personal list of must-have resources at the end of Chapter 1.

This book contains additional differentiators from other technical books on failure analysis:

- Inclusion of real-life examples and specific tactics rather than general admonishments,
- Inclusion of practical, actionable methods to develop good communications and plans,
- Inclusion of lots of hints and tips (this book is both the instructor and the textbook),
- Inclusion of exercises to cultivate clarity of thought – the foundation of competence!

Inclusion of Real-Life Examples and Specific Tactics rather than General Admonishments!

Targeted tactic: I highlight different aspects of the failure analysis process (with their relevant conceptual basics) in the context of that specific step of the analysis.[6] Together, we will tackle the nitty-gritty aspects of failure analysis such as how to figure out where and how to cut specific types of parts, how to preserve and photo-document evidence, and case studies. These topics have been left out of most of the literature.

> *Early in my career, I was involved in a litigation project where well over 1000 photos were taken of the Subject Component. This was well before the days of digital photography. There were rolls and rolls of film shot and developed and copied and shared. Not one showed a view of the curvature of the deformed tubular component at the damaged area before it was removed from the service environment. This turned out to be critical in determining how wide the crack*

[5] Michael J. Gelb, *How to Think Like Leonardo DaVinci: Seven Steps to Genius Every Day* | ISBN 978-0007323821

[6] For working engineers doing failure analysis work, this book is supplemental to Failure Analysis references publications with a more detailed technical foundation.

was in service. How wide the crack was in service was critical to understanding the consequential damages due to the presence of the crack and how much of the toxic material leaked through the crack.

Most references simply command the reader to preserve evidence. This is what I mean by an "admonishment". This book presents specific techniques to preserve evidence and how to perform a visual examination. I won't tell you to, "Look carefully at all aspects of the environment and take pictures from multiple angles." That is way too vague. It's not actionable for a young engineer. It won't reliably produce a desirable result! It is an admonishment that only top experienced people could potentially implement.

Inclusion of Practical, Actionable Methods to Develop Good Communications and Plans

Another tactic: lay a foundation with communications skills so you are able to negotiate the resources for an investigation appropriate to the circumstances. This involves helping everyone ask better questions.

I just prepared a project proposal where I offered a ***fracture*** *analysis (which would provide information about the physical causes of the cracks) and a* ***failure*** *analysis (which could provide information about preventing future fractures). This explained the value of the more expensive failure analysis over the lower cost, but perhaps less economical, fracture analysis.*

In Chapter 3, we explore a method to describe and document the appearance of items that are the subject of our analysis. We do this with specific photographic skills and we do it with words. But words are always approximate and even photos can fool the eye. By choosing certain words to describe parts, we can maximize the chances of *Getting to Yes*[7], building support for the project and its conclusions, and doing so from the earliest stages of the investigation. Words that are truly descriptive of features that are visible to anyone with functional eyes enables starting with agreement. By beginning with a shared reality, we inspire faith in our commitment to facts. We may even reduce the fear that we're favoring one side over another.

This Book Is Both Instructor and Book

Another tactic: this book is a dialogue with the reader. I try to convey the hints and tips we failure analysts need to watch for and avoid as we go about our jobs. I am going to provide the types of pointers I wish I had been given when starting out and even later in my career. Even after thirty years as an independent consultant performing failure analysis, I often work on products for the first time.

It wasn't until I had worked on a few piping corrosion failures that I realized how important it is to record the orientation of the pipe. Was it horizontal or vertical? If it was horizontal, the 12 o'clock position (or whatever is accessible) should be marked.

Knowing how to preserve evidence is a lifetime task; especially when we are given a small component from a larger investigation. We have to work with the people who were on the scene to try to get the best background information we can.

Actionable Practice for the Cultivation of Clarity of Thought: The Foundation of Competence!

Another tactic: I am annoyed by the common advice given these days, "Be sure to use critical thinking." It's annoying to me because I have never seen or heard that advice followed with any useful strategies for accomplishing it. In this way, this book is different.

There are many resources on critical thinking available. However, I don't know of any specifically for engineers, except *Pragmatic Thinking*

[7] Roger Fisher and William L. Ury, (Bruce Patton was a co-author for later editions.) *Getting to Yes: Negotiating Agreement Without Giving In*, Penguin Group, emphasizes use of objective criteria. Starting with basic facts that would make a disagreement look silly is a good way to set the stage for the rest of the investigation.

and Learning: Refactor Your Wetware, by Andy Hunt[8]. This was written by a software engineer for software engineers. It's an excellent book and I recommend it for other engineers as well.

Maybe critical thinking books for engineers have been lacking because we already believe we know how to think critically. Isn't that why most of the critical thinking text books out there these days are used in freshman English courses?

I've been working with David Levy's *Tools for Critical Thinking: Metathoughts for Psychology* for over twenty-five years. It's a great book because we all must use our minds in order to do failure analysis – the content is definitely applicable to what we do. But there's a lot of jargon that engineers don't necessarily absorb very easily. And there are a fair number of examples that he uses that might be a bit embarrassing for the typical engineering personality type to discuss!

I liberally draw from these sources and my own discoveries in thought clarity to provide you with insight.

Constructing Competence in Failure Analysis represents a distillation and simplification of critical, rational, creative, and systems thinking techniques as well as decision science, communication hints, and advice to commit to a lifetime of learning, specifically as they apply to analyzing failures and interpreting the data for the purpose of resolving a problem with a commercial product or civil structure. These techniques can be used within a single company, in a dispute between companies, or a litigation matter that has entered the public arena.

Wishing all readers enjoyable and challenging use of this practical guide to failure analysis.

---

[8] Hunt, Andy; *Pragmatic Thinking and Learning: Refactor Your Wetware* (2008); Pragmatic Bookshelf; 251 pages, Paperback. | ISBN 9781934356050

CHAPTER 1

What Is Failure Analysis?

An Omni-Disciplinary Approach to Failure Analysis

1.1 Failure Analysis Knowledge and Professional Competence

In the simplest sense, failure analysis is a process to determine what went wrong. In this book, we'll be studying engineering failures[9]. What stands out in this book, compared to other engineering textbooks and guidebooks, is that we will work together to facilitate the integration of engineering knowledge with your knowledge of what it means to participate in the human search for professional competence.

I find that stories are powerful as a communication and teaching technique. A few of my colleagues have told me through the years, "Cut to the chase. We don't need all these irrelevant stories." While I understand their impatience, a technical person rarely achieves robust competence without also evolving other human capabilities.

Albert Einstein[10] believed in the power of stories as the foundation of developing intelligence. He said, "If you want your children to be intelligent, read them fairy tales. If you want them to be more intelligent, read them more fairy tales."

Over the years, my evolution as a failure analyst has depended heavily on learning from experience, practicing particular skills so that I could learn to think more clearly, and framing the lessons that I learned into stories.

One young failure analyst told me, "I skipped all the stories." Well, this person missed the whole point of

[9] Analysis of component failures and failures associated with mechanics of materials

[10] Albert Einstein, half-length portrait (1931) created by Doris Ulmann. 1882-1934, photographic platinum print, Reproduction Number: LC-USZC4-4940; Call Number: Unprocessed in PR 13 CN 1975:127 at the Library of Congress Prints and Photographs Division Wash., D.C. USA; https://www.loc.gov/pictures/item/96522310/

the book then. Don't let that be you. Each of these stories illustrates a key point that I learned. Even if you are not going to (or did not) learn it the same way, **the story reveals that real knowledge comes in the context of experience**.

This issue of context and its importance is addressed at length in David Wootton's <u>*The Invention of Science*</u> *(2015), where he tells the story of a conversation that the great astronomer Kepler had with his wife about a salad she was making as he was trying to interpret his findings regarding planetary motion.*[11] *The early scientists all told stories about what led them to do the experiment that they did in the way they did it. Later, stories were stripped from scientific reports. I find that helping people understand how progress in thinking was made is a very important ambition. Maybe it's time to turn the tide and welcome stories back into technical learning! Humanity needs imaginative scientists, not only those who don't mind dry, boring, and repetitive work.*

Furthermore, I doubt that many math teachers in elementary schools or even engineering teachers in college believe that word problems, which are essentially stories, are useless.[12] *Why would failure analysis stories be useless then?*

One of the most popular events at ASM International Failure Analysis Society Conferences turns out to be the late-night discussions at the bar where we share "war stories". Some of it is essentially gossip; we talk about who made this or that stupid or genius argument in a court of law somewhere. But much of it is useful. It serves as a lesson. A peer's story about a subtle physical metallurgy fact helped us understand physical and mechanical metallurgy in a new way. He helped us become better experts. Stories teach at multiple levels.

Therefore, each of the stories in the book illustrates a key point that I learned, and more importantly, how I used it to think more clearly. The stories frame the lessons that I learned and you will be the beneficiary of this life long effort.

The other main difference between this book and other books is the integration of my new seven step method of Thinking Skill Optimization. These techniques are keys to compounding your competence in any field.

Since failure analysis is so interdisciplinary, even the most skilled practitioners have never mastered the entire discipline. Those of us who choose to work in this field have many opportunities to challenge ourselves to gain more complete levels of clarity of thought so it is easier and easier to learn and apply new facts and theories as we gain maturity.

Do you want to avoid avoidable dementia? Use your brains. Reinforce your thinking skills every day with the seven techniques that will be demonstrated at as many points in the text as I can.

This is a large book. It is likely more content than most students will read in a single semester. But keep it with you. Reading and thinking about the content will continue to pay dividends throughout your career.

I have been working with and teaching Thinking Skill Optimization for 25 years now. After about 10 years, I started noticing big bumps in understanding how to use the techniques in more and more powerful ways. The lesson: practice brings accumulated gains that have continued to compound over the years. In a way, I am glad that I procrastinated completing the book for 20 years. If it had it been completed earlier, it would not have this evolved and powerful methodology of Thinking Skill Optimization presented in such a useful and integrated way.

---

[11] Kepler's wife pointed out that random events happen but do not always achieve the most desirable outcomes. *The Invention of Science: A New History of the Scientific Revolution*. By David Wootton. 2015. Harper Collins, New York. 784 pp. (pages 263-265)

[12] Math educators use stories to improve word problem outcomes. Web search: word problems and the relationship to stories | education leaders deploy story-based problem solving to improve outcomes.

There's no substitute for practice; the practitioners of this method, which focuses on critical, creative and systems thinking, will be rewarded.

Some professors and mentors bemoan, "The youth of today do not having critical thinking skills." What do they really mean? Most people who have answered when I asked them this very question have understood my unspoken point: They mean that they disagree with the results of the thinking activity performed by the "non-critical thinker". But disagreement over outcome does not prove bad reasoning.

Here, we'll give specific definitions for critical, rational, logical, and systems thinking accompanied by exercises to refine the use of each of these related mental practices.

As for traditional technical content, data collection and evaluation, these are still key activities and will be covered in this book. Setting up a plan to ensure that the needed data will be produced is the first of two underappreciated skills required for excellence in failure analysis. The specific data collection methods used will definitely affect the analyst's ability to evaluate or interpret the data.

If I am constrained to emphasizing two guiding principles for failure analysts, mastering scope negotiation and evidence preservation might be useful choices.

- Learning how to **preserve evidence** is a lifetime task. It is important to work with the people who are on the scene to try to get the best background information we can. And sometimes information is unnoticed by the on-scene experts. That is when we need to know how to preserve evidence as well as how to explain to others how to preserve evidence when we can't personally supervise the work.
- **Scope negotiation** is difficult if you are at the bottom of company hierarchy. We provide some help on this topic in Chapter 7. Studying the provided guidelines will help you understand what is involved in getting the resources you need
- Experienced failure analysts will find that Chapter 7 presents a way of thinking about what to ask for, and maybe even what to demand, if you are in a position to do so.
- If I can't get the resources that I need to do what I think will be a competent job, I have occasionally turned down the work. Not everyone has this option.

1.1.1 Evidence Preservation is Crucial. Don't Be a Robot! Be a Human!

Since evidence preservation is such a key aspect of failure analysis, let's start with a deeply important, yet simple example of how using a *methodology* is a better approach in failure analysis than using a *recipe*. We will come back to this topic several times.

> EXAMPLE OF A METHODOLOGY RATHER THAN RECIPE: *I was pleased to hear the word methodology on the radio during the final editing stages of the draft of this book. Ketanji Brown-Jackson was being grilled by the United States Senate Judiciary Committee members in anticipation of her confirmation as the newest Supreme Court Justice. As the senators asked her for her judicial philosophy, she reportedly stated that she did not have one; but she had a methodology.* [13]
>
> *A few days after I heard her describe her methodology, I realized that by its very nature, a methodology is philosophy in action. After all, logic is a part of philosophy, and so is epistemology. Even reasoning has been considered a valid topic of philosophy. As all recent US Supreme Court nominees do, she avoided saying that she would favor one* **political** *philosophy or another.*

A methodology means that you are not following a set of simple rules. Rather, the existence of a methodology means that you have considered both logic (If A then B) and a theory of knowledge that

[13] We look at David Hume's thoughts on reason in Chapter 9. Evaluation of data is an important activity for Supreme Court justices as well as failure analysts.

allows you to ensure that there is a good chance that A is correct in the first place. Without that theory of knowledge, logic is worse than useless. As they say, "Garbage In, Garbage Out." But with the cover of logic, we can feel and act like we are justified, whether we are or not.

A methodology is philosophy in action!

PHOTO-DOCUMENTATION - PART OF EVIDENCE PRESERVATION: *I was at a litigation "party" and watched other engineers take photographs of a nominally cube-shaped component. One engineer made all six views directly perpendicular to a different one of the cube surfaces. There's no way that such an oversimplified procedure had a chance of revealing the details in areas that have different reflectivity, for example. And this is one of the most important things to look for when performing an initial visual examination. Sometimes a small change in the angle of the incident light will make a feature go from invisible to glaringly obvious.*

See the difference between the same image taken at different angles (Figure 1.1A and B). It's pyrite crystal (fool's gold) with other black crystals intermingled.

Since failure analysts are often seen as the outside-brains hired to help others determine what went wrong, those who cultivate clarity of vision and clarity of thought will have more credibility. What is the philosophy behind such a mechanical set of actions? Honestly, there was none. The human brain will default to automatic thinking (what has been done before) unless you are directing your actions with conscious forethought. This is one of the many ways people can act like robots – through automatic thinking or copying. So, don't be a robot! Be a human! Learning to see the reality of what is in front of our eyes is a lifetime process. I'm still cultivating and compounding my competence in this specific skill myself. I will share some of the techniques that I have developed to help you attain higher skill levels sooner rather than later in this book.

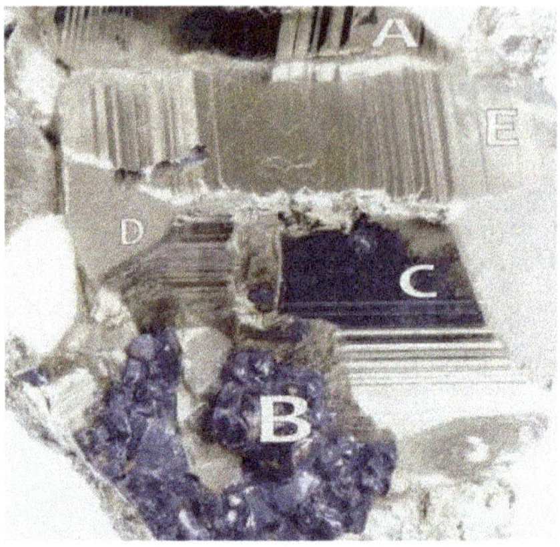

Figure 1.1A: *Pyrite crystal, with other crystals intermingled. First of two views.*

In fact, I'm not much of what some experts call, "a visual thinker", at least not by nature. Until the last ten or fifteen years, it was much easier for me to learn something new if someone explained it in words, rather than showed me. Maybe because it was difficult for me to learn what is natural to many who choose this field, I remember what it took to learn to see properly, and it's easier for me to explain it.

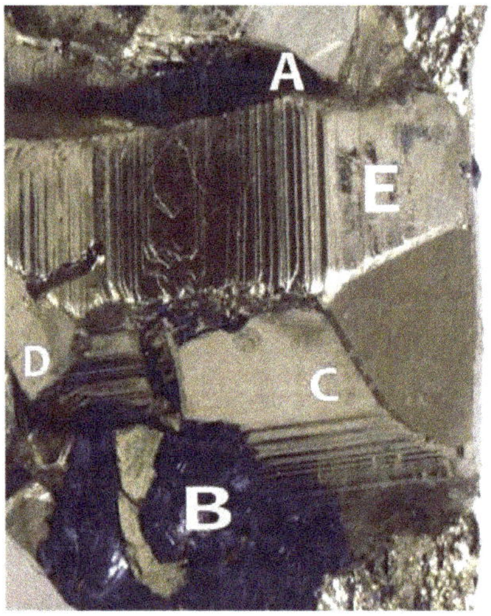

Figure 1.1B: *Second view taken at a different angle and exposure. Same pyrite crystal, with other black crystals intermingled.*

In Figure 1.2, several views of a broken tensile bar are shown. Is it obvious to you that these images are all of the same objects? What else could have been done to create a relevant change in view?

After 25 years of working deliberately to improve my basic thinking skills, I have come up with a catchy short list of ways to promote good thinking. Some of them are hardwired into the structure of our brains. Others require more attention.

NOTE: Since this book is both text book and instructor, the steps need to be presented in a linear order (text book) while some of the most difficult, foundational, and complex steps will be repeated multiple times (instructor) in different ways during each portion of a failure analysis example. Some steps will even be repeated several times within a single stage of a failure analysis example. Remain focused here!

Figure 1.2: *Note how each view from a slightly different angle reveals a different aspect of this broken tensile bar.*

1.2 How to Construct Your Competence as a Failure Analyst - The Foundations of Human Cognition in Steps

I wrote this book with the working title "Cultivating Competence in Failure Analysis". But since Cultivating Curiosity is the first step in Thinking Skill Optimization, I sought a different word to avoid repetition. After telling my sub-conscious to find a new word, I awoke two days later to Compounding Competence. Cultivating sounds like harder work, so having a title with the word Compounding, which sounds like you can sit in a chair and let Father Time himself do the work, might be more attractive to some people.

I shared the interplay between cultivating and compounding with a client who said, "I prefer 'compounding' since it seems to relate to industry while the term 'cultivating' seems to relate to agriculture." I asked my long-time friend and engineering associate, Peter Ried, the same question. He suggested: How about 'Constructing Your Competence'?

My publishers know many people on the path of mastery, and among masters, the analogy of cultivating or constructing are perfect metaphors for the story of evolving your competence. And, compounding is the result you will earn if you do the work of cultivating or constructing! Although I chose the metaphor of constructing for the title, whichever word you choose as a personal metaphor, whichever field in which you work, both effort and the passage of time are required to allow your skills and experience to ripen into greater competence, and the justifiable confidence that accompanies it.

Now, let's get back to my catchy short list of steps to promote good thinking in the failure analysis process and to enhance all cognitive skills.

In Table A, each thinking skill practice is preceded by a common term used when discussing cognitive skills. Following the common term, please find my (usually) three-word description of how that skill is carried out in practical terms by a failure investigator. Let's look at them one by one as they apply to failure analysis.

1.3 Be Human: STEP 1 - CULTIVATE CURIOSITY

STEP 1: It is really difficult to do competent failure analysis work if you don't have a strong desire to learn and know and understand. As adults, life can become overwhelming in aspects related to earning a living, taking care of a family, etcetera. Curiosity is natural in young, healthy children, but as adults, many of us can benefit from reminders that there is always something new and interesting to learn.

For many years, I have walked my rural property to enjoy nature. We've had several very severe storms in the more than three decades that I have lived in Central Lower Michigan. There are many trees that have been blown down by the wind. I have spent some time trying to see whether there were weaknesses in the trees that made it easier for the wind to knock them over.

In 2020, an electrical engineer asked me for help explaining how we could tell that it was wind, not lightening, that damaged a large number of trees that were insured against damage from lightening, but not wind. My curiosity led to a hobby of observing and explaining downed trees, hopefully future food for edible mushrooms (another hobby), which eventually, happily, also led to some billable work!

TABLE A: Catchy Short List of Steps to Promote Good Thinking for Failure Analysts

Step	Common Term	Thinking Skill Optimization Distillation
1	Being Human	Cultivate your Curiosity!
2	Critical Thinking	Clarify the Context (or Question the Context)
3	Rational Thinking	Compare-Contrast
4	Creative Thinking	Connect-Create
5	Systems Thinking	Contemplate Complexity
6	Decision Science	Calibrate your Confidence
7	Communication Skills	Converse or Compose Capably!

1.4 Critical Thinking: STEP 2 - CLARIFY CONTEXT (Question the Context)

STEP 2: Clarify (or Question) the Context reminds us to examine the frame of reference in which we find ourselves related to the project at hand. What specifically are we being asked to do or determine? The negotiation of the project scope provides an opportunity to educate the requestor about what is and is not possible.

Once I had a potential client ask me to carbon date some galvanized steel. That's just not possible. But it did open the door to explain to them what information might emerge from a materials engineering-focused failure analysis. In this case, I was clarifying the context of what a metallurgical investigation entails, and may provide, for the benefit of my potential client.

1.5 Rational Thinking: STEP 3 – COMPARE-CONTRAST

STEP 3: Compare-Contrast. This is the foundation of rational thinking. The word ratio refers to a comparison of two numbers. When we are performing rational thinking and we compare, it involves a broader set of ideas (not just numerical values). In that case, we are *looking-for* similarities; and when we contrast, we are *looking-for* differences.

This mental practice is repeated during every step of the investigation. We *look-for* differences and similarities between the subject of the investigation (the failed part) and specifications, engineering drawings, past production, later production, and our mental (or private) ideas. We compare to published (or private) libraries of damage appearances, microstructures, and mechanical test results. In all of this, we're comparing and contrasting.

The term "Rational Thinking" itself must be put within a context. The common usage within the USA implies fact-based rather than emotion-based thinking. In other places around the world, it is used to imply logic. However, as already noted, logic alone does not ensure agreement with reality. In this book, we'll be referring to the compare-contrast process or practice when we use rational-thinking. Compare and contrast is truly the foundation of building useful knowledge. It is the way our brains are wired. We do it naturally all day long.

1.6 Creative Thinking: STEP 4 - CONNECT & CREATE

STEP 4: Connect & Create. Connection is the essence of how creative thinking works. In our failure analysis methodology, we first prime the cognitive pump with Step 3 (Compare & Contrast). Then, by a mysterious process, our brains suddenly glimpse new combinations, new possibilities, and new ways of understanding. Rational thinking thus lays the groundwork for intuitive breakthroughs to something new.

Connect & Create is also part of how our brains are wired. When we give the brain (our mind) increasingly rich material to work with, we will make note of increasingly creative potential solutions. When we feed our minds with quality, relevant reality and knowledge-based data (Step 1), we will have a higher chance of producing useful, creative, practical and economically viable solutions.

1.7 Systems Thinking: STEP 5 - CONTEMPLATE COMPLEXITY

STEP 5: Contemplate Complexity. Before we finalize our recommendations to the client, a meditation on the myriad of factors that influence any situation is in order. We do this despite the fact that some people just want "an answer". We must mull over the reality that every action has a reaction.

Contemplating complexity is often key to a client perceiving your competence.

Change always entails costs as well as benefits. Engineering is a complex balancing act. Our credibility will increase as our clients and colleagues see that our improvement proposals have a strong chance of achieving the intended consequences as opposed to doing something careless, or doing nothing and setting a chain of catastrophic events in motion.

Contemplating complexity, in my mind, is a key to data evaluation. It is one of the ways we can look back at the frame of reference we established for the project and get an idea if it was comprehensive enough.

Contemplating complexity is key to data evaluation.

Practicing contemplation of complexity will mean that your brain gradually becomes better able to hold more factors in view simultaneously. This is a powerful engine of intellect. We'll focus more on these ideas in Chapter 9 when we look at data evaluation, reporting, and client communications.

1.8 Decision Science: STEP 6 - CALIBRATE YOUR CONFIDENCE

Step 6: Calibrate Your Confidence: We've collected all the data, turned it into information, and now we are in a good position to choose a potential solution to the problem. We need to remind ourselves, and all stakeholders, that sometimes unknowns remain, and our confidence is low. In these situations, we say to ourselves and all stakeholders, "I may need to revise my advice at a later time, as we learn more." We need to train ourselves to remember that *confident* does not always mean *highly confident.*

1.9 Clear Communications: STEP 7 - CONVERSE OR COMPOSE CAPABLY!

If you perform all six of the earlier steps, **Step 7** will likely be natural. However, every person brings a different skill set and some people have to work harder at some skills than others. In any case, you will have thought over what happened, how you ruled out other possibilities, and how to explain your choices of method and the information revealed; you will have done this long before you write the report. You will much more naturally and capably converse, compose, and describe all of the considerations that need to be described for the record. But it never hurts to seek out opportunities to hone any important skill, especially something as fundamental as communications.

1.10 Learn from Success as well as Failure!

Many of these techniques are also directly applicable to success analysis.

It has been said that it is impossible to do a success analysis. The school of hard knocks continues to prove that fact to people who thought they knew how to do something. Because they remained unaware of some hidden factors that were important to their earlier success, they did not include those factors in their recipe for success. This may then lead to failure.

Remember, the part doesn't know what the specification is!

In general, we humans tend to take credit for our own hard work and smarts when we succeed. When we fail, we blame outside circumstances. If we are going to benefit from success analysis, we need to be sure that we don't fall prey to this trap of our own making! Again, this is a matter of seeing reality clearly, whether in success or in failure.

Almost all of us professional *Homo sapiens* have had this experience at some point or other in our lives. To avoid the traps, we use analytical techniques that are the foundation of failure analysis. For example, it is extremely useful to characterize the materials of structural (load carrying) components that passed a durability test. And that disciplined method can help us avoid delusions.

Success Analysis: The analytical techniques that are the foundation of failure analysis are extremely useful to characterize structural (load carrying) components that passed a durability test.

Obviously, for large civil structures such as bridges and buildings as well as machinery used to manufacture consumer goods, there is no durability test phase. For civil structures, engineers are compelled to base their designs on codes. It is often, but not always, straightforward for the properly trained individual to determine whether there was code (specification) compliance or not. And occasionally, failures reveal that the codes or specifications were inadequate.

Even if the applicable codes are adequate, if nobody is verifying that the materials and construction methods being used are in compliance, no code can prevent failure. Investigations into failures of great consequence often include a review of how the required oversight was carried out. Lack of proper oversight is often a finding in large disastrous failures.

Hurricanes and earthquakes sometimes reveal that construction codes were widely ignored. Ignoring codes and oversight breakdown cannot happen when properly-specified communication channels are functioning. Of course, for oversight to catch wrong actions, they have to be noticed by someone. Often, the people with opportunity to notice are not high up on the corporate ladder. Thus, they have to be brave enough to report the wrongdoing.

The chain of bravery can break at any link.[14]

> *Many, many times over my career I have been faced with this challenge. If you are a young professional, do you feel confident that you will be able to report ethical problems that you find during your investigations? To whom would you report the problems? What if they don't act? What if they say, "There is nothing to worry about," but they don't explain to your satisfaction?*
>
> *This happened to me years ago, and I ended up writing a memo that gave my boss a metaphorical heart attack. I wrote that I would not want to be driving anywhere near the vehicle that had this perceived safety problem. It took years for me to understand the full nature of engineering risk. To my knowledge, nobody ever was injured as a result of ignoring my advice. My boss and his boss did not have the communication skills to explain to me why I did not need to worry about the issue at the time the decisions were being made.*
>
> ***BOTTOM LINE:*** *Sometimes you will find tribal knowledge is more robust than it appears. Even if you are wrong about the specific concern, speaking up is the foundation for the development of a reputation for having strong ethics.*

We see that empowered employees are a key factor in failure prevention. Empowered employees are different from employees who are admonished to do the right thing, but are not coached or mentored so that they know how to do the right thing. Knowing happens at multiple levels. The bigger the perceived potential personal negative consequences, the harder it may be for most people to take correct actions.

For many mass-manufactured items, especially those that have extended warranties, a company often performs some sort of durability test or a sequence of such tests. Perhaps there will be a simplified corrosion resistance test, such as salt fog exposure. While such simplified tests do not represent the real environmental exposure that the product must endure, survival does provide some information about the suitability of the product. This is especially true if the company has experience with using the test for similar products over time.

Evaluation (materials characterization) of durability test pass specimens, properly performed, will provide a basic description and documented record of what survived the known conditions of the test. With this data in hand, it is much easier to evaluate the significance of deviations from specification. This is a key to understanding failures of consumer products.

With an actual known-good Test Pass component in hand, there are even more opportunities to perform tests that were not originally required or to test features of the component that were not evaluated originally. I am always very happy when a company trying to understand a failure can provide an original, known-good from the Production Part Approval Process (PPAP)[15].

[14] One such example of a break in the chain of bravery may be reviewed in the Texas incident involving a gunman at an elementary school. As he was killing 19 students and 2 teachers, the police appeared to spend their time handcuffing a mother trying to go in and save her child. From what I saw, the police stood around watching the handcuffing (they dawdled, in my opinion, in a cowardly manner). This incident reinforced for me that bravery, even among those who are well armed, is not automatic. (Robb Elementary School, Uvalde, Texas, 24May2022)

[15] PPAP is a structured process and organizational system that helps manufacturers and suppliers communicate and approve production designs and processes before, during, and after manufacture of a physical part.

Having this is often the key to understanding failures of production manufactured components.

Many real-life examples of disastrous failure happened as a result of ignoring the technique of Failure Modes and Effects Criticality Analysis (FMECA). It is useful in helping us prevent preventable catastrophic failures, even those whose likelihood is thought to be vanishingly small. There are books on this subject. Here's my personal FMECA story:

> FAILURE MODES AND EFFECTS CRITICALITY ANALYSIS (FMECA) STORY: *My father is getting on in years and his eyesight is deteriorating. His eye doctor informed him that he would not be able to renew his driver's license after his 85th birthday. After it expired, my father turned down his mechanic's offer to buy his old car because, even though he had no better offer, he thought it was worth more. A few weeks after he had lost the right to drive, he nonchalantly told me, "Hey Debbie, I took the car out today."*
>
> *"Dad," I exclaimed, "you know you aren't supposed to drive."*
>
> *"Yeah, yeah," he muttered, "but I was going stir-crazy." He could tell that I was very upset so he continued. "The risk of an accident is so low. I'm a good driver and I went out in the daytime so I could see just fine."*
>
> *"That's not what your eye doctor said," I explained, "and the consequences of an accident while you are unlicensed and uninsured are extremely worrisome. What if someone was hurt or killed? Senior citizens have been sent to prison for vehicular homicide in the United States. You could lose everything you own, even your freedom."*
>
> *This is the essence of the concepts underlying a failure mode effects criticality analysis. When I shared the story of my dad's risky behavior, a colleague who used to work for a company that made home kitchen appliances told me about one of their FMECA projects. It was related to their oven's Sabbath-Mode and it was marketed to Orthodox Jews. If the Sabbath Mode failed (the oven turned on during the Sabbath), this would entail the owners of the oven breaking one of the Ten Commandments. This is surely to be considered a very critical marketing problem for that segment of the customer base.*
>
> *... Back to the story about my dad. (I'm interweaving stories here to work at multiple levels of your brain systems.)*
>
> *"Dad," I cautioned him over the phone. "In two days, I'm going to come for a visit."*
>
> *"That's nice, dear."*
>
> *"And when I leave your house," I said firmly, "I'm taking the car with me."*
>
> *"Whaaaa... ?"*
>
> *"I love you, Dad. But you have got to make peace with walking and using taxis. You can do it." He did hand me the keys.*

The next sections cover the thinking skill *considerations* applied to failure analysis: the basic steps of Thinking Skill Optimization. Please review again the catchy names in shown in Table A (but included below for your convenience) to practice constructing your failure analysis competence and confidence.

- Being Human - Cultivate your Curiosity!
- Think Critically - Clarify the Context (or Question the Context)
- Think Rationally - Compare and Contrast
- Think Creatively - Connect to Create
- Think in Systems - Contemplate Complexity
- Use Decision Science - Calibrate your Confidence
- Communicate Clearly - Converse or Compose Capably!

Over the next sections, we will explore how each of these mental processes applies to the basic aspects of performing failure analysis work. These considerations are presented out of order to show they are interrelated and iterative.

1.11 Rational Thinking Considerations in Failure Analysis

The Compare & Contrast step is the essence of rational thinking; and it is really one activity. When we find a common trait between two things, we call it a comparison. When we find a difference, we say, "We're contrasting." The word "Rational" is derived from the word "Ratio". What is a ratio? It is a comparison of two numbers. So, when we talk about rational thinking, it is just an extrapolation of comparison and contrast to a broader field than that of numbers. In your role as a failure analyst, what questions would you ask to form the basis of an excellent comparison and contrast? (If you are experienced, or brave, stop and find an answer before continuing to read.)

Fundamentally, Compare and Contrast would be to leverage the existence of a known good part. If you don't have a known good part, how can you say for sure that a different part is bad? Part of failure analysis work is determining whether the part was bad, or whether something unusual happened to it. It is common, but incorrect, to reason that "Because the part failed, it was bad."

Think carefully before selecting your "Control" or "Reference" or "Exemplar" specimens for a failure analysis. Skipping a properly chosen control specimen, when one is available, is almost always a failure of the failure analysis process. Often it leaves you with insufficient information and you may lack the ability to provide confidence-inspiring action plans. Rational thinking is so fundamental to failure analysis that it is woven throughout the entire book.

Remember, the part itself never fails; it simply responds to its specific situation. The failure is always to be found in the human arena. When thinking about a part's role or the part's processing or environment, were any of these inadequate? The part doesn't "know" the specification. We have to rule out (or in) the existence of unusual weaknesses or vulnerabilities in the Subject Part. Comparison to the known good often allows us to do this in a clearly demonstrable manner.

1.12 Critical Thinking and Creative Thinking Considerations in Failure Analysis

One of the first things I learned when I started studying critical and creative thinking is that one of the basic rules of *critical* thinking is to practice *creative* thinking! And here's an interesting truth; one of the basic rules of *creative* thinking is to practice *critical* thinking!

Clarifying the context is the essence of critical thinking in the way we are using the term in this book. Defining the frame of reference, another way to say this, is one of the primary goals of the initial project scope negotiations. But because we need to understand all the available steps of the failure analysis before we can take a lead negotiator's role on the investigation team, the negotiation process is placed toward the end of the book (Chapter 7). Understanding how test parameters differ from the in-the-field experience of the subject under investigation is also an exercise in clarifying the context.

Connecting the dots between ideas and concepts that were not previously related in other people's minds is the essence of technical creativity.

> *Years ago, I heard an interview of a rock star from the 1970s explaining to the National Public Radio host how he got new ideas for songs. He tuned two radios to different stations, and put one on each side of his head! The result was hearing and connecting new relationships between lyrics and musical elements.*
>
> *In reading Michael Gelb's wonderful book* "How to Think Like Leonardo da Vinci: Seven Steps to Genius Every Day", *I especially appreciated the chapter on learning to more readily see connections. One of the ways to do this is to look at two different things and find five (or more) features in common. We've been calling this technique "rational thinking" in this book. It multiplies your neural connections. The greater the difference between the two items, the more powerful the approach.* ***This compare and contrast to connect and create technique is truly a basic technique that anyone can use to***

improve their mental abilities in any field.

1.13 Systems Thinking Considerations in Failure Analysis

Contemplating complexity is a way to remind ourselves to stop before we draw all of our conclusions. It is a way to take the time to consider the interactions between all of the factors that we have identified as having played a role in the failure. Systems Thinking is a whole discipline unto itself. It provides its own structured methodology that reduces the chances that we will overlook a relevant factor.[16]

Years ago, when working on a fastener failure, the team found 82 potential factors that might have produced a higher stress level in the fastener than was originally anticipated. Systems Thinking is a structure that can reveal such oversights!

1.14 Decision Science Considerations in Failure Analysis - Hey! Can you say E-pis-te-mo-logy?

As we noted earlier, calibrating our confidence doesn't necessarily mean we have high confidence. Calibrating our confidence involves knowing how confident the reasonable person would be – a person with relevant expertise – a person experienced in the field-in-question – how confident would this person be? **When used effectively, the thinker will be more likely to more frequently generate multiple attractive recurrence prevention alternatives from which the client may choose**. I'll address the first one first.

Knowing how confident the thoughtful, aware person would be is a big part of the philosophical discipline known as epistemology. I love that word. It's so much fun to say! Epistemology is also known as Theory of Knowledge. Epistemology has to do with examining how we know what we think we know.

Democratic societies in the second decade of the Twenty First Century CE have experienced a serious deterioration in "buy-in" to a shared reality. Using the methods presented here can ensure that our knowledge is sound so that it accords as well as possible with observable facts. Epistemology has positive potential to help overcome the divides in perception of our world.

Many people are now claiming to do critical thinking. They explain that critical thinking is thinking-for-yourself and rejecting conventional understanding and wisdom. This is not critical thinking. Critical thinking must be geared to minimizing the need for belief when validated knowledge is readily available. Critical thinking must be based on the foundation of epistemology, which means seeking redundant, independent sources that point (metaphorically, at least) in the same direction, support a common understanding, and conform to reality.

In 2012, I presented a paper at the Machinery Failure Prevention Technology (MFPT) Society's annual conference. The topic was epistemological approaches to negotiating project scopes in failure analysis. At that time, I could not find any other publications on this subject. Generally, only philosophers, theologians and hard scientists bother with this arcane, yet highly relevant academic field that is a backdrop for gaining reliable knowledge. Those who developed epistemology were either trying to prove that God does or does not exist, or they were interested in proving that one scientific theory is more or less valid than another. You will find updated content from my MFPT presentation in Chapter 7 of this book. When you want to rely on your judgment as a failure analyst, it pays to have confidence that you actually know what think or believe that you know.

---

[16] Donella Meadows, *Thinking in Systems: A Primer*, Chelsea Green Publishing, 2008. This is an excellent book to get started in the understanding of complexity.

EXERCISE 1-A: What Do You Really Know for Sure?

Please stop and think for at least one minute before deciding. Once you have an idea of what you really know for sure, set a timer for 1 minute and write your list in the space provided below. Do all of the exercises below before you look at my lists, which I provided in Section 1.29.

Here are some of the things I am most sure of:

Please stop again and think for at least one minute before writing. Once you have an idea of some of the things that you have been fairly confident of, set a timer for 1 minute and write your list in the space provided below.

Here are some of the things I am / have been fairly confident of:

Please stop again and think for at least one minute before deciding. Once you have an idea of some of the things that you don't know, set a timer for 1 minute and write your list in the space provided below.

Here are some things I do not know:

After generating your own lists, now read my lists (Section 1.29 at the end of this chapter). Contemplate how they compare.

Have you contemplated and compared our lists? What are some of the points, that on reflection, I might be trying to make with these lists? Write your ideas in the space below. Don't turn the page until you have some ideas written down.

____

____

____

Now, think about this: did your parents ever tell you "No, you're not angry / hungry / sad!" or "No, you don't like ice cream!" or "No, that didn't happen"? Would that tend to confuse a young person?

The term "gaslighting" is used to describe this confusing education. Gaslighting is a form of psychological manipulation intended to alter behavior by sowing seeds of self-doubt or confusion so that the person is easier to control. Gaslighting is the Merriam-Webster word of the year for 2022.

In addition to using validated knowledge, we must learn how to recognize what our bodies and memories are telling us, since this is the basic human detection system on which our confidence rests. Without a basic ability to recognize and name events and experiences, our ability to develop an accurate moral or ethical compass will be severely inhibited. This is disastrous for a failure investigator, as well as for running our type of business. Knowing how I feel allows me to extrapolate how others might feel in similar situations. Without this, I am lost. I am not unique in this consideration. Humans receive critical information during this examination process. Empathy is the foundation of both ethics and morals. If you can't recognize your own sensations and emotions, it is hard to be empathetic or moral. Note that morality is more of an issue when evaluating the actions of a powerful person.

1.15 Going Beyond "I'm the Expert!" Epistemology Continued

Epistemology helps us understand the strengths and limitations of what we think we know. If we are performing failure analysis, whether it's to fix a problem or assign blame, we better have some way of explaining to ourselves and others how we reached the conclusions we reached. It is not good enough to say, "I'm the expert." Appealing to authority without providing justification for the claim is an age-old way of manipulating people and is often used to push them to act against their own interest.

Francis Bacon (1561-1626) was the first in the western tradition to make a noticeable dent in the armor plate of traditional knowledge. Francis Bacon did not invent and did not have at his disposition a formal epistemology; but his writings (*Novum Organum*[17]) laid the groundwork for early modern scientists to create an epistemological framework for evaluating truth claims. Bacon implored his readers to observe for themselves rather than relying exclusively on revealed truth.

If we are cultivating competence in a technical field, we would do better to avoid the appearance of "I'm-the-expert, self-serving communication

[17] Novum Organum, is formally named in Latin, *Novum Organum, sive Indicia Vera de Interpretatione Naturae* (New organon, or true directions concerning the interpretation of nature). It is considered to be a foundational work for the progression of scientific thinking and a dramatic improvement over Aristotle's work, *Organum.*

styles".[18] While such approaches may have worked in courts of law in the past, American Federal Law (try putting "Daubert" in your search engine) generally now prohibits such tautological (circular reasoning based) claims of expertise based solely on the reputation of an individual.

Epistemology is very useful, and easier done than explained! Most of you are probably already doing it to some extent already. Following the practices outlined in this book will help you develop a procedure to evaluate how confident you should be in a given technical situation and how to communicate that confidence in a professional manner.

Please note that even when epistemological principles point toward great confidence, you can still be dead wrong. Never forget this. It is usually a bad idea to claim certainty, rather than confidence. Certainty is more useful to people deciding how to interact with a religious community, than for those in the middle of an engineering project.

Finally, providing a real choice means that there is more than one option. If you have scoped a project to provide options, then you must do that. If you want your clients to respect you, you need to be clear about backing up your claims of expertise. That simply gets back to the creativity aspect of our constructing of competence. We have to generate enough options during our analysis so that there is something to choose from. When all else fails, one option can always be, "Maintain the status quo." Sometimes, sadly, after all of our competent and careful work, we realize that solving the problem will cost more than the problem is currently costing the client. If there are no safety implications, doing nothing may be a legitimate business decision.

1.16 Systems Engineering Considerations in Failure Analysis: Contemplating Complexity!

Everything we deal with in failure analysis is part of a multi-layered multi-system system! The part we are analyzing is a result of a design system; it is a result of a manufacturing system. In fact, it is a result of multiple levels of design and manufacturing systems. Some relate to the raw materials. Some relate to the machinery used to make the raw materials and components. Some relate to the components; and some relate to the sub-assemblies and assemblies and to the integration with the ultimate products.

When evaluating something as simple as the microstructure of a piece of hardened steel, we may be confronted with the effects of hundreds of factors that ultimately created the specific arrangement of atoms that the microstructure specimen reveals. If some of these factors varied far enough from the usual range produced by the process, we may have to perform a lot of work in order to determine where and what the deviation was. This may (or may not) be necessary to reduce the risk of recurrence.

And then, there is the consideration of the "Law of Unintended Consequences". So, let us say that a particular investigation has gone well. You have sussed-out what the problem was, and suggested a fix, as promised.[19] But that fix often, if not usually, comes with higher cost, more restricted availability, or other drawbacks. Often, for high-volume production products (or products made for military or aerospace applications), re-certification is required for process or specification changes.

While it is often not the job of the failure analyst to evaluate all of the potential problems associated with a fix, these issues sometimes result in a

[18] My publisher and many advisors recommended presenting my expertise in a straightforward style to a greater degree than I wanted. I preferred to stress my own struggles with self-doubt and how I overcame them to gain confidence. Hopefully, we struck a balance because this journey is critical to competency in FA.

[19] Another excellent resource with advice on how to check the likelihood of benefit from the recommendations is, *The Logic of Failure* (4th Edition), by Deitrich Dorner, Basic Books, 1977, 240 pages. | ISBN 978-0201479485

decision to live with an occasional minor problem (or, frequently, a major problem).

1.17 Communicate Clearly: Considerations in Failure Analysis

While most universities offer degree programs in communications, a basic skill level is a necessity for any professional activity. There are different types of reports appropriate for different types of organizations or individuals.

Going through all of the previous steps to constructing competence and communicating along the way with the project team and stakeholders is challenging communication work. And that work is a valuable resource for a final report before you even select your font or format.

What you learned about the team along the way will allow you to customize your expressions to ensure that the people who requested the work understand it in a way that they will be able to make good use of the data.

Converse and Compose Capably reminds us that the way we both speak and write has consequences; it can clarify or mislead - intentionally or unintentionally. The number of courses and consultants who provide guidance for technical communications is large. A key point that many of these entities make: know your audience. That takes us back to confirming the context. Of course, the context can change as we uncover new information and newcomers can enter the picture over the course of the investigation.

Many technical communications specialists recommend extensive use of visual aids. As the old Chinese saying goes, "A picture is worth a thousand words." The report preparation content in Chapter 9 advises, when possible, giving the client the entire relevant data and photo record.

1.18 Failure Analysis by Methodology Rather Than Failure Analysis by Recipe

A recipe implies that there is a fixed set of basic steps that anyone can follow. Now, there are very few things in life that are so simple that a recipe could be all inclusive of everything you need to know.

> BAKE-A-CAKE-BY-RECIPE EXAMPLE: *In order to bake a cake, a recipe writer may assume that you know how to:*
>
> - *grease and flour a pan.*
> - *measure flour using a cup in the United States or know how to use a scale to weigh the flour.*
> - *cream butter (beat with the sugar until fluffy)*
> - *read the test stick (a toothpick or wooden skewer) that has been inserted into the cake after a certain time to see if the cake is done.*
>
> *Once you have these basic skills, you can usually reliably bake a nice cake - assuming your oven is level and the temperature controller is calibrated, and that you are there to remove the cake from the oven when it is fully baked.*

A procedure is usually more detailed than a recipe.

> BAKE-A-CAKE-BY-PROCEDURE EXAMPLE: *There might be a procedure, for example, following the cake recipe example above, to grease and flour the baking pan. There might be a procedure to cream the butter, which details how to tell when it has been softened enough, and a procedure to add the sugar bit-by-bit, rather than all at once. So, there might be 10 procedures to be learned if the baker had never baked a cake or watched anyone else bake a cake.*

One of the important aspects of communication is having cultural context. Below is a story that highlights this.

> SHOE-LACE-FAILURES EXAMPLE: *In 2008, I was in India as a tourist. The Pune, India ASM International Chapter kindly invited me to give a few presentations. During my first presentation to the students at the Government Engineering College, I brought up a key idea about how components fail.*

I resonated with a section in Don Wulpi's book[20] *about shoelaces. So, I asked the students, "Where does a shoelace break? This is a good example," I continued, "because all structures that are prone to break typically break in a specific way." Then I waited. And I waited. But the students were silent. Then I looked around the room at what was in front of me. (Clarify the Context!)*

It was February 28th in Pune, India (about as far south as Jamaica). All the students were wearing sandals. Besides me, there was only one person in the room with shoelaces – the department head! I was lucky he was wearing shoelaces. Many people of my generation started wearing Velcro shoe closures after high school. Sure, these students could have performed calculations, but they did not have the experience to whip out an instant answer to where shoelaces break. Even in the USA, my guess is that most young readers of this book might not choose to wear shoes with laces often enough to have an instant answer to where shoelaces break. That's why I am still seeking the new universal story to explain failure analysis.[21]

If you have any ideas, I'd love to hear them.

Okay, I've quickly covered recipes and procedures and now it is time to discuss another type of instruction set - a methodology. A methodology may be thought of as a mental toolkit based on a structured knowledge foundation.

Using a methodology requires more than knowledge of the steps a failure analyst would need to take during an investigation. It requires an extensive knowledge of the object being investigated as well as the object's use, its design history, the methods of manufacture, the raw materials used, etcetera. Details of any inspection techniques used prior to placing the object/ assembly into service may also be relevant to some investigations. Effective use of a methodology in failure analysis requires a lot of knowledge, both broad and deep. It also demands knowledge of how to perform the failure analysis using the correct steps in the right order.

Here's another story in the form of an old joke from decades ago.

HOW TO KICK THE TV: Back in the day televisions were heavy. Consumers really did not want to carry their heavy old black-and-white TV set to a repair shop. Joe was having TV problems and called his handy repairman, Mike. (Yes, it was highly likely a repair man).

Mike knocks; Joe opens the door and lets Mike into the house.

Mike looks at the TV to be sure it's plugged in and turned on. It is. He walks around to the side and kicks the TV. It starts to work. Mike announces, "That will be $100^{00}, please."

Joe is outraged, "$100^{00} just to kick my TV? I could have done that!"

"Sure," said Mike calmly. "But only I knew where to kick it, and how hard!"

Joe ruefully opens his wallet.

Metaphorically speaking, competent Failure Analysts have learned where to kick the TV and how hard to kick it.

Preparation for effective visual examination is an important part of learning failure analysis methodology. Step-by-step, the skills required are covered in this book.

No matter how long we have been doing failure analysis work, we have more to learn. All of it adds to our mental toolkit – our methodology.

No matter how long you have been doing failure analysis work, you have more to learn. At one point, I worried about the quality of work I did early in my career. Fortunately, I have not often been horrified

[20] Donald J. Wulpi, et al, *Understanding How Components Fail*, 3rd Edition, ASM International. | ISBN: 978-1627080149 |Chapter 1, Page 5

[21] Thanks in advance for sharing your universal story to explain failure analysis. | daaliya@itothen.com | Email Aliya Analytical |

about big mistakes. That was because I was fairly meticulous to follow the procedure that I knew, which meant going from less to more destructive, and document, document, document what I did. That documentation is a lot easier now than it was in 1986, when I started doing failure analysis, before the days of digital photography.

So, you will be the judge. I've been using the techniques specifically to cultivate clarity in my life and work for 25 years. There was not a lot of noticeable improvement that I could directly attribute to the practice for quite a while. Then, the techniques started to become second nature. According to a free online IQ test I just took; the benefits do not show up in an IQ test. However, in a practical way, I find myself being able to understand my personal and professional failures, as well as those my clients pay me to analyze, in more useful ways. None of this means I live problem free. New challenges are always arising. But these techniques have gradually helped me to become really efficient and effective in most of the routine failure analysis work that I do on a day-to-day basis. Hopefully you will have a similar experience.

1.19 The Importance of Understanding Materials, Mechanics, and Human Factors

If we are studying failure analysis of structural components, we absolutely must have knowledge of basic mechanics.

We must understand that forces and loads on the component result in stresses and strains within the component.

We must also understand that temperature changes may result in shape changes that may result in additional stresses and strains, forces and loads. Temperature changes are accompanied by material property changes. Shape changes may also result in, as well as be caused by, temperature changes! And shape changes can influence the mechanics!

We must have, or gain an ability, to visualize how the stresses are distributed within the components as they are subjected to common loading geometries.

The chapter in this book on macro-fractography will help you learn this skill if you don't have it, and will help you refine this skill if you do. For many fracture events, it is adequate to do this in a qualitative way. You don't need a calculator. Even people who are experts in computer simulation of stress fields benefit from intuitive understanding of the stresses that develop in different simplified loading geometries. It was surprising and disturbing to encounter two FEA experts who did not know the term "hoop stress". Without such basic understanding of how engineering has been performed for hundreds of years, the ghost of failure will hover over the expert's shoulder.

I had been doing failure analysis for many years before I learned the loading geometry aspects of macro-fractography. However, I always practiced basic epistemology even before I knew the word. So usually, I was able to help my clients gain some understanding of their problem. But once I became skilled at visualizing the stresses in simple shaped parts, my utility to my clients increased noticeably. **The basic mechanics aspects of macro-fractography are the link that allows materials and mechanics specialists to be able to communicate with each other in a meaningful way.**

For example, knowing every detail of the microstructure of Material X is irrelevant if you don't understand the stress and strain experience of the part.

If you are already knowledgeable in mechanics, it will still benefit your practice of failure analysis to learn the simplified loading geometries presented in this book. They facilitate communication with non-engineers. Everyone can learn to visualize the stresses if someone explains the loading geometries.

1.20 Ductile and Brittle at the Macro- and Micro-Scale (a Valuable Approach)

This book will present an accepted alternate approach to classification of mechanical behavior characteristics than the one historically taught in universities by materials science or engineering professors.[22] Understanding the difference between ductile and brittle behavior is usually taught as a highlight of any introduction to materials. Brittle behavior is easy to understand and is contrasted with ductile behavior. Glass breaks in a brittle manner. If you can find all the pieces after a fracture, you could put them back together and they would more or less fit like a puzzle. Aside from the crack itself, the shape doesn't change much. Ductile fractures are taught as having (usually visible) deformation before breaking.

Ceramics usually break in a brittle manner, and metals and polymers sometimes break in a brittle manner. (Did you notice that brittle and break start with the same sound?)

Figure 1.3 shows a steel part that broke in a brittle manner. There is no change in diameter at the break. The two parts would fit back together easily if we tried. We did not try to fit them back together because that would damage the features on the crack surfaces that might otherwise provide information to help us understand what happened.

Figure 1.3: Note the brittle crack at the macro-scale.

But many metals and polymers are inherently ductile, so when they break in a brittle manner, it indicates that something went wrong. This is where the alternate approach to understanding brittle and ductile behavior demonstrates its power. Let's look at Figure 1.4 (next page).

The upper section of Figure 1.4 shows an example of a classical ductile steel test bar, similar to what professors usually present to their students. Note the pinched down area at the center. The diameter of the test coupon was originally constant in the entire area between the red dots within the purple ink.

The middle section of Figure 1.4 shows a higher magnification view of the pinched down (deformed) area at the center of the length of the test bar. There's a reason the pinching occurs at the center of the length in test bars. It is because it has been machined (despite what I just said!) to be a little bit smaller diameter (1% maximum) than the ends of that constant length. It forces the break at the center of the length.

As I said, this is an image of the "classical" ductile test bar. But let's look at it a bit more closely. The lower image in Figure 1.4 shows the actual fracture surface of one of the fragments. We see some complicated features. The center is perpendicular to the axis of the cylinder - just like that classical brittle crack shown in Figure 1.3. It is pretty easy to imagine that a crack will "want" to break a long, skinny part at its smallest area, and in an orientation which is perpendicular to the load. There was definitely something strange going on in Figure 1.3 that caused the crack to propagate in the larger diameter.

In Figure 1.4, we see that the crack started to propagate perpendicular to the load, at the center of the smallest area, which was the area previously pinched-down by the deformation processes associated with the ductile behavior. So, the center of the crack surface is defined as macro-brittle because the center where it broke is flat. We're choosing to define ductile or brittle at the MACRO-scale entirely by whether the atoms were sliding at an oblique angle (ductile) or the atoms

[22] The method described in ASM International Metals Handbook, Volume 11-2002 approaches fracture based on brittle and ductile behavior; *"Fracture Appearance and Mechanisms of Deformation and Fracture"* by William Becker, page 557. The Constructing Your Competence in Failure Analysis method is consistent with this handbook.

were being pulled apart (brittle). The first results in a "shear lip" and the second in a crack perpendicular to the maximum principal stress.

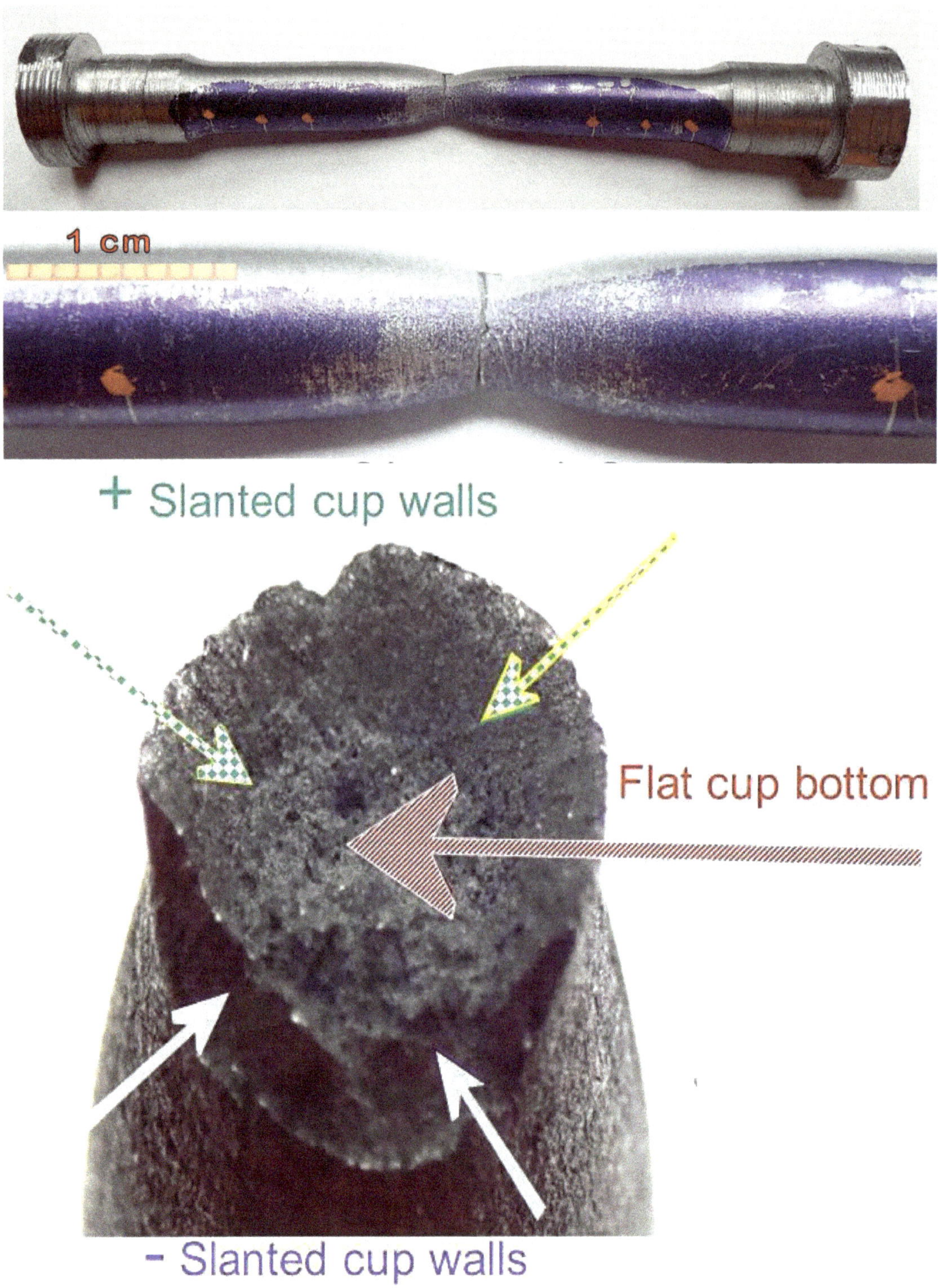

Figure 1.4: *Broken Tensile Test Coupon. Necking (upper, middle) or pinching during stretching of the length is a classical ductile behavior; a crack often results that is partially perpendicular to the load (part axis). The crack edges are at a slant angle and are thus macro-ductile, but the flat center of the crack is macro-brittle.*

We're choosing to define the ductile and brittle cracks this way for multiple excellent reasons, perhaps the biggest one being that most people can learn to visualize stresses and loads this way, and mechanical engineers learn how to analyze stresses consistent with this method

This insert is specifically for those who learned Ductile and Brittle by "the old method" and want more explanation and justification.

For the rest of the readership, these concepts will be picked up later.

The stresses that result in brittle cracks are called "normal stresses". The stresses that result in ductile cracks (at some oblique angle to the normal stresses) are called shear stresses. Mechanical engineers often work with the von Mises Stress Criterion when developing their designs. This forces them to analyze both principal (or normal) stresses and shear stresses. So, once we explain to a mechanical engineer that this is what we are doing, we usually have fairly clear sailing to explain the rest of our findings.

In Figure 1.4 (lower), we see that the outer edges of the break are at a slant angle. The outer edges of the crack, the "cup walls" of this "cup and cone" fracture, are at a slant angle. The protruding portion of the cup wall between 6 o'clock and 10 o'clock is on the mating piece. The cup side wall feature is ductile at the macro-scale. So, we see that this classical ductile fracture, in the old days defined by its obvious change of shape leading up to the fracture event, is a mixed mode brittle (flat center) ductile (knife edge side walls) crack.

Perhaps some experienced people will object to this naming convention. I can hear them now, "I know that the material is ductile so why is she calling it brittle?"

And I respond, "Well, I understand that the MATERIAL itself is behaving in a classically ductile manner prior to breaking. But the break itself is flat. It is not a classical slant fracture at the center."

Yes, it has noticeable shape change. But again, I want a method to describe the crack surface itself, the way we see it with our eyeballs (at least when the part itself is big enough to be seen by our eyeballs). I want a MACRO criterion that fits with the VISUAL examination.

NOTE: I'm using a lot of CAPITALS here because so many people (myself included when I first learned this method) have an extreme emotional reaction to the cognitive dissonance associated with calling a crack BRITTLE when it has a pinched down cylinder with a cup and cone.

It is sometimes even more difficult for people to want to try this system if they are metallurgists who know that the micro-scale features are going to look like the classical ductile micro-scale features. It is cognitive dissonance on top of cognitive dissonance.

"WHAT?" they say. "You want me to believe that my beloved professors who taught me all that I use to earn my living WERE WRONG?"

"No. I'm not saying that at all." I answer back to them. "I am saying that we are going to refine our system of naming and describing mechanical behavior as it relates to fracture features of the Subject Components in our investigations. We're going to do this to align our macro descriptions with the stress states that we want to see from the Finite Element Analysis models that often are helpful in understanding the observed damage. "

(continued on next page)

We're also going to do this to be able to speak to non-specialists in a way that they can easily link their sensory impressions to what specialists are saying about the damage. We are making the language and thus communication about failure universal!

And we're also going to do this so that we can readily discern two aspects of mechanical behavior, macro- and micro-scale, so that we can compare and contrast them to gain a fuller understanding of what happened. Sometimes macro and micro behavior is self-consistent (ductile at both macro- and micro-scales or brittle at both macro- and micro-scales) and sometimes it is not.

Fatigue is (usually)* macro-brittle and micro-ductile. Again, that is by definition. Fatigue cracks propagate when the local stress is above the tensile strength but the nominal (average) stress is below the yield strength (so, no plastic deformation can occur).

Finally, we are going to define MATERIAL behavior as either ductile or brittle, and CRACK behavior, independently at the macro- and micro-scales, as either ductile or brittle. For example , a part that cracks in a brittle manner, at both macro- and micro-scales, could potentially deform (behave in a ductile manner) after the crack event.

This will be reviewed in detail in Chapter 8. This naming convention allows you to be able to communicate across engineering and business functions in a clear, accurate, and concise manner. Even as it can admittedly be tongue tying.

* *When torsional fatigue occurs in some geometries, the crack can propagate on a macro-scale shear plane. This means that in this case, fatigue is ductile at both macro and micro-scales (and that will give us a headache as well).* ***So, here's my hint to you.*** *We just need to remember: All classification systems are to some degree arbitrary. We humans have forgotten to convince Mother Nature that these processes were supposed to conform to our beautiful theories! (Theoretically, bumble bees can't fly!)*

1.21 Materials Science and Engineering as a Foundation for Failure Analysis

Materials science and materials engineering are disciplines that, like every other modern field of study, offer many different work experiences requiring different sub-sets of knowledge. Materials scientists wear white coats and work in basic research labs. Materials engineers wear green canvas fire resistant clothes to make steel. We work in aluminum foundries. Materials engineers sit on their rear-ends and write specifications. Materials engineers help manufacturing engineers tweak process details to optimize polymer injection molding operations. Materials engineers often end up working as quality engineers, having the final say as to whether a lot of powder, pellets, billets, bars, castings, stampings, or I-beams are fit for use. Materials engineers may determine welding process parameters or define inspection criteria. Many activities are included in the field of materials science and engineering.

The essence of Materials Science and Engineering is the knowledge to apply an understanding of how a piece of *material*, subjected to a *process*, will develop (often at multi-levels) a *structure*, which will determine its response to physical, mechanical and chemical aspects of its environment. We refer to this response to any environmental factor as "the properties" of the material. See Figure 1.5 .

Properties are relevant because they determine performance. Every material is selected, specified, and inspected in order to achieve a set of properties! There is a whole discipline, based in both science and art, called metallography, which is the major method (but not the only way) that materials engineers evaluate microstructures. A brief description of metallography is included at the end of this chapter, and a more detailed explanation is included in Chapter 6.

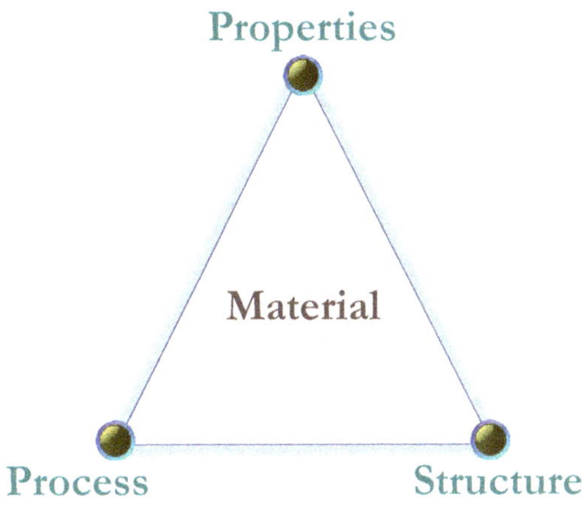

Figure 1.5: *The Process - Structure - Properties Triangle*

Armed with the knowledge of the role that structure plays in determining the behavior of the material, the materials engineer is uniquely suited to help shed light on *why this one failed*, and *why this one failed now*, and *why this one failed here.*

The materials engineer is the one who is prepared to examine a microstructure and consult the literature to gain understanding about how the component is likely to behave in different circumstances. The materials engineer is the one who is suited to examining and interpreting the features in microstructure specimens to shed light on corrosion failures. Materials engineers examine and interpret micro-scale features on fracture surfaces, to shed light on the speed of the loading that resulted in the separation, and the environment that facilitated the damage.

The materials engineer is uniquely suited to help tighten a materials or process specification so that the same undesired event is less likely to recur.

Every material is selected, specified, and inspected in order to achieve a set of properties!

Before leaving the technical aspects of failure analysis, we're going to delve a little bit deeper into the mechanical behavior aspects of materials. It should be less controversial than the discussion above. This discipline is key to understanding deformation, fracture, and wear damage. Since corrosion damage often leads to deformation or fracture, due to loss of the material that is helping to transmit a load or force, mechanical behavior is an important foundation of failure analysis in general.

1.22 Brief Introduction to Mechanical Behavior: Is Brittle Bad?

Let's take a little more time now to introduce ourselves to the key concepts of mechanical behavior that will facilitate understanding the rest of the book and failure analysis work in general.

"Ductile versus Brittle" is arguably the most essential characteristic of what is called mechanical behavior. Ductility, rather than strength, is important because, even if the material is very strong, if it is also brittle, it may not get to act strong. A faster-than-anticipated load event, or traces of corrosion, may prevent the full strength from "manifesting" if the material is not able to behave in a sufficiently ductile manner.

Charpy V-Notch (CVN) testing is one popular way to evaluate toughness in steel products.[23] Toughness, or crack resistance, is a result of both strength and ductility. The CVN test allows us to make multiple measurements of different characteristics that allow us to tease out the contribution of the ductility to the apparent strength.

Figures 1.6A and B show two sets of broken steel Charpy V notch test coupons. The untested

[23] If you have not seen a Charpy test, this 2019 YouTube video link shows the initial coupon at 45 seconds as well as the machine and the test itself: Charpy Impact Tester Per ASTM E23 Notched Bar Impact Testing of Metallic Materials | https://www.youtube.com/watch?v=qeR6Lyyrotg

coupons are 10 mm square bars 50 mm long. A "v" shaped notch is cut parallel to the small ends in the center of the length.

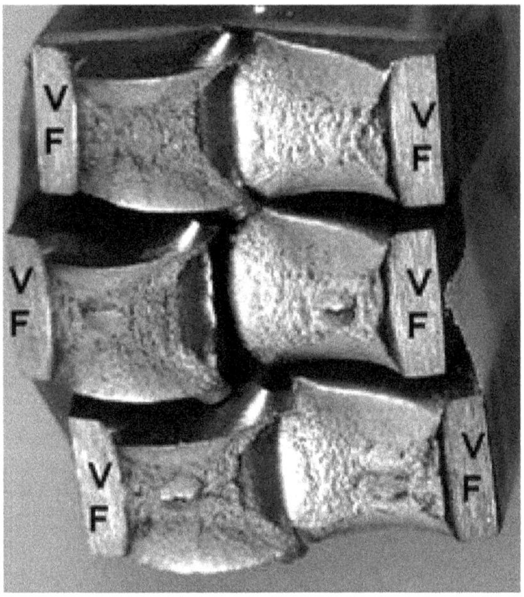

Figure 1.6A: *Ductile Behavior – "VF" means V-notch face; this is the original machined surface of the stress concentration notch.*

We see the faces of the "v" notches on the outer edges of all of the broken fragments (marked VF). The bars were broken by hitting them with a freely swinging hammer on the surface opposite the notch. We can see that the ductile coupons have noticeable deviations from the symmetrical 10 mm square bar described as the starting point for these tests. The brittle coupons have very smooth crack surfaces, and very little evidence of deformation. They still appear symmetrical and square.

On completion of the impact test, we read how much energy was absorbed by the deformation and fracture processes by noting how much lower the hammer swung back up after falling down to the position where it encounters and breaks the test coupon. The value is shown in foot pounds or Joules. But we're not finished!

The technician may measure the widest portion of the deformed bar. That allows calculation of a "lateral expansion" percentage, which is one measure of ductility. Zero lateral expansion is as brittle as it gets.

Finally, the technician may examine the fracture surfaces to determine the "percent shear". Shear technically implies ductility at both macro- and micro-scales, although the test method does not require anything more than a visual check using the eyeballs. Shear is evaluated by the presence of "shear lips".

- The shear lips on the Brittle coupons are non-existent. The percent shear would be zero.

Those on the Ductile coupons are very large (although not as large as the cup sides on the cup and cone fracture of Figure 1.4). They involve over 50% of the fracture cross-section.

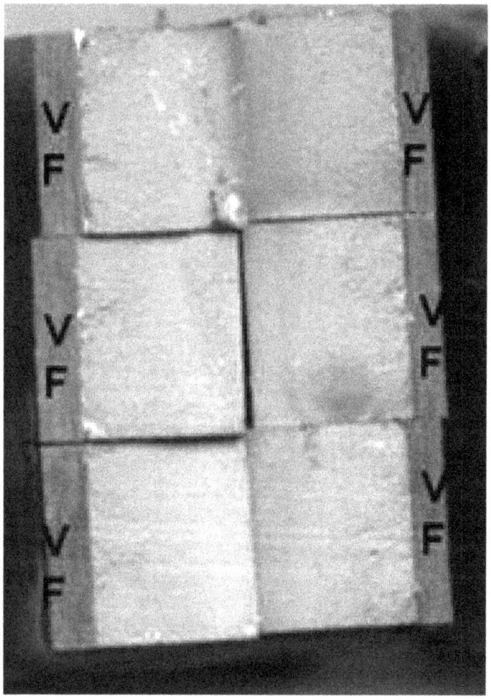

Figure 1.6B: *Brittle Behavior – "VF" means V-notch face; this is the original machined surface of the stress concentration notch.*

Figure 1.7 highlights another important feature of ductility. This figure depicts two sets of coupons cut from the same part. They have intermediate ductility values compared to what is shown in Figure 1.6. Ductility is facilitated by higher temperatures and impeded by low temperatures.

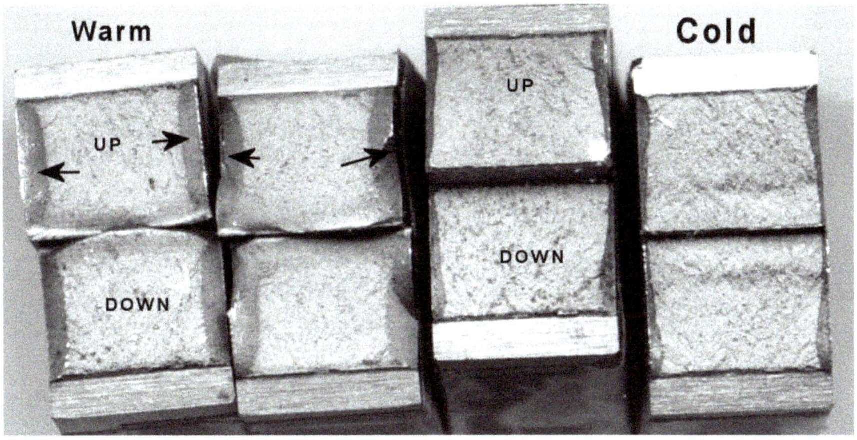

Figure 1.7: *The coupons tested at a warm temperature have about 25% shear. Those tested cold have only 10% shear. The protruding ("Up") shear lips are matched by recessed ("Down") shear lips. Note that the second coupon from the left has mixed "up" and "down" shear lips.*

Charpy tests are one type of mechanical testing. Tensile testing is another. Tensile testing, like Charpy testing, provides numerical values that characterize both the strength and the ductility.

But the tensile test results are easier to use in a design, because the tensile strength in pounds per square inch, or MPa, directly helps the engineers calculate the loads and forces that the structure will be able to support.

But if something goes wrong, such as a loading spike, there could be an issue. Even though computer modeling is being used more often as computer power increases, it is still difficult to reliably predict the behavior of a large structure or complex shape. If the specified structural steel had appropriate minimum Charpy requirements, assuming the material met the requirements, a bit of a safety factor could be ensured.

> *This was not the case, it was determined, for a very long series of castings made for the US Navy. The metallurgist at the foundry decided that she knew better, and when the test bars failed to meet the requirements, she lied and stated that the castings met all requirements.*[24] *She is now facing prison time, and she and her employer will have to come up with a chunk of change to cover the fines associated with this criminal activity.*

Understanding the nature of ductile and brittle behavior is the key to understanding the job of mechanical and civil engineers involved with design of structures. Materials that display ductile behavior surprise us with disaster less frequently than those that are inherently brittle.

To return to the header question of this section: Is Brittle Bad? The answer is that for most of the commonly used structural metals and alloys (iron, aluminum, copper), ductile is generally highly preferred. For some applications, a sacrifice of ductility is required in order to gain some other property, usually strength.

1.23 Human Factors

The failure analyst must also have some basic understanding of human factors. If we are the person who will provide the answers about "why it failed", **we must be able to prevent as many failures in our own thinking processes as we can while we work.** We must also be able to communicate our findings. We must be able to negotiate access to the resources we need to get the job done in an appropriate manner.

Our skill with these human factors will, to some degree, determine the complexity of the projects we will be asked to work on.[25] As we work on

[24] Read the U.S. Attorney's Office press release: Web Search: *Former metallurgist lab director pleads guilty to major fraud on USA* | *https://www.justice.gov/usao-wdwa/pr/former-metallurgist-lab-director-pleads-guilty-major-fraud-usa*

[25] Our willingness to work on projects that we have ethical concerns about also influences which projects we work on.

more complex projects, we gain additional skills that will be useful in other complex projects.

1.24 Materials Engineering Informed Failure Analysis: A Process Overview

Let's review the basic steps that are usually a part of the failure analysis work. Especially for those who are not engineers, it is important to have an idea of the basic activities that are included. It is also important to gain a general idea of how we use the data generated by these activities. Our goal is to create the most realistic and detailed picture possible of how the damage happened.

The preliminary steps are glossed over in much of the existing literature and teaching programs for failure analysis. This book will differ on that count, but for now, I will follow that trend, and gloss over the details of background data collection and visual inspection. The main goal of the present section of the book is to set the stage for beginners and non-technical people to get introduced to the common lab work that is performed as part of a failure analysis, and the general ideas that are used to interpret the data. People who have been practicing materials engineering informed failure analysis for even a short time will naturally be familiar with these basic steps.

This book takes a **four-step approach to failure analysis**. It starts with Step Zero and proceeds to Steps One through Three.

STEP Zero thing. If you do not know how to do nothing, count backwards from twenty and breathe deeply. Then consider your frame of reference. Or, as we said earlier, clarify the context in which the investigation is to take place. Ask yourself what, exactly, are the goals of the investigation. Focusing on goals will facilitate a higher chance of determining the answers to the relevant questions.

STEP 1 is all the remaining activity that precedes any obvious destructive testing of the components.

What is the result of Step 1? You must prepare a detailed but flexible plan with defined steps to answer the specific questions of interest to the stakeholders.

We start with background data collection and visual inspection. Photography, and any microscope work that may be done without changing the physical evidence in any way, follow.

NOTE: for big disasters, moving the fragments of a broken structure might be considered destructive.[26] In this case, there might be a "Step One-Half!" As the medical first responders know, if a person might have a broken neck or back, they are not to be moved, without first ensuring that the spine is supported. Part of becoming a competent failure analyst is learning to cultivate the mind-set that moving a fragment might be considered "destructive testing" in some cases.

Step 1 includes deciding how the part will be cleaned for additional inspection, and how the part will be cut, if needed. In other words, Step 1 includes detailed specification of the protocol, focusing on getting consensus for any destructive testing. Step 1 may include radiography, but many of the activities classically called nondestructive testing or evaluation (NDT or NDE), are nondestructive to components as manufactured, or in service, but not necessarily to evidence required later in the investigation.

Many NDT methods are thus actually part of Step 2. Step 1 includes optical microscopy and electron microscopy, as well as surface hardness testing, if they can be done relatively nondestructively. For many small parts, those that can be easily manipulated for optical or electron microscopy, Step 1 can be very

[26] While there is much overlap of these steps for all types of damage and failures, we'll focus on the case of fracture for portions of the rest of this introduction.

informative. For the very experienced person, it might provide information that can be used to unravel the mystery of the sequence of damage. But truly gaining data that demonstrates a chain of steps that led from being intact (usable) to being damaged (not usable) often requires additional work.

STEP 2 follows Step 1, with cleaning[27] when needed, and cutting to allow microscopy, both optical and electron. Step 2 also includes the classical testing that most people probably think of as the heart of the analysis, cutting sections for strength evaluation, core hardness checks, and metallographic cross-sections.

Step 2 often reveals the microstructure and characterizes the response that the component had at the time of the damaging event. In the case of mechanical damage, mechanical test coupons may be extracted, or removed from, the component. Specific loading parameters will be chosen to simulate some aspect of the loading. These same general ideas may be applied to corrosion damage evaluation.

Step 2 usually requires cutting. But if we could find an equally cost-effective but nondestructive way to get the required data, we would. Whoever is doing the cutting must work to produce the required test coupon from the relevant position within the component, without modifying the structure. This is a real challenge for some materials, and for some sizes and shapes of components.

Metallographic cross-sections are the primary tool that materials engineers use to reveal the microstructure of a component. This, along with document review, is what allows the analyst to determine whether the part was manufactured properly.

Casting, forging and heat-treating processes leave tell-tale signatures of their conformance to recommended guidelines, even if it is not possible to re-perform the tests that were done for initial certification. Reading these patterns, which are revealed by metallography, is what Materials engineers learn to do. Analysis of the microstructure is required to determine whether the part was made from the proper material, using the specified procedures.

A microstructure evaluation is commonly performed by preparing one cross-section or more (for thick sections) of the part. The goal is to allow the variation in the microstructure from the near surface layers to the depths of the core to be documented.

GOING DEEPER

Metallography Steps

1) Cut the cross-section
2) Deburr as appropriate
3) Mount as appropriate
4) Grind and polish to mirror finish*
5) Examine
6) Etch
7) Examine

****This usually involves 5 to 10 steps of increasingly fine abrasive.***

NOTE: Materials engineers learn to reveal, read, and record hidden microstructure evidence through the use of metallography.

Once documented, the microstructures of the subject of the investigation may be compared to known good microstructures of the same

[27] Sometimes the Subject Components are covered with grease, or corrosion products or deposits that make it impossible to make much progress. The cleaning process must be negotiated in the same way that every step of the protocol is honored.

material or published microstructure images made from known materials using known processes.

See Figure 1.8A (below). It depicts a piece of a weldment that has been cut in preparation for metallographic work. Three cuts were made to facilitate having a short piece for electron microscope crack surface examination (which is the left cut) and optical metallography (which is the middle cut). Because the right-end piece was very tall, a final cut was made to accommodate a metallographic mount.

Figure 1.8A: *This part had a weld running across what is shown as the top (center left to center right). The weld concentrated stresses along the toe. A crack formed and grew to the back surface, then extended left and right.*

Figures 1.8B and C show the final mounted, polished and etched condition of the cross-sections. The trained observer may learn much about the details of the process history of a component or assembly from a properly prepared metallographic mount.

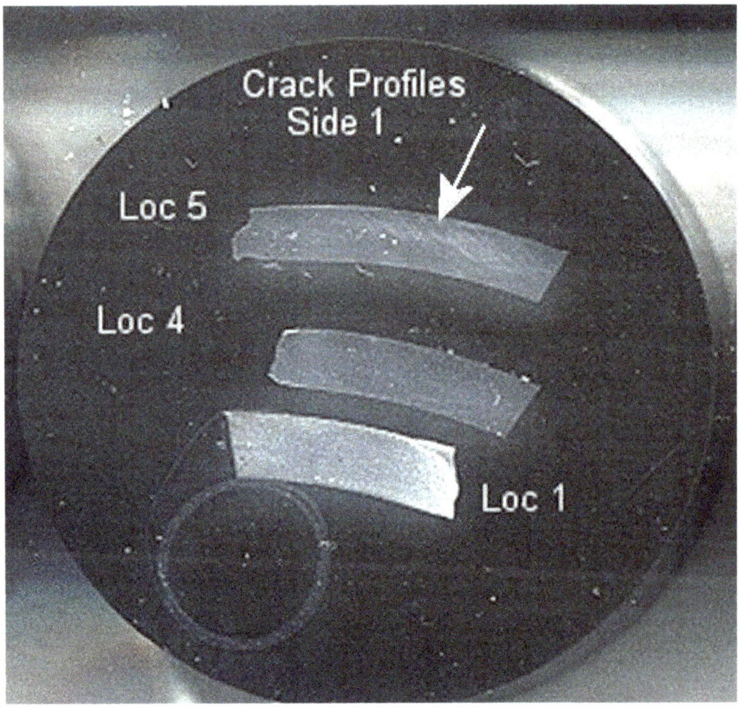

Figure 1.8B: *The upper section (white arrow) is the cross-section of the far-left fragment of 1.8A. The middle section below it is the left side of the second cut (slice) of 1.8A. The small pieces of 1.8A have been stood on end, surrounded with powdered plastic, hot pressed, polished to a mirror finish and etched to reveal the microstructure. The notes were marked on the images to let the client know where the higher magnification images were in relation to the individual sections.*

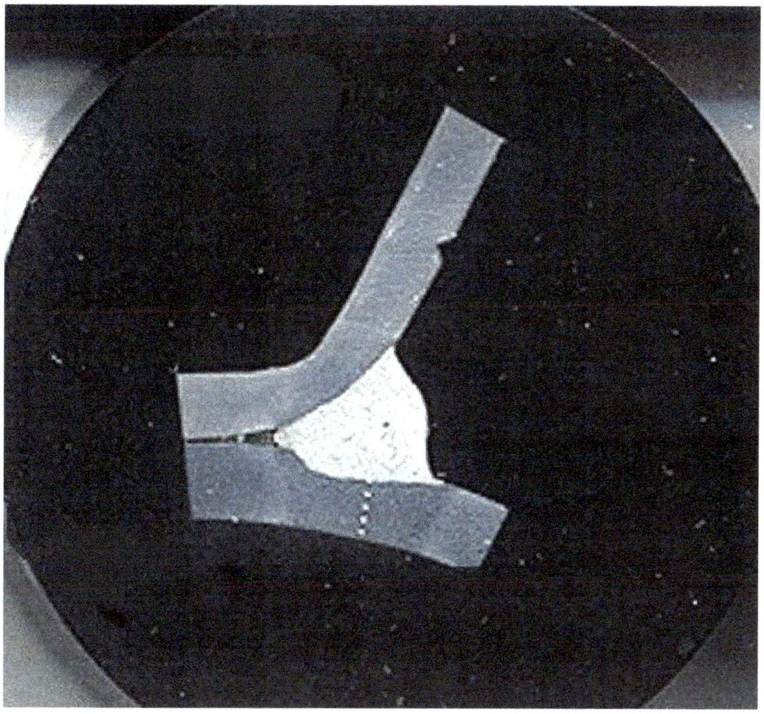

Figure 1.8C: *Cross-section of a weld, made as described for 1.8B, from a different piece. The etching clearly reveals the base material as one color and the weld metal as a different color. The heat affected zone of the base metal is also clearly visible (a little darker than the surrounding material). A Knoop 500g hardness profile was obtained in the dark weld heat affected zone (column of bright spots).*

The first step of the process to prepare the extracted segment as a metallographic cross-section usually includes embedding the material in some sort of polymer resin, usually in the shape of a disk.

If the material being evaluated is polymeric, usually a cold mounting process is used, often a two-part epoxy, similar to many materials used as adhesives. Cold mount is colder than hot mount, and exerts less pressure on the specimen, but it may still get warm. Polymer parts can be challenging to mount without damage.

For metallic parts, the more common procedure is to use a thermosetting plastic that is cured under pressure in a heated mold.

If the material is thin, or of a shape that could deform during mounting, cold mounting without pressure may be used for metallic specimens too. In either case, the cured disk is ground to clean up the exposed cross-section, and then ground with successively finer grits, to remove the damaged layer from the previous grinding, and leave a thinner and thinner layer of damage. Usually, the final step is polishing with a grit that is approximately one micron in diameter (or less). This produces, when properly done, a mirror finish, free of deformation that could obscure the features of interest. The cross-section may then be examined with a microscope.

Some features are visible in this as-polished condition, while other features require etching the mirror surface with various substances, usually acidic liquids, to reveal the groupings of the different types of atoms that were used to make the alloy in question. This is what microstructure indirectly reveals: the groupings of different alloy elements. The groupings that are revealed result from the processes that the material experiences.

Chapter 6 describes three main types of microstructure features created by the basic processes used to manufacture components out of metallic materials. The features that are tell-tale for polymer processing are usually more straightforward to interpret. These include such characteristics as uniformity of mold fill, alignment and distribution of reinforcement fibers and mineral fill, when present.

GOING DEEPER

Quality engineers and technicians spend a lot of time ensuring that the manufactured component meets the dimensional requirements. But it is important to remember that there is more required from a "good part" than meeting the dimensional requirements. While dimensional conformance is the foundation of a part being able to be properly assembled, especially for highly engineered products, the mechanical and physical properties of the material from which the part is made will influence its performance in service.

Many people are aware that there are many materials that look alike. There's no way to visually distinguish a piece of sheet steel that has 0.08% carbon from one that has 0.16% carbon, or 0.80% for that matter. There is no way to tell the difference between high and low molecular weight polyethylene simply by looking at it. Thus, there is more to a "good part" than the fact that it meets dimensional requirements and is made from the "correct material". Many common metallic materials may have a range of potential strength values that vary by five times (500%)!

The major additional factor that is required to make a "good part" is to ensure that it has an appropriate microstructure. Following the recommended processing guidelines is crucial to producing a part with characteristics suited to the specific application. Casting, forging and heat-treating processes (time at temperature, deformation history) leave "signatures" of their conformance to recommended guidelines.

Materials engineers learn to reveal, read, and record these patterns (hidden microstructure evidence) through the use of metallography or materialography.

While visual inspection in all of its plain and aided (microscopes, etc.) varieties reveals how the damage happened, usually microstructure analysis is required to determine whether the part was made from the proper material, using the specified procedures. This latter evaluation is usually what we mean by why a failure happened, if the failure is due to improper manufacture, rather than, or in addition to, improper use.

STEP 3 of the failure analysis is evaluation of the data, and preparation of the report. The key activity here is determining if the data sets from all the different tests that were performed, on the different parts or different areas of the component, are all generally self-consistent. If the data all "holds hands", then all is well. But if one set of data (macro appearance) says, "We have a brittle fracture," and another set (micro appearance) says "We have a ductile fracture," then it's time to take a break and come up with an explanation of why you think that the macro- or micro- data take precedence, or, alternatively, perform research to determine if there is a type of fracture that is brittle at the macro-scale and ductile at the micro-scale. (**HINT:** There is at least one. And there is one that appears ductile at the macro-scale and brittle at the micro-scale.) Barring such discovery, it may be necessary to go back to square one, and re-examine the part, or photographs, if possible.

As you write the report, using whatever format the situation calls for, keep in mind that intelligent people usually do not like to be told what to think. Wise analysts show enough data, and "translate" the engineering story the data tell so that the interested reader may follow along, and not feel bamboozled.

The higher goal, in important cases, is to explain in enough detail that the interested reader may reach the same conclusions as you; or at least, having been swept along by a well narrated mystery story, say, at the end, "Aha! I knew it was Mrs. Peacock (Stress Corrosion Cracking) all along! Colonel

Mustard (Creep) was completely innocent!" *(The analogy is a reference from the board game Clue by Hasbro. There was also a 1985 movie.)*

1.25 Are You a Goody Two-Shoes?

What is a "Goody Two-Shoes"? This term refers to a person who is on the top end of the population distribution when it comes to unselfish, helpful behavior. There is a negative connotation; Goody-Two-Shoes can be a term associated with self-righteousness. Great individuals, at least in historical context, can be considered great humanitarians rather than Goody Two-Shoes: Mother Theresa, Swami Vivekananda, Pema Chodron, Martin Luther King, Jr., Harriette Tubman, Desmond Tutu, Princess Diana, Mother Jones, and Siddhârtha Gautama. Of course, great people are often criticized even when they are respected.

Why am I bringing this up now? Failure analysis is about figuring out what went wrong; so, it is important that it be done correctly. Done right, we can avoid multiplying the damage.

Sometimes it is difficult to do the right thing. I mean doing the right thing because it's the right thing, even when it's difficult, even when there is personal cost, not because you are seeking to be praised or feel self-righteous. Do this and you will keep moving toward the humanitarian end of the scale and away from the self-righteous types who annoy most of humanity.

EXERCISE 1-B: List one or two people who you consider to be overly self-righteous, and one or two who might be considered humanitarians. These can be personal acquaintances, rather than famous people. What are the details of what they do and, perhaps more importantly, how they do it, that lead you to put them in the specific categories that you choose for them?

____

____

____

How you are perceived - as an annoying self-righteous jerk, or an unselfish and helpful individual, or as a humanitarian - depends on your true sincerity, commitment and intent.

1.26 Personal Costs of Doing What Is Right

S.L. Young is an award-winning social entrepreneur. In his blog on the HuffPost (*Unexpected Cost of Ethical Behavior*), S.L. Young comments about the personal costs of acting in accordance with higher ethical standards. His point is well taken; it is often not obvious what the right decision is, unless you have clear insight, clear guidelines and strong moral fortitude, because there can be repercussions and pushback. If humans had more of these qualities on all levels, we would naturally behave more ethically. The fact is, much wrongdoing occurs and goes undetected for long stretches of time. See the reference below for details of a situation happening as the text of this book was being finalized.[28] A detailed case study of this situation is provided later in this book.

Failure analysts are involved in a wide range of work-related activities. Some fields are obviously critical to public safety and some might seem like they are not. Even if it doesn't appear to be a safety matter, it might become one later. For example, what about maintenance? When does lack of maintenance due to hard financial times become a safety hazard?[29] When do overstressed employees, not given time to cool down in heat waves, or

[28] Ibid: "*Former metallurgist lab director pleads guilty to major fraud on USA*" | For more on the costs of ethical behavior to: https://www.huffpost.com/entry/unexpected-cost-of-ethica_b_6027034

[29] Web Search: Surfside condominium collapse | https://en.wikipedia.org/wiki/Surfside_condominium_collapse

delivery drivers not given time to walk while carrying the packages, start to have safety failures? Or die? Who caused the failures? The employee? The management team? The business environment? How can we sort these issues out? I'm not offering answers here. Texas made it illegal for municipalities to require water breaks for outdoor workers and several died in the early part of the summer of 2023.[30]

Investigations into the Florida condominium collapse of June of 2021 revealed to the public that structural inspections for new condominiums were required only 40 years after construction was completed. Because 98 residents died in the collapsing building and engineering recommendations were ignored, everyone could see simply meeting minimum legal requirements can be unethical and deadly. In a special session, the Florida legislature created a more rigorous milestone-based inspection plan based on proximity to the coast and age of the building to prevent future tragedies.[31]

Earthquakes, tsunamis, floods, fires, hurricanes – all natural disasters expose examples of maintenance-related public safety issues. Hurricane strength winds have revealed corrosion so severe it allowed a structure to be torn apart. Some structures released toxic chemicals as a result of the damage. Are these ethics failures if no specific law was broken? Or are they then simple results of the difficulties of the human condition? In 2017, the Michigan Civil Rights Commission determined that the root cause of the lead poisoning of many citizens of the City of Flint was systemic racism.[32] The Governor apologized, and the ramifications were still multiplying in 2022.

Engineers who obtain a professional license to perform and approve work directly affecting public safety are obligated to maintain their behavior within certain ethical standards. Usually, these engineers are working to reasonably well-defined structural codes. In these cases, the right ethical decision is usually clear, even if the parties choose to do otherwise.

1.27 Suggested Reference Books - Failure Analysis

1) ASM International and other organizations have published many books of case studies. These are always good to have to consult. Maybe it is the material that you can't find an example of anywhere else; or maybe the fracture mechanism. Or maybe some detail of the shape of a component with which you are unfamiliar. The problem with case study books is that they usually don't do a good job of describing the specific issues I have commented on above, such as "How did they know where to cut the part so that the discontinuity was revealed?"

2) 2002 version of the Metals Handbook, Volume 11, on Failure Analysis and Prevention. This book presents the mature thought of many people who had been working in the field for decades using the then relatively new technologies of the scanning electron microscope and computer simulations of stress fields and accident scenes. The impact of these technologies on how the active and contributing members of the profession approach the work can hardly be overestimated. This version did not have the components and industries section in the 1985 and earlier versions.

3) *The Metals Handbook Volume 11A: Analysis and Prevention of Component and Equipment Failures*, was published in 2021.

4) There is also the older, but still very valuable *Volume 12 of the Metals Handbook*, whose main content is an atlas of fractographs of many different materials, mostly metallic. There is also a new version of this book.

---

30 Texas Governor Outlaws Life-Saving Water Breaks | https://www.democracynow.org/2023/7/11/texas_prisons_heat_water

31 Ibid; Web Search: Surfside condominium collapse | Impact - Legislation Section.

32 'Systemic racism' at root of Flint water crisis | https://apnews.com/article/us-news-race-and-ethnicity-mi-state-wire-flint-michigan-df42de2ec4424193866467a2981ccb51

5) Then, there is the classic work, still one of ASMI's bestselling books after many years, *Understanding How Components Fail*, by Don Wulpi, revised by Brett Miller. This book does an excellent job of helping the reader learn to identify and classify different types of damage in metallic components. The emphasis is on mechanical damage.

6) Mars Fontana's classic, *Corrosion Engineering*, is an important work. Fontana was head of the Metallurgy Department at The Ohio State University. I think he deserves credit for creating a system of classification of corrosion types that is essentially still the foundation of the field today. One of my colleagues who has spent many years doing corrosion analysis is not a fan of the Fontana method. The complaint is that Fontana's method confuses characterizations, mechanisms, and modes. After thinking about my colleague's complaint, I understand what he is getting at. I might say it differently. Fontana's method alerts us to different circumstances in which we might expect to find increased risk of corrosion due to different types of geometries, corrosive substances, environments and material combinations. Being a generalist, I find it useful to know that pits, crevices, pairing of dissimilar metals, low or high pH solutions, and high temperatures in the presence of oxygen or sulfur can degrade structural materials. I find it useful to know in advance what evidence is compelling when I am inspecting a component or assembly that appears to have corrosion damage.

In its black cover and handsome copper lettering, it makes a beautiful addition to a book shelf.

7) In *Plant Failure Analysis* by Neville Sachs is geared toward people who don't necessarily have a metallurgy lab. It helps the general maintenance and operations employee gain an understanding of what information they can extract from a visual examination of a "failed" machinery component. It also gives guidance on when a more detailed investigation would be helpful. This is a good book to help the failure analyst with a materials background understand more about the lives of various machinery components.

8) Periodical journals of failure analysis are also helpful to many people. *Engineering Failure Analysis*, published by Elsevier, and the *Journal of Failure Analysis and Prevention*, published by Springer, and edited by a committee from ASM International's Failure Analysis Society, are both useful resources. Of course, there are many virtual journals that publish on topics that may be of interest to failure analysts.

1.28 Suggested Reference Books - Clear Thinking

1) *How to Think like Leonardo da Vinci: Seven Steps to Genius Every Day*, by Michael J. Gelb, has seven chapters with short exercises to improve your creativity. He makes the point that creativity is a cornerstone of clear thinking.

2) *Thinking in Systems*, by Donella Meadows, is a whole book that explores specific methodologies for contemplating complexity. Systems Thinking was developed to understand environmental degradation. Now there is a complex problem for humanity!

3) And finally, the indispensable *Tools of Critical Thinking: Metathoughts for Psychology*, by David Levy is composed of thirty short pithy chapters with descriptions of how people think erroneously. It contains short exercises to overcome these limiting thoughts and the behaviors they can produce.

On my path to cultivating competency, I consider these to be my core references. If you are just starting out in the field, build your library. Invest in yourself!

1.29 My Exercise Answers for Compare/Contrast

My answers are below listing some decisions where I am most sure I am correct.

- When I am angry or ashamed, I KNOW it.
- When I am hungry/ thirsty / want sex / need the toilet, I KNOW it.

- I do not like to drink Guinness or any stout.
- When a new pain arises, I KNOW it.
- When I feel lonely / want love or companionship, I KNOW it.
- When I feel annoyed, I KNOW it.
- When I am happy, satisfied, proud, I KNOW it.
- When I feel good about having helped someone, I KNOW it.
- If I am on a plane that starts losing altitude too rapidly, I will experience anxiety, fear, or terror.

My answers are below listing some of the things I am (or have been) fairly confident of.

- If I jump out of a window, I will fall, not fly.
- If I walk into a closed door, I will not end up in the room beyond the one I started in.
- If I find myself in a building with a large uncontrolled fire, I may grab some things but I will immediately start planning an escape.
- I will be hungry when dinner time arrives.
- My dad would call me to sing Happy Birthday on my birthday.
- I will never eat tomato hornworms, even though I have seen some recipes and some people think they are tasty.

(Image by Doug Smith from Pixabay)

- I generally enjoy the flavor of the fruits and vegetables that I grow in my garden.
- The methodology presented in this book will improve the failure analysis and general thinking skills of those who practice it.
- That room temperature unalloyed iron near the surface of Earth is usually found in a bcc crystal structure.

Below is my list of some things I don't know:

- When I will die.
- How many years of healthful living I will have.
- When and exactly how the next war will erupt.
- Who will win the next election for congress / president / township clerk.
- When my failure analysis business will close its doors.
- Exactly what I will eat for dinner tonight, or any night.

Below are two insights I can draw from my lists:

- The things we KNOW most surely are almost all related to feelings and sensations.
- If we can't acknowledge our own feelings and sensations (at least to ourselves), we will have no foundation for recognizing ANY TRUTH.

CHAPTER 2

A Conspectus of the Failure Analysis Process

Humans Doing Failure Analysis

2.1 Overview

Why did the part fail? Isn't that what we all want to know?

Luckily, we always have a true answer that we can provide to the question of why something broke or got deformed. It's because the stress exceeded the strength.

We always have a true answer to the question of why something has lost some portion of its original mass through wear or corrosion. For wear (mechanically induced material loss) the answer is the same as for fracture and deformation. That's because wear is essentially a complex process that involves either deformation or fracture, or both of these damage types, at the microscopic scale. For corrosion, it's because the environmental conditions were more severe than the component could resist.

Of course, these answers are an example of tautological reasoning. We're not really providing any new or useful information beyond what the asker already had stated.

For all failure analysis, we need to remember that the strength and corrosion resistance characteristics depend on the microstructure. Certain environmental factors can change the microstructure over time. Usually, these changes result in a degradation of useful characteristics when compared to the carefully optimized properties that the manufacturer first produced.

In a sense, this entire book and all of the practice of failure analysis of engineered components and structures may be viewed as an attempt to get more specific, and thus useful, answers to the questions of why something is no longer serviceable. In order to answer that question in terms of the basic physical changes that led to the diagnosis of damage or failure, we must determine the specifics of how the damage happened.

The core of that technical process starts with

classifying the basic damage category into mechanical, environmental, or both. Environmental damage may result from thermal exposure or impingement by other types of physical energy (light, radiation, etc.). The other main category of environmental damage is classical corrosion: exposure to incompatible chemical compounds or elements or incompatible concentrations or combinations of chemicals.

When we contemplate complexity, we see that all of the interactions of the component with its environment are complex. Even what we think of as simple damage always takes place within an environment including chemical, physical (energy impingement), and mechanical (load or force) interactions. The temperature affects how forces are transformed into stress fields and how aggressively a corrosive substance may attack a material. Even air is a different environment from a hard vacuum. Air is not necessarily the default of no consequence.

The answers that we are able to extract, like a confession from a criminal, will be more useful if we are thoughtful about how we go about getting the information. Just as torture often leads to incorrect information from suspected criminals, damaging the evidence from a failure event can also mislead.

Failure analysts need to cultivate multiple skill sets in addition to knowing how to apply technical knowledge. This includes learning about the domain of human factors. Sticking to a methodology (human factor of discipline!) will go a long way toward gaining confidence from others who may have natural tendencies to mistrust our findings or conclusions. Understanding when someone might be tempted to hide something is a crucial skill in some situations. Knowing how to tell that someone is unaware of the significance of an important factor is also critical. These are all so-called soft people skills that produce valuable hard outcomes, and the interested reader will need to study some of these more subtle human behaviors with other resources. Asking questions while listening for self-consistency in the responses is one way to move toward improving your skills in this area. Listening for discomfort, and trying to interpret whether it is hesitation due to simple ignorance, or discomfort and hesitation due to deceit is another skill set that may be cultivated. Obviously, being sensitive to such matters will help the cultivator of competence harvest their crop!

The most powerful and useful failure analysis methodology is flexible enough to modify when a roadblock to the investigation appears.

Understanding why a failure took place must be accompanied by understanding how it happened. Both of these questions (why and how) have many layers from which they may be approached. But that is still not enough.

2.2 Ethics in Engineering

The link between basic competence in any professional endeavor and ethically produced work product is very strong. It is often difficult to distinguish ignorance from incompetence, or even a slightly more benign failure to keep up with new developments. In many professions, continuing education is a requirement of ongoing personnel certification or licensing. Investigations of corruption in the public eye remind us of how difficult it can be to achieve a consensus about whether someone really doesn't remember, or is acting in an intentionally deceitful manner.

That said, I have observed the most common cause of incorrect statements in failure analysis reports is the combined effect of time constraints and intellectual laziness. It is admittedly difficult to explain the technical details of a failure event in a manner readily accessible to a person completely lacking a background in science or technology.

> *In early 2022, one of my neighbors informed me that science is nothing but a bunch of people sitting around, talking, deciding what is the most popular theory, and calling it truth. He claims he got this definition from a dictionary. I wonder what dictionary he used because my experience with dictionary definitions of "science" is not remotely similar to his. If it was, nobody should use the word science. But his definition of "science" is not what I am basing this book*

upon. Hopefully, this book will be a useful "trail of breadcrumbs" for anyone who wants to start on a path of gaining clarity rather than sowing confusion.

A slightly less distressing example comes from a November 6, 2019, public radio talk show (1A.org) guest. He tried multiple times to explain the difference between an annual quantitative calibration of a breath alcohol spectrometer and its "before each use" calibration verification. In radio, every second counts; and they had to move on from these boring technical details before everyone participating in the conversation understood.

Why would that be? The participants were presumably all educated people. Maybe the state of science education in the USA is a factor. Sadly, the United States ranked 25th out of 72 countries that participated in the 2015 OECD test of science knowledge among 15-year-olds.[33] While that doesn't necessarily correlate directly with adults' ability to use scientific knowledge in their daily lives, my personal opinion is that it does not bode well for communications between lay and technical people.

NOTE: *"Beta readers" of this chapter suggested that they wanted to know more about case studies in this book, especially those where only a small snippet is provided. There are multiple reasons for keeping many of the examples to this condensed meta-level. First, I don't want to make anyone unnecessarily unhappy with me by sharing their recognizable stories. This is a business ethics issue.*

Second, it's often the meta-point (or overarching concept) that I am trying to convey rather than the specific facts of the case. The meta-point applies in more situations. As an old saying goes, "I'm pointing to the moon. Don't look at my finger." Identifying which metaphorical level is most relevant can be challenging and is another cultivatable communications skill.

Thirdly, the original title for this book was Cultivating Competence in Failure Analysis: Materials, Mechanics, and Human Factors. The human factors that are emphasized in this book are ones that are related to how individuals perform failure analysis, not the human errors that result in failures we are paid to investigate. There are other books on that subject. Of course, there is overlap. But one of the biggest errors people make in performing failure analysis is "jumping to a solution". People may do this multiple times at different steps of the investigation. Learning to tolerate ambiguity and uncertainty is one of the most powerful ways to lay the foundation for useful results.

Many lay stakeholders do not care to spend the time to understand the details of the investigation we have just completed on their behalf. How do we decide that the explanation was close enough to enable the stakeholder to take appropriate action, and when it was oversimplified? The thoughtful, competent engineer may struggle with these communication and data interpretation decisions over the course of a long career. For failure events with devastating consequences in terms of human or financial costs, it is incumbent on the lead investigator to take whatever actions are required to ensure that the stakeholders are adequately educated.

Competent failure analysis requires getting data from multiple sources (tests, observations, review of documents, etc.), When the data are self-reinforcing, the analysis is proceeding toward confidence inspiring conclusions. Conflicting data tells you that you have more work to do. Ignoring conflicting data is easy to do. Most people are not contemplative by nature. Most people accept what they are told when they are young. Most people do not question revealed truth, whether it is about religious issues or secular. Most people go through life with a single world view. People may gain additional factual knowledge, but moral values are generally persistent and usually unquestioned.

[33] See the National Science Foundation's statistics at https://nsf.gov/statistics/seind14/content/etc/nsb1401.pdf

"Of course," you may be thinking, "Murder and kidnapping are bad." But most of the situations that failure analysts work on are not that black and white. People who come from a culture that discourages questioning authority may find it more difficult to speak up when something is wrong. It's important to cultivate several unrelated relationships with people who you think would be supportive if you find yourself in an ethical quandary.

> **WRITE THIS DOWN IN YOUR OWN HANDWRITING!**
>
> **Keep it in your wallet or as a graphic on your digital device!**
>
> **Ethical questions sometimes have obvious answers on what needs to be done, even if the individual struggles with how to find the inner strength to do it. But often, answers to the big questions are not so clear. For example, how much technology is good for humanity? Is a particular short cut acceptable or not?**
>
> **The human condition is difficult because we are always balancing different hoped for "goods" against potential feared "evils". There is no easy-to-follow recipe to ensure right and avoid wrong.**
>
> **When an individual faces ethical questions as a consequence of participating in a failure analysis investigation, it is important to seek out a trusted colleague, mentor, or a trustworthy person (when available) if a technically savvy person can't help.**
>
> **No single person has the all-encompassing vision that would be necessary to judge every ethical quandary that might arise.**
>
> **Think about it. There is less shame in asking for help than there would be dealing with the responsibility of a disaster.**
>
> **BALANCE IS THE KEY!**

Here's one of my own personal examples: Recently, I assisted a mechanical engineer on a small damage claim he was investigating for an insurance company. His conclusion was that Party A made a mistake. I agreed with his conclusion. It was fairly obvious. But he and another engineer involved invoked a specification that they noted was not met by the Subject Non-Functional Assembly. That is an easy way to explain it. "It failed to meet the specification." However, it was obvious to me that the specific failure that happened could still have happened even if it had met the specification. And, it might never have come to anyone's attention for failure, even if it did not meet the specification.

Inert mechanical devices do not know the specifications to which they are made.

It bugged me (it still bugs me) that this engineer was too lazy (as I saw it) to explain the actual situation to their client. However, the client was probably happy not to have to strain their brain to understand complexities when they really "didn't matter". The client was probably happy not to have to pay another 500^{\underline{00}}$ to cover the engineering time to document the more complicated reality about the product. This "lazy" engineer may be seen to have better communication and business skills than I do. He's certainly more financially successful than I am!

In contrast to the personal costs ("micro-ethics") paid by the individual who is acting with higher ethical standards than the rest of those in the environment, employees of companies that hold themselves to relevant ethical standards are able to do better work ("macro-ethics"). Fortune Magazine noted costs of $170 million for just two of its featured ten top corporate scandals of 2017.[34] In addition to fines assessed by the governments, many of the other ethics scandals caused loss of market share, accusations of

[34] See Fortune.com for the whole article: The 10 Biggest Business Scandals of 2017 by Lucinda Shen | https://fortune.com/2017/12/31/biggest-corporate-scandals-misconduct-2017-pr/

bribery, loss of position for top leadership, and criminal charges.

Corporate ethics has long been seen as affecting the relationship between the company and the customer. New business trends consider both outside and inside relationships. The way companies treat their employees is part of the new, more holistic corporate ethics and customers are more likely to punish companies for ethical lapses, even when they like the product or service, than they were in past eras.[35]

2.3 The Process of Failure Analysis

Failure analysis for components is a series of activities that unfold in a flexibly sequential manner, intentionally planned to reveal how the damage that led to loss of function occurred. Many people are not clear on the distinction between how something happened and why it happened. This is true even for people in manufacturing and construction. Describing the "how" must focus on the physical evidence. The "how" is the foundation of the "why".

The skilled failure analyst must have background knowledge in relevant disciplines of science and engineering. Being able to determine why the damage happened often requires a strong background in human factors (psychology, behavioral science, sociology, economics, etc.) as well as the so-called "hard sciences". Both physical and social sciences (primarily in the area of cognition) are considered in the body of this book.

Failure analysis as presented in this book is an engineering discipline. Engineering is based on science but is not identical to it. Many scientists seek knowledge to inform curiosity. Many scientists say that they are interested in truth. Engineering is less concerned with ultimate truth, and more concerned with using the knowledge that scientists and engineers have revealed. The engineer's goal is human convenience or financial gain. Engineers happily use empirical knowledge that has been gained in the absence of any theoretical foundations.

The endeavor of science is inherently concerned with understanding how the physical world works.

The human world, including psychology and other social sciences, is part of the physical world. The human experience of consciousness, including our ability to experience aspects of the physical world through our sensory apparatus, is part of science. Human behavior is part of science. See "Understanding human perception by human-made illusions" by Claus-Christian Carbon[36] to understand the importance of perception as a human factor and our powers and limitations to observe.

If we, people who work as failure analysts, are the ones who are "figuring out" what others can't, we need to understand the limitations of how our brains and sensory organs collect and interpret data. If we know what mistakes we tend to make as members of the human race, we have a greater chance to avoid some mistakes over the course of the investigation, and over the course of our careers. Francis Bacon was the first Western thinker to write about how our cognitive biases affect our ability to see truth. This was an advance over Plato's simple mistrust of sensory data. Chapter 3 takes a look at some of these issues in greater detail.

Humans perform further information processing in additional areas of the brain prior to experiencing normal awareness. In addition, humans have many heuristic thought pathways (short-cuts) that allow rapid analysis of data, which is often below the level of conscious awareness.

[35] This was written in 2018, before the ravages of COVID 19, which has, at least momentarily, only accelerated trends of employees in the USA refusing to work for companies that they find unethical.

[36] Carbon, Claus-Christian. (2014). Understanding human perception by human-made illusions. Frontiers in Human Neuroscience. | https://doi.org/10.3389/fnhum.2014.00566 | https://www.ncbi.nlm.nih.gov/pmc/articles/PMC4116780/

Humans are also able to perform conscious rational thinking. We can compare two, or more, things, simultaneously. We can compare a whole range of "things", from simple physical items to complex theories. We know how to look for both similarities and differences.

Just What Is Science, Anyway?

Modern science is a joint endeavor by a community of scientists who are able to understand the methods used in experiments well enough to potentially reproduce the finding or to speculate on why it was not reproducible. True modern science is inherently epistemological in nature, by which I mean that it works forward and backward. It works from experimental data in order to formulate theory; and it works from theory to try to predict new findings. The best new scientific theories suggest ideas for new experiments, which can be used to support or weaken (or even kill) the new theory. Modern science brings new knowledge to light in what could be viewed as an ever-expanding helix of understanding.

Figure 2.1: *The opening of this snail's shell gets larger as the snail grows. As time passes, the volume of the shell also grows in three-dimensional space.*

Rational thinking is a foundation of critical thinking. The most essential function of critical thinking is the ability to question the context, to frame the situation explicitly, and to communicate the relevant background information. Failure analysis is most successful and useful when it discovers all of the relevant data needed to improve the understanding of the stakeholders. Even if the "failure" was a one-time event on a unique structure, a properly evaluated catastrophe will eventually shed light for the community at large.

Failure analysis is most successful when the context of the design, manufacturing environment, and service conditions are all considered.

2.4 Business Decisions as Human Factors

Let's discuss another subset of human factors. They are broadly related to corporate culture. By studying these big issues, we may learn why the problem that led to failure arose. These might include greed, rushing to market, etcetera. These are factors that are currently thought to have contributed to two Boeing 737 Max 8 airplane crashes within less than six months of being placed into service. The crashes happened within a few years of the first of these models being delivered to airlines. How did the company balance profit with safety? For the majority of small, routine failure analysis projects, we are not asked to investigate to this level.

After the collapse of the World Trade Center in 2001, the main take-aways were the need to improve evacuation in an emergency. The stakeholders in the business of designing, building, regulating and occupying skyscrapers determined that it was not feasible to require skyscraper designs to be able to withstand the heat from a fire initiated by a full load of jet fuel.

What does the term safety mean in this context? All of engineering, all of life, may be viewed as balancing the tradeoffs between safety (or risk) and the rest of our activities. Up to a point, it's safer to stay in bed than to walk or get in a vehicle and go to work, school, or vacation. It is always possible to make a more robust structure. But what price is the customer willing to pay? Working to "generally accepted industry practice" is one way to decrease the chances of catastrophic failure; or at least of being excoriated, blamed or fined if things do go wrong.

These business decisions have been drawn into public view as a result of some high-profile engineering investigations. The terrorist attacks that destroyed the World Trade Center in New York in September, 2001 were primarily seen as having political roots. Yet twenty years later, humanity is no closer, and perhaps farther, from addressing the conflicts that result in violence. Is that the result of a failure in political scientists' work or of the way human affairs are generally conducted? Do we lack education and opportunity to flourish? What considerations do we miss?

EXERCISE 2-A: Compare and contrast the how and the why of an object's lost usefulness.

In the space provided below write how one of your possessions lost its usefulness. Also, make note of why it lost its usefulness. *(Note, there may be multiple levels at which you can explore this assignment.)*

Going back to the time of Plato, philosophers argued that we contain, within us, knowledge of the structure of reality. Plato taught that ideal forms were more real than the models we make of them by sensory interaction. The interested reader may wish to contemplate how our knowledge of specific exemplars of Plato's ideal forms (the ideal horse, the ideal man, the ideal river, etc.) relates to what is commonly called intuition.

EXERCISE 2-B: How would you compare and contrast knowledge of Plato's ideal forms with knowledge presented by the holistic brain function known as intuition?

Dreams have a long history of providing new scientific and mathematical theories; I have occasionally awakened with fresh insights into failure analysis puzzles.

Humanity benefits when engineers use their knowledge for the general good. Is the backlash against scientific knowledge in the West related to the fact that technological progress and the enlargement of the human population has come at the cost of the rest of Earth's biome? Of course, some of the backlash is related to the same types of reactions that Galileo experienced from the religious organizations hundreds of years ago.

Is the backlash today related to the use of technology to create weapons of mass destruction? What is the obligation of the individual engineer to contemplate the effects of her or his work, beyond the effects on a personal bank account?

> *I have had the chance to work on projects where salvage is being considered for out-of-specification products found before shipment to the customer. Instead of scrapping an entire production lot, I like to encourage salvage when appropriate. Mother Earth has paid the price for the material to be extracted. Energy has been used to make the product. If waste can be avoided without excess risk, salvage may be the right choice.*

2.5 The Process of Science: All Theories Are Not the Same

As we said, engineering is informed by science. Science advances through the formulation of hypotheses, which are then tested multiple times by multiple investigators. At some point, a consensus may emerge among those working in the field. Either the hypothesis is strengthened, or it is weakened, or even thrown out. At some point, a fact-supported hypothesis may be used to successfully predict phenomena that have not yet been observed. Then, the hypothesis becomes a strong, reality-based theory. Some recent examples are:

- Success of quantum physics in predicting and facilitating observation of the Higgs boson.
- Success of Einstein's work, which allowed planning for (and eventually observing) a gravity wave.

Therefore, scientific theories are inherently different from personal theories in that they have been independently verified by multiple individuals and methods on the way to becoming an accepted theory. It is generally not possible to document the development of personal theories in the same way. One reason that personal and scientific theories are inherently different is that it is impossible to totally separate the observer from the observation.*

> *I may have a personal theory that I should wear red lipstick on Tuesdays. Lucky charms used by many athletes result from such personal theories. Unless I keep a verifiable record of people's response to me when I do and do not wear red lipstick on Tuesdays (and compare to some other day to evaluate the significance of Tuesday), my theory is merely a completely subjective opinion. Yet, we use the same word – theory – to describe a subjective personal opinion and a widely demonstrated and vetted set of scientific observations.*

It is impossible to totally separate the observer from the observation.*

**** The first to document this concept was the Hindu sage Kanada in his Vaisesika Sutra.***

It is important to emphasize again that a scientific theory is very different from a personal theory. While true, pure objectivity does not exist, as it is impossible to totally separate the observer from the observation, the process of scientific advancement inherently works to minimize the effects of personal subjectivity. Modern science is now working to minimize the effects of cultural subjectivity. Women and people from ethnic groups that have been under-represented in scientific fields will provide critical insight into how to continue to root out cultural subjectivity.

EXERCISE 2-C: Separate personal theories from scientific theories. List three personal theories you have and three widely accepted scientific theories.

____

____

____

____

____

____

Now list three debated scientific theories.

____

____

____

Despite the multiple layers of verification underlying any valid scientific theory, most scientists make no claims of eternal, absolute truth. Those who do are engaging in scientism rather than science. Similar to the definition in the "*Merriam-Webster Dictionary*", I'm using the term "scientism" to mean an exaggerated trust in science, and an unquestioned belief that if it isn't visible or detectable with instruments, it doesn't exist.

The proponents of scientism tend to be dogmatic materialists. They usually do not believe in anything they can't see or measure. They fail to understand that absence of evidence is not necessarily evidence of absence. Take a moment to absorb this.

Scientistic scientists believe in a WYSIWYG (What You See Is What You Get) world.

> *Anyone with any remote sense that they are more than the contents of their bag of skin is hereby invited to reject a WYSIWYG world.*
>
> *Science was founded as a reaction against conservative thought patterns of ancient thinkers. It always remains open to improved explanations of reality.*

Scientific theories are, at their root, models of how the world works.

- There is not yet a universal theory of everything.
- Scientific theories are generally grounded within a particular discipline.

Unlike a scientific theory, logic may be performed in a domain devoid of physical reality. Logic is not the same as critical or rational thinking. However, understanding the limitations of different types of logic is an underlying skill needed by people who hope to practice critical and rational thinking. And that includes professional failure analysts.

Here is an example of logic: If A, then B. If B, then C. According to pure logic, this would imply that if A, then C. In our very human minds, we usually automatically attempt to compare A, B and C. In fact, logicians love to trick their students:

- by making up nonsense words to see if students can follow a logical argument.
- by using real words to try to confuse the student, and
- by providing definitions of words that make the sentence false even though it would normally be true.

In physical reality, logic may not be clear cut at all. Logical thinking does not inherently include clarification of the context (critical thinking).

> **Proper use of logic ensures that the statements on which the argument is made tally with reality. Statements must be clearly defined and relevant to the given circumstances.**

We've touched on scientific theories and logic so now we must discuss another branch of study that is very useful to failure analysts: epistemology.

Epistemology provides a way for us to demonstrate that we really know what we think we know. It also helps us come to peace with the sadness of having to discard a beautiful, but false hypothesis or theory.

In short, epistemology involves an investigation into what distinguishes justified belief from opinion.[37]

When we take the time to evaluate the self-consistency of multiple data sets associated with the same event, we have the right to greater confidence in our conclusions, be they our originals, or newly modified as a result of epistemological confirmation.

Again, scientific theories are inherently epistemologically demonstrated. A hypothesis may not become a theory until the observations have been verified by additional properly trained individuals. **Failure analysis data evaluation is also inherently an epistemological activity** even if most failure analysts do not use this terminology.

For example, let's look at a failure investigation as an epistemological activity. A fracture investigation would normally include evaluation of the ratio of a component's fracture strength to its supposed service stress. That protocol might include measurement of the tensile strength, yield strength, and elongation as well as hardness, microstructure, and impact resistance. This information is accompanied by a background document review. The document review preferably includes a finite element analysis that predicted local stress levels. Composition analysis (to confirm the correct material was used) adds to the strength of the eventual conclusions. During our investigation, the fragments' appearances will certainly be subjected to macro-fractography and possibly micro-fractography. The failure analyst will seek multi-faceted evaluations of different "If A, then B, and then C" methods of possible preliminary theories about how the damage logically happened. Then the failure analyst evaluates and compares the individual theories with each other and examines them for coherence. Below are some details of coherence examinations from an epistemological viewpoint.

[37] Epistemology is a major subfield of philosophy, comparable to ethics, logic, and metaphysics. Epistemology is concerned with knowledge and how we become confident that we know something.

- If the hardness is consistent with the tensile strength, the heat treating might have been properly performed.
- If the yield strength and elongation are within specified ranges, then the heat-treating might have been properly performed.
- If, on top of all of the above, the microstructure is consistent with the specified heat treatment, the heat-treating is starting to look like it **was** properly performed.
- If the composition was within specified ranges, then we are right to have even greater confidence in our preliminary conclusions. Our hypotheses are starting to look like legitimate technical theories that might underpin the creation of an accurate causation scenario.
- If the elongation of a common structural metal was 45% in 50 mm, and the event supposedly happened at room temperature, and the macro-fractographic inspection reveals macro-ductile behavior, then yet greater confidence is warranted, because of multiple levels of self-consistent data.

NOTE: the original logical statements were based on relevant scientific and engineering knowledge; most (if not all) of that knowledge was originally founded on inductive logic.

In turn, **inductive logic is based on observation.**

Originating a relevant branch of knowledge means that someone, or a group of people, made observations. Then she/he/they compared and contrasted her/his/their way to a preliminary hypothesis about the relationship between the observations.

EXAMPLE OF THE ORIGINS OF INDUCTIVE LOGIC BASED ON OBSERVATION: The Englishman, Isaac Newton, noticed that the tides were related to the position of the moon and developed an explanation of action at a distance. This new theory of gravity revolutionized scientific thinking at the expense of many other ideas about the movement of the waters covering the earth.

Newton's Theory of Gravity united findings of the motion of the planets and what was then called "The Law of Fall". Newton's work gave one explanation that explained aspects of the world that had previously been seen as unrelated. That's the power of a great scientific theory.

That the same theory predicting tides was later applied to closely predict the motions of the heavenly bodies showed the power of another type of primary logic that has benefitted humanity over the millennia. **That is deductive logic.**

The progress of science thus inherently relies on inductive reasoning followed up by epistemologically driven deductively-guided experimentation. Or the other way around! Either direction gives us epistemological confidence.

Once a theory is accepted and moves into a wider population's thinking, the line between deductive and inductive becomes blurred. Practically speaking, the wider population accepts the new theory as a revealed truth and uses this framework to structure their data collection and they eventually subject the data to new rounds of deductive reasoning. This is true of intentional data collection as well as the sub-conscious data collection that has evolved for those in possession of a normally functioning human brain.

At least in theory it works that way; however, *in practice*, most failure analysis relies on a **third type of logical reasoning, called abductive logic.** [38]

Abductive logic starts with a set of observations and then seeks the simplest and most likely conclusion from the observations. Abductive logic allows people to maximize their resources by planning a pragmatic project protocol. Another way to see abductive reasoning is as a "SWG". What is a "SWG?" A "Scientific Wild Guess!" Occam's Razor, or the idea of "parsimony", is another way to understand abductive logic[39]. All of these ideas overcome the problem of limited data.

Properly performed abductive reasoning results in the creation of several scientifically informed explanatory scenarios, from which the simplest that is consistent with all of the observations may be selected. Abductive logic, unlike the inductive and deductive types, incorporates epistemology in at least an informal manner. Abductive logic also relies on the creative thinking function, a holistic way of evaluating all the data simultaneously.

Quality abductive reasoning in failure analysis further relies on the results of earlier deductive and inductive reasoning performed by others to build the scientific and engineering bodies of knowledge on which the investigation team depends.

Reliance on the properly collected data, and just enough creative speculation to create credible scenarios as just described, is the most robust foundation for the development of failure recurrence prevention. You will never have all the data you might wish for. Get used to it!

Another concern of philosophy is how we interpret the data that comes in through our senses. Experienced engineers are often quick to jump to conclusions based on their in-depth knowledge. When performing failure analysis, it is important to minimize such leaps of faith, where fact and opinion are not clearly identified. Describing color, texture, patterns, reflectivity, shapes, and density on their own terms allows all parties, even those lacking the detailed experience of the long-time practitioner, to see the important features that may be used to build understanding

[38] Abductive reasoning was formulated and advanced by American philosopher Charles Sanders Peirce.

[39] Occam's razor: with competing theories or explanations, the simpler one with fewer assumptions is usually preferred. If you have never heard or Occam's razor, check it out here: https://en.wikipedia.org/wiki/Occam's_razor

of what happened. This will be covered in greater detail in Chapter 3.

The study of critical thinking is commonly understood to include familiarization with common thinking errors made by almost all humans. One of the most well-known is confusing correlation with causation. Correlation on its own does not allow the direction or even existence of causation to be determined. Logic must supplement the correlation data with theories that explain the proposed causation. Exercise 2-D (below) will help you explore these issues.

David Levy warns against falling prey to "The Spectacular Explanation Fallacy".[40] Although the specific issues he discusses differ from those of interest to a failure analyst, it is important to remember that careful investigations of truly catastrophic (multi-casualty) events usually reveal no lack of precursor warnings. The thorough investigator of a catastrophic event usually finds evidence of longstanding failures to adhere to well established safety protocols.

Another problem that humans may run into when trying to use logic to investigate a failure event is perhaps the most difficult to avoid. David Levy calls it by several related names in his section on Investigating Phenomena.[41] **We must learn to see how our beliefs influence our perceptions.** Our underlying beliefs color our experience, and are difficult to identify. If we don't know that we're not seeing every relevant fact, we will be more likely to miss a relevant fact.

Now is a good time to make a note to get a copy of Levy's book. Put it in your "wish list" at your favorite on-line bookseller.

As we have already noted, failure analysis is an engineering discipline founded on science. Science is fundamentally different from religion, which is the other human activity that is concerned with truth. How are they different?

The scientific method is founded on asking questions; that is of upmost importance. Religion, on the other hand, is based on answers being provided by the founders; that is of upmost importance. Science, also, does not make claims of providing eternal truth about the subjects of investigation.

Good scientific practice eventually leads to reliably useful scientific knowledge. Older knowledge that conflicts with new knowledge is sometimes slow to be expunged from the memories of working scientists. Quantum physics, for example, was formulated approximately one hundred years ago, yet most scientists today still struggle with what it might mean on the level of everyday occurrences. In fact, most lay people are still comfortable thinking in terms of Newton's Laws of Motion, which, at the time of writing this book, was 333 years old.

EXERCISE 2-D: List three pairs of things that are correlated.

Try to find one that is a true causal pair, one that is apparently a coincidental correlation, and one that is really hard to determine. Justify your examples of each type.

____

____

____

____

____

____

____

[40] Spectacular Explanation Fallacy: extraordinary events do not require extraordinary causes. | Source: *David Levy, Tools of Critical Thinking: Metathoughts for Psychology*, Chapter XIX

[41] Ibid, *Part Four, Investigating Phenomena,* Chapters XXII – XXV, pages 133 - 172

If you can't think of something on your own, you could explore health or nutrition advice columns. Do wine drinkers enjoy better health because they drink wine? Do vegetarians suffer from poor health because they are vegetarians? Do viruses cause disease?

Engineering knowledge is slightly different than scientific knowledge in that it contains more empirical knowledge. Thus, it is perhaps more prone to error due to application of experimental results; especially when the experimental results are outside of the context of those in the original data collection.

EXAMPLE: I viewed one of the first editions of the ASM Failure Analysis Handbooks, which listed corrosion rates of carbon steel placed in about a dozen different types of soils. I viewed a later edition of that handbook, which stated that the earlier publication had been shown to be misleading and unreliable. The new edition clarified that tests on the specific soils in question must be used for any quantitative corrosion predictions. Yet entrenched engineers, in possession of the older references, continued to use the overly simplistic recommendations for corrosion rate calculations, even after the publication of the new data.

2.6 Definition of Failure

Up to now, we've been discussing failure analysis as well as the nature of scientific and engineering knowledge, both of which support the practice of failure analysis. Now we'll move on to contemplate the term "failure" itself.

Many people think that fracture is the same as failure. However, it is not. Fracture is not even a true, completely overlapping subset of failure. Sometimes, failure-to-break could be a failure.

- Think about the case of a shear pin intended to protect a machinery component.
- Think about a pop-top can that must fracture in order to let the beverage escape and serve the purpose of quenching thirst.
- If a laboratory test coupon intended to measure the fracture strength did not break, the test would be considered a failure.

Also, consider Figure 2.2 (below). It is an image of a children's toy. To use the toy a child would hold the small circle and throw the animal shaped feature at a wall (or another object). It stretches and sticks. If the leopard's leg broke off, the toy could still be used. The same is true if the small ring at the far end of the animal's head broke off *(In fact, this happened right after Figure 2.2 was photographed.)* With the loss of the end ring, the fun a child might have would be reduced because the distance the animal-feature could reach would be shorter, but it would still function as designed.

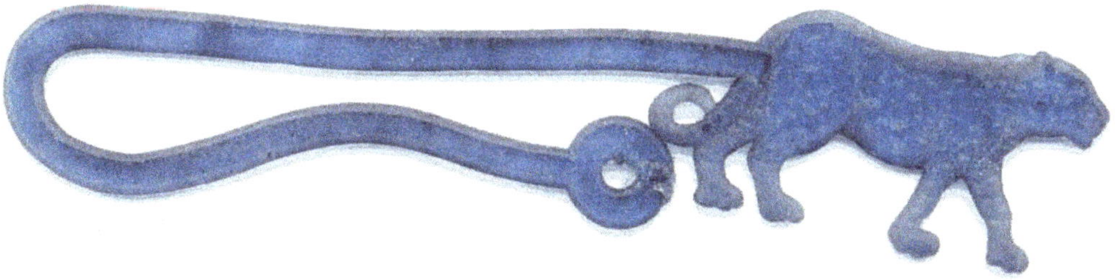

Figure 2.2: *Children's leopard toy with the potential for fracture and non-fracture damage categories*

Understanding how the toy is used allows the reader to predict another failure mode, which is completely unrelated to fracture. Dust adheres to the originally sticky surface of the leopard and the leopard is no longer sticky. Then, the fun is gone and the toy is no longer fit for purpose.

For the purposes of the engineering-informed failure analyst, failure is an inability to perform the intended function. Just because we have what others are calling a failure of a particular physical object, it does not follow that the object itself failed. Furthermore, the purpose of the investigation is to determine what went wrong. Improper use may cause unanticipated stresses which damage the object. At the risk of appearing to anthropomorphize the inert subjects of our investigation, this seems unfair! We must remind ourselves that objects respond to their environment in a nominally predictable way. **The failed object is not at fault.** It has no free will. Despite the fact that American (and perhaps other culturally Western) children are taught otherwise[42], it could not have tried harder. When used to refer to objects, the word failure has strong human factor overtones that confuse rather than enlighten. Furthermore, such word use allows the speaker or writer to avoid responsibility, and eliminates opportunities for learning, understanding, and prevention. For these reasons, I almost never use the term "the failed part". I also object to the popular scientific, passive voice, writing style, for similar reasons.

2.7 Materials Engineering Informed Failure Analysis

The central concept of the materials engineering informed failure analysis process is sequentially performing the least destructive portions of the analysis first, followed by the more destructive portions. High level failure analysis for complex failures, however, must be flexibly sequential. Some activities will necessarily be performed in parallel, and some planned activities will need to be modified as data is revealed by the initial work.

Failure analysis is not a process that may be performed "blind". There are some stakeholders who think that they are promoting an objective analysis if they hide background data. This is mistaken. The potential list of scenarios that might set a damaging chain of events in motion is virtually unlimited. Background data and initial visual inspection must be used to ensure that the data required to evaluate the likelihood of any credible scenario is properly collected. If the part is broken, but the part surface looks pristine, it is often a waste of time to evaluate the corrosion response of the material.[43]

EXERCISE 2-E: Imagine a failure of an engineered component. What responsibility would the failure analyst be avoiding by blaming or implicating the part, rather than the people who designed and/or built and/or sold, used, or performed maintenance on it?

____

____

____

____

____

____

____

____

[42] https://en.wikipedia.org/wiki/The_Little_Engine_That_Could
Note: I am not being a grouch and implying that this morality tale should be eliminated from the children's entertainment and instruction repertoire. I am simply urging us to distinguish physical from metaphorical truth.

[43] Hydrogen embrittlement of steels and stress corrosion of copper alloys due to ammonia or other nitrogen compounds are potential exceptions to this rule. They may look clean.

If a casting is broken, and a large void is visible, but the void is not at the crack initiation position, it is probably a waste of time to evaluate the porosity, beyond a basic documentation of the fact of its presence and position. In these specific cases, the analysts would be falling prey to the so-called red herring fallacy. Following a predetermined one-size-fits-all failure analysis protocol would waste time and might be interpreted as an attempt to divert attention from the true cause. If a broken hardened steel part has a sparkly fracture surface, and background information indicates that it had been subject to a sustained load, it would be important to consider ruling in or out the possibility of hydrogen embrittlement. That may require analysis of the hydrogen content, but it must include scanning electron microscopy to confirm that the sparkly surface is intergranular and not cleavage.[44] Metallographic work[45] to confirm that there is no other reason for an intergranular fracture (should that have been found to be the case) is also required. If, for example, the microstructure work had revealed the presence of an intergranular network of ferrite, there might be an alternative scenario to the hydrogen embrittlement. It is possible that closer inspection of the intergranular facets on the fracture surface might reveal some evidence for cleavage, and then the likely scenario would shift to a sudden impact load facilitating the separation, despite the documentation, and perhaps reality, of a previously sustained load.

2.7.1 Damage Or Failure?

Consider using the word damage, rather than failure, when describing the physical facts surrounding an incident. Fracture, wear, deformation, corrosion, creep (microstructure degradation), and molecular breakdown in polymers are all types of damage. Creep and molecular breakdown are not always visible directly to the eye in the way that fracture, wear, deformation and some other forms of corrosion are. Rather, creep is generally diagnosed as a result of additional testing following visible deformation or fracture, or as a result of an intermittent in-situ nondestructive microstructure evaluation, when it is an expected consequence of the specified service conditions. Radiation and ultraviolet light damage would also usually be diagnosed following deformation and/or fracture, or in the case of some polymers, following a change in appearance (such as cloudiness in amorphous materials).

Visible damage to the Subject Component is usually something that all stakeholders can agree on. The Subject Component may have changed shape or color, broken, been partially dissolved away or have missing material due to mechanical abrasion.

2.7.2 Root Cause Analysis

It has become popular in both industry and government to talk about finding the root cause of an undesired event. The problem with this phrase is that it has so many different meanings. Some refer to physical causes, but many refer to human factors. Wiktionary[46] defines 'root cause' as "An initiating cause of a chain of events which leads to an outcome or effect of interest." One problem with this definition is that it overlooks the fact that chains of unusual events rarely proceed in a pre-determined, linear manner, as implied by the analogy to a chain.

The Rootician's Dictionary[47] gives "A root cause of an adverse (harmful) phenomenon (effect, result, outcome) is a single one of the harmful conditions, behaviors, actions, and/or inactions that harmfully affected one or more important attributes (nature, magnitude, location, timing...) of the specified effect and was not in turn the result of more important deeper underlying conditions, behaviors, actions, and/or inactions"

[44] Intergranular and cleavage will be covered in Chapter 8.

[45] Metallography will be described in Chapter 6.

[46] Wiktionary is a free dictionary on the internet: https://www.wiktionary.org

[47] Corcoran, William, The Rootician's Dictionary 2015.06.15, https://app.box.com/s/3absbz3xbsm917v1q8ns or https://www.dropbox.com/s/0t1kkvi8lqcdg76/The%20Rootician%27s%20Dictionary%202015.0 6.15.pdf?dl=0

as the first and best of 51 total definitions. That sentence is a mouthful!

Each definition in The Rootician's Dictionary is accompanied by listings of the benefits of the particular definition, its drawbacks, and some examples of how the term has been used in that particular manner. Definition 3 in the same document is "The deepest underlying cause or causes of positive or negative symptoms within any process, which, if addressed appropriately, would result in the elimination, or substantial reduction, of the symptom."

This definition gets to the key aspect of a common reason for performing most root cause analysis, which is prevention. Root cause definition number 6 is "A basic underlying cause that management has control over." The compiler (or editor) of the Rootician's Dictionary suggests the following very practical question to focus the investigation when using this definition: when an organization uses this definition it should ask, "What are the factors that directly resulted in the nature, the magnitude, the location, and the timing of this "root cause?"

2.7.3 Causation - Multiple Levels

It is the job of the team specifying the investigation protocol to plan the investigation so it has the greatest chance of revealing the complex interaction of factors that allowed the subject event to happen. KEY LESSON: **it is rare that a failure has a single, simple cause**.

Even with something as simple as dropping a pencil, and having it fall toward the ground, there are multiple factors involved beyond simple cause and effect. If someone holding a pencil between two fingers increases the space between the contact points of their fingers and the pencil, the pencil will fall. But the finger motions are only one part of the background factors that allow the pencil to fall. Gravitational effects certainly play a role. If this event is happening outside, and the wind is blowing, it might change the position where the pencil ultimately contacts the ground. If there was a baby in the pencil's path, the pencil could have landed point down in the baby's eye.

With complex events, such as a car accident, which is perhaps being blamed on the fracture of a steering component while the vehicle was being driven down the road, it is even less likely that all of the specific consequences that resulted from the fracture did so from a single cause. Factors which might have made the accident result in more or less severe consequences include alertness of the driver, pre-existing damage to mating components, road conditions at the time of the accident, and road conditions over the life of the vehicle. The specific timing of the final separation event would, in the case of a pre-existing fatigue crack, depend on the entire stress history that led to the specific size and shape of the fatigue crack at the moment that the final load cycle divided (or fragmented) the component. Individuals participating in a thorough investigation will do their best to lay the groundwork for the creation of multiple credible scenarios for the sequence of events which led to the damage and subsequent failure. The scenarios must then be checked against the available physical evidence for self-consistency. This means planning to look for evidence for all of the factors that might, or could have, played a role.

EXERCISE 2-F: Here we are contemplating complexity. Practice makes perfect. Take the time now to think of fifteen contributing factors to what you had for your last meal.

Classically, failure origins (or causes) have been sought within specific aspects of the product design life cycle. This includes design, manufacturing of raw material, manufacturing of the Subject Components, service conditions and maintenance operations. See Figure 2.3 (below). This is a helpful scheme to review at the time of defining the scope of the project. It could help to minimize the chances of overlooking a potential contributing factor to the damage event. If a formal Failure Mode and Effects Analysis document exists for the product in question, consulting it during the documentation and background review might further minimize chances of failure to collect the data needed to rule a factor in or out of the proposed scenario.

2.7.4 Damage Category Determination

NOTE: Details of the major types of damage will be covered in Chapters 5 and 8. Some of the concepts listed here are fairly advanced for the lay person. If this describes you, consider using targeted web searches and persistence to begin absorbing the advanced ideas that matter the most to you.

The competent failure analyst makes plans to determine the damage category an early priority. Sometimes it is completely obvious to the most casual observer that the Subject broke, got deformed, or wore out. Sometimes corrosion damage is "self-evident". In other cases, the Subject Component may have evidence of multiple types of damage, and it may not be obvious which happened first or which was most damaging. Despite the prioritization of damage categorization, in complex damage mechanisms, the specific manner and contributing factors may not be determined until the end of the investigation, after the destructive portions of the testing have been completed. Environmental and Complex Damage categories are described in Chapter 5.

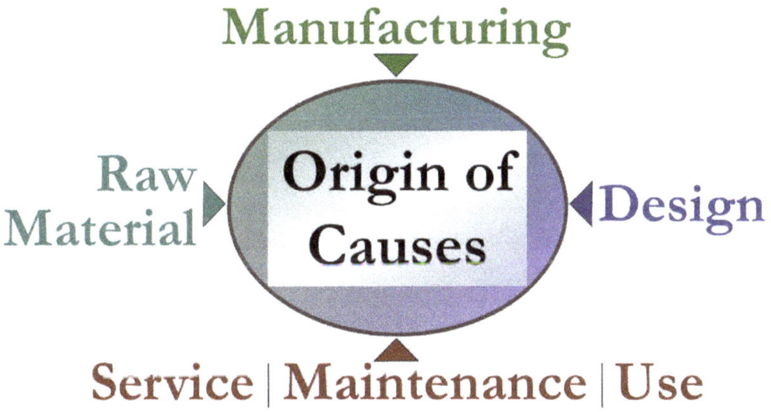

Figure 2.3: *Origin of causes schematics help avoid overlooking a major category of potential factors facilitating damage.*

Let's stick to Mechanical damage for now. The "Big 3" include deformation, fracture and wear. The major factor in mechanical damage is stress. Deformation, fracture and wear arise along with stress. One way to explain mechanical damage is as follows:

The Stress Exceeded The Strength.

Much of the rest of the investigation into the physical causes will center on defining which types of stress and strength were involved.

Many types of mechanical damage proceed in a progressive manner, including environmentally influenced damage processes, such as stress corrosion cracking in metals, and stress cracking in polymers. Pure mechanical damage processes, including fatigue or bearing wear resulting from Hertzian stresses and friction, and creep are also progressive.

Other types of mechanical damage happen essentially all at once. Loads or forces that exceed the tensile stress at the relevant temperature and strain rate fall into the monotonic or single load fracture category.

Creep may be due to a monotonic load, but it is progressive. This is because the load must be sustained for an extended time (often, decades) before the damage accumulates sufficiently to interfere with fitness for service.

EXERCISE 2-G: Question: If the damage happens all at once, due to a single load, is it more or less likely to have been due to an unanticipated event than in the case of progressive damage? Explain your thought process.

____

____

____

____

Now, let's consider Environmental damage. These processes include corrosion (chemical attack by an aggressive substance, usually in liquid or gaseous form), and energy impingement. Heat, ultraviolet light and nuclear irradiation are examples of energy impingement. Most of the environmental damage categories proceed over an extended period of time. If the visual evidence indicates that environmental conditions, rather than stress, were the primary facilitator of the damage, we may explain that:

The Environment Was Too Harsh.

WAS THE DAMAGE MECHANICAL OR ENVIRONMENTAL?

Sometimes it is not easy to tell. Here are some clues for when it is.

If the part is broken or deformed, a good working theory is mechanical damage.

If there is discoloration, especially when accompanied by material loss or build-up of foreign material, a good working theory is environmental damage.

Material loss with no discoloration could be wear, although wear in the presence of heat (think about a bearing with a lubricant problem) could be accompanied by discoloration. And some types of corrosion have no discoloration.

IT'S COMPLICATED! LOOK AND THINK!

Just as we must look at the variables that affect the stress when evaluating mechanical damage, in environmental damage, we search for factors that may have led to increased susceptibility to particular substances or conditions in which the part served. Alternatively, maybe the subject was used for longer than the originally planned life cycle. Maybe it was damaged, but never failed! Maybe, indeed it demonstrated exemplary skill and care by all involved, from the designer and nested manufacturers, to the maintenance crew and the owners.

GOING DEEPER: When the thorough analyst reviews the details of the life of an investigation Subject, the thought might occur that all types of damage are truly complex. The simplest fracture event is influenced by the temperature and the strain rate. It is still helpful to have a basic categorization of damage types.

In March of 2021, in the wake of the global societal disruptions related to the COVID19 pandemic, a major global trade disaster occurred in the form of a giant container ship being blown by strong winds approximately 45 degrees out of alignment with the sides of The Suez Canal, causing a complete stoppage to shipping traffic for 6 days.[48] Was this failure caused entirely by human factors?

Figure 2.4: *This satellite view, approximately 2.64 km wide, shows the ship named "Ever Given". It is stuck in the Suez Canal.[54] (Image found on Wikipedia; contains modified Copernicus Sentinel data; processed by Pierre Markuse)*

2.8 The Spectrum of Failure Analysis Work

Both damage and failure are best understood as multi-level processes that have inputs and consequences. If the consequences were not negative, we would not use the terms damage and failure.

Human factors that contribute to the likelihood of failure include simple mistakes due to distractions, lack of training, inadequate sleep, being in a hurry, carelessness, and use of psychoactive substances. Many human initiated factors (events) result in mechanical stresses that are beyond the design envelope.

[48] For more information, see the Wikipedia article: https://en.wikipedia.org/wiki/2021_Suez_Canal_obstruction | also you can read about it in the web article: Why we shouldn't blame the pilot of the container ship stuck in Suez Canal by Tyler Dawson | https://www.saultthisweek.com/news/world/suez

Business factors overlap with human factors. A bad economy may push some decisionmakers to delay needed maintenance. This could result in injury to employees, or inability to hold tolerances on the parts being manufactured on the improperly maintained equipment. The improperly manufactured parts may or may not be found prior to use.

Sometimes the consequences of an in-process failure are minimal, such as an assembly problem that momentarily slows production. Sometimes, the consequences are catastrophic, such as the collapse of a building under construction.

Issues that interfere with global and local societal harmony and justice may be viewed as the root of failures such as some terrorist attacks, and destruction of the environment.

A 2019 investigative radio drama called "When Lighting the Voids" was created to disclose the Center for Investigative Reporting determination that a multiple casualty 2009 explosion at a shipyard in Mississippi resulted from the illegal use of lights lacking explosion resistance features, in combination with extremely high (illegal) concentrations of solvent vapors. Listening to the drama, anyone with a basic knowledge of industrial safety can predict the consequences. At least one survivor had extremely severe injuries, which were due to working in a confined space without proper safety gear or training.

The wheels of justice work slowly. In 2017, the Senate overturned new rules that would have made it easier to hold shipbuilders accountable for employee injury and death. See the article by Jennifer Golan, "Congress kills rule forcing contractors to disclose safety problems" (https://revealnews.org/blog/congress-kills-rule-forcing-contractors-to-disclose-safety-problems/).

Nuclear energy companies, oil and gas producers, chemical plants, and airlines that have disasters are usually forced by government policy to perform in-depth human factor analysis, in addition to the engineering analysis of the damaged components. Human and business factor analysis are almost always required if a sound foundation for prevention of recurrence is to be created. The investigators from the Center for Investigative Reporting, mentioned above, found that the company locked away the ventilation fans that would have removed the explosive fumes from the work area of the ship, because using these fans increased costs and reduced profits.

More legitimately, many companies make the business decision, either explicitly or implicitly, that component failures in consumer products are too expensive to prevent. When the fix would require a significant change in how components are designed or manufactured, and the costs of the failure are low, many businesses will accept such failure as a part of the cost of doing business. This helps keep a wide variety of products and services available to a larger potential group of buyers.

EXERCISE 2-H: What is meant by implicit and explicit business decisions?

____

____

2.9 The Spectrum of Approaches to Failure Analysis

There is a spectrum of approaches to failure analysis. (This sentence will be repeated several times in this book!) One way of looking at this subject is to consider "The Template versus The Toolbox".

- A template is a list of activities that must be accomplished.
- The toolbox approach is more flexible, and the team of stakeholders spends more time up front deciding which steps to take, and which to skip.

Using the toolbox tends to require at least a minimum amount of discussion about which tests and evaluation procedures will shed the most light, and why. If there is some back-and-forth, even informal, negotiation of the protocol, some background information will be brought to light. When properly used[49], The Toolbox forces the failure analysis team to think about how the data will be interpreted before the destructive testing starts. This approach is superior to a list of activities (The Template), because a given activity may or may not be relevant.

The most basic failure analysis protocols start with steps that reveal the physical factors within the damaged component that facilitated the damage. This will naturally lead to data which may be evaluated in order to reveal one or more potential sequences of conditions and events which allowed the part, assembly, or structure to become unfit for purpose.

For more complex failure events, a more comprehensive approach is needed. Human and business factors must be considered. Doing failure analysis as an outside consultant will provide opportunities for comparison of engineering design philosophy in different industries. I've told multiple clients that their companies have problems because they are using a guess-and-test design method. Yet, unless they have suffered greatly (great financial loss, or injury or death), most do not welcome advice on improving their design process, even if they do take advice on improving the design of a particular component. It is difficult for a company to move from a guess-and-test method founded on "tribal knowledge" to a comprehensive, more quantitative engineering approach.

No matter what the complexity of the investigation, odd (unusual) data points "cry out"

[49] The proper use of The Toolbox will be detailed in Chapter 6 of this book.

for thoughtful contemplation. Figuring out how to explain outlier data is often the clue that reveals what happened. If this is not the case, it may reveal what might have happened. When postulating what really happened, a story that does not have any facts in conflict with science and engineering knowledge, even if it is not certain, is preferable to one that has documented conflicts with established theory (Abductive logic, Occam's Razor, Parsimony!).

But before we move on, now is a good time to introduce the concept of what failure analysis is not![50]

- Failure analysis is not process-troubleshooting.
- Failure analysis is not design of experiments.
- Failure analysis is not statistical analysis.
- Failure analysis is not putting the part in a CT scanner to see what you can see.
- Failure analysis is not cutting up damaged parts to see if they met the specification.
- Failure analysis is not thinking about what might be wrong based on history and logic, and diagnosing the cause.

All of these activities might be part of a failure analysis. But on their own, they are not failure analysis. Let's take a closer look at the differences.

Process troubleshooting and Design of Experiments (DOE) are related activities whose practitioners' primary goal is to make a manufacturing process more reliably produce an in-specification product. The evidence is the stability of the process; established by understanding variation and the target specifications through controlled experiments that manipulate multiple variables at the same time. This differs from true failure analysis in that the DOE's primary goal is to eliminate product variation – not to reveal failures. In DOE it is easier to use a template, and the criteria for success are generally predetermined (product meets specification more reliably). **Outputs of a DOE may help or inform a failure investigation if processes are suspects.**

Statistical analysis may be helpful in determining what happened to cause a harmful microstructure or dimensional deviation. It may be helpful in understanding how widespread a problem is, and thus how comprehensive an investigation is warranted. Statistical analysis may even be helpful in understanding what physical processes or conditions contributed to the undesired event. But alone, statistics can't answer the questions that create the narrative that describes the problem and the factors that allowed it to happen.

A Computed Tomography scan may reveal a dead insect in an electrical switch box that resulted in a frail, elderly woman's death due to a long train of interconnected events.[51] But doing the scan doesn't reveal how the insect ended up in the switch box, or whether that was truly the cause of the switch failure, or the control system of which it was a part.

Failure analysis is most definitely not cutting up parts to check the composition, hardness and microstructure, to see if they meet the spec. Failure analysis may include those activities, but the specific location where the evaluations are made is as critical as the results themselves.

When I speak at engineering education events, I like to use the phrase "putzing around in the lab" to refer to this type of unfocused activity. It can bear fruit, and sometimes it is a reasonable business decision. But it is not failure analysis, which must be based on some foundation of a formal process, even if the practitioners have

---

[50] The concept, of what Failure Analysis is not, will be detailed more in Chapter 4.

[51] Larry Hanke, PE, FASM, of Materials Evaluation and Engineering gave a presentation on this case study at West Michigan ASM International local chapter dinner meeting in October, 2017.

never gone to the extent of using a formal methodology, such as this book promotes.

What are some limitations of failure analysis work? In 2020, one of my colleagues convinced a technical committee of an industry association to fund some failure analysis of additive-manufactured tooling. There was a lot of excitement around this new technology, because some hidden features could allow the process times to be reduced, and that means higher profit margins for the manufacturer.

It was known that these hidden features could promote the early onset of fatigue cracks. But there was nothing in the technical literature to document this.

We received a tool that had five visible cracks, while the verbal description of the problem noted only two. Only the edges of the cracks were visible, but not the crack surfaces. We needed to decide how to cut the part to reveal the crack surfaces in order to determine where the cracks started.

There was an obvious stress concentration visible, but the crack did not follow that notch.

Finally, the people who had designed the tool gave us a 3D model of the geometry. At least some of the cracks were aligned with an internal feature that was clearly a stress concentration. Now we had a data-based reason to choose the cut location. Had we used our first plan, we probably would have left the crack initiation buried in the closed portion of the crack.

It was very difficult to explain to the group of mainly manufacturing engineers WHY they NEEDED to give us the engineering prints of the parts. We were not just putzing around in the lab. We were trying to do a real failure analysis. We had to explain the benefits of the real thing, when compared to application of tribal knowledge.

Failure analysis is not: going over all the possibilities in a brainstorming session or devising a fishbone diagram to reveal potential factors that contributed to the damage/failure, and then picking an answer by a process of logic based on how the process is supposed to run.

Failure analysts, who rely on actual physical evidence, may get useful input from people familiar with the processes used to make the subject of the investigation. Failure analysts may get useful information from people using their intuition about what is possible or not possible regarding a process with which they are intimately familiar. Yes, using problem solving techniques where thoughtful people knowledgeable of the processes used may be part of the failure analysis. No, this is not the entire failure analysis.

At heart, troubleshooting methods that rule possibilities in or out on logic alone are, in a way, inherently opposed to the basic failure analysis philosophy of keeping an open mind and relying on the physical evidence.

One of the primary features that makes failure analysis different from other technical activities is the tenacious adherence to the practice of keeping an open mind until all the evidence is in. That doesn't mean we can't speculate when that is useful. But we speculate with a caveat: until we have studied and evaluated all the data, we don't know the answer. New facts might lead us to change our interpretations and conclusions at any time. That's the nature of the failure analysis endeavor.

2.9.1 The Importance Of Paying Attention To Anomalous Features

What is an anomalous feature? This is another very important concept for those who are doing failure analysis work.

In casual conversation, "anomalous" often refers to an outlier. We've reviewed this earlier in this chapter. The root of the word itself has to do with something that "has no name". The word can also refer to something that doesn't make sense; something that doesn't fit with the rest of the data. An anomaly is difficult to fit into one of the standard categories of a relevant classification system.

An anomalous micro-scale fracture feature might be intergranular features covered with smaller, shallow micro-voids. Is this an intergranular fracture, which is usually considered brittle? Or is it ductile fracture? In general, when two features point in different directions, the one acting over a larger scale would be given greater weight. In this case, that would be the intergranular features, and if the damage happened at room temperature, (rather than elevated temperature), it would be considered micro-scale brittle. But that doesn't mean we ignore the smaller scale feature. The fact that there are micro-voids on the apparently intergranular features may indicate something of importance. Maybe the crack surfaces are parallel to, but not exactly at, the grain boundaries. Additional work (likely including microstructure analysis) will need to be performed in order to figure it out. These can be some of the most interesting projects. The ones you later share happily at technical conferences or publish for the enlightenment of your fellow engineers!

Sometimes, the anomalous features will be in evidence from the earliest parts of the visual investigation, while other times, a subtle factor that is inconsistent with the rest of the data will be revealed toward the end of the investigation.

> *It was shown in the middle of the 20th century that up to 88% of humans are unable to use negative findings properly. This great majority of humans may thus be presumed to be unaware of the significance of a negative finding. Could a major factor in whether we are considered smart or not be related to this ability? I remember being surprised on multiple occasions, early in my career, when others failed to see what was obvious to me, even after I tried very hard to explain it., Although, my IQ does not remotely meet the threshold to become a member of MENSA, it is possible I have a rare talent.*
>
> *I had a very strong Aha Moment when I read the Thompson book.*[52] *The cited study by Bruner, et al demonstrated that the vast majority of people were unable to accurately work the negative data into an evolving theory. This frequently prevented them from reaching a correct conclusion.*
>
> *My personal biggest Aha Moment was related to my assignment to review a huge investigation that I had not been involved with, and that was considered a failure. That is, the investigation was considered a failure as the investigators could not determine the cause of the irregular deformation of the sheet steel their employer was rolling. Certain areas of the steel ended up with objectionable ripples in the final product. These ripples were parallel to the direction of rolling.*
>
> *A big part of the investigation was a focus on documenting evidence of excessive non-metallic inclusions. To the dismay of the investigation team, the excessive concentration of inclusions was not present in the heats of steel they had followed. They dejectedly reported their defeat to management. Some months later, when I looked over the data, I discovered that this low inclusion steel in fact did not have the ripple problem. Since two thirds of the heats had the problem, this simultaneous finding of low inclusions and no ripple problems should have brought shouts of joy. Yet, apparently nobody on the team, or the managers who reviewed the findings, were able to see that the investigation had been successful in moving the inclusion hypothesis toward becoming a legitimate theory! No inclusions and no ripples were at least as good a result as inclusions and ripples. By good, I mean good in terms of supporting the original hypothesis.*

Well, let's go back to the people with thought patterns represented by the 70 - 88% of the population who don't grasp the importance of a negative finding. In any of their life activities, they would be less likely to include any comments on the negative finding in the report.

As Lucy Goodchild van Hilten notes in her editorial on the benefits of publishing negative findings by scientific researchers, at the very least,

[52] Robert Thompson, *The Psychology of Thinking*; Pelican, 1959. | ISBN: 978-1446272473

it could save a lot of time for others who get the same "great" idea.[53] The 1897 publication of the "failed" experiment by Albert A. Michelson and Edward W. Morley led to a Nobel Prize for Michelson, the founding physics professor at the University of Chicago[54], and the first American to win a Nobel Prize in a scientific field. The work debunking the existence of the aether helped Albert Einstein reshape his theory of Special Relativity into the more comprehensive General Relativity.

In reporting on any significant failure investigation, the professional failure analyst will note what common features were not found, or what scenarios were ruled out, and why.

2.9.2 *Use Of Reference And Control Parts In Failure Analysis*

Fitting all of the data into a coherent framework that explains how the physical damage happened, and what background conditions facilitated it, is much easier if the initial scope definition includes thoughtful selection of reference specimens.

Rational Thinking:
Compare - Contrast

Selecting a Reference Part for a failure investigation is a key way to practice rational thinking! Explaining to others why Reference Parts are necessary helps you to master the practice of rational thinking.

Skillfully selected Reference Specimens routinely provide the key to creating a believable damage scenario. For investigation of a routine in-process manufacturing failure, a current production part, or current production parts from multiple lots prior to the manifestation of the current problem, are often beneficial reference parts. Part of the purpose of these specimens would be to acquire a sense of the normal variation in the production part characteristics. The best reference parts for durability test failures or field failures returned from service are durability test passes.

Unfortunately, many companies do not retain test success components or assemblies. Nor does their testing protocol require that they ensure that the prototype "Test Pass" was representative of the product they expect to be making in the future.

A potential substitute for a reference part is a field return for an unrelated concern, perhaps peeling paint, when the problem of interest is a fatigue crack. Such reference parts are more worthy of the name "Control" part than a randomly selected part of unknown durability. Of course, it is important to gather any available data on the actual service conditions experienced by a part that is going to become a "Control".

2.9.3 Troubleshooting Manufacturing Process Failures

Troubleshooting manufacturing process failures in some companies can be summed up as, "Turn it on and turn it off."

A sphere of knowledge that is related to failure analysis is manufacturing troubleshooting. Stamping plants may want help determining why their parts are splitting, or not ending up with the desired final shapes. Forging plants may want help determining why their parts are developing higher scrap rates due to indications found during NDE. Plastic molding plants and foundries may wonder why the parts develop blemishes after plating or painting. Assembly plants may wonder why threaded fasteners are breaking as they are installed.

Because manufacturing processes are very complex, some plants have adopted a Design of Experiment model for troubleshooting. Various factors that can be tweaked to modify the process are grouped together, to reduce the total number of trials that need to be performed in order to

---

[53] *Why it's time to publish research "failures" by Lucy Goodhill van Hilten* | https://www.elsevier.com/connect/scintists-we-want-your-negative-results-too

[54] Their experiment was designed to detect the relative motion of matter through the "luminiferous aether".

sort out the parameters that might be helping or hindering the desired production and scrap rates.

Some of the process conditions that might be tweaked include temperature, lubricant application, and timing of sequences. If the plant is using purchased raw material, and they have more than one supplier, they may switch out the stock to see if one company's product runs better than another. They may sort raw stock by some noticeably variable dimension, and run, for example, thicker or thinner sheet metal. They may try re-calibrating their torque wrenches.

They seek sets of parameters that they can modify in order to "turn the defect on" and then "turn the defect off". If they can turn the defect on or off, they falsely believe that they know what is causing the problem. Sometimes companies do this to meet the demands of their customers to demonstrate that they know how to turn it on and turn it off.

The problem with this exercise is that it is usually impossible to prove by this type of manufacturing optimization alone that the original defects were caused by the same set of parameters that were used to produce the intentional defects.[55]

For example, many people believe that the clamping force created by a threaded joint is determined by the installation torque. While it is true that the clamping force created by a threaded joint is related to the installation torque, there are many factors that can result in higher or lower clamping load even if the torque used is tightly controlled to a narrow range. A small modification in the amount or type of lubricant or surface finish can drastically change the clamping load attained. Smoother surfaces or more effective lubrication can readily decrease the amount of friction that has to be overcome during installation; this results in a much higher clamping force. The force can be so high; maybe it could break the externally threaded fastener or strip the threads from the internally threaded fastener. If all of the good fasteners have been used up and only the problem lot remains, how can it be determined which of these two factors (lubrication or surface roughness) might be at fault?

It is possible that an externally threaded fastener can break because the hardness of the internally threaded fastener has increased. If the internal threads are too hard, then the externally threaded member of the joint could be loaded in torsion instead of tension. The fastener is weaker in torsion than it is in tension. In any case, we should be happy if the fastener breaks, because then we know there is a problem. If instead, we simply have a low clamp load, the joint may be subject to premature demise via fatigue in service, usually a more expensive problem to fix.

Furthermore, maybe there is something wrong with the fasteners themselves. Especially if they are hardened and/or electroplated, they may be suffering from some type of heat treating or plating process anomaly. In-plant troubleshooting is not likely to be able to determine the true physical cause of damage.

Personally, I am not a fan of the idea of turn-it-on/turn-it-off as the sole way of demonstrating the existence of a robust manufacturing process. When manufacturing problems arise, if they have costly potential future consequences, it is helpful to ensure that the actual problem has really been dealt with in an appropriate manner.

Turning it on and off doesn't really do that. The entire idea here is that knowing the cure proves the cause. If I take aspirin for my headache and the headache goes away, does that prove that the cause of my headache was lack of aspirin? Much of the turn-it-on/turn-it-off method of proving the cause is of similar logic. There may be multiple causes that result in the same problem.

[55] Another approach is to seek the key factors that drive variation and sort them in order of theoretical importance.

In *Tools for Critical Thinking*, author David Levy notes that "*The Cure Doesn't Prove the Cause.*" Turning it on and off doesn't prove you know the specific cause. It only proves you know one potential cause | ISBN 978-1577666295

Years ago, shortly after I started working as the corporate metallurgist for a manufacturing company at their corporate headquarters, I was sent to our foundry and then to a heavy steel fabrication shop.

First, let's talk about the foundry. This foundry had problems. It was using a new high-tech process to pour aluminum castings without gas porosity. The scrap rate had hovered near 30% since the line had been installed. All of the sudden, the scrap rate plummeted to 5%. The vice president of the company found out that the R&D department had just hired a new, young, educated, and supposedly smart metallurgist. No matter that the company had already spent $2000^{\underline{00}}$ (1985 money) for me to attend an out of state conference on aluminum welding metallurgy. This was a subject that the company knew I would require training in. Time to pack up and go to the foundry.

Not that I knew much about aluminum foundry practice either for that matter. All of my limited experience to date had been in steel. But my boss informed me, the Veep had spoken and I had to go. So, I did. I spent two days there. I asked a lot of questions. I got some data. I gave the people at the foundry some ideas on other data they could collect. I asked more questions. I observed. I wrote a report and went to my conference on welding.

The next day, I was informed that the scrap rate went back up to 30%. The bottom line is that proper process control must be based on disciplined operation of the product line and an understanding of the key process drivers. What are the chances of someone who knows virtually nothing about a particular process being able to immediately eliminate a problem that the permanent plant personnel had not been able to solve in two years? I am glad the Veep had such confidence in my skills. Unfortunately, I could only disappoint him.

I had better luck with the steel fabrication shop problem. In that case, the purpose of my visit was, first and foremost, to familiarize myself with the operation so I would know who I was dealing with when they called to ask for help. But while I was there, the plant engineer told me that they were having a problem with some high-strength, low-alloy (HSLA) structural steel.

The material had to be bent in a few places in order to make the flange of a specialty beam with a variable height web. They had been having problems from day one, as the high strength material tended to crack. So, they decided to warm it up with a torch before they formed it. But then they ran into a batch that formed easily, without being warmed up. So, they cut off two pieces for me to take back to my lab, one from the formable and one from the problem lot.

It turned out that the problem steel had the required yield strength value, slightly over the minimum specified 550 MPa. The good material, the stuff they liked, had a yield strength value of only 344 MPa. So, we ended up doing some work to formalize the pre-forming, warm-up procedure and documented the fact that there was no detrimental change in strength or ductility properties resulting from the process modification.

Such basic applications of materials engineering to manufacturing troubleshooting are useful and legitimate. The use of turn-it-on/turn-it-off methodology is to be thoughtfully evaluated before presuming that a problem has been eliminated for good.

CHAPTER 3

The Visual Inspection

Seeing Through Our Mental Filters

3.1 Foundational Concepts of a Failure Analyst's Visual Inspection

What is a visual inspection? Is there anything we need to know about visual inspection that is specific to failure analysis?

> *Years ago, my mother was shopping for a quality step ladder. Of course, pretty much everything available was made offshore by people without the decades of manufacturing history that informs quality product manufacturing in well-known brands. I was visiting her from out of town, so she decided to take advantage of her daughter's metallurgy and failure prevention knowledge before making her selection.*
>
> *But what could I do? I could not exactly perform a durability test there in the store. There were several models that met her specifications for height, appearance, price, and rated load. Some looked more flimsy than others. Some had mechanical fasteners holding the steps to the frame, some had welds. None had obvious pre-existing discontinuities. There was one sturdy looking model, the welds looked uniform, and they appeared to be sized in line with normal industry requirements for proportion to the members being joined. I unfolded this ladder and stood on it in the store aisle. I could feel how well it supported my weight. It felt solid, not wobbly. This would help to prevent my mother from losing balance.*
>
> *As a failure analyst, I would note that I don't have x-ray eyes and that I did not have access to the weld qualification and welder certification documents. As an engineer, I read the labels looking for data and considered cost; but, what about the safety of the ladder? From the perspective of a daughter, I recommended the one I selected for the physical (stand on it) test.*
>
> *Let's review my inspection procedure:*
>
> 1. *I looked at the ladder to get a general idea of the appearance.*

2. *I looked at the labels to see the load ratings, warnings, and country of manufacture.*

3. *I looked at the welds to gage their size and uniformity of shape.*

4. *I looked at any other fasteners present to see whether they appeared adequately sized and properly installed.*

5. *I stood on the ladder to feel how well it supported my weight.*

Thus, we note that the visual inspection may be viewed as two major activities. First is LOOKING-AT. Most of this chapter is devoted to this activity. Looking-At is open-ended.

The second is LOOKING-FOR. Looking-For requires that we know *what* we are looking for.

On my ladder evaluation exercise, four of the tasks were looking for something specific.

Later sections of the book present detailed information on the looking-for aspect of visual inspection. At the end of this chapter, we'll discuss perhaps the most significant feature that we must *look for*, to get a head start on the important concepts of ductile and brittle general behavior versus ductile and brittle cracking.

LOOKING-AT: Of the information previously published on visual inspection for failure analysis, most of it is in the second category (LOOKING-FOR specific things). Due to the open-ended nature of LOOKING-AT, it is, in some ways, more difficult.

EXERCISE 3-A: Compare and contrast LOOKING-AT with LOOKING-FOR

Look at your hand and find five features that you have never noticed before. This is a result of *looking-at* your hand.

Now, count your fingers, to see if you have the expected number. Look to see if there is any dirt under your fingernails. Estimate the distance between your knuckles. Look at the pattern of lines on your palm and see what an internet search says about your life based on this pattern. Or skip this if you are offended to find such a ridiculous suggestion in a technical book. But first, cut me some slack and try to find a good pedagogical reason why this idea might have occurred to me. Then find one or two more features to *look for* on your hand.

3.2 CASE STUDY: LOOKING-AT Application

This past year, I had the opportunity to pay someone to look at the insides of my eyes. My first vision incident was occasioned by the red numerals on my digital clock briefly turning pink, when my left eye was covered. The second time was a result of flashing lights appearing in front of my office building, as I was backing out of my car's parking spot at night. I was impressed to note that the two different eye doctors looked at my eyes and then at the insides of my eyeballs after the technician put drops to dilate my pupils. They looked to see what was there, after listening to my "complaint" about HOW my vision had changed.

The second time was when I was working on this chapter. This eye doctor used three different magnification devices, each stronger than the previous one, and had me look up, to the upper right, to the right, to the lower right, down, to the lower left, left, to the upper left, and up, to examine both the affected eye and the other eye, while she determined the extent of potential damage to the back of my eyeball. So, she used a low magnification device first, and followed by a higher magnification device, and then one more with a really bright light and higher magnification device.

DISCUSSION: Did you catch my opinions creeping in to this description? How did I know the subsequent magnifiers were higher magnification? If I were making factual statements, I would have said the lenses had more pronounced curvature! The truth is that I just "assumed" that was what she was doing!

3.3 What Affects the Visual Inspection

Looking-at things is natural. If we are not visually impaired, we do this all the time and have been doing it since we were born. We open our eyelids and look around. This original state of visual data reception, what I like to call LOOKING-AT immediately starts getting lost in favor of LOOKING-FOR. And all of our looking, whether AT or FOR, is, for better and worse, also, looking THROUGH!

EXERCISE 3-B: LOOK-AT versus LOOK-FOR versus LOOK-THROUGH

Once more into the breach: as soon as you read this paragraph of instructions, I want you to look closely at your hand for all of the lines that cross over each other. You are LOOKING-FOR "Xs" on your hand. Look for only 5 seconds, then close your fist and look at the book. Do not re-open your hand to look for the answers – keep your eyes on this page. The 5 seconds start now!

In the box provided, draw a quick sketch of the lines of your middle finger's middle knuckle (inside and above the palm). Don't open your hand to look. Does it look the same as the lines in the first (top) knuckle? Do you remember seeing those knuckle lines? If not, that is because you were *looking through* a filter set by your brain to look for "Xs on your hand".

We are *looking through*[56] our mental and neurological filters. We humans have many automated operations related to visual processing. In fact, even newborn babies are programmed to look for human faces so they are not capable of pure *looking-at*. Of course, babies don't have words to describe what they are looking at, although researchers have shown that babies in utero do learn the rhythms of their mother's language. As we age, the relationship between the words we frequently use and how we see the world gets stronger. Most people do not realize how many layers of interpretation come along with our everyday vocabulary. Note the Compare-Contrast pairs in the following paragraphs.

Whether you are a beginner or are very experienced, the first exercise for the failure analyst trying to improve their skills is to learn to disentangle the words from the reality and the opinions from the facts. Of course, this is not a black and white (dualistic or dichotomous) choice. The idea is to promote use of a vocabulary that clearly distinguishes between basic fact, things that anyone can see or notice, and interpretations based on experience.

There are three major reasons for doing this.

- The first is that it helps the analyst to explain what they see to people who lack a technical background.
- The second reason is that it helps the analyst question their assumptions. This in turn improves skill and reduces the tendency to jump to erroneous conclusions.
- Thirdly, when we, the lead failure analyst, start our communications by stating the obvious, we make it easier for other interested parties to agree with us. This sets a trend of agreement, which makes it easier to gain the confidence of other stakeholders.

EXERCISE 3-C: Fact-based description. Briefly describe the object pictured below.

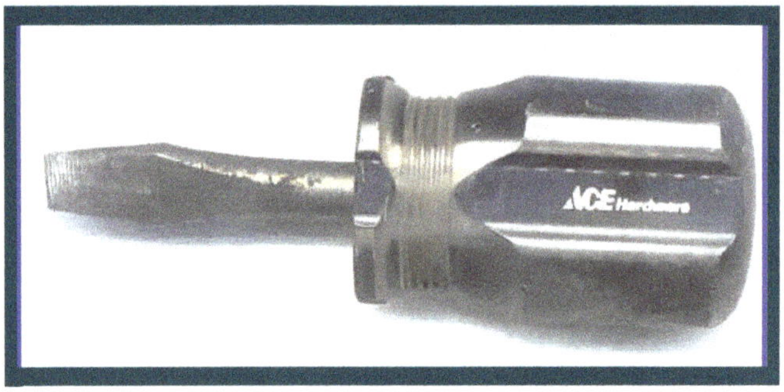

After generating your own description, read my brief description (Section 3.8 at the end of this chapter).

NEXT try to describe the object using only words related to the shape and other features of the appearance. And after generating your own description read my description (Section 3.8).

[56] Because the human brain makes heavy use of neurological and mental filters, we are always LOOKING-THROUGH them (at least partially), even when we are LOOKING-FOR data or are LOOKING-AT something with fresh eyes. The classic example of neurological and mental filters is the exercise to look around a room and identify all blue items in the room. Then after you close your eyes, you are asked to identify all of the red items in the room. The strength of our filters is demonstrated by the difficulty remembering red items because our task is set by the brain's agenda and filters. This was just demonstrated by EXERCISE 3-B.

NEXT observe an actual screwdriver (go get one); add details relevant to your screwdriver. (If you have two screwdrivers, then compare and contrast them.) Use no words that require your specific engineering knowledge. Therefore, your list won't contain the words steel, plastic, rust, blade, heat-treated, forged, etc. We're going for simple descriptions of shape and color and other features that may be characterized as a result of direct sensory input.

SUGGESTIONS FOR APPEARANCE DESCRIPTIONS: You could comment on the feel as well including how dense the two ends feel relative to each other: consider the colors, the reflectivity, etcetera.

When you are finished, repeat these exercises with a few more objects.

FINALLY, to test your skill, read your descriptions to someone else and ask them to sketch what you described and to guess what it is.

This definitely gets easier with practice. The key words to have in mind are the basic shapes of three-dimensional objects and surface and bulk characteristics such as color and density. The table provided on the next page contains suggestions of descriptive words.

Table B: Descriptive Words

Shapes	Surface Characteristics & Appearance	Bulk Characteristics	Relationships
Volumetric (3D) or Superficial (2D)	For 2D- True or Nominal[57]	Size	Relative size with reference
Cylindrical, Disk Shaped, Spherical (avoid round for 3D items)	Red, Green (other colors)	Opaque, Transparent, Translucent	Opaque, Transparent, Translucent
Spherical	Shiny, Dull, Speckled, Spotted	Heavy for size, Light for size	Thick, Thin
Prismatic (Triangular or other shapes)	Flat, Rough, Bumpy, Pitted	Rigid or flexible (how shape responds to loads)	Nominally 2D (as in the sides of a box)
Grooved, Perforated	Uniform (or not)	Solid, liquid, gaseous	How it sounds when struck
Centered, Offset	Slippery	Solid, Hollow	Cold, Warm, Ambient
Lens (cross-section of a magnifying glass)	Faceted	Compressible, Incompressible	Smell
Conical, Tapered, Helical	Iridescent, Sparkly	Hard, soft	Concentric

What have we just done? We have primed our minds with visual description terms. This is an important practice for failure analysts. I practice this every day, all of the time. Now that we have primed our minds with visual description terms, see what you make of the following exercise.

EXERCISE 3-D: Imagine and draw identical lenses (draw in the box provided on the next page)

Imagine and draw two identically-sized lenses. There is nothing unique about the shape of a lens, unless you look at it in cross-section. So, what we are going for here is a pair of doubly pointed ellipses, lined-up end-to-end. Take the opportunity to draw the whole sketch before turning the page after the stop sign!

One of my guinea pigs (I mean test subjects) for this exercise drew the ellipses side by side, and the experiment still worked.

[57] An example of a nominal 2-dimensional object is a piece of fabric. Another example is the surface of a three-dimensional object that is not completely flat

NEXT, mark the center of each ellipse with a dark spot.

NEXT, imagine a large oval surrounding the two, pointed ellipses and with the short axis of the oval aligned with the long axes of the ellipses. Can you name what was just described?

Have you finished?

The pictures on the next page show a hand-drawn sketch and software generated version, using the preceding instructions.

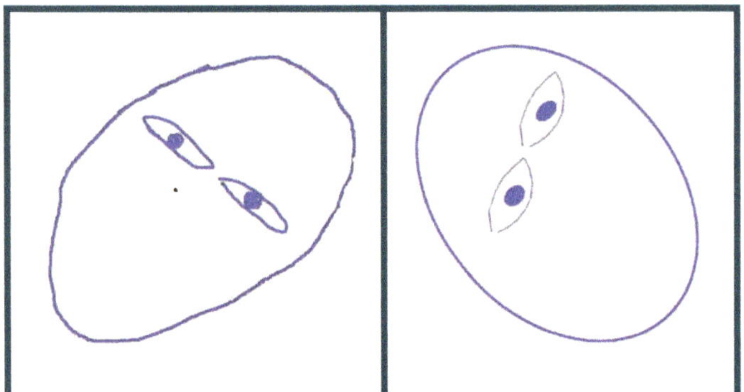

Hand Drawn Attempt | *Software Aided Version*

I found it interesting that the rough sketch was more recognizable as the eyes in a human face than the more "precise" (if unskilled) computer drawing. Of course, a face is rarely a perfect oval. And most of us have ears!

Humans have an inborn Go/No-Go gage for the presence of human eyes. Newborn babies are programmed to seek out the eyes of their care givers. The eyes may not focus well at a few hours after birth, but light and dark and bright colors register on our retinas and an impression is carried to our brains for interpretation.[58] As we age and interact with our widening world, we gradually lose more and more of our ability to see much of what is in front of us in that open-ended way.

Even as infants, we look for our caregiver, our toy, our food, our blanket. And when we look at and recognize something new, we go toward it. We want to touch it, smell it, listen to it, and taste it. This curiosity-driven process is hard-wired into human infants.

From an early age, we lose the ability to simply see the images in front of us in favor of filtering for what is of use. We start to make associations between data that is presented individually by each of our senses. We learn to see, for example, that something is heavy by noting how another person interacts with it.

[58] This is an oversimplified description of seeing.

This sensory data is processed in the mind and usually happens under the level of our awareness.

Usually, this processing is helpful. But when we are the experts who are supposed to be figuring out what happened and why it happened, we need to be able to see all the evidence and not just the features that make it through our usual filters. We need to constantly train ourselves to perform these tasks.

We need to be able to see what is present, beyond what everyone else notices.

Challenging ourselves to stop and look at an object or situation for longer than usual, and trying to find words to describe it at the simplest level, helps us remember how much of our seeing is influenced by what we expect to see. The added benefit is that slowing down gives us time to notice anything unusual.

I have found it helpful to prove to myself that I really don't normally see "what is there", but see a processed version of what is there. I hope you will take the time to do the same thing and practice it frequently. This awareness is critical to avoid self-deception.

3.3.1 The Four Idols

You may wish to investigate what Sir Francis Bacon called "The Four Idols"[59] in relationship to automatic visual data processing activities. In summary, Sir Bacon instructs us to devote ourselves to studying all sides of a situation because our natural impressions, the influence of society, and our communication (to ourselves and others) are flawed. In fact, he says that our senses impede a true understanding of the world.

[59] The treatise, The Novum Organum, by Sir Francis Bacon was first published in 1620. The full name is *Novum Organum, sive Indicia Vera de Interpretatione Naturae or Instaurationis Magnae, Pars II.* It was written in Latin and the title references Aristotle's work *Organon.*

"But by far the greatest impediment and aberration of the human understanding arises from the dullness and inadequacy and deceptions of the senses..." – Sir Francis Bacon-

Think of his idea in terms of light and color perceptions influencing our individual judgement of relative sizes and positions, and the linguistic style of our communications - all could be affected – possibly distorted. Previously we discussed how language strengthens bias and influences our perceptions. Likewise, these automatic visual data processing activities could result in wide variations in perception and descriptions of what is nominally the same scene.

In the nature-nurture equation, some distortions are built-in and some are learned; so, how do we learn to study all sides of a situation equally and objectively as Sir Francis Bacon proposed?

Below is a scientific way to practice these skills. Studying optical illusions can also demonstrate for you the specifics of humanity-wide, automatic, and otherwise invisible data processing activities. Figures 3.1 – 3.5 (below) are some well-known optical illusions to get started.

Figure 3.1A: *Simultaneous Contrast Illusion*[60]

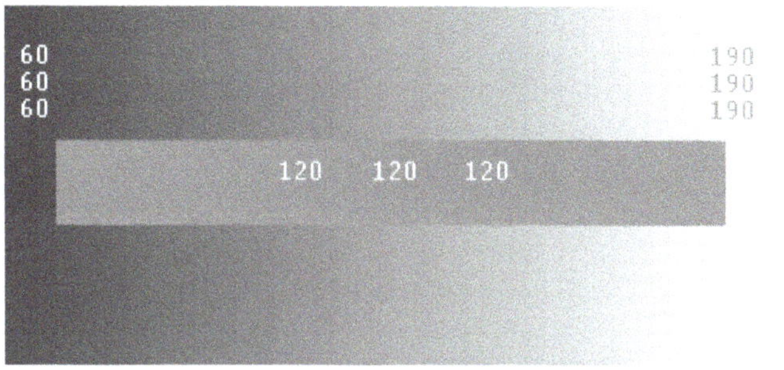

Figure 3.1B: *Simultaneous Contrast Illusion with color values*

In both of the images above, the background is a color gradient and progresses from dark grey to light grey. The horizontal bar in the center appears to progress from light grey to dark grey, but is in fact a single color across the bar.

It's difficult to believe that Squares A and B are identical colors in Figure 3.2 (below). This demonstrates that our brains correct for shadow.

Figure 3.2: *Squares A and B are the same color.*

This was originally created by Edward H. Adelson, vectorized by Pbroks13. - Own work | CC BY-SA 4.0 Creative Commons

You can prove this to yourself with an eyedropper tool in graphics software that shows the RGB or other color system values. Now look at the illusion in Figures 3.1 and 3.2 (above). See Figure

[60] Attribution: Understanding human perception by human-made illusions by Claus-Christian Carbon, 2014, Figure 4 | https://www.frontiersin.org/articles/10.3389/fnhum.2014.00566/full#B25

3.3 (below). This figure demonstrates the same perception that the moon is larger when it is low in the sky and adjacent to trees and buildings whose size is known to us.

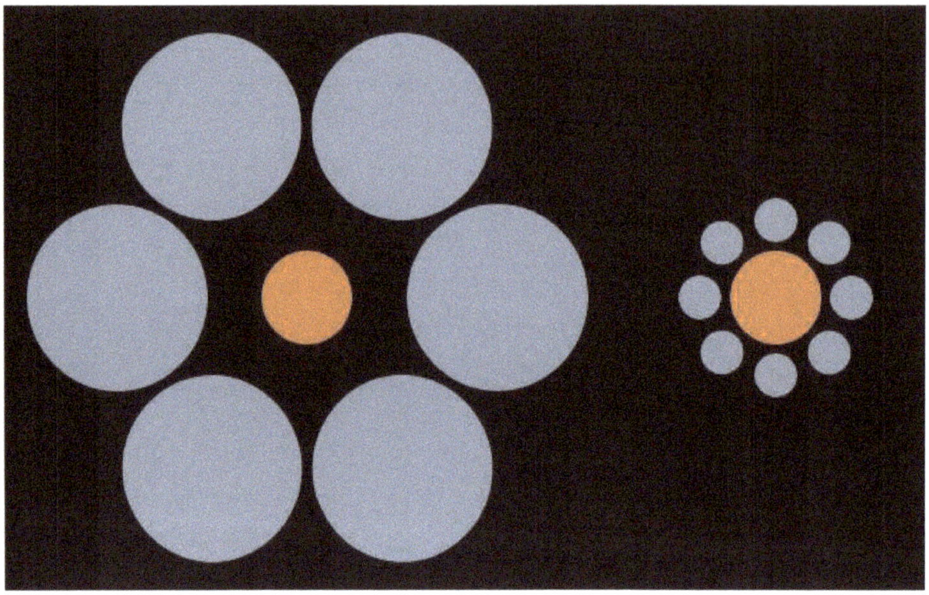

Figure 3.3: *The two orange circles are the same size.*[61]

EXERCISE 3-E: What do you see first?

Describe what you see in two words. Even if you have seen the image, Figure 3.4 (right), a number of times, what do you see first? Do you see an old hag or a young woman?

Once you have seen both, can you switch back and forth? Have you noticed it's impossible to see both at once? Try it.

Now, make a list of the features that contribute to each image. DO IT NOW before you continue. Space is provided below and on the next page.

Image Source:
https://commons.wikimedia.org/wiki/index.php%20p?curid=16926542

Figure 3.4: *Picture by William Ely Hill – Public Domain*

[61] Attribution: *http://en.wikipedia.org/wiki/File:Mond-vergleich.svg*

By making the list, you practiced seeing and distinguishing both images. Once you created the list, were you able to shift more easily from crone to lady and vice versa?

See Section 3.8 for my list of attributes that helped me switch from old woman to young one.

The next interesting concept to explore is the possibility of seeing both images simultaneously. It's impossible; that is interesting. Why is it impossible? It is because when we see a face, we see a face. And in our frame of reference and social conditioning, a face has two eyes when looking from the front and one eye when looking from the side. The ears are in a different relative position to the edges of the face than that occupied by the eyes. We simply can't see a single feature as an eye and an ear at the same time.

In other words, our mental processes are automatically determining the frame of reference. **That's a huge concept for the performance of failure analysis!** Our decision to see an eye or an ear is happening at a sub-conscious level. Here is an opportunity to reflect on the necessity of Confirming our Context.

Let's pause to stop and think for a few minutes about what this means when we are performing visual inspection. Our eyes and mind fill in features based on patterns that we have seen in the past or that are hard-wired into our genetic programming. Now, let's change the conditions of the same image slightly. See Figure 3.5.

Is it just as difficult to find each face as the first time you tried? Did you adjust quickly after the previous practice? If you saw both images faster, is that good for a failure analysis professional or bad?

Could lots of practice with the faces allow you to see features more quickly on a fracture surface? Consider that our tendency to see what we expect to see is well documented.

Figure 3.5: *Alternate version of Hag-Lady illusion by William Ely Hill*[62] *(Cropped black and white image).*

David Levy says, "Ye shall find what ye shall seek."[63] And what we expect to see is often below our level of direct awareness. Let's apply this directly to failure analysis.

---

[62] William Ely Hill (1887–1962) ("Puck", 6. Nov 1915) [Public domain], via Wikimedia Commons; Also see https://en.wikipedia.org/wiki/My_Wife_and_My_Mother-in-Law

[63] David A. Levy, *Tools for Critical Thinking: Metathoughts for Psychology*, Chapter 24; Print Book, English, ©1997; Publisher: Allyn and Bacon, Boston; 262 pages; ISBN:9780205260836

If we are *looking-for* a crack origin and can't determine it definitively, we're going to have to give our best scientific guess. Some people are more inclined to be intellectually honest and say, "The designated origin is presumed." Then they state their reasoning. But others do not. Over the years, I have seen several *final* fracture features identified as the *origin*. The analyst knew they had to identify an origin and so they did. They found what they sought; whether they identified it correctly was potentially less important to them.

If you are not armed with a reliable methodology to determine the position of an origin, it's easy to make this type of mistake (especially if the fracture surface is dirty, damaged or if the material is brittle). See Figure 3.6A (below) displaying grey iron casting fracture surfaces (this type of surface is notoriously featureless). The example in Figure 3.6A is characteristically so. Can you determine the origin? It's pretty difficult to tell.

Figure 3.6A: *Cast Iron Fragment*

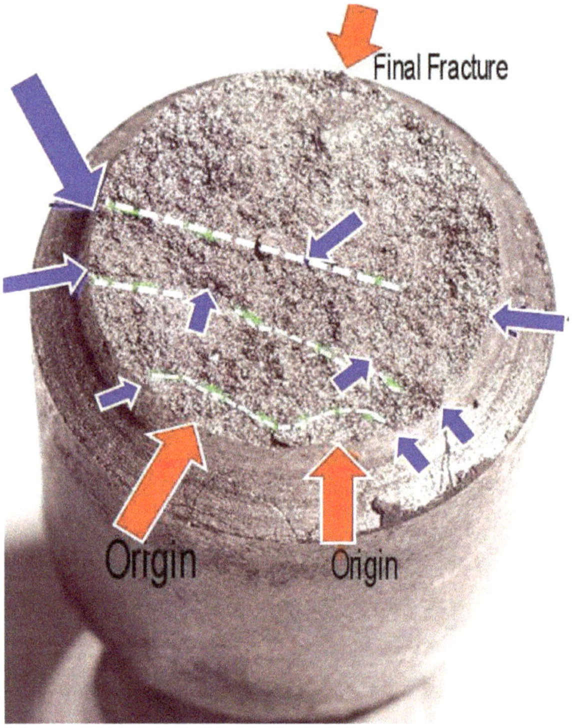

Figure 3.6B: *Presumed fracture features highlighted on grey iron casting fragment; contrast was enhanced from the original photo.*

Only by examining the mating part, which is not shown here, was I able to speculate on a loading geometry. And then I further speculated on the likely initiation position.

After staring at the fracture surface for a long time, in conjunction with the knowledge of the loading geometry, my eyes found some very faint beach marks. See Figure 3.6B (above), where the speculated beach marks have been highlighted. Without the support of the information from the loading geometry analysis, these faint features would have been easy to ignore.

It's possible that I fell prey to "seeing what I sought", but the client found the analysis useful. The damages in this case were small and the client was simply seeking to rule out an obvious case of insurance fraud. The data I was able to provide from this low input exam was all that they were willing to pay for.

If you lack a methodology to determine the position of an origin, it's easy to make a mistake, especially if the fracture surface is dirty, damaged, or the material is brittle.

We often begin the process of making a mistake or of resolving the mistake by linking the inputs from individual sensory experiences with other sensory inputs.

- We link similar visual experiences with each other (as I did with the separate fragments).
- We link similar sound experiences with each other.
- And, we link simultaneous sights and sounds to each other (and to other sensory input data) in order to create multi-dimensional experience.

For the last few months, I have heard my heart fluttering, or beating much faster than usual, but only when I lie down to sleep. Very strange. Finally, I tried to take my pulse at my wrist. It was normal. Hmmm. So, I stopped worrying so much, but still felt confused. Finally, I realized that there is a vibration related to my furnace blower fan which is making my bed vibrate. It seems strange that an all-foam mattress would transmit these vibrations; so, I paid even more attention and I noticed that the "heart problem" was happening at the same time that the furnace fan started up. In fact, it happens a few minutes after I hear the oil burner kick in. For a long time, it didn't occur to me that the sensory input was from an external, rather than internal, source.

Editing this in July, I realize that my heart problem is gone. I have not needed the furnace for a few months. Maybe there's a bearing wearing out in the furnace fan. Time will tell. For the moment, my confidence in my root cause analysis of my fluttering heart has increased.

What else shapes our experience?

Emotions affect the way our memory shapes experience. Emotional tagging of sensory memories explains why we can see trash on the road as a dead small animal or see a dead leaf blowing across the road as a jumping frog. It explains how hypnotists are able to make people who are *looking-at* a rope believe they are seeing a snake.

When we start to learn about failure analysis, among the first things that we encounter are explanations of how to distinguish *brittle* from *ductile* fracture events (see inset text box) and *fatigue* (cracking due to repeated loads) from *monotonic* (single load) crack propagation. Once we know how to recognize these four categories of features on broken parts, it can be very difficult to look at a broken part to simply see what it reveals.

Even after ten years of being very clear in my own mind about the necessity to LOOK-AT, most of the time when someone shows me a fracture, the first thing I am going to notice is the result of my judgement about the ductile or brittle nature of the crack and my determination of whether it happened all at once or during repeated load cycles. This is not seeing.

This is not LOOKING-AT.

This is a result of LOOKING-FOR items on a list of features, some of which I "know" must be present.

Obviously, complete beginners who don't know how to tell a fatigue crack from one that happened as a result of a single load cycle are better in some ways at *looking at* the visual data. Sometimes beginners see details that a more experienced person misses. Many of the details beginners see and point-out result in the comment "Oh, that doesn't matter." This mentoring is both good and bad. It helps build experience, but is also begins distortion in the mirror of observational skills.

KEY WORDS: CLASSICAL USE OF THE TERMS DUCTILE and BRITTLE *(AT THE MACRO OR VISUAL LEVEL)*

Ductile behavior resulting from the loading event(s) that culminate in fracture causes permanent shape change (permanent deformation) in the area of the part immediately adjacent to the crack, in addition to fragmentation. **Brittle cracking** results in fragmentation with extremely limited or no visible permanent shape change in the immediate vicinity of the crack surfaces.

Technically, we must specify whether we are referring to macro- or micro-scale ductility (permanent shape change). This will be embellished in Chapters 4 and 8. For now, since we are in the chapter on Visual Inspection, we'll be referring to macro-scale ductility or shape change that can be seen with the unaided eye - if you know how to look!

Brittle cracks may happen in materials that are engineered to behave in a ductile manner. When this happens, they are often unexpected. Their consequences may be more dire. Thus, "Ductile or Brittle" is often the first technical question that comes up regarding a failure that is associated with a fracture.

Brittle cracks don't require anywhere near as much force as ductile cracks, all other factors being equal. That's why brittle cracks are often greeted with surprise.

In Chapters 4, and especially Chapter 8, we are going to move beyond this common, but oversimplified description of the way to distinguish ductile and brittle behavior.

We will study the benefits of adopting a mechanical engineering approach to distinguishing macro-scale ductile from brittle cracks. We will also distinguish macro-scale ductile and brittle cracks from macro-scale ductile and brittle behavior before and after the actual crack event.

EXERCISE 3-F: Cultivating and Compounding Competence in Visual Inspection

What does the image in Figure 3.7A (below) represent? If you are not at all familiar with metallurgy and metallography, you might have no clue. If you are, what does it look like? Slow down and see.

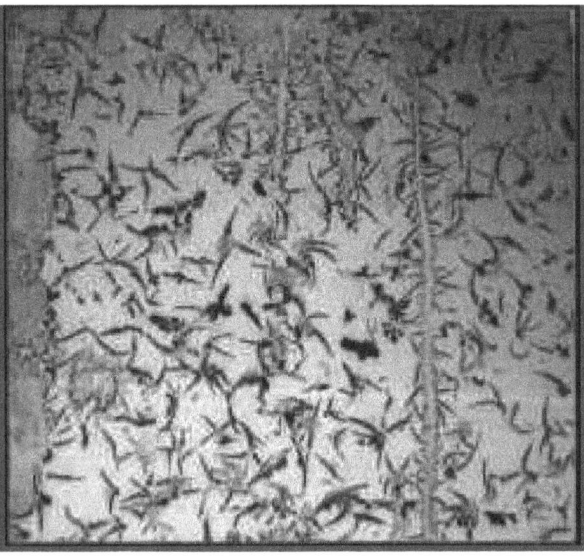

Figure 3.7A: *Mystery Image*

Well.... to me this looks like a grey cast iron microstructure, a metallographic coupon in the as-polished condition (unetched). But is it really? There are a few suspicious features that might say it's not a grey iron microstructure. I am primarily suspicious of the long white (or light grey) skinny feature that runs top to bottom along the edge of the right third of the image; although, that could be a ferrite vein. Figure 3.7B (below) shows a wider view in its original surroundings with its original colors.

Figure 3.7B: *Mystery Image – original surroundings with original color*

Although I can tell you that this is the frost covered doorpost of my car with multiple reflections, I seriously cannot figure out what all the stuff to the immediate left of the frosted door post is, even though I took the photo.

Learning to recognize patterns across different categories of objects is beneficial to holistic brain function; and helps you become a more effective failure analyst. A little later in this chapter we will delve deeper into recognizing patterns across different categories (such as the graphite in grey iron and the frost patterns on a car window).

EXERCISE 3-G: One More Eye Test (Visualization)

Look at Figure 3.8 (right – *The Michigan Fish Test*) for 5 seconds and then write what you saw in the space below.

Figure 3.8: *Michigan Fish Test*

This test is called, "The Michigan Fish Test".[64] If you came from a Western cultural background, you likely noticed the three big fish on the right. Those from an Eastern cultural background more likely noticed the large fish and the seaweed, the little frog, the snail, and the smaller fish.

I'm not sure why they made the water appear such an unnatural shade of green. I've looked at this image many times, and I am just now noticing the bubbles rising from the furthest left of the three large fish.

An important point is that the background helps us understand the situation. If we don't pay careful attention to the background, whether it is the physical details of a scene in front of our eyes or the bigger situational picture surrounding the event (metaphorical background), we have a greater chance of misinterpreting the data.

NOTE: Experienced analysts see the world through stronger filters.

These stronger filters are useful. They can also be problematic if the analyst screens out features that could reveal something we need to understand about the failure. One of the most frequent types of features where this happens are those related to post fracture damage that results in shiny areas. In fact, sorting out the significance of post fracture damage, in many cases, can be the most difficult part of a fracture analysis.

Here is a story from the early days of my career, where I struggled with an unrecognizable feature. The fracture was clearly a result of a fatigue crack. I was the only one at that company who knew anything at all about fracture analysis, so, I was on my own with the Metals Handbook, 8th Edition, Volumes 11 and 12. I started poring over the images in the Atlas of Fractography in Volume 12.

Based on my one semester course in college on fracture mechanics and fracture analysis, it was clear that the crack had initiated at the surface of the part, in fatigue. There were classical fatigue beach marks.

I also noted a weird (the technical term would be "anomalous") feature that I did not understand, in the middle of my fracture surface. It looked like a volcano protruding from the center of the fracture surface. It had a slightly oval shaped base. Even more strange, in my opinion, the volcano had a slit down the center, along a vertical plane. (By vertical I mean if the feature had been a real volcano, it would be oriented radially on the face of the earth. Vertical was not related to the service orientation. It was vertical on my desk at this point, and that was a consequence of how the part had been cut down to a reasonable size for analysis.)

It's been a while, but I think I spent at least two hours poring over the reference images. At last, I found something that looked just like my feature. Aha! I must have smiled. The hunter at last glimpses the prey! Now, for the caption, that would surely explain all.

Well, so much for that. The caption did not even take note of this strange feature. The caption only described the details of the fatigue portion of the crack. I finally gave up, since I knew that the crack initiation was more important to identify.

Years later, I realized that this strange split volcano was essentially a shear lip (final separation feature) that was confined to the core of the original part. It was a result of the last portion of the load-transmitting, cross-section to be

[64] The attribution for the Michigan Fish Test: Nisbett, Richard & Masuda, Takahiko. (2003). Culture and point of view, Proceedings of the National Academy of Sciences of the United States of America. 100. 11163-70. 10.1073/pnas.1934527100. Other interesting links include the following: http://www.cnn.com/2011/OPINION/03/06/iyengar.fish.freedom/index.html and the accompanying file at https://i.cdn.turner.com/cnn/2011/OPINION/03/06/iyengar.fish.freedom/story.three.large.fish.courtesy.jpg

pulled apart. The split down the center was probably a result of centerline shear stresses, perhaps acting on a plane of weakness due to centerline segregation of low melting point, low strength impurity alloy elements during the solidification process at the steel mill. I have never been able to confirm it because the part and the company where I did the investigation were both long gone.

I'm usually much more efficient now than I was during that first difficult fracture analysis. But I have to make a concerted effort to look at the part for features that are not necessarily on my list of what to look for. Sometimes it is just these unlisted features that provide the key to understanding the situation. I also try to remind myself that the feature pointed out by my presumably less experienced or totally inexperienced client may actually be very significant. What looks insignificant to me may tell the manufacturer or designer critical information.

3.4 CASE STUDY - Visual Inspection of a Corroded Stainless Steel Pipe

Client: Here is a corroded stainless steel pipe. Why did it corrode?

Discussion: Early in my career, like most people, I was often given some "failed" component or sample, but little or no background information, and had little to no likelihood of getting any. Now I know, to a much greater degree, how to ask questions when the inquiry first comes my way.

Preaching about background information collection does little good if your understanding of engineering concepts is immature. We all start that way, and the only way to acquire the knowledge of what questions to ask is by asking questions, and realizing later that you didn't ask enough questions, or didn't ask the right questions. No simple list will ever be able to include all the relevant questions for every situation.

The leaker pipe of Figure 3.9 (shown below) is one of my favorite case studies. I was given a piece of pipe from an industrial facility. It had been used to convey a chemical said to be extremely hazardous to human health. Before I could even open the plastic packaging in which it was wrapped, I had to take it to a chemist for analysis and decontamination of trace residues of the dangerous material. Fortunately, he found no problems. That was fifteen years ago, and I am still alive, so, I guess he was probably right.

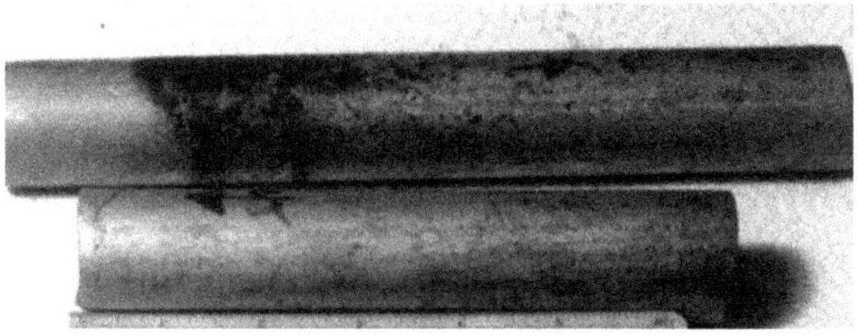

Figure 3.9: *300 Series Stainless Steel Pipe with corrosion associated leak in dark the area – the black arrow on lower pipe indicates flow direction in service.*

Now with the Subject Tube in front of me, I could begin my visual examination. It was two cut lengths.

Nobody had taken any photos before they cut the short pieces out of the piping system at the factory. So, I did not have any information about whether the pipe was horizontal, vertical, or at some slanted orientation (this is a different case than the one mentioned earlier for the litigation project). I also had no record of the radial orientation in service. Which side was up or facing away from the wall? Was it next to a hot or cold line? None of this had been recorded. Someone did think to indicate the flow direction. And I can't now remember if I got the pipe as one or two pieces.

So, here we are with a dark stain in a chunk of stainless steel. What is the next step?

Well, take the advice of a fountain of experience; turn it over. See Figure 3.11 (below). Hmmm.

Figure 3.10: *Same pipe as in Figure 3.9 showing detail of hidden surfaces.*

Looking at this today, I see there are some very interesting dark circumferential streaks. At the time, I didn't pay much attention to them. Although the orientation of the pipe section had not been formally recorded by the people at the plant, these natural features did the job. At this point, it is now completely obvious to me the pipe in Figure 3.10 shows the top position and Figure 3.11 depicts the bottom side in service. The dark circumferential streaks are drip marks. Well, so be it. Live and learn.

Note that when I say it is now completely obvious, actually, there could be other "causes" of these features. Maybe the specimen was subjected to some environment other than the service environment. Had the piping been stored outdoors before installation? Did the damage occur then?

At some point, if we are trying to come up with a scenario of what happened and how, we have to give ourselves at least some working hypothesis. Here at least, the working hypothesis is that the dark stain facing up in Figure 3.11 was facing upwards in service. In a more thorough investigation, we might find reason to re-evaluate that assumption (contemplate complexity) later during the investigation. In any case, it is important to note in the project record that we made an assumption about the orientation of the pipe in service.

Nobody had been injured in this case and I was not made aware of any prohibitively expensive cleanup efforts. Evidence preservation is rarely on the primary agenda of people who are faced with an immediate safety concern. This is a fact of life that we failure analysts have to learn to live with.

Now, let's remember to examine the inner surface (Compare and Contrast the Outer and the Inner). Below, Figure 3.11 shows the only inside view I have left in my files.

The dark stain on the upper surface is foreshortened. If you only had this view, it could look like simple surface roughness.

The inner surface appears to have some reddish material in a deposit of irregular thickness. As soon as we choose the word "deposit", the implication is that this is due to gravity. The natural, automatic (unconscious) decisions I made about how to photograph this artifact were possibly helpful.

However, I am missing the mating view, which might reinforce the conclusion that the bottom in the image was the bottom during service, or it might weaken it.

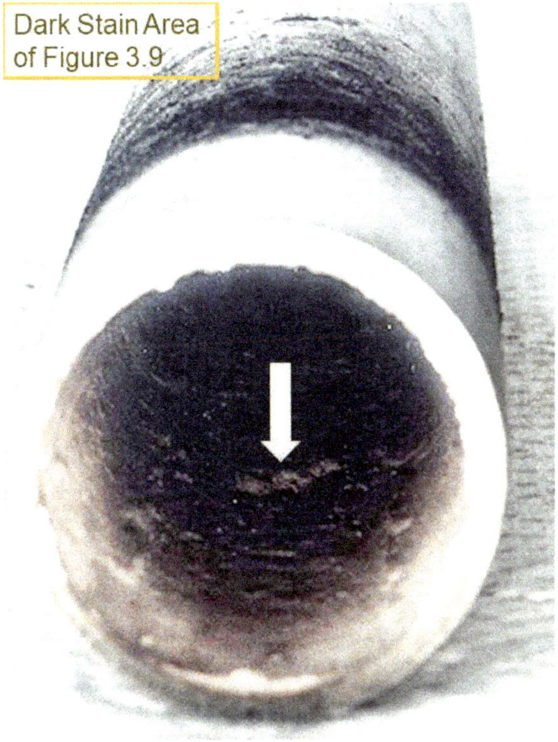

Figure 3.11: *Inside of the same pipe as in Figure 3.10 with dark stain*

Since I selected this image for the slide show from a dinner talk I have given many times, my guess is that there was a reason. However, it's possible that I never looked at how much deposit there was on the inner, upper surface. I will cut my younger self some slack and assume that there was less or no deposit on that inner, upper surface: that there was nothing interesting to photograph. It's preferable to take photos from multiple directions.

Deposit, as a technical term, is more generic than another commonly used term that we might have chosen: corrosion product. Corrosion product implies that the solid material on the bottom of our pipe is a chemically combined compound that includes atoms that used to be part of the metallic structure, as well as atoms that used to be present as individual atoms, ions, or molecules from the service environment.

A deposit could be the solid portion of a corrosion reaction, hence, corrosion product.

But a deposit could also be the remains of particulate material that was simply carried in with whatever was in the pipe. An example could be suspended dirt.

A deposit could be material that was in contact with the pipe as individual ions from a source other than an intentional part of the system itself. An example of this type might be calcium carbonate that has exceeded its solubility limit, and precipitated from solution.

Here we start to see that even the most neutral and dry technical terms carry a lot of assumptions.

KEY POINT: This case study is an example of an application of epistemology.

Drip lines and deposit position are both pointing to the same orientation of the pipe in service. We can now be more confident that "we know what we think we know" about the orientation of the pipe.

Well, the time had finally come to make a decision about how I was going to cut the part. One of the big questions was whether or not the dangerous chemical caused the corrosion or not. That meant I had to open the pipe along its axis to get a better view of the inside.

Figure 3.12: *Silicone replica used to capture the loose material at the dark spot – this replication material is applied as a liquid and solidifies in place.*

Because I did not want to contaminate the stain and deposit areas with metallurgical wet saw coolant, I decided to cut the pipe lengthwise with a dry band saw. However, band saws tend to create a lot of vibrations in lightweight parts. Therefore, I decided to "capture" the loose material in the stain area with a silicone replica (as shown in Figure 3.12 above).

When it cured, I peeled back the silicone. The loose particles were now held in the same relative positions (as when they were on the outside of the pipe). Figure 3.13 (below) shows the central damage area after removal of the replica.

Figure 3.13: *Detail of corrosion product or deposit filled hole in stainless steel pipe*

At this point, we still didn't know for sure whether the dark stain was really the source of the leak and the reason for the investigation in the first place!

Finally, I cut the part, along the assumed horizontal, diametral plane. At that point, all the remaining central dark stain area was shaken loose and the presence of a through hole was revealed. The moment of truth had arrived! Confirmation that we were indeed *looking at* the location of interest - the leak position had been obtained.

Figure 3.14 shows a metallographic cross-section through the edge of the hole. It is clear that the outer surface is rougher and the corrosion is more severe there than on the inner surface. Thus, it is most likely that the corrosion did not initiate inside the pipe from the dangerous chemical, but from another source outside of the pipe. This was very good news for the owner of this piping system. It's usually a lot easier to inspect the outside of pipes than the inside.

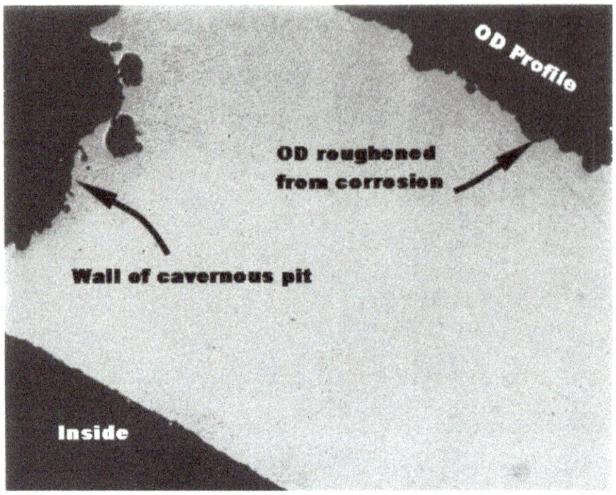

Figure 3.14: *Metallographic cross-section of stainless steel tube with through wall hole*

Another feature is prominent in Figure 3.14 (above). The corrosion is actually most severe in the core of the pipe wall when compared to either the inner or outer surface. That allows us to classify this corrosion process as pitting corrosion.

Figure 3.15 (next page) shows an EDS spectrum of the dark material that fell out of the hole during the band saw cutting operation. Note the high concentrations of both iron and chromium, most likely from the stainless steel itself; also note the oxygen, chlorine and sulfur. We now have an idea that this material is most actually deserving of the name "corrosion product".

However, looking at the data for the hundredth time (or thereabouts), it jumps out at me that there is no nickel and this material was an austenitic stainless steel. Well, this is not the first time that I have looked at old data and noticed something glaring for the first time. I will also comment that I have presented this case study on at least one occasion to an audience of probably over fifty people who work in the failure analysis field, and I was not questioned by my colleagues about the lack of nickel!

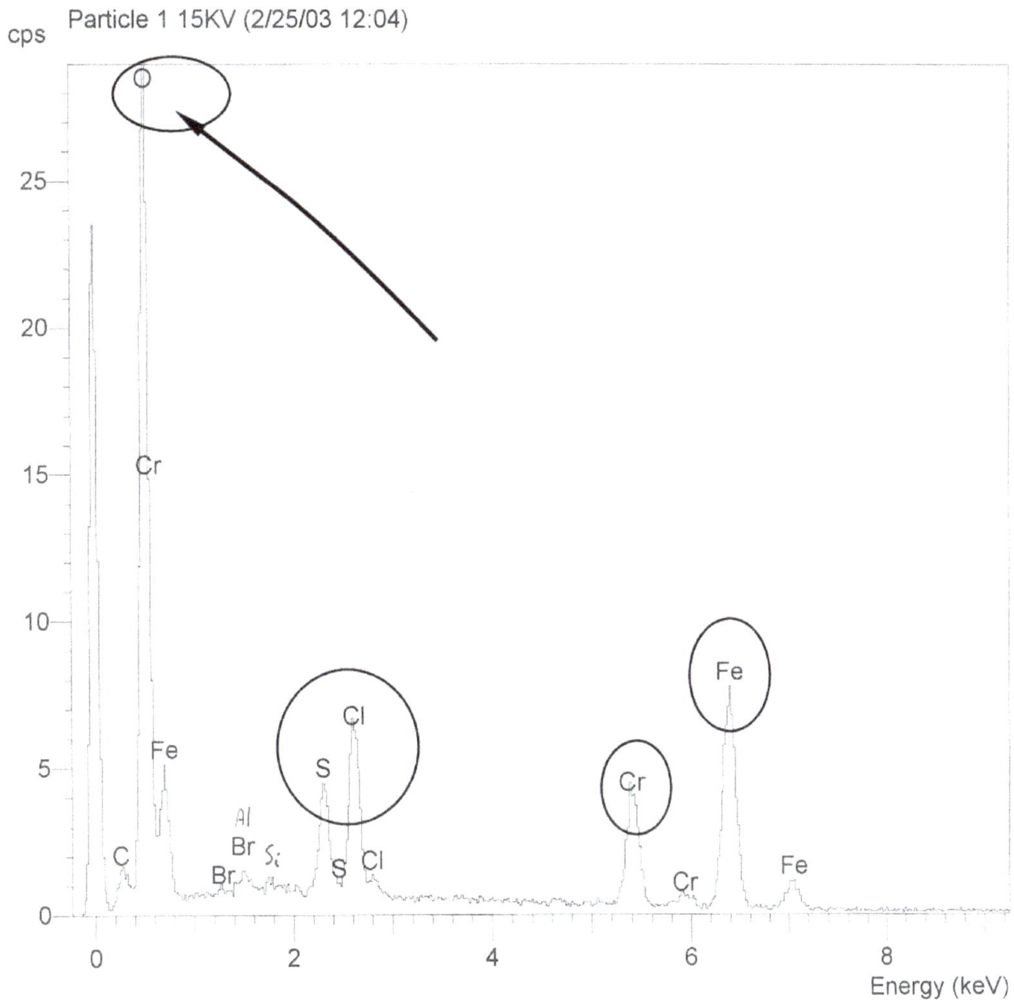

Figure 3.15: *EDS (15kV accelerating voltage) Spectrum of a particle of what used to fill the hole in the subject stainless steel pipe*

Of course, nickel is highly corrosion resistant in comparison with iron and chromium. Since there was a hole in the pipe and presumably nickel was dissolved along with the rest of the alloy, what happened to the nickel?

This is a great time for you to try to think of some possible answers.

One idea that jumps out at me is that the corrosion product is not from this corrosion event. I don't think it is likely, but at this point, I can't rule it out.

This brings up another point. The human brain is wired to notice what is present and ignore what isn't. According to Bruner, Goodnow and Austin, only one in seven humans routinely notice the significance of negative data.[65] Formal use of epistemology and basic critical and creative thinking tools encourage us to make it a habit to "Consider the opposite." Learn to ask "What don't I see?"

At last, we are in a position to put the data all together. Let's go back to Figure 3.11 and ask ourselves anew what we see. It may be helpful to read the following points aloud.

1. We see a dark stain that surrounds a dark deposit that appears to be all that is left of what was once a portion of the upper service surface of an intact austenitic (nickel bearing grade) stainless steel pipe.

2. The hole, that now breaches the pipe wall, was filled with material that is most likely a true corrosion product. It contains elements found in the tube itself as well as chlorine (Cl) and sulfur (S), which are known as elements that are aggressively corrosive to the subject pipe's base material.

3. The corrosion happened by a type of pitting corrosion. The size of the pit is larger in the core of the wall than it is at the surface, presumably in direct contact with the aggressive material.

For those who know some corrosion science, this last "piece of information" points toward a micro-biological corrosion process. Bacteria may cause this type of damage. It's often called "MIC" or microbe induced corrosion.

WHAT PROTOCOL WAS FOLLOWED?

- *Visual Exam*
- *Scanning Electron Microscopy with EDS microchemical analysis*
- *Optical Metallography*

FEEDBACK

Well, I was still scratching my head, but finally called my client. He had been at the plant for other reasons, but at least he had been to the plant where the pipe was in service.

So, I told him what I had found, and told him about "MIC". He was quiet for a moment. Then he said, "Birds."

"Birds?" I asked.

"Pigeons. There were pigeons flying around inside the plant."

"Aha!" Finally, I was able to see the dark, drooping stain for what it was. The corrosion product and residue from a pigeon plop. This was a combination of corrosion product and "deposit". But probably the deposit itself was long gone. If there had been some left, with the new high resolution EDS instruments now available, I might have been able to detect some nitrogen between the carbon and oxygen peaks. Animal manure in general has high nitrogen. But the EDS unit that I used back in 2003 was not able to detect nitrogen unless present as the major constituent, especially when present with oxygen and carbon, which bracket the nitrogen peak position.

This case study is one of my favorites to show that you really have to teach yourself to be able to see what is in front of your eyes. The difference between "seeing a bunch of light and dark and more or less colored patches created from interactions with light within our retinas" and "seeing the whole situation and the surrounding story" depends on just this ability to see.

Seeing is not one thing. Seeing is a whole multi-level process that involves our entire mind as well as our sensory apparatus.

[65] Bruner, J. S., Goodnow, J. J., & Austin, G. A. (1956), *A Study of Thinking*. Oxford, England: John Wiley and Sons, discussed in Chapter 4 of Thompson, Robert (1959), *The Psychology of Thinking*: Penguin, Edinburgh, Great Britain. This was mentioned earlier in the book.

Back when I was in kindergarten, my teacher assigned our mothers the chore of feeding us three different foods while we kept our eyes closed. The mothers were supposed to put the food directly into our mouths.

I only remember one of the three foods my mom chose. It tasted like nothing I could remotely recognize. Having the actual food in my mouth, with its accompanying texture, didn't help. My mother informed me that it was a slice of strawberry. This would have been in 1963. That was before the days of tasteless supermarket fruit. Without seeing, my taste buds just couldn't work properly.

Some people who are permanently impaired in one of the sensory functions learn to develop others to compensate. But a one-time trial of "guess the food" left me guessing.

EXERCISE 3-H: Cultivating and Compounding Competence in Visual Inspection

Take 10 minutes to write a list of features that you find by looking around in the room where you sleep. Write it on a separate piece of paper. You could start with the normally expected furnishings. A bed, perhaps some hooks for hanging clothing, the flooring material, any posters or wall decorations, a window, etc. Then, take at least half of the time to list features that you have never noticed before.

Repeat this exercise with a small familiar object, making sure you focus on the two different processes of LOOKING-AT and LOOKING-FOR.

Finally, repeat the process with something that you have never really studied before. When you complete the lists, try to write a description of one of the objects, that someone in a different place could read to create an accurate image of the object in mental space. To make this challenging, avoid using any words other than those that define shape, size, color, reflectivity and texture, using the words in Table 1 as examples of description, rather than interpretation. Read the description to someone and see if they can name what you just described.

3.4.1 GOING DEEPER (With A Slice Of Intellectual History)

In the heyday of the Greek philosophers, Plato scorned sensory input as being an unreliable way to learn about the world. Pure thought and the concept of meditation on the ideal form were promoted as being superior ways to know truth. As you might imagine, this put a damper on scientific progress; because if we can't trust what we observe, then we can't really do the basic type of classification that forms a foundation of modern science. And if we can't trust what we see, hear, smell, feel and taste as being reliable guides to actual characteristics of the world around us, then there is no point to setting up an experiment in the first place. To benefit from the experiment, you have to do it. Someone must collect and interpret the data. Such was the strength of the philosopher's love of conceptual ideals and belief in the unreliability of data collected from fallible human sensory organs, that it took almost 2,000 years before someone was born who was able to break out of this mental prison.

Galileo performed a series of experiments (1589 of the Common Era) using actual falling bodies to test Aristotle's long-standing theory that heavier things fall faster than light things, in direct proportion to the weight. Aristotle had taught that a 10-pound weight would fall ten times as fast as a one-pound weight.

Galileo's series of experiments[66] showed, at first, that light things actually fell FASTER than heavy things. He also showed that the heavy things eventually caught up and passed the light things. But as he refined his technique, he eventually

[66] For more in depth information, please web search this very interesting project: The Galileo Project funded by Rice University | or follow this URL: http://galileo.rice.edu/sci/theories/on_motion.html

concluded that heavy and light things fell at the same rate.

> ***THE GREAT KANADA***
>
> ***Philosophers of the Indian Subcontinent were likely ahead of the west. The great Kanada who possibly lived about 100 years before Plato (and possibly before the Buddha), wrote his inadequately studied work called, "Matter and Mind: The Vaisesika Sûtra of Kanâda".***
>
> ***Kanada started from his observations of the world around him and built his theories from there. Kanada beat Newton to a description of the laws of motion by possibly up to 2000 years. Perhaps this is because science remained tied to philosophy and wisdom for a longer time in India and there was no early schism between science and religion, which was caused in my mind by a priestly hierarchy.***
>
> ***About the same time as Kanada, the Jain religion emerged. One of the "Three Gems" of the Jain religion is "Right Knowledge". Sensory knowledge is the first of five types of knowledge identified in Jainism. The key aspect of the Jain system of using Right Knowledge is that the individual is taught to consider every data point from multiple viewpoints. The Gem of Right Knowledge acknowledges that some aspects of reality are indescribable.***[67]

Plato's teachings were one of the main reasons for science getting such a late start compared to art, politics and other fields of human endeavor. Because educated people knew that experiments with real objects were unreliable, experiments with real objects were left undone.

One of the main forces, other than Plato, that held western science back was the Catholic Church. There was a big mental investment in the interpretation of the picture of the universe that said that the Sun orbited the Earth. We still live with the results of the painful science-spirit divide today. It may be shocking to some Westerners to discover that Genesis doesn't actually SAY anywhere that the sun circles the earth or that the earth is flat. But the post dark-age thinkers in Europe had lost a lot of the knowledge gained by the observations of earlier peoples. And if the "Crown of Creation" is humans, then it followed LOGICALLY (to them) that we must be the physical center of the universe. Thus, the sun must circle the earth. This is a prime example of the danger of thinking that all we need is logic. We must always have a way to question our assumptions. Logic alone won't keep us from error.

Likewise, humans were not viewed to be as perfect as the heavens. The heavenly bodies were thought to be perfect. Aristotle was reportedly bothered by the fact that the moon was not perfectly uniform in brightness when full.

Thus, we can see that these ideas of a perfect heaven predated Christianity.[68]

Leonardo da Vinci drew pictures of birds in flight showing the motion of individual feathers. Da Vinci's accuracy was confirmed only in the early 1900s, with high-speed zoom photography. But Leonardo, a product of his time, apparently never drew an accurate picture of the moon. People's beliefs that heavenly bodies are perfect, along with the knowledge that a smooth sphere is the perfect shape for a perfect body, prevented Europeans in the late Middle Ages from seeing

[67] Web Search: "Subhash Kak Kanada" for a translation of Kanâda's work or go to the following link at Google books for a new translation of the nearly 2,500-year old work | Mount Meru Publishing, 2016 -Vaiśeṣika - 222 pages | *https://books.google.com/books/about/Matter_and_Mind.html?id=l8oHvgAACAAJ&source=kp_book_description*

[68] Aristotle died in 322 - Before the Current Era. The first Roman Catholic Church in Rome apparently dates to the 4th century of the Current Era, which is 600+ years after Aristotle.

the "Man in the Moon" or any shadow features on our satellite.

And I just realized one more reason (besides the Rorschach effect[69]) that not all cultures acknowledge a man in the moon. See Figure 3.16.

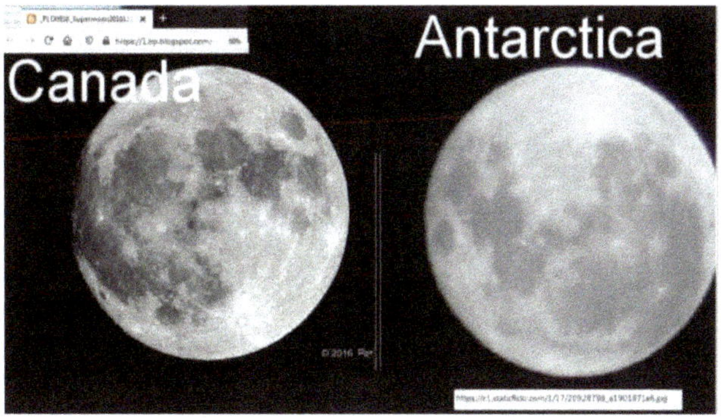

Figure 3.16: *What is our frame of reference when we look at the moon? We see the same features all over the earth, but the image is rotated.* [70]

There was a revolution when people actually started using telescopes to look at the moon, including Galileo, who came to the intellectually honest conclusion that the moon was not smooth. This was one argument that the church apparently did not make a big fuss over, compared to some other observational evidence... Perhaps it was not TERRIBLY painful for the church officials to admit that the heavenly "bodies" might not all be as perfect as the ultimate divine being itself.

Of course, we moderns are beyond all that. Even as a reasonably knowledgeable practicing engineer, it just took me about an hour and a half to gather the data (web surfing on a DSL connection) to be able to report the preceding historical information to you. We have different things to worry about. The idea that experiment and observation is inherently unreliable has LONG been put to bed by scientists. We RELY most soundly (whenever possible) on a combination of theory, logically thought out, and experiment, carefully performed. This is how we advance our knowledge today. We know that logic and reason are at least as prone to error as observation and experiment. Of course, I should not leave out the role of the hunch or inspiration as additional inputs, which in their turn, may provide paths to a more complete understanding of a situation, or alternatively, be in conflict with anything that we can demonstrate scientifically.

TO OBSERVE IS TO DISTURB!

Chapter XXI of David Levy's *Tools of Critical Thinking* book is entitled, "Reactivity: To Observe is to Disturb".

This insight perhaps comes to us most forcefully from quantum physics via The Uncertainty Principle. But metaphorically, this principle has describable effects in psychology. Whether we are observing people, or measuring the speed or position of an electron, we influence the behavior that we are trying to measure.

When we saw the moon as the pitted rock that it is, it changed how we started seeing many other aspects of our world. Cultivate your curiosity and read up on Werner Karl Heisenberg and The Hawthorne Effect.[71]

Probably the most shocking technical event in my life was when I discovered that there is a form of iron oxide that has a lower melting point than metallic iron. The first day of Materials Science 101, back in 1975, included a lecture on the general characteristics of ceramics (including iron oxide), metals and polymers. One of the primary differences is that ceramics have higher melting points than metals. How could my professors have

[69] The Rorschach effect occurs when person describes what they see in ambiguous inkblot images based on our experiences.

[70] To see more details, go to the following url: https://upload.wikimedia.org/wikipedia/commons/e/e9/Modeled_moon_appearance_for_same_longitude_30_minutes_after_moonrise.jpg

[71] Werner Heisenberg – Facts. NobelPrize.org | https://www.nobelprize.org/prizes/physics/1932/heisenberg/facts/

lied? I was shaken! Of course, there are exceptions that I knew about even then.

Mercury obviously is molten at temperatures where polyethylene is solid. But this was different. This was an oxide of the same metal! That day was important for me as a still relatively young practicing engineer (38 years old). I had to start to come to terms with the fact that even the things that I knew most surely might be wrong.

The way most of us learn much of the information we use is neither by observation or logic, but by believing someone else of greater authority. Obviously, we are not all as smart or curious or driven as Galileo and Aristotle. So, we have to take a lot of the knowledge that we get for granted, at least until we are unavoidably faced with the fact that it might not be correct. Maturity and skepticism have encouraged us to question the sources of our information, and not take everything the experts say as truth. Today, we have available so much information on how our minds work and how our sensory inputs are evaluated by our minds that we would do well to spend some more time looking at the origins of our knowledge. If we are the failure analyst, we better have something to offer those we're supposed to help.

3.5 Understanding The Eye-Mind Connection

Human beings were not the first to see. The history and evolution of the eye is long. The earliest eyes developed from bundles of light sensing organs in organisms that evolved long before humans. It's easy to understand that even a single-celled, photosynthesizing life form would have an advantage if it could somehow sense and thus have a foundation to follow or seek light. Whether we are thinking of the first organisms that got their energy directly from light or the organisms that fed on these, light is a foundation of life as we know it.

Untold numbers of engineering changes later, the eye emerged, sending detailed patterns of light, dark, and in some cases, color, to a bundle of nerves. Eventually, patterns meaningful to the survival of the owner of this eye-mind combination emerge and guide.

Experiments done at the Research Laboratory of Electronics at MIT in the 1950s based in the anatomy and physiology of a frog's vision (species Rana Pipiens) indicated the presence of a fairly primitive binning system for the results of analysis of the field of view for the frog, essentially consisting of only two categories: Big (predator) and Small (food). Frogs only perceive these big or small things if the things are in motion. After a few years of thinking about this, I suspect there's more to the story.

However, there was an entertaining video on You-Tube some years back, showing a poor frog in the corner of a room painted white. The frog kept bumping into the wall. In the wild, the frog clearly must be able to sense rocks, trees, and other nominally motionless objects in its natural environment. Being in the colorless room, devoid of any naturally shaped features, was clearly a confusing experience for the poor frog. How smart a test animal appears has been shown, in recent years, by many experimental biologists, to be strongly related to how smartly the experimenter has set up the experiment.

Despite this, there is not a lot of ground to stand on when trying to argue that frogs are just as mentally flexible as humans. Clearly, we have the advantage. Even before the development of writing and the consequent explosion in human civilization, we had the advantage in being able to adapt to a wider range of environments than any single species of frog. And before rational thought developed, we did it with genetically transmitted pattern recognition systems based in the eye-mind.

The existence of these systems can be easily demonstrated with the use of some well-known optical illusions; for example, the maiden-hag illusion, which has been previously described.

Now, look at the picture of Figure 3.17. It's the face of a man, wearing glasses. But, do you notice anything else? STOP and LOOK before reading.

***Figure 3.17:** What do you see?*

What if I had instructed you to look at the word written in Figure 3.17? What do you see now?

Again, we run into a cultural divide here. The young generation is no longer taught how to write in cursive and individuals who are not fluent English speakers may be at a disadvantage. But for a certain group, the word "Liar" will stand out, once we get into the mode of "reading" rather than *looking-at* a picture.

Now, *look-at* Figure 3.18.

***Figure 3.18:** What do you see? Frog?*

Clearly, it's a frog. We were just discussing frogs so please don't be shocked to find a frog in your failure analysis book. The frog is looking off to the left. It's sitting on a partially submerged log. Nothing more straightforward, right? Now, look at Figure 3.19. What do we see now?

***Figure 3.19:** Same picture of the frog!*

What? A horse? The frog's eyes have become the horse's nostrils. The partially submerged log has become the horse's mane. A dark streak between the front and back legs now becomes the horse's eye. The frog's back knees have become the horse's ears. The coloring is ambiguous, and helps along the illusion that the shape could become either animal's profile.

It is probably impossible, or at least it would be extremely difficult, to overestimate the importance of the pattern recognition system in how we see. It's just not correct to think of the eye as seeing and the brain interpreting. Seeing is inherently combined with interpreting. Otherwise, we should just call it visual sensing.

Getting back to a more direct connection to pattern recognition, the failure analyst uses these basic skills at multiple levels to interpret features of the subjects of the investigation.

First of all, we use pattern recognition to fit the subject of the investigation into a category of objects. Is it a screw? A shaft? A housing? A bracket? A gear? An engine valve?

Each of these recognitions brings with it a slew of both reliable information and preconceived notions. For some of the items listed, the experienced engineer will have an immediate basic understanding of the loading geometry, and more generally, the service environment and its fluctuations. This will sometimes allow access to an immediate insight into whether the failure is of an expected (well-studied) type or not. For example, the bases of screw threads are loaded in shear, but the body of the screw is loaded in tension, when properly installed. If the threads strip, we can tell something about what happened from that evidence. If the body of the screw is broken, and the fracture faces are smeared, that tells us something else. If the body of the screw is broken and the fracture surfaces have clean fracture features, that tells us yet something else. Pattern recognition in this case is a combination of interpreting features using the genetically encoded programs we all carry, and the specialized knowledge that we bring due to our particular educational backgrounds.

Secondly, we use pattern recognition to evaluate wear and deformation patterns, and fracture features at the macro- and micro-scales; and if we perform metallography (or its associated practices for non-metallic materials) we use pattern recognition to evaluate microstructure.

One of the simple and effective ways to improve your pattern recognition skills is to learn to see patterns across a wide range of size scales.

For example, Figure 3.20 (below left) shows the sky on a cloudy day across the street from my house. Figure 3.21 (below middle) shows some fatigue striations in an austenitic stainless steel. And Figure 3.22 (below right) shows pearlite in medium carbon steel (this pearlite-microstructure feature is often found in annealed or hot-rolled or normalized steel).

Figure 3.20: *Note the striped cloud pattern.*

Figure 3.21: *Fatigue striations in stainless steel – scanning electron micrograph*

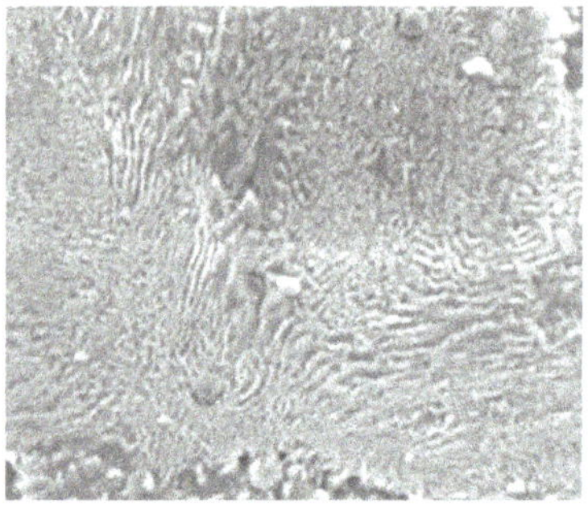

Figure 3.22: *Pearlite-microstructure (original magnification 1000x, nital etch, prepared metallographic cross-section)*

The alternating ridges of the altocumulus clouds are a result of a phase transformation just like the pearlite is a result of a phase transformation. And yet, the fatigue striations look more similar to the cloud pattern than the pearlite.

Hmm, this interesting observation demonstrates a number of things, but the most important is…

… Simplistic pattern recognition is *not* the best way to perform failure analysis. We need to couple pattern recognition with an understanding of how the patterns form. For fracture feature and microstructure interpretation, that understanding comes from the interaction of topics in mechanical behavior and physical metallurgy.

Finally, maybe it just seems like entertainment to notice that Figure 3.23 (next page) shows heat treat scale on a leaf spring that was coated with a clear glossy paint. It is so beautiful, and maybe it looks like there is a funny owl in it. But if you are enjoying your work, maybe you will be happy to spend the extra couple of minutes that it takes to let your neurons connect and create a new piece of information from what was previously simple data.

Figure 3.23: *Heat treat scale on a steel leaf spring – these springs were to be eventually upholstered so the most economical way to keep the scale from flaking off during the following manufacturing steps was to paint them with the "lowest cost available paint".*

This steel leaf spring happened to get clear coated. This image was obtained from a flatbed scanner, and the colors were brightened. I realize that this is a type of Rorschach Test and not everyone may see the "owl" with its wings slightly outstretched. It's a funny owl. One of the proofreaders saw a blue fish and another called it a grey whale. I'm ashamed it took me so long to see the fish/whale, swimming below the owl.

Others may see different patterns. Different cultures notice and describe different things all of the time as we've seen throughout this chapter.

IMPORTANT: Again, simplistic pattern recognition is not the best way to perform failure analysis. We need to couple pattern recognition with an understanding of what forms the patterns. For fracture feature and microstructure interpretation, that understanding comes from the interaction of topics in mechanical behavior and physical metallurgy.

3.6 Aids to Visual Examination

While this chapter has focused on the mental aspects of visual examination, we should not overlook the physical aspects.

As wonderful as our natural eyes and brain are, sometimes a little magnification will reveal features that are otherwise hidden. Never underestimate the value of a quality magnifying glass. Especially for large parts that are difficult to move and turn, a magnifying glass can be very helpful and save time. Magnifying glasses from 2x to 10x are readily available; and they are less expensive than microscopes.

Sometimes, microscopes are necessary. Microscopes reveal images that we process with our eyes. While there is no hard boundary between images that are perceived directly from object to eye and those that come through a microscope, there are things that we need to keep in mind when we are using a microscope.

- *Are we looking at a two-dimensional representation of a three-dimensional surface? If so, what do we need to keep in mind when interpreting the data?*
- *This caution is especially true when using a scanning electron microscope because the cues that our brains use to determine up & down, in & out, and protrusion or pit may result in incorrect mental images of the object.*
- *In the world in which we evolved, the sun is above us and the earth is below us. At high magnification in general, we lose context. Shadows may look the same, but when the angle between the source of the "light" and the thing detecting it is different from what our brain thinks it is - this is when we can easily reverse pits and bumps.*
- *As soon as the magnification is high enough that we lose our usual frame of reference, we need help to understand the images. The following guide may help you with "Visual Examination through the Microscope".*

3.7 Key Concepts for Microscope Based Documentation

There are four keys to performing and documenting a microscopic examination in FA.

3.7.1 KEY 1 - Provide Context For The Image

Overview: You are on the failure analysis team. You have been working with the Subject Part. You know what you are looking at. Show compassion for other people who may not have strong technical backgrounds. Help them understand what they are seeing. Especially if the final documentation is to be distributed to a wide audience, take the time to provide a few extra images to help people make the bridge from one image to the next.

Context: A well-documented failure analysis report allows the interested reader to reach the same logical conclusions that were reached by the expert investigators. Of course, the lay reader will not have the same technical background to criticize your technical findings and raw data interpretation. But they should be able to follow your basic train of thought. The photo record is key to achieving this level of clarity.

How To: When creating documentation of the damage event, provide an overview of the entire object or a feature that will be recognizable from a few overview images of the entire object. For example, Figure 3.24A (below) doesn't show the entire screw, but it shows the entire threaded length. Note also the addition of the scale, which provides a reference to the size of the object.

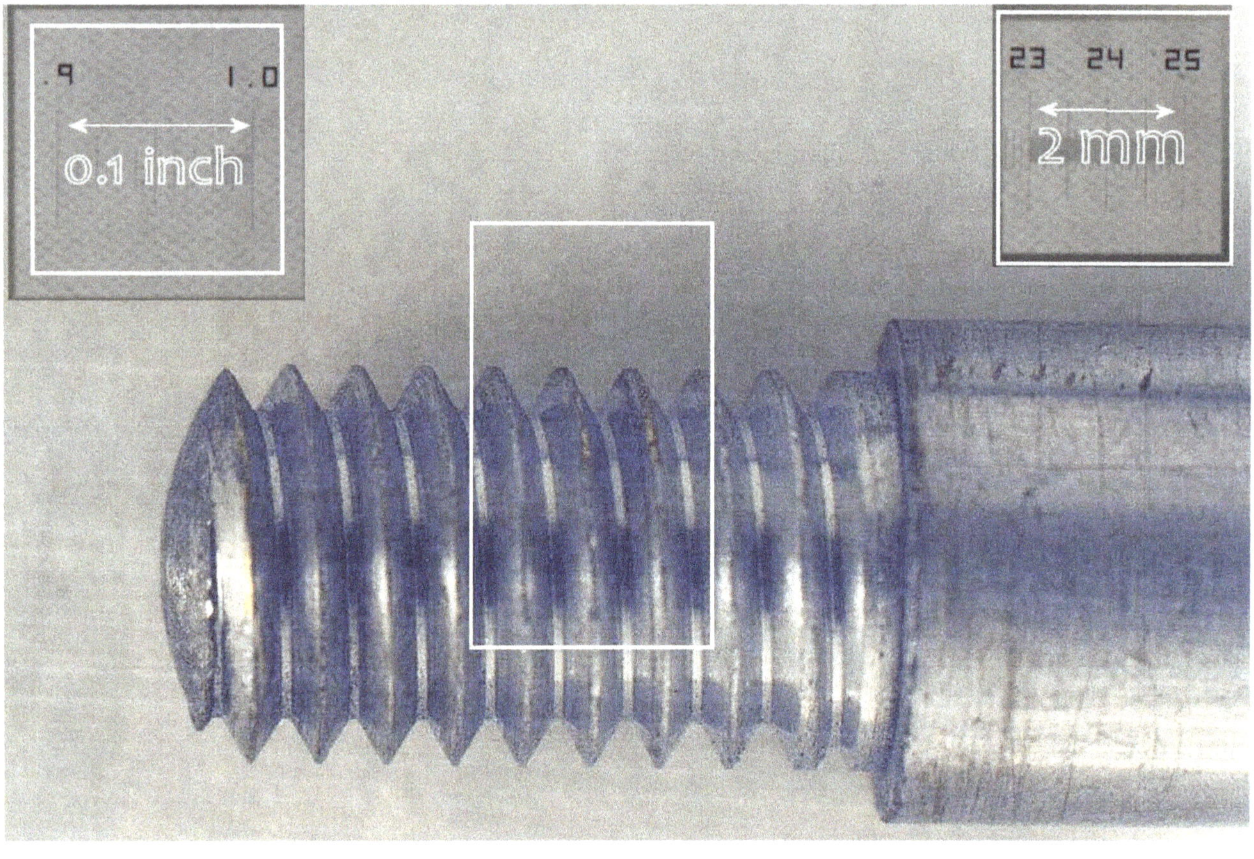

Figure 3.24A: *Overview of a fastener's threaded length*

Figure 3.24B shows damage to the thread crests.

Figure 3.24C depicts the same fastener, but a viewer may have problems reading this image; it is problematic. Only by referring back to Figure 3.24B can a viewer understand that the shiny areas are the thread roots. If we didn't have that view, the eye could easily be tricked into thinking they are the crests.

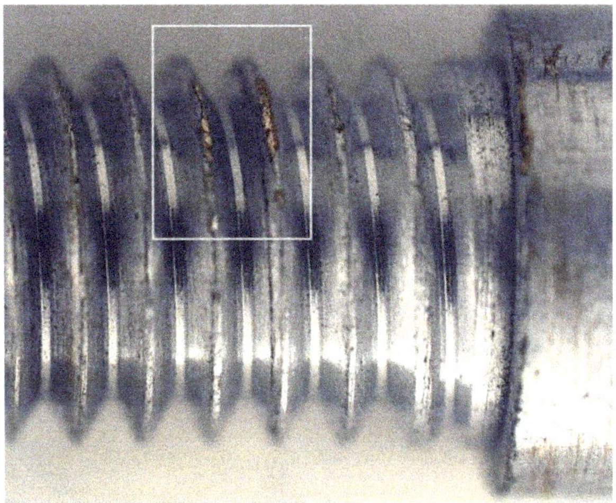

Figure 3.24B: *Detail of the damaged thread crests*

Figure 3.24C: *Notice how the lack of context makes it difficult to understand this image.*

Readers in my non-fiction writing group could not see that the two images below were of the same bug sitting on the tinted glass of my office door, Figures 3.24D and E. Some engineering fields must learn to rotate views of objects in their minds in order to perform their work. This is also a valuable skill for failure analysts. Figure 3.24F shows another type of moth wing from a magnified point of view.

Figure 3.24D: *Moth sitting on tinted glass*

Figure 3.24E: *This is the same moth shown in Figure 3.24D. Notice how the point of view rotation changes the perception of the moth's appearance.*

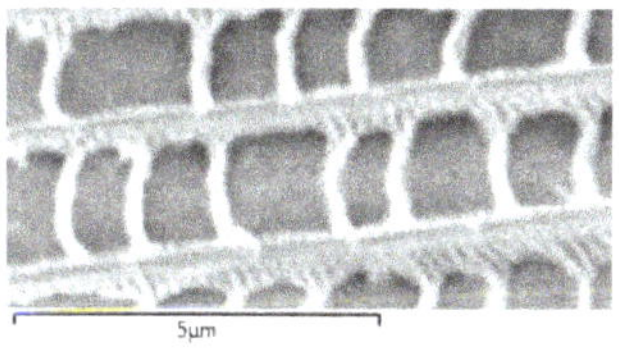

Figure 3.24F: *This is a picture of another type of moth's wing at 6000x. Notice how the lack of context makes it difficult to understand this image.*

3.7.2 KEY 2 - Depth Of Field

Overview: Part of the reason that 3.24C is difficult to interpret is that much of the image is out of focus. Optical microscopes have limited depth of field. This is also true for cameras when they are used to take extreme close-ups. Compare the magnification of Figure 3.25 (below) to that of Figure 3.24A (above). Figure 3.25 is higher magnification, but the entire image is in focus. This is what a scanning electron micrograph provides.

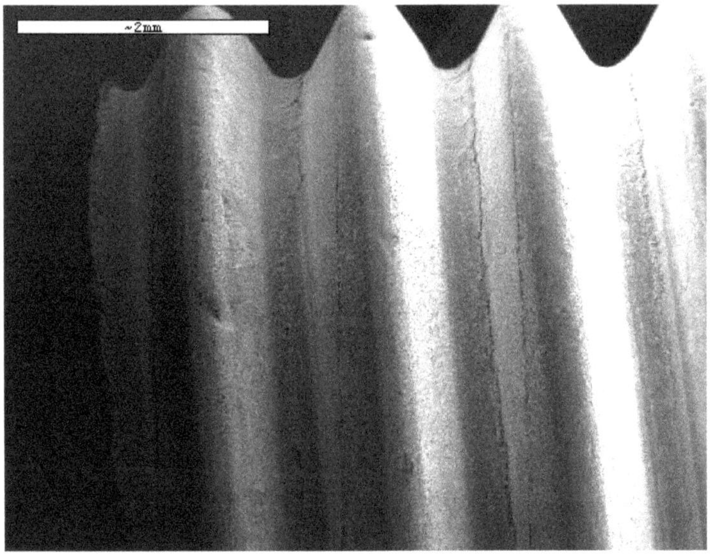

Figure 3.25: *This Scanning Electron Micrograph has double the magnification of Figure 3.24A but it is easier to read.*

This is a different screw.

Scanning Electron Microscopes (SEM's) have greater depth of field, but they are inherently greyscale.

Some optical microscopes have software enhanced focus, where the computer captures multiple images and stitches the portions that are in focus together to create a color image with good depth of field.

Note that sometimes, poor depth of field is beneficial. When the image is showing a lot of busy details, intentional selective de-focusing can help to tell a story. Every tool has its strengths and weaknesses.

Context: Know the strengths and limitations of the image capture devices you have available. Depth of field is a primary variable among different technologies.

How To: Use optical devices if you need to capture large areas. Electron images have greater depth of field at high magnifications, but are limited to small areas. Scanning electron microscopes often come with an attached microchemical analyzer. New microscope types with different capabilities are being invented.

A microchemical analyzer can be used to create color element maps, which show relative concentrations of different elements at each position of the image. If you are looking for some pizzaz to interest business people, or others without technical backgrounds, this can sometimes be effective. This is not useless "pizzaz". It's informative pizzaz! The subject of microchemical analysis will be elaborated later in this text.

3.7.3 KEY 3 - The Tricky Nature Of Incident Light

Overview: The angle of the light may be used to hide or highlight different features. See Figures 3.26A-C (next page). Figure 3.26A was taken from a flatbed optical scanner. Figure 3.26B has two views of the same area: a detail of Crystal B (between the paper clips) but at different angles and brightness levels. Note how different features jump out in the different views.

Figure 3.26A: *Overview of four pyrite crystals – the one on the far left (above the large paperclip) comes from a mine in Spain. All of these are pyrites / iron sulfide / Fool's Gold.*

If I told you are LOOKING-FOR pyrite, you might not even notice the crystal from Spain. Note how the large face looks very dark, and the smaller crystal, at an oblique angle, is very bright. A similar range of dark to bright is visible in the other pyrites.

Note also how the SEM image of an area with black-and-gold-colored crystals (as viewed in scanner image 3.26B), appears all grey in 3.26C. The EDS microchemical analyzer detected manganese and sulfur in the visually black areas (dark colony of crystals labeled "B" in Figure 3.26B Left).

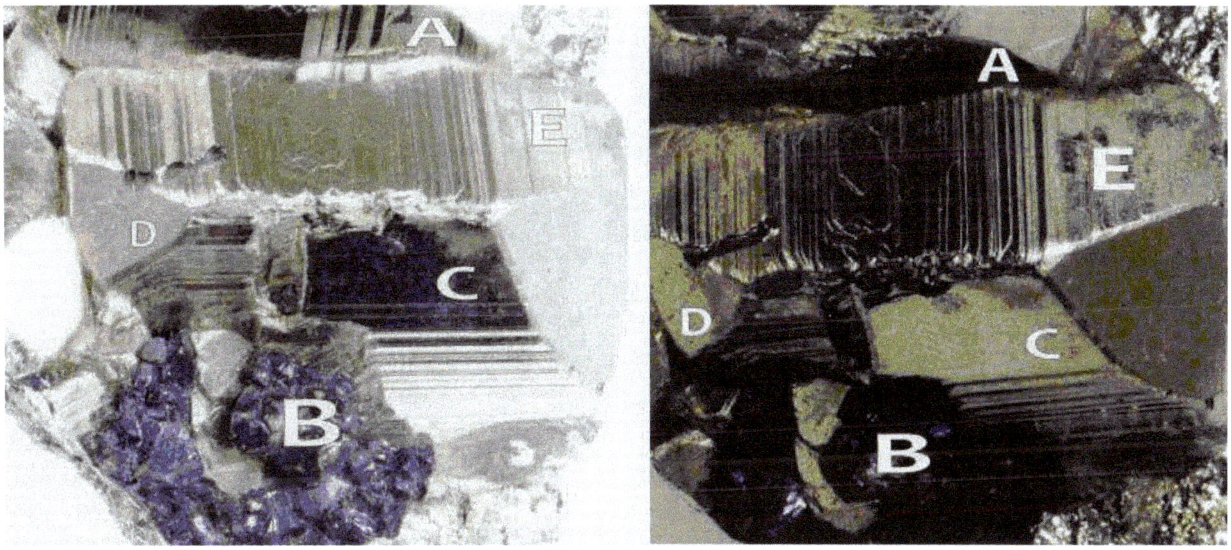

Figure 3.26B: *Note that some features are easier to see in one view than another, and other features are easier to see in the other view.* ***NOTE:*** *the scanner artifacted colors; e.g., hauerite on the left (B) looks like amethyst, but it's not.*

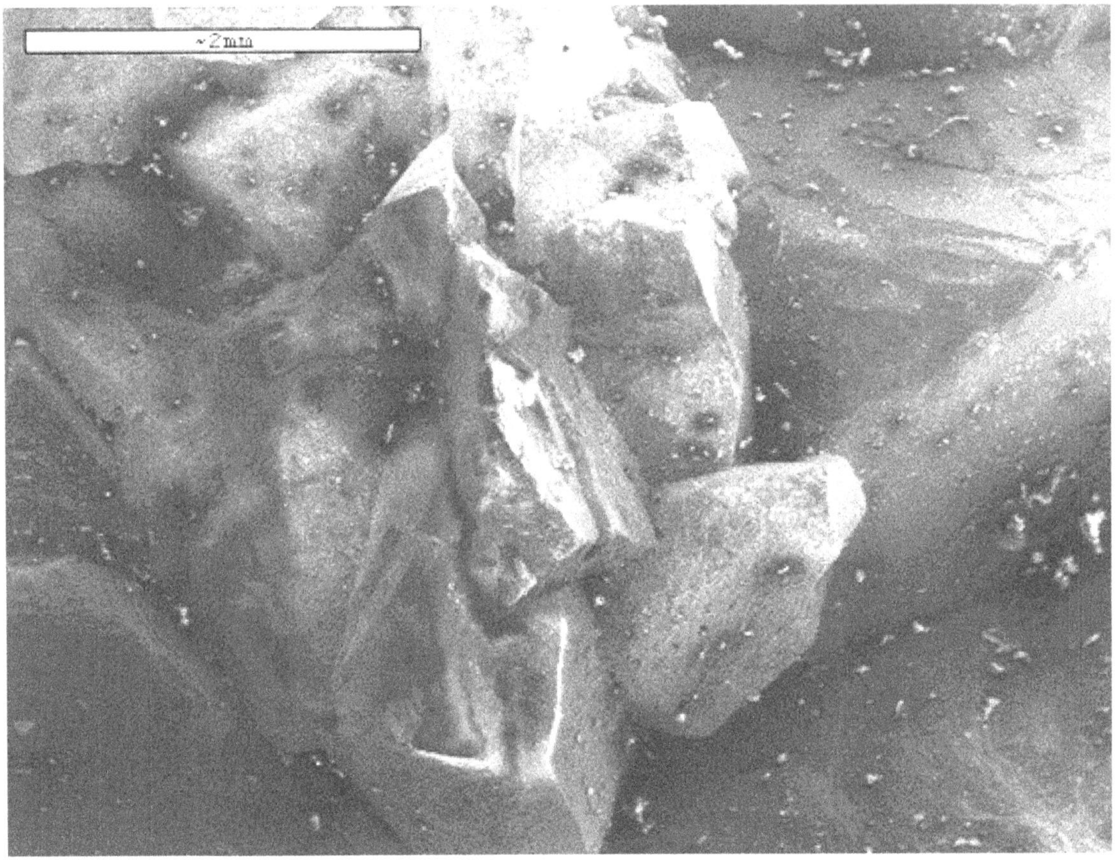

Figure 3.26C: *This image is a scanning electron micrograph of a portion of the crystal shown in 3.26B. The readily distinguishable black and gold crystals are all depicted in greyscale in this SEM image.*

3.7.4 KEY 4 - Using And Explaining Cross-Section Images

Overview: We covered cross-sections very briefly in Chapter 1. Microscopic views of metallographic cross-section images may be presented for different reasons.

Context: While the general public rarely has an intuitive understanding of these reasons, they may readily grasp the concepts via "compare and contrast" through pattern recognition. Give them a chance to compare and contrast two parts (Subject and Reference, for example) at the same magnification and image capture settings. What is important is for them to see that they are different. Even if they don't recognize it's because of a different grain size, or aspect ratio, or micro-scale fracture features, they will be able to see differences. Or, they will get the gist that you think they are the same, even if you did not perfectly match the brightness and contrast of the two images.

How To: When you are documenting for a lay audience, it is worth a few words to explain what a cross-section is. This will be detailed in Chapter 6, but was introduced in Chapter 1. If the material has a different structure in different directions, such as wrought metals or extruded or injection molded plastics, it is further worth a few words to explain that the same location of the object may appear different depending on the orientation of the cut. People may not be used to words like "orientation". But that is more precise than "direction", which implies a line instead of a plane.

The appearance of a cross-section view can be misleading if the specimen is not prepared properly. Someone must have enough knowledge to be able to tell the difference between the actual microstructure and damage from poor preparation technique.

The appearance of a cross-section view can change drastically if the lighting method is changed. Use of polarized light, for example, can turn white things black, and vice versa. Interpreting such images properly requires knowledge of what type of lighting was used to capture the image.

As-polished cross-sections will generally be used for inspection of geometric features such as: thread forms on small screws; the presence of voids or cracks; corrosion; flow lines in polymeric parts and cold headed parts and forgings; intended or undesired second phase particles; and coatings. This may include glass fiber orientation in nylon. Full documentation of microstructural features such as grain size and evidence of proper or improper heat treating generally requires etching with an aggressive chemical solution.

When presenting images of cross-sections for documentation of microstructure, color sometimes adds information, but often is provided primarily for aesthetic value. Compare Figures 3.27A (below) and 3.27B (next page). The greyscale image actually has more information about the iron matrix surrounding the graphite nodules. 3.27A has been etched, and 3.27B has not.

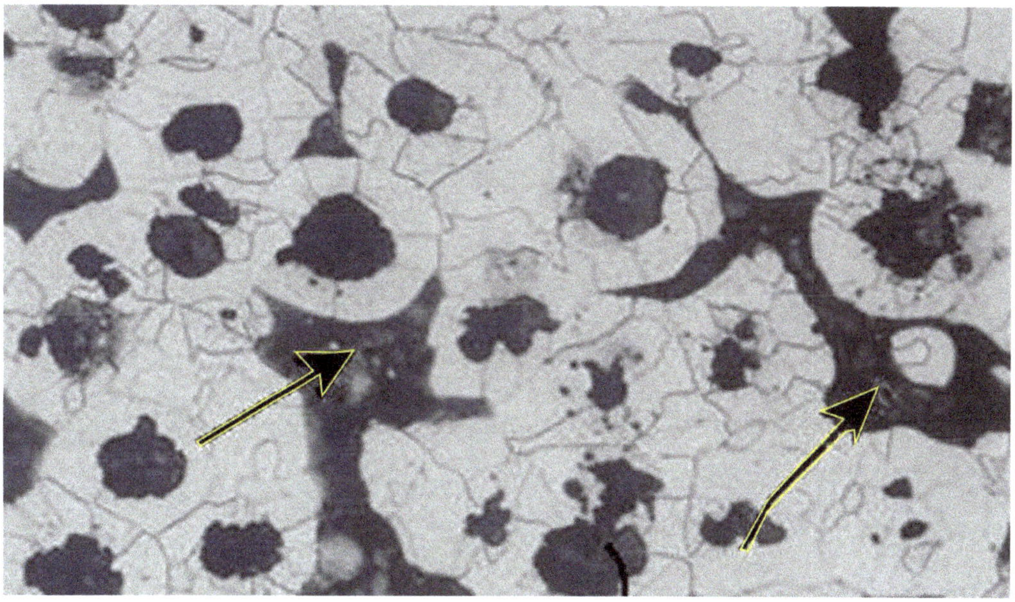

Figure 3.27A: *The rounded dark features are graphite nodules. The irregular dark features are pearlite as shown by the black arrows. The white is ferrite (nearly pure iron). This is an optical micrograph of a ductile iron casting.*

If you are not familiar with physical metallurgy, you will need to find help with an atlas of microstructures or a friendly materials engineer. Figure 3.27B has more information about the graphite nodules, but these have less effect on the material behavior as long as they appear more or less circular. The 3.27A nodules (they are spherical or spheroidal in 3 dimensions) would generally be considered to meet the more-or-less-round criterion even though they are obviously less round than those in 3.27B.

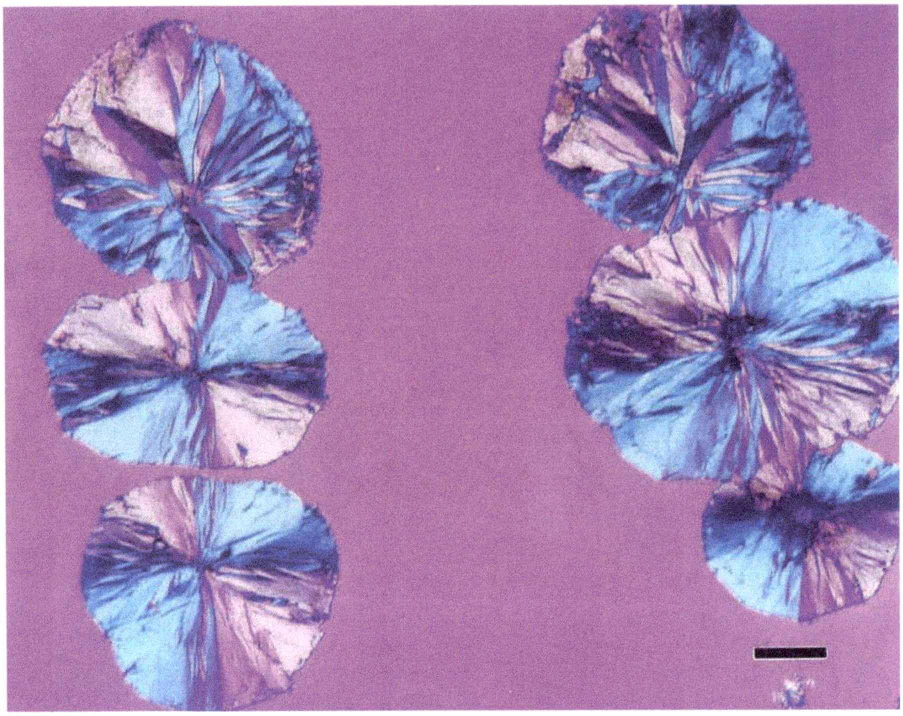

Figure 3.27B: *Optical micrograph of a ductile iron casting – image courtesy George VanderVoort*

Some common materials have recognizable structures; and there are a lot of structures that repeat in nature. Widely different alloy systems have similar looking features. Just because two features in a micrograph appear to be the same color or seem to represent the same pattern, it doesn't mean the material has the same composition.

More information on microstructure interpretation and its uses in failure analysis will be provided in Chapter 6.

With mastery of these "Four Keys to Microscope Images", you are on your way to being a capable communicator - at least for the details of your investigation.

3.8 My Case Study Exercise Answers

Here's my answer for EXERCISE 3-C: Brief, fact-based description of a screwdriver.

It's obviously a screwdriver. It has some discoloration on the steel blade and a few scratches on the plastic handle.

Here is my answer for EXERCISE 3-C: Description of the screwdriver using words related to the shape and other features of the appearance:

A small diameter cylinder, a rod shape, flattened at one end, surrounded on the other end by a cylinder about three to four times the diameter of the smaller one. The larger diameter cylinder has grooves running parallel to the axis of the flattened cylinder, and a single wider, circumferential groove toward the end from which the rod protrudes. The rod-shaped portion has a light grey surface, with many red-brown irregular spots on it.

Here is my answer for EXERCISE 3-E: My list of features that contribute to seeing an old hag or a young woman (Figure 3.5).

- Old hag: big nose, looking down, staring eyes, a bulky collar, and protruding chin...
- Young woman: slim neck, looking away, with a stylish feather in her hat, dark necklace ...

Here is my answer for EXERCISE 3-E: My list of features that contribute to switching back and forth between the old woman and the young one (Figure 3.5).

When I asked myself what contributed to be seeing the hag drawing and distinguishing the young woman, I noticed:

- above the young woman's nose, see the crone's eyelashes, on the hidden eye;
- the crone's upper lip is a necklace around the lady's neck;
- the lady's left ear is the crone's eye.

CHAPTER 4

Musings on Critical Failure Analysis Concepts

Insights from Life as an Engineering Failure Analyst

4.1 Overview

This chapter is an invitation to think deeply about critical concepts in Failure Analysis (FA) and avoid the traps waiting to nab the unsuspecting. We examine atypical ideas for an engineering text. They will be particularly fruitful for understanding the human factors in running a business, in the performance of quality engineering work, and in the communication of concepts. The entire life of a part is considered from the earliest inkling of the employees of the company that designed the unit to the Failure Analysis (FA) of any damaged component.

I suggest reading this chapter, one or two sections at a time; perhaps between the reading of other chapters. The topics below in the text box are highlighted in other chapters when they appear, so that you can consult Chapter 4 as needed as you progress through the book.

Chapter 4 Section Topics That Are Also Highlighted In Other Chapters

4.2 How to Think About Failure. What Exactly Is Failure & Failure Analysis?

Grammatically, failure is a noun; yet, there is no such *thing* as a failure. There is nothing that is manufactured under the category of failure. You cannot go out and purchase a failure. In fact, failure is only a concept. When Failure Analysts think of a concept as a physical thing, they can run into difficulties. Failure is a concept not a physical thing.

David Levy writes, "It is tempting but misleading to regard abstract concepts as if they were concrete objects. This error in thinking is called *reification*. To reify is to invent a concept (or a 'construct'), give it a name, and then convince ourselves that such a thing objectively exists in the world." [71]

The sentence, "This bearing failed" is never meaningful on its own. "Failure" must be defined within a context before the sentence becomes a tool of clarification rather than obfuscation.

Why is this important to the Failure Analyst? Because the Failure Analyst must use a specific methodology to perform their analysis and they must clearly and concisely communicate their findings. They must use language that clarifies the situation for business leaders, designers, manufacturers, or maintenance personnel.

The problem with the reification error is that the valid criteria for judging an object differ from those appropriate for concepts or constructs. Here is an example. The statement, "The bearing is worn," may be readily verified by visual examination. Compare it to the statement, "The bearing failed." *The bearing failed* is a concept statement. In order for the statement to be actionable or verifiable requires more extensive analysis.

For example, if we report, "The Space Shuttle Challenger and the lives on board were lost due to an O-ring failure," then how do responsible parties understand what happened or know how to prevent it from happening again using this information? They cannot; there is not enough data contained in this communication. That is because *criteria used to judge* the state of an object are different from *those used to evaluate concepts.*

In fact, the proper manner to judge concepts is by how useful they are. They simply should not be judged to be *true* or *false*. This is because concepts are essentially *models of reality*. They are *not direct features of physical reality* itself. They may, at best, closely approximate immediately relevant aspects of reality.

So, if there is no such *thing* as a failure, then how is the term "failure" useful to the professional analyst? It clarifies that something went wrong. For physical objects, failure is the bad thing that happened and that is exactly the subject of our investigations. And the bad thing that happened is always a result of *physical* damage. That may not be the case for medical and software failures!

The *construct of failure* impels us to begin our journey of investigation because something bad happened to a physical object; but once we start our work, the Failure Analyst needs to focus on and understand the sequence of the progress of the damage before communicating the results.

> *When I started doing failure analysis work, the concept of "a failure" seemed pretty obvious. A failure was what I was analyzing when I was assigned to perform a failure analysis. A failure analysis was trying to determine what caused the failure.*
>
> *Most people starting out in failure analysis are simply given components that someone else removed from an assembly.*
>
> *At my first failure analysis job, the components on my desk or the ones in the shop (too large to sit on my desk) were truck components. Typically, the*

[71] *Tools of Critical Thinking: Metathoughts for Psychology* by David A. Levy | Page 9, Chapter II, entitled "Reification Error: Comparing Apples and Existentialism"

"failure" was a crack, or corrosion damage, or wear. We didn't really have to deal with deterioration due to high temperature exposure.

Similar to most beginners in failure analysis, I started out with some version of the obvious questions: "Did the part meet the specification;" "What was wrong with the part;" or "Did they use the specified material?"

The obvious implication of these questions was if specifications were followed and the part failed, then someone must have done something bad to the part. Usually, that someone was either the customer or maintenance personnel. With this approach, the only available outcome was blame; however, there should be many other possible outcomes, other than blame, in a real failure analysis.

If a beginning Failure Analyst is simply given components that someone else removed from an assembly, then it is essentially impossible to perform a true failure analysis.

What do I mean by this?

Without background information (the context) or an understanding of the environment the "failed part" lived-in, it is difficult to reach useful conclusions and nearly impossible to give reliable advice for prevention. If you are working for the company who makes and sells the part, then you are indeed in luck! You can at least get access to the specifications.

If you work at an independent testing company without access to specifications, you are in a more difficult situation. If you are familiar with the type of component at issue, then you can compare the characteristics of the part with others in your personal memory bank, or those of your colleagues. If not, you will need to procure another part – a comparison part. In this way, we gather "reference data". In this way, we can judge whether the material of the "failed" part was obviously wrong.

In short, the quality of the data gained is strongly related to the quality of our choice of a reference part or the availability of different types of reference parts. We will circle back to this later.

There's a big difference between damage characterization and failure analysis.

- Damage characterization is always possible to some degree if you have the damaged part. The goal of damage characterization is to describe the difference between how the part was at the time of manufacture and how it is now (after it stopped doing its job). Sometimes it is possible to get a sense of how the damage happened.
- Failure Analysis goes much further. Failure Analysis aims to understand how the damage happened, the details of what unusual circumstances allowed it to happen, and if possible, why it happened. The answer to "why" usually requires multiple answers from multiple viewpoints: quality of the raw material, processing parameters used to manufacture the object, and service conditions are some of the most common.
- Also, Root Cause Analysis is often conflated with Failure Analysis. We discuss this further in Section 4.11.

Even with no specifications or reference parts, we will still use our eyes and mind and available instrumentation to characterize the damage. But this in itself is not a failure analysis. In the case of cracks or corrosion we can often determine where and perhaps how the damage started and how it progressed over time. For example,

- someone familiar with the basic fracture types can often tell at a glance if the crack happened over time or if it happened all at once.
- someone familiar with basic loading geometries can often tell at a glance how the part was loaded when the crack initiated.
- someone familiar with the specific material may be able to make a reasonable scientific guess about whether the material was embrittled.

But it is rare that this type of information is adequate to determine why the part failed. In the end, if we are truly doing a failure analysis, there must be an attempt to determine *why* the part failed, and more importantly, whether the part itself was the problem. More information is required to answer the question of why it happened. As previously mentioned, getting the required data from multiple viewpoints, including environments the part experienced, is necessary. Greater cooperation is required for data gathering in Failure Analysis than for Damage Categorization. True Failure Analysis allows you to make an informed inference about how and why the event happened.

But failure is a philosophical judgement, a concept. We need to remember that the parts themselves are always responding to an environment. They can break, wear out, get overheated, etcetera. They can't fail. Only humans can fail.

At some point, the facts and the context and the evidence from many viewpoints may be persuasive enough to build a credible story supporting the philosophical judgment of "FAILURE". This is another important concept. The fact that the component is broken, corroded, worn out, or bent is a physical fact that is often, but not always, easy to see. You may need to dig deeper. There may or may not be direct physical evidence that the part used was originally in fit condition to perform its intended function; but now, it is no longer fit for use. If it is broken, deformed, corroded, worn, or has degraded property characteristics, these are essential facts. This is the first step to arrive at a valid philosophical judgment.

Figure 4.1: *How does the American judicial system's sentencing policy compare to "failure as a philosophical judgment"? What damage may underly it? (Photo Credit: jraffin; https://pixabay.com/es/users/jraffin-1935103/)*

Discernment is required. Sometimes it is not completely obvious whether the damage is a result of mechanical loading or environmental exposure. But the bottom line is, if the part were still fit for service, nobody would be doing a damage characterization - let alone a failure analysis.

With this said, it is the job of the Failure Analyst to judge whether it was truly the part itself that failed to perform as expected or to determine whether the conditions in which the part found itself were beyond the design envelope. If we are *looking-at* a component from an assembly, which is usually the case, determining that conditions were beyond the design envelope leaves us with at least two options.

- First, maybe someone really did something that was clearly outside of the operating parameters.

- Second, maybe it was a different component from the assembly that "failed" but the damage was most obvious in the part given to the analyst.

We are often faced with a situation where someone did not notice the significance of "missing evidence".

Years ago, a client performed a durability test on an assembly and experienced a miserable failure. The part did not last even half as long as usual. They decided to send me the section of the assembly for evaluation. This was actually an assembly with which I was extremely familiar. I "chopped it up" and did all my tests. I use the term "chopped" to imply a casual approach to the work. This was nominally a routine job!

I found that the material characteristics were right in the middle of the usual statistical range. In addition, there was not even a single feature that was "out of specification".

So, the nagging fear was in my mind, "Will they think I'm stupid because I can't find a problem? Will they fire me and find a new consultant?"

I wrote my report. I told them "I looked and looked for something wrong but found nothing wrong with this assembly. The microstructure, hardness, composition (etc.) were all perfectly normal."

They were confused. About a month later, I got another call, "Whoops!" The plant engineer told me, "We sent you the wrong part. The one that actually failed the test is still here in the plant!"

I cut-up and evaluated this assembly and found large deviations from their usual good practice.

Anyone who does failure analysis and materials characterization on a regular basis knows how easy it is to find "out of specification" conditions in a manufactured part. The fact is, for many manufactured items there is no specification for the final strength of the product. There are only specifications for the raw material or for hardness at certain locations. Anyone who truly specified the characteristics of a part at every location would run out of room to store data on the corporate data base. It's an unrealistic task.

For me, facing my fear and doing what was right had high merit. I am grateful that I did not write the "safer" report, listing all the findings that were unusual with this particular assembly. I couldn't do it. It turned out well for my future work. I did not get fired for being stupid, and I was likely privately acknowledged for being talented and careful, while they dealt with my inadvertent revealing of their own mistake. But there are no guarantees in life.

Here's the hard business truth: it is not that rare for Failure Analysts to be fired for **not** finding what the client thinks should be found. It's your decision, of course, what you will decide when your bank account is low or the pressure is high. But the great Failure Analysts are trustworthy; their reputations build their long-term careers. Failure analysis is **not** first and foremost a service industry and the customer is **not** always right. There are always ethical considerations in our work.

Failure analysis is, first and foremost, a professional activity. A person is asking you to help sort out what they cannot. But on occasion in some routine failure analysis work, as long as you have negotiated the scope of the project clearly, there's nothing wrong with diving-in and "seeing what jumps out", even without a detailed protocol proposed. **But realize this is not failure analysis.** I call this "putzing around in the lab". It is easy to miss clues to the real problem with this approach.

If the Failure Analyst is not given the instructions or resources to review the whole situation, then it is not a failure analysis.

Depending on the circumstance, there may be value in doing a restricted scope damage analysis. If there are multiple failures and no significant financial or health or safety ramifications, it's a perfectly legitimate decision to go ahead and destructively test something to see if an obvious clue jumps out. Most of us who have been in the field, for any length of time, have performed a "quick and dirty" or "economy failure analysis" (damage characterization). This is fine as long as all parties are clear about the scope of the project and they know it is not true failure analysis. It's a good idea to get confirmation in writing of the scope and purpose of any destructive testing prior to initiating the work.

Saying it again, if the Failure Analyst is not given the instructions or resources to review the whole situation, **then it is not a failure analysis**. It would be legitimate to call the activity a certification check or a damage analysis, fracture analysis, corrosion analysis, etc. But it is not a failure analysis; **it is not a failure analysis.**

Part of the purpose of doing a failure analysis is to determine how the failure happened. If we start out assuming that the only relevant details have to do with the part and its specifications, we're usually going to miss most of the picture. I recently worked on a project and came to a conclusion. After I issued the report, another interested party gave me a piece of background information that had not previously been provided. As I listened, that background information seemed to be in conflict with the physical evidence that I had reviewed.

> *Had I known about this fact at the beginning of my work on this project, my report would have been different. I would have noted that I had found it impossible to come up with a scenario that was consistent with both the background information and the physical data.*
>
> *So, what could this mean? What couldn't change were the physical facts available to me. Those were solid. There could have been an additional component from the assembly that was the primary failure and nobody thought important to provide. Or, someone was telling a "fish story". The new information was ambiguous and conflicting so this would have been a reasonable avenue of further investigation had it been revealed earlier.*
>
> *Maybe it's just as well that I did not have to put in writing the fact that I suspected one of the parties might have been dishonest or that one of the other investigators was trying to cover something up. I am not clear that the full story of the failure was revealed, but I didn't rewrite the report after the job was closed because they achieved their business objectives and were ready to move on.*

Based on my experience, I would estimate well over three quarters of the failures that happen in durability test labs are due to inadequate specifications. The part does meet the print, but it failed. And if it failed in a durability test, usually it's a surprise to the designer.

> **Often, we can't find a specific physical cause of the failure. Nothing lasts forever. Risk is inherent in life. For that reason, I prefer to label my reports as engineering investigations rather than as failure analysis reports.**

Obviously, these stats will vary by industry. I work mostly in what I call medium-technology industries. Not much of what I do involves aerospace or nano-materials or micro-electronics. My clients only rarely have a Materials Engineer on staff. So, it's the design engineers who select materials for a new design; and usually it's because the materials are similar to previous designs that worked.

There are a lot of companies in the industries I serve who specify "hot rolled steel". They don't even realize that "HRS" is not a particular composition with a specified strength range. It may take up to twenty engineering change orders (ECOs) to the component before they use up a safety margin. That safety margin may have been determined decades previously. So, what failed in this case?

Saying that the part met the specification may be all that can be done. The failure was not with the part, but with the business decisions that didn't give the designer the tools to do a better job.

But wait! This is often the most economical way to design a part. I call this design methodology "Guess-and-Test". In my experience, it requires a huge effort to take an engineering design group and transform the way it works from Guess-and-Test to semi-quantitative design methods.

It's time to apply what we've discussed about the concepts of failure, damage and failure analysis.

EXERCISE 4-A: COMPARE FAILURE ANALYSIS TO DAMAGE ANALYSIS.

In both of the following activities, you are looking for background information and examining the Subject Component. In both activities, we use our eyes and minds. In both activities, we may take a lot of photos, perform composition and microstructure analysis, and do mechanical tests. Now it's your turn. What is similar in Failure Analysis and in Damage Anaylsis?

It can be confusing to think about damage as detrimental physical changes and keep failure as a separate category. You need to find a way to keep these two categories separate and distinct. You need to burn the features of the physical world – the ones that facilitate damage – into your knowledge structure. Figure 4.2 (below) is my way of keeping these categories separate and distinct.

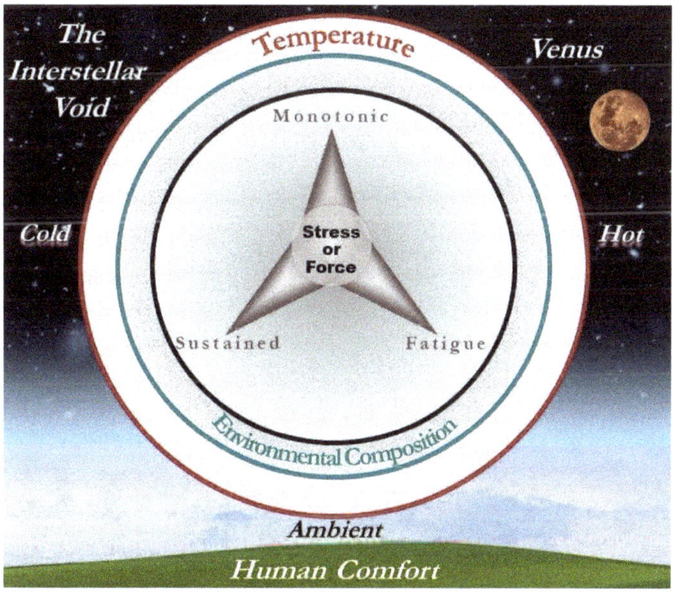

Figure 4.2: *Features of the physical world that facilitate damage*

Imagine the two outer rings rotating to align Hot with various environments (vacuum, salt water, air in an industrialized city, etc.) and the relevant sustained, monotonic or cyclic (fatigue) loading conditions.

EXERCISE 4-B: CONTRAST FAILURE ANALYSIS TO DAMAGE ANALYSIS.

Fracture Analysis is subset of Failure Analysis involving fracture. In Fracture Anaylsis, we must attempt to directly link the loading and environmental conditions to specific details of the damage.

In this exercise, contrast failure analysis to damage analysis, in which it's acceptable to simply describe damage using standardized damage classification methods. Below is my take on the exercise.

Damage Analysis	Failure Analysis
It is legitimate to simply provide findings when the allocated project resources are exhausted.	We must define specifically what we are trying to achieve; *i.e., assign blame, prevent future failures, or optimize design.*
It requires a discussion between the requesting party and the one performing the project.	The public is familiar with the term failure analysis; it rolls off the tongue.
Analysis of structural components is typically performed according to a generic methodology.	This analysis almost always requires creative effort.

Now it's your turn:

Damage Analysis	Failure Analysis

4.2.1 Damage Is Physical and Failure Is a Concept

One of my favorite teaching aids, for the discussion of the meaning of the term failure, is a little stretchy toy, such as shown in Figure 4.3 (below).[72]

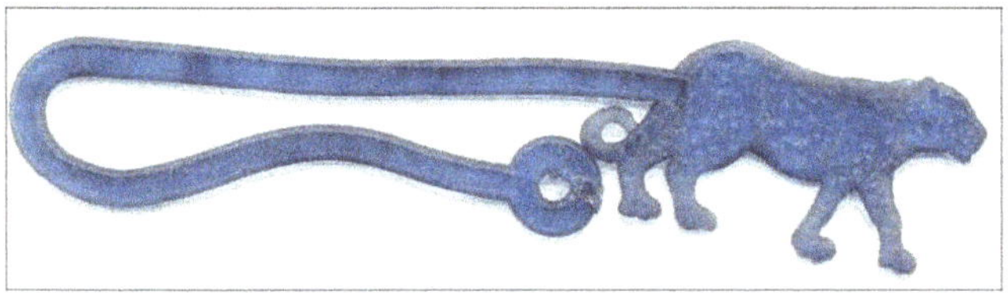

Figure 4.3: *Children's stretchy toy*

How might this toy fail? Usually, people answer that it will break. By now you should have an inkling that this answer is very simple and does not encompass the understanding that failure is a philosophical judgment.

Here is the way the toy is used: hold on to the ring and throw the animal toward something with a flat surface like a wall, a refrigerator, or a car. The animal will stick to the flat surface. It's fun. Kids love it.

When I demonstrated this at a friend's lab, the animal toy left a colored stain the size and shape of itself on the wall. So, my friend considered this a failure that required the wall to be repainted.

Note here that the context of its use takes us in a different direction from just imagining that it will break. We can imagine the ring breaking off due to the stress concentration at that location. Or the "tail" could break if the "kid" stretches it too far. Or if there is a void in the "casting" then it could break at that weakness. But another "failure category" is that the toy's sticky surface will eventually pick up so much dust, that it no longer sticks to whatever the kid wants it to stick to! Check "sticky toys you throw" in your favorite search engine for more ideas.

Many Failure Analysts get stuck in a false equivalence between fracture and failure!

72 This example was presented in Chapter 2 but it will be discussed a bit differently here.

Damage is different than failure. The stretchy-toy example highlights that false equivalence. Be sure you retain a broader and differentiated frame of reference in life and in work!

4.3 Frame of Reference (Confirm Context!)

Warning: this is a new subject. We have beaten failure versus damage to death! Now we are going to talk about critical thinking.

Whenever we set out to do any real-world work, as opposed to an exercise in school (where presumably the professor already knows the answer), the most important thing to do is to define our frame of reference.

In the real-world, it's the obligation of the Failure Analyst to define the frame of reference. By clarifying the frame of reference and thus the tasks, all parties have a layer of protection from later disagreements.

In the real world we can only manage complex problems: we cannot solve them for all time.

In failure analysis work, which is conceptual work, background data collection is how most professionals put together a frame of reference.

- It is important to understand how the Subject of the investigation was intended to be used, and whether there is any evidence of it having or not having been used that way.
- It's important to know if the Subject of the investigation is considered to be broken, deformed, corroded, or worn (etcetera).
- It's important to know whether the design is unique or a common production item.
- It is also important to identify and understand the service environment.
- Assembly and in-process manufacturing "failures" have their own requirements.[73]

It's extremely unlikely that test conditions will be identical to the service conditions.

Failure analysis work often includes testing to determine the material characteristics. The lab-measured properties rarely match those displayed in service in a direct manner. Strength and many other characteristics are dependent on strain rate, temperature, humidity, and geometry. These are controlled in the lab, but may vary in the field, and are often difficult to pin down with any quantitative confidence.

In the automotive suspension realm, for example, did a large pot-hole in the road surface contribute to a severe load? Did the driver hit a curb? Did the driver have a habit of turning the steering wheel through a large arc without the tires rotating? The test conditions will rarely be identical to the service conditions. This is why extreme care is important during data interpretation. This is a major subject of Chapter 9.

Confirming the frame of reference includes understanding the application of the Subject Part and paying attention to the details of our tests as we attempt to see if the Subject Part was properly manufactured.

Frame of reference also involves a basic familiarity with the industries and companies that made, installed, and used the Subject of the investigation.

When you are starting out as a new Failure Analyst, you will not have the background to thoroughly understand all of the implications of the interactions among the manufacture, assembly or installation, use, and maintenance history. And

[73] An assembly "failure" is a failure that happens during manufacturing before the end-user ever gets the item. If we are investigating an assembly failure, we must understand assembly conditions.

the implications of the choices you make as you proceed to do the failure analysis are also difficult to grasp. But as you gain experience, if you cultivate your curiosity, if you truly want to understand why "this one failed and why it failed now", you will naturally start to gain the ability to see how the factors interact.

Years ago, I worked on a fastener failure. At one point, the team had generated a list of over eighty factors that they thought might have contributed to the early demise of the bolt. In the end, when they finally determined the cause, the main process problem had not been on the list of eighty potential problems!

Define and Refine the Frame of Reference. Negotiate the Scope!

One of my main goals in writing this book was to allow the student to gain practice using "critical thinking skills" in order to enhance their ability to perform failure analysis. Now is a good time to emphasize the importance of one of the main goals of critical thinking, which is to:

1) Define your frame of reference (Confirm the Context!). Get in the habit of stopping to think. Stop and *do* nothing. Stop to clear your mind. Stop to ask questions so you understand:

A. what technical questions need to be answered; and

B. what needs to be done so the people who experienced the problem will be able to use the information produced by your investigation.

This last item is usually left off of the list of considerations when defining the scope of a "generic" failure analysis. In fact, this is the key feature of doing a real failure analysis rather than a generic testing protocol. Generic testing protocols are created in the hope that they might provide some useful information. A real failure analysis doesn't leave the scope in the hands of hope!

2) Refine your frame of reference. Get in the habit of figuring out how to acquire reference specimens that will allow the most definitive conclusions.

3) Negotiate the scope of the work. Some of the questions you might ask yourself and ask your client are:

A) Is this a routine job (this means no injuries or extensive property damage)? If this is the case, it might be a legitimate business decision to do just enough work to check the box to say someone looked at the problem.

B) Were there injuries, deaths, extensive property damage, or an incident that needs to be reported to the authorities? If so, you will probably need to document every step you take.

Don't leave the scope in the hands of hope!

EXERCISE 4-C: Think of some examples where it is okay to "get it done" quickly and then move on.

____

____

____

____

The rest of this book will help you learn specific times during the investigation where you can apply critical and creative thinking as well as a simplified form of decision science.

Learning to address the issue of defining the frame of reference is a key skill.

NOTE: As soon as you get the inquiry or are informed of the possibility of doing the work, it's time to start mulling over the frame of reference!

What obligation to the client do you have as a professional Failure Analyst. What does the client want?

GOING DEEPER: The term "frame of reference" is a difficult concept to apply to our lives as opposed to specific work related tasks. In fact, "The True Frame of Reference" is invisible.

> *Our personal frame of reference is all the subconscious ideas and beliefs that we have simply by virtue of growing up in the particular environment (perhaps family) that we did.*
>
> *Years ago, I had a discussion on the Yahoo Groups Root Cause Forum, where it seemed to me that my frame of reference was so different, nobody else in the 1200+ person group could respond in a meaningful way.*[74] *I was trying to convey the idea that if humans believe they are inherently sinners (as some religious groups believe/teach/preach), then that puts an obstacle in front of them when they run into difficulties trying to improve safety systems or anything else. I asked, "Why should we bother to do all the difficult work to convince others to do things differently, when we're all too tied to the comfort of our own flesh to really care? There are other religious teachings that promote the idea that humanity has divine aspects. Even Christianity teaches this idea, but in many denominations, that teaching is quieter. How would a stronger belief that humans have some divine qualities inspire someone to try to make things better for everyone?" Responses came from very different frames of references but none provided new insights to me.*
>
> *What did I learn? Large-scale sets of cultural beliefs do affect how we go about our work lives, even if only at a sub-conscious level.*

[74] Or maybe everyone else was smart enough not to argue with an avowed Born-Again Christian regarding a potentially negative consequence of his belief system.

4.4 What is a Methodology?

This book promotes a methodology for performing failure analysis. This is different from a recipe. While it is perhaps easier to memorize the steps of a recipe, or the page number of the book the list is found on, knowing how to expertly use a methodology will ease the way for more challenging projects. There are plenty of experienced people who have never really thought about the difference between a methodology and a recipe, or checklist. They learn a checklist approach, and use it for an entire career.

A recipe may detail how to bake a cake, for example. It is usually a fairly straightforward list of steps that, when properly followed, will result in the desired outcome – a yummy cake. A methodology is required when a simple list of steps will not routinely produce the desired outcome.

Of course, there are simple cakes that even an inexperienced cook can reliably make from a recipe. Other cakes require specialized knowledge, such as recognizing the difference between beating egg whites into soft or stiff peaks. I ran into a recipe that stated, "Make a roux" as one of the basic steps. If I didn't know what roux was or how to make it, I would be stuck. What I would hope to find in the cookbook is detailed information for inexperienced chefs to make a roux - more information on the advanced steps. I would call this a procedure.

If a laboratory desires to receive ISO accreditation to perform failure analysis, they need to prepare detailed procedures for each major step in the failure analysis process. But they don't necessarily need to wrestle with the idea of a methodology.

We now understand recipe and procedure. So, what is a methodology? A recipe and a procedure are mostly linear, whereas a methodology is best seen as a branching decision tree. There are open ended "either-or" decisions, in addition to "go-no go" decisions.

A methodology is required to describe the process by which every step is to be performed. It also helps you to decide when to move on to the next step[75].

In Chapter 3, I described a methodology for performing visual inspections. I explained how to determine what to look for, and then I explained how to get out of the mind-set of *looking-for* in order to use *looking-at* to notice data that might otherwise have been missed. Notice the *looking-for* is more like a recipe (although this can be readily expanded), but the *looking-at* process involves mindset, observational principles and description techniques that make the combination a methodology.

Later in this book I will provide a methodology for how to select areas for destructive testing. It will not be a recipe. It will not even be a recipe enhanced with procedures. It will be a methodology, because failure analysis is a complex task that benefits from a more structured approach. Many people find the recipe approach sufficient to do the work expected of them. But for those who want to develop excellence in failure analysis work, for those who want to be the ones who can get the answer when others cannot, for those who choose to cultivate competence just because that's who they are, there is no substitute for developing the discipline of the methodological approach.

Practice, practice, practice the methodology until it becomes a habit. This skillset will be the Failure Analyst's foundational tool to apply when a difficult project presents itself.

Getting into the habit of using a methodology to characterize damage before you perform any destructive testing will mean you always do it this way - and people will be less likely to rush you into making a mistake.

[75] Methodology Definition (modified from Collins Dictionary): A system of methods, tools, processes, rules and principles for doing something, for example for teaching or for carrying out research.

Maybe it's enough for some to look at a broken shaft and say, "Oh! That's a fatigue crack." Then take some pictures. Maybe that's all that is needed in some cases. Other times, cutting off the fracture face and making a cross-section to look at the microstructure will be required. But just stating the damage type barely tells the story of failure. Don't let this be you. A skilled Failure Analyst would get into the habit of using the following technique *(below)* to characterize the damage before any destructive testing is performed. A skilled Failure Analyst will exercise the discipline to use a methodology whenever possible. The added benefit of relying on a structured methodology is that people will be less likely to rush you into making a mistake.

Here is the basic process that has become an integral part of my broader methodology. There will be a longer section on how to apply it in Chapter 6, *The Failure Analysis Process.*

1) Make a preliminary determination of the primary damage category (deformation, fracture, corrosion, property degradation, wear, or complex).

2) Make a preliminary determination of whether the damage initiation is localized.
   A. If it is a fracture, it will be localized even if there are multiple initiations.
   B. If it is corrosion or wear, it might be localized.
   C. If it is localized, make a preliminary determination on the initiation location or the position of the most severe damage.

3) If the primary damage category is fracture, determine whether the crack is macro-scale ductile or brittle. Also, determine the loading geometry.

4) Perform photo-documentation in light of the findings of 1-3.

5) Review the background information, select reference specimens, and plan the rest of the required testing to understand what happened.

Note the initial part of the method may require multiple iterations. For example, you may end up changing your mind about your initial assessment that the damage category was wear. You may conclude on closer examination that the wear happened after the part broke. Which was first? Wear or fracture? Sometimes it's hard to tell.

> *During the workshop portion of a seminar that I facilitated, someone brought photos of a large forging with a crack in it. It was large enough that it was extremely inconvenient to cut. They obviously didn't want to spend the time, money, or effort to cut the part to reveal the fracture surface. They wanted to review the process data to see if there was any evidence for process deviations that would promote a quench crack. Then they would report it was a quench crack and congratulate themselves for a job well done.*
>
> *There is a problem with this approach. This is not failure analysis. This is guess-and-report. Without evaluating the appearance of the revealed fracture surface, there is no way to ever prove that it was a quench crack, or a forging discontinuity, or a raw material discontinuity.*
>
> *It is understandable that it may be difficult to justify the cost of cutting the part in the complex manner required to reveal the crack surface appearance. In those cases, process troubleshooting may be the most realistic source of information. But please, do not call this failure analysis. Failure analysis must be driven by data collected from the physical objects themselves. Thus, "cyber-failure analysis" performed from photos sent by email from around the world is also not real failure analysis.*

Competent Failure Analysts, and those working to compound their competence, rely on a disciplined methodology.

Having a robust methodology is strengthened by the concept of a toolbox. To some people, the toolbox is the machine shop, the microscopes, the cameras, and the hardness tester. But to me, the first things in the toolbox are my brain and my mind[76]. Next are my eyes, which are connected to the first two. Following that is my voice, so I can ask questions. Then I have all the hard instruments.

Tools from a regular carpenter's toolbox (for example, a hammer or a screwdriver) are inert objects. It is obvious that it is not the hammer that selects which nails to hit, and it's not the screwdriver that selects where to put the screws or which screws to use. Yet, we tend to take our eyes, voice, brain, and mind for granted. In Chapter 3, we learned how to compensate for the tricks our brains play on our eyes when performing a visual exam – the tricks that lead us to think that we saw one thing when something else was present.

Once we get in the habit of cultivating clarity through developing our rational and creative thinking skills, we'll be on the way to becoming competent in our work. Once we get in the habit of contemplating the complexity of the situations before us, we'll be less likely to fall into the trap of helping to create undesirable unintended consequences.

> *Once a colleague told me, "We don't need philosophy to do failure analysis! All we need is logic. That's not philosophy!"*
>
> *Well… a few people overheard this rather loud portion of our conversation. Immediately a discussion broke out. Most disagreed with my colleague, "Yes, it is!" Meaning "Yes, logic **is** part of philosophy!"*
>
> *In fact, the earliest development of philosophy was logic. But if you read Plato's recollections of Socrates' teaching sessions, you will often find that the result of using logic was an understanding that*

---

[76] The brain is the physical organ composed of nerves cells and blood vessels. It has a distinctive shape. The mind has no shape, is non-local, and includes the summation of our many thinking activities and mental processes. Brain and mind functions are affected by emotions, feelings, their associated hormones, sensations and physical health.

the original premises were either incorrect or at least poorly defined.

There are two important types of logic. One is deductive and the other is inductive. One starts with data. After mulling the individual data points, a theory is formulated that provides a new way of understanding the data. This is inductive logic. Deductive logic starts with a theory (that someone else formulated) and predicts an outcome. Logical statements originating with data can never prove anything absolutely for all time.

For example: in Europe, for millennia, it was an established fact that all swans were white, because every swan anyone had ever seen in recorded history was white. Then a European traveled to Australia and glimpsed black swans living there. The Europeans' exemplar of logic, "If it's black, it's not a swan," was destroyed with a single viewing. See Figure 4.4.

Let's look at an example of incorrect reasoning:

Bookshelves hold books.
The screwdriver is on the bookshelf.
Therefore, the screwdriver is a book.

Figure 4.4: *When Europeans first saw black swans, they had to revise their notion of a logical impossibility.* https://pixabay.com/es/photos/cisne-negro-ave-acu%c3%a1tica-elegante-3340932/ | *Photo Credit: Alexas_Fotos;*

Huh? It's easy to see that this is incorrect.

But what if you are not familiar with books, screwdrivers or shelves? It might not be easy to catch this error. What if you are not familiar with a part, assembly, or material presented to you to analyze? Without context it might not be easy to catch an error. This is why it's so important to gain an understanding of the context of the failure analysis work you are planning to undertake. This is why it is important to understand the context of every step of your work. You must be highly confident of your facts and context before you proceed with the work.

The purpose of logic is not necessarily to singlehandedly come to a final, ultimate and absolute truth. Deductive logic cannot reveal ultimate, absolute, and eternal truth either. Remember that logic assumes its premises are correct. Logic paired with epistemology and other thinking tools is more powerful than logic alone.

Logic may be based on inductive or deductive mental processes. It's possible to start with data and facts and try to see a pattern. If our data are too limited, we may see a pattern that is not actually present in the broader world. The pattern may be an "artifact" of the artificially limited boundary. It has been said that humans are pattern seeking animals. Perform an internet search on "pattern seeking in humans" and you will see how seeing patterns creates opportunities and mischief!

If data and facts are not available, knowledge may be advanced by starting with a theory and working to prove or disprove it.

While I was researching this topic, I came across a new category of logic. It's called abductive logic.[77] This type of mental activity is useful when there is insufficient information for either inductive or deductive logic to create a direct chain of intermediate conclusions that reveal "the answers". In my experience, when we are talking about making real decisions in our work and in our personal lives, the lack of sufficient

[77] The term "abductive reasoning" was first reported by Mirriam Webster in 1877 | https://www.merriam-webster.com/dictionary/abductive., | Wikipedia attributes this to the American philosopher Charles Sanders Peirce (1839–1914) at https://en.wikipedia.org/wiki/Abductive_reasoning. I prefer to use the term logic. We will discuss reasoning in Chapter 9.

information is usually the situation. We make an educated guess. I think abductive logic should be given more attention.

Abductive reasoning is akin to Occam's razor. This is a principle that encourages us to adopt the simplest theory that correctly explains a set of observations. Most people will embrace the simplest available explanation as a good working theory, until something appears to contradict it.

The famous philosopher Plato preferred inductive over deductive reasoning. He didn't trust any information coming from the senses. He placed his trust in ideal forms, which he thought resided in some imaginary realm. He felt and acted like he knew that this imaginary realm was more real than the physical world. Plato's distrust of sensory data held science back for centuries.

To my mind, logic without epistemology is not useful. We need both!

Living at approximately the same time as Plato (around 2500 years ago), a philosopher from India, Kanada, held a different idea. He acknowledged direct perception as a valid way to know reality.[78]

Most of us are probably not as clear thinking as Plato, Socrates, or Kanada. Note that I still credit Plato with clarity of thought, even if I "know" that he was mistaken in some fundamental ways. It is an important lesson that a self-reinforcing story with data supporting your logically integrated ideas does not mean you are correct – you still can be wrong. That said, I have found that we produce our best work when we carefully use our trained senses and properly functioning instruments to collect data, turn it into information using a validated thought process, and then draw conclusions; but that is not the end. We then double-check the reliability of our conclusions using epistemology (a concept we met back in the Chapter 1, which was reinforced in Chapters 2 and 3).

To sum up, failure analysis is a methodology - a system of methods, tools, processes, rules and principles for truly understanding what happened to the Subject of an investigation.

4.5 Subject and Reference Specimens

The gold standard process used in a medical study to determine the efficacy of a new treatment is a randomized, double-blind, placebo-controlled study. This type of medical trial is specifically geared towards human participants, because our minds might otherwise influence the outcome. The essence of the double-blind, placebo-controlled study is that neither the medical staff nor the patients know who is receiving the treatment and who is receiving the placebo (inactive substance).

In the Failure Analyst's world, the "Reference Specimens" play the role of the people who receive the placebo in human clinical trials. Some new clinical trials are adding a third group of "specimens". These are people who are selected like the first two groups, but not given any treatment. They become the "natural history" group.

EXERCISE 4-D: Can you think of a corollary use for such natural history specimens in an engineering failure analysis?

The situation is different regarding inert objects. Obviously, we do not have to worry about the thoughts and emotions of the component having an influence on the microstructure, strength or composition that we determine based on our test protocol. What we do have to worry about is how to interpret the results and what to report to the client or our boss.

78 This is explained in the translation of Kanada's work by Subhash Kak. The book is called *Matter and Mind: The Vaisesika Sutra of Kanada* (Mount Meru Publishing, 2016). The book delves into perception and its relationship to understanding reality.

4.5.1 Current Production Parts

Those who are beginning their career as a Failure Analyst often think that it is enough to compare the characteristics of the "Subject Component" to a new part ("control part") that has never been used. But we need to pay more attention to the provenance, or origin, of the "compare and contrast" part.

Some people use a random current production component. But is that current production free from problems? A Subject Part that has been out in the field may be "the canary in the mine". Perhaps the current production is also problematic but hasn't been used enough to manifest the problem.

> **Did You Catch the Term "Subject Component"?**
> ***What Exactly Does That Mean?***
>
> - **It is the term that is highly preferred and more professional than "Failed Part".**
> - **The Subject Component is the main "subject" of our study.**
> - **The Subject Component's undesired behavioral response is what caused us to need to perform the failure analysis.**

Sometimes, we may obtain a set of specifications against which we may compare both the Subject and the Control specimens.

When working on large pieces of machinery or civil structures, we are unlikely to get someone to pay for (or even find) an "exemplar" or "reference" to be sacrificed to science. In this case, finding an appropriate reference specimen from another area of the Subject of the investigation, or the written specification, will have to suffice. Just as care is required to select a reference part, care is also required to select a reference specimen from the Subject Part.

If we are dealing with a high temperature application, choosing an area that would have been expected to have similar exposure should be considered. However, even this may be difficult to obtain because sometimes the "failure" is slated for repair rather than replacement. It is not uncommon for the teams working on high temperature damage to have only the published alloy data to rely on.

It is important to consider not only the specification, which is usually based on minimum characteristics, but to look for typical data as well. Over time, the Failure Analyst develops a gut-feel from years of working with design engineers, as well as learning about manufacturing and construction practices. That gut-feel will warn you about atypical behavior. There is no substitute for the combination of knowledge and experience.

Beginners: Take heart! Some day you too may become the expert who has internalized the proper methodology and practice!

Whether we are dealing with a manufactured component, a machine, or a civil structure, such as a bridge or building, we need to remind ourselves that:

> **1) The part doesn't know the specification.**
>
> **2) Many specifications have fairly arbitrary requirements.**

So, what is the answer for the Failure Analysts? When dealing with manufactured parts and when it is available, the best choice is a *successful durability-tested part* to use as a Control or Reference part.

Obviously, if the Failure Analyst is confronted with a current production assembly failure and the failure rate is very low, then using *randomly pulled* or even *any available component* might be acceptable.

This discussion thus far has been concerned with making sure the Subject Part can be properly compared to one that we have reason to believe does not have the same problem that led to the failure.

4.5.2 Materials and Processes

But, what about a project using raw material? Let's say we have a metal stamping that cracked during assembly. Can't we simply compare it to the flat stock that we used to make the part? No, because the stamping process changes the characteristics of the material.

4.5.3 Size

But what if the part is very small and it's not possible to extract a test coupon that is big enough to perform a standard test? Well, that's even more reason to use a known good part. You will then do as good a job as possible to match the test location to the same part feature on both Subject and Reference.

It is important to realize that changing the specimen size often changes the test results. While yield and tensile strengths are relatively stable, elongation and impact energy are strongly dependent on specimen size and shape.

4.5.4 Service

If there is environmental damage due to service conditions, then that needs to be accounted for in the data interpretation.

> *I have several clients who are using the BOB WOW technique of specimen selection for failure analysis support work. "BOB WOW" means comparing the "Best of the Best" of all material to the "Worst of the Worst" of all material to make a product. This is also useful in evaluating assemblies experiencing intermittent or occasional durability test failures.*
>
> *One of the example-projects that I worked on was annealed flat sheet steel that formed consistently and gave proper dimensional tolerances on the finished stamped parts, versus material supposedly purchased to the same specification that resulted in warped stampings.*
>
> *One of the other BOB WOW projects had me comparing strength of miniature tensile test bars cut from the Best of the Best group of durability test components to those that had very short fatigue lives in the same durability tests. These are excellent "compare and contrast" protocols. If the mechanical properties measured are essentially the same in both groups, it tells you there is some other problem.*

4.6 Macro, Meso, Micro and Atomic Scales

This section introduces a powerful system to help readers understand the different levels of structure present in a Subject Part: MACRO, MESO, MICRO and ATOMIC scales.

> *Some years ago, two colleagues and I were teaching a three-day failure analysis workshop for the Minneapolis Chapter of ASM International. Dr. William Becker, one of the editors of the 2002 Edition of Volume 11 of the Metals Handbook, was the lead instructor. A workshop participant asked him a very insightful question, "What is the most misused word in failure analysis?"*
>
> *Bill didn't blink before responding. "Ductile and Brittle," he said. Then Dr. Becker went on to explain, "These terms should only be used when the context is specified."*
>
> *It's entirely possible for a component to fracture in a brittle manner at the macro-scale, yet have ductile micro-scale features. Likewise, it's possible, if less common, for a component to fracture in a ductile manner at the macro-scale, yet present what are usually considered brittle features at the micro-scale.*

> **FUNDAMENTAL IDEA**
>
> **The terms "ductile" and "brittle" should only be used when the context (macro or micro) is specified.**

We'll spend quite a bit more time on this subject of scale in Chapters 6 and 8 because there is hardly a more important single factor in physical analysis to be considered.

But macro-, meso-, and micro-scale features refer to far more than just fracture (we will deal with atomic-scale immediately below). For a materials engineer, every component must be described at multiple levels. The levels are nested within each other and are not mutually exclusive or independent. If we truly want to understand all the characteristics of a component, we need to understand that materials are made from atoms, which are, in turn, made from sub-atomic particles. Those contained in the nucleus are primary determinants of some of the characteristics of the material, while the electrons surrounding the nucleus are primary determinants of other characteristics.

In many basic failure analysis projects, if a component is nominally made of an isotropic and homogeneous material, we may be able to get away with considering only the micro- and macro-scale features. The micro-scale features of interest in a particular situation surrounding a steel component might include grain size, distribution of phases or constituents in the microstructure (ferrite, martensite, pearlite, etc.), and the presence or absence of second phase particles, non-metallic inclusions, or, for steels, a surface layer of decarburization.

There are other features requiring consideration of the entire component, or at least a cross-section representing all layers of the part from surface to core. This might include the

- depth of an intentionally hardened case[79],
- weld characteristics such as penetration and fusion,
- grain flow lines in a forging, and
- macro-segregation of low melting point alloying elements or impurities in castings.

While not in widespread use, I find the concept of meso structure to be helpful in my own thinking. Meso structure is especially useful when I consider an entire structure where the Subject Component was a member. In this case, I re-classify the features just listed above in the "macro" section as meso. Then I put macro to use as a placeholder for the entire structure and the situation surrounding it, including its manufacturing and service history. See Figure 4.5.

[79] One type of case hardening is the process of hardening the surface of steel by adding carbon to the near surface layers. The underlying metal retains its original composition.

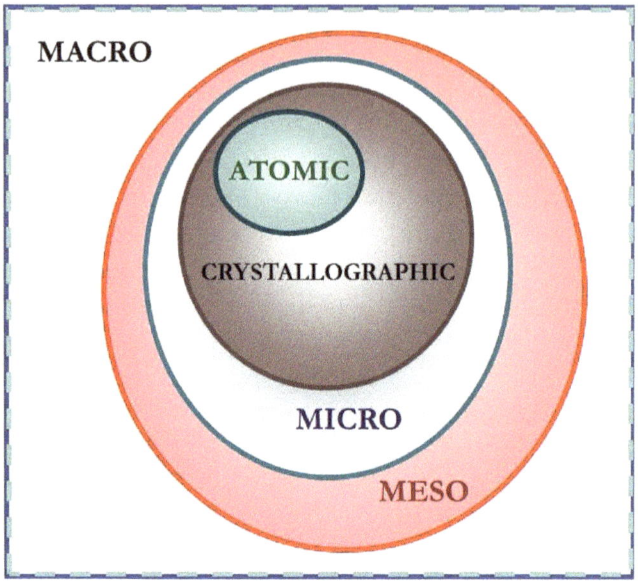

Figure 4.5: *This image, although not to scale, shows the Levels of Structure from atomic to the visible.*

4.7 Musings on "A Structure"

In this book, and generally in my work, I use the term "structure" to refer to any component or assembly that must sustain forces in order to perform its function. A structure is a thing. The word is also used to refer to "micro-structure" and "macro-structure", which are constructs based on concepts, and differ from "a structure" (a thing).

A micro-structure is a *characteristic* of "a thing", that influences its behavior (strength, corrosion, resistance, etc.) and these characteristics are a result of its thermo-mechanical history.

The thermo-mechanical history is a way to refer to the manufacturing process from the point of view of the "the object".

A macro-structure can be observed with the naked eye, or slightly higher magnification, for features such as the presence of welds, a coating or geometric features on parts.

Examples of structures include the beams in bridges. Fasteners, which hold together a housing, also fit the bill without much deep thought. Soldered-in-place lead wires in electronic components may not come to mind, but if the wire form is not designed with enough slack to account for movement due to heating and cooling, the wire or the soldered joint may inadvertently become a structural component, subject to force or stress. A housing just sitting in place may not seem to be a candidate for the title of structure, especially if its purpose is simply to keep out dust or moisture. We might not think of this as being particularly stressful. However, if the temperature changes or if the material degrades, or both, we may find field-returns due to cracks. The presence of cracks means there were high stresses. The stresses may be due to the mass of the structure itself.

Even a change in appearance might be a cause for concern and require a failure analysis. Imagine a transparent plastic housing whose job is to keep out moisture and dust and allow visual inspection of the internal mechanism. If an incompatible chemical, or sunlight, or extreme temperature changes cloud the material, it has basically lost its ability to perform one of its major functions. Even though there may be no cracks, the structure has failed.

In the real-world, temperatures vary. Sometimes unanticipated stress levels result from that variation.

A client called me about a section of aluminum hand railing. He told me the railing cracked only a few weeks after installation. "Another party," he said, "determined that the problem was a cold snap that hit the region shortly after installation of the handrail."

My first response was, "That's ridiculous." However, when the section arrived at my shop, I was shocked to see what looked like a broken tubular tensile test coupon. The large diameter vertical tubes had been set in concrete and the smaller diameter horizontal tubes welded in place.

Originally, the design called for galvanized steel railings, but at the last moment, the end customer decided that aluminum was prettier. The engineer of record for the project was not informed of the change.

I performed a simple calculation of the relative length change due to thermal contraction in the horizontal member had it not been locked into place by the vertical posts embedded in concrete. The value approached the percent elongation in a two inch (50 mm) standard tensile test coupon.

The percent of elongation usually decreases as the total stressed length increases. The length of the hand rail was much longer than a standard tensile test coupon. Suddenly what had seemed ridiculous moved to the top of the list of most reasonable explanations!

Indeed, the thermal expansion co-efficient of aluminum is approximately triple that of steel. To ensure structural integrity in an area that experiences temperature variation from -20 to +35 degrees Celsius on an annual basis, the characteristics of the material must be accounted for. Changing from steel to aluminum would require a significant shortening of the length of the restrained members.

In this case, it's obvious that a hand rail must sustain the stresses of people bumping into it. It must sustain the stresses of kids (or adults) sitting on it or leaning on it. Maybe it needs to be able to resist a sidewalk sweeper or lawn mower bumping into it. But it was a surprise to all but the original engineer that the thermal stresses would be the demise of the beautiful aluminum structure.

There is a monumental sculpture located in Belgium that beautifully depicts nested structures.[80] See Figure 4.6 (next page) for one view of it.

[80] Sculpture in Belgium. Photo Credit: superdirk0144983; https://pixabay.com/photos/brussel-atomium-landmark-958408/

***Figure 4.6:** Sculpture depicts nested physical and metaphorical structures*

Now it's your turn.

EXERCISE 4-E: Structure versus structural component.

Look around the room you are in or a room that is familiar to you. What might you now label a structure or a structural component that might not have seemed to fit the category before you read this section? Explain how service stresses could affect the function of the item.

ON IMPLEMENTING NEW IDEAS

"You have free will, but have one foot nailed to the ground." ~ Gurdjieff (paraphrased)

Your path to exercising your free will is The Seven Steps to Constructing Competence.

4.8 Culture

Hidden in plain sight and often difficult to pinpoint, are very powerful factors that influence a Failure Analyst. The summation of these factors we often call the company culture.

Cultures vary from company-to-company and industry-to-industry. A successful company that makes or operates safety-critical, high technology items such as nuclear power plants and airlines must have a culture that ensures the integrity of the product. Typically, this requires people to look over each other's shoulders and procedures with redundant functions. In this way, mistakes are usually caught.

However, modern companies are under great pressure to cut costs. Accountants may not see the value of performing multiple design reviews or paying for engineering oversight when purchasing critical components. Therefore, one aspect of culture that is relevant to failure analysis is being able to classify the degree to which the workers or contractors at a company are inherently double-checking each other's work.

Another aspect of corporate culture is the social environment. Some of the companies I work with have people who are unfailingly polite. Others seem to promote a culture of antagonism whenever a problem arises. It is important to discern and to understand which situation you are in or if there is another mix of practices from which management tends to address problems. When you are investigating a fatality, extreme caution is required. Knowing how to interview people in a way that improves the chances of getting the information you need is a critical skill for failure investigators.

Although I have helped to investigate several fatalities, I've never been the main investigator interviewing bystanders or witnesses. I have also never had to interview someone who was the subject of a criminal investigation, so I have no detailed advice for that type of work. I found that reading the transcript of the interview of a quality engineer who got in trouble with the U.S. Navy was helpful to shed light on how to conduct an interview with a suspected criminal. See Case No. 3:20-mj-05145 dated June 15, 2020 of the United States District Court for the Western District of Washington at Tacoma.[81]

With that in mind, I can suggest you examine the social culture before any interviews. During interviews, try to help the person being questioned feel as comfortable as possible. Developing rapport and understanding the person and the context of their social position and work environment is helpful. This is likely the only way you can infer what happened, which is the Failure Analyst's primary goal. Blaming people or criticizing them will generally interfere with the person's willingness and ability to provide information. Torture has been shown to provide bad and generally unreliable information[82].

A few years ago, I rashly accused a new client of withholding information. My ignorance of the facts that he withheld led me to propose ineffective testing. He was very offended that I saw him as being deceitful. He claimed that he simply didn't realize that the information was important or even relevant.

I remember having a hard time believing that I hadn't asked him the relevant questions that would have brought out that information earlier in the investigation. Whatever the details of the situation, I won't have to worry about it again with that client. He "voted with his feet" and chose not to provide me any more business. It was an important lesson.

It is often hard to tell the difference between ignorance and corruption. Ignorance is not usually a crime, unless it's intentional on the part of someone in a position of relevant responsibility. Even a company that prides itself on integrity may have situations where individuals feel defensive and want to hide information. Learning to recognize when this is happening is an important communication skill to develop. If you are just beginning in the field, finding a successful mentor with these skills will help. There are also courses on this topic.

In my personal experience, the best path to figuring out if a person is keeping information from you is to ask a lot of questions. If the answers all contribute to a self-consistent story, then you have evidence (but not proof) that the facts provided could be accurate. In other words, you have no cause to be greatly suspicious.

Unfortunately, companies occasionally send a messenger who really isn't involved in the situation or who is not high enough on the responsibility-ladder to know the answers to all of my questions. In this case, I will follow up with someone who knows more. If the company cannot be persuaded to provide basic information to enable me to create a scope of analysis for a professional result, I usually turn the work down. That's why I am not financially wealthy! Alternatively, I inform the company in writing, as part of the quotation for work, that the work is considered preliminary and may not provide the information they seek and may even destroy the chances of figuring it out later.

Many factors create a company culture and the specific event of being investigated may bring out the dark side of any corporation's practices. Even a straightforward approach in failure investigation can create antagonism and enmity toward the investigator.

[81] You can view the case for free if you sign up for a governmental Pacer account and download less than $30 per quarter | https://pacer.uscourts.gov/find-case | Or, contact the author at daaliya@itothen.com

[82] Shane O'Mara, *Why Torture Doesn't Work: The Neuroscience of Interrogation*; Harvard University Press; Nov. 30, 2015 | ISBN 978-0674743908

Early in my career I worked as a quality assurance technician. It was my job to determine whose fault it was when the steel mill mis-marked the composition grade on a billet. Was it the continuous casting department? The bar mill? The transportation group? Whatever my determination, somebody was unhappy with me! In that position you were damned if you did and damned if you didn't. Someone was going to be unhappy.

High level failure analysis is not a job for the faint of heart or for the heartless! Trying always to achieve a balanced nature and relying on basic physical science and the engineering body of knowledge will serve the investigator well. Consider these power questions as you investigate:

1) Am I doing what is right?
2) Do I need a mentor for guidance?
3) Will I be satisfied with my work?

4.9 Complex Systems

What is a system? There are many definitions. Oxford Dictionary's first two definitions seem to be the most useful in this context[83] (below).

1) "A set of things working together as parts of a mechanism or an interconnecting network; a complex whole."

2) "A set of principles or procedures according to which something is done; an organized scheme or method."

Even these two definitions are lacking, in my opinion, despite the use of common fields of knowledge to explain the meaning of system. The first definition (above) goes on to provide an example of fluid being pushed through the pipes in a plumbing system. While that is not wrong, I think it's helpful to see the fluid itself as part of the plumbing system. Also, we shouldn't forget the aspect of the system that feeds the fluid into the plumbing system. If it is not a recirculating (closed) piping system, we need to consider how the fluid gets back out of the system.

Is there heat provided to the system? If so, how? Is whatever mechanism that provides the heat considered as a part of the system? I think that as Failure Analysts, perhaps trying to understand why the plumbing system has leaks, we absolutely need to be aware of how heat is provided to a system.

Is the fluid always completely pure? Again, if we are investigating plumbing system leaks, small amounts of solid debris dropping to the bottom of a pipe in a horizontal run may get stuck and produce nucleation of crevice corrosion. If the plumbing system is part of a boiler heated recirculating water heating system, there will usually be corrosion prevention chemicals added to the water. The addition of chemicals means there must be a monitoring system to ensure the desired concentration range is stable.

Back in 2014 in Flint, Michigan some government officials wanted to save money in the operating budget for the drinking water system. They refused to approve the $37^{\underline{00}}$ per day expense to prevent corrosion of the piping system. The result was that large flakes of lead oxide began to break off of the decades long build-up of scale. Piped water carried the flakes of lead oxide into the homes of the people of Flint. The water was used to prepare baby formula and baby food. The water was used for drinking and bathing and recreational activities.

Flint's children started showing up at their doctors' offices with indications of lead poisoning. The lead from the piping system ended up in the physical systems (bodies) of the residents of Flint and damaged the nervous systems (brains) of several young children. Young children are more susceptible to lead poisoning than adults. Lead poisoning causes long-term neurological damage and personality changes.

From the information discussed above, we can now revisit the first definition of system: a set of

[83] If you do not have an Oxford Dictionary, you can go online at en.oxforddictionaries.com to view definitions.

things working together as parts of a mechanism or an interconnecting network; a complex whole.

What we see, in reality, is that any system must be defined. The boundary of the system may be somewhat arbitrary, but it should contain any subsystems that may be relevant to the Subject of the investigation. The Flint, Michigan example shows us that multiple systems may be nested inside each other, or overlap - sometimes neatly and sometimes not so neatly.

EXERCISE 4-F: Diagram a specific system.

One of my favorite exercises is to try to diagram the system that brought YOU into the world. Obviously, there needed to be a mother and father and an event or situation that allowed the potential mother and father to meet and connect. Then the mother had to be in adequate health to gestate the baby you. Then you had to be nourished and protected to grow up to be someone who could be studying failure analysis. Stop READING NOW and take a clean page or set up a new document. Try to fill in at least 20 levels of events or situations that enabled you to reach this moment. Write the name of the document below.

Name of Document: ______________________________

- The Failure Analyst who wants to go beyond routine work will take every opportunity to cultivate the ability to imagine the entire life cycle of the Subject Component.
- The Failure Analyst who wants to go beyond routine work will take every opportunity to cultivate the ability to understand the function of the Subject of the investigation and the effects of the larger environment on its durability.

Before leaving this short (very short!) musing on the term "system", let's look back at the second Oxford Dictionary definition: "a set of principles or procedures according to which something is done; an organized scheme or method".

Most of the "objects" (things) that are the subjects of our failure investigations are constructed or manufactured according to a set of specifications or codes. This leads us back to the "principles or procedures according to which something is done". A manufacturing company or a construction contractor is supposed to work according to defined procedures, and when deviations occur, usually nothing bad happens.

Usually, deviations happen because someone is in a hurry or doesn't want to stop and repair something that broke / wore out / corroded, etc. So, deviations usually save money - at least in the short run. But occasionally, damage occurs. Often it is because multiple deviations built up.

Because deviations usually save money in the short run, experienced people who do not understand the system-level effects may believe they can get away with infractions of the defined procedures. Self-reinforcing feedback loops, physical and mental, are another part of a complex system. Just as a psychologist might need to provide medication and talk therapy, beliefs and practices must be aligned to ensure proper manufacturing operations.[84]

Getting away with a deviation leads to a reinforced belief that it's okay to operate with deviations (plural). When management doesn't have a system in place to prevent ongoing deviations from the known robust operating procedures, eventually problems are the result. Eventually, safety margins are eliminated and an undesirable event happens.

[84] David Levy, *Tools of Critical Thinking*, Chapter III, "The Simultaneity of Physical and Psychological Events".

Sometimes, the deviations are even approved by management. Such was the case in Flint, Michigan.

And that was the case with a batch of steel that one of my clients could not get to harden to specification after heat treating. For years, they had reduced the alloy content of the material they bought to lower and lower levels in order to control costs. Then one summer about five years down the road, there was a heat wave and they could not keep the quench water cold enough to chill the steel fast enough to ensure that it transformed to the required hard martensitic condition.

The steel composition met the current purchase specification, but the laws of physics will almost always prevail, regardless of what the plant manager thinks about the situation.

Real systems *are* complex; and there are always subtle interactions between factors that are imperfectly understood. It's rarely possible to specify proper ranges for every factor in a system. System factors, by nature, are interdependent - generally in ways that remain partly understood and partly hidden.

4.10 Design Practices

In this philosophically-oriented book by a materials engineer, I won't presume to tell the reader how to calculate whether a design is structurally adequate or if it will perform within the desired risk category. What is important to all people who are involved with structural component failure analysis work is a basic understanding of the actual requirements of the structure. What, exactly, does the part have to *do*? What strength levels are required? What corrosion properties are needed to resist relevant environmental stressors? And so on...

People who are involved with structural component failure analysis work need a basic understanding of the actual requirements of the structure.

As already stated, a simple answer to the question "Why did the part break?" requires knowledge of stress and strength in the given environment and a known service history. These must be evaluated at a specific position within the component (where the crack initiated).

It is also important for the Failure Analyst to have an idea of the level of detail of the requirements embedded in the design. In other words, did anyone ever properly specify the raw material used to make the part? Is there a corporate policy to review certification documents? Is there someone who knows how to interpret the data on these "certs"?

Just recently, I reviewed documents provided by a client for 8620 steel. The mill had not reported the chromium, nickel or molybdenum content. The paperwork was two years old! I was informed by my client that there was nobody at this "lean company" who knew what the grades of steel meant.

It is important for the Failure Analyst to have an idea of the level of detail of the requirements embedded in the design.

Without knowing what the technical sophistication of the client is, it will be difficult to provide meaningful and practical advice on how to avoid repeats of the failure. In forensic engineering, without this knowledge, it will be impossible to assign responsibility in a manner that deserves the term fair.

In the real world, most components and assemblies are designed without a realistic and quantitative knowledge of the relationship of stress-to-strength at every point in the assembly. Even in the aerospace industry, the capacity for computer modeling that would be required to account for every possible situation does not exist. That is why, to a certain extent, the field is always a test lab.

In some product families, the *only* test lab is the field.

Years ago, I was given a small screw-machined part that held a molded plastic chair seat onto the frame. This chair was meant to be placed in a public space so that people could sit down while waiting their turn in line.

There are some types of seats that have quite strict test requirements. Movie theater seating is one of these. The City of New York, for example, requires that the manufacturers demonstrate that the seats return to vertical positions after being folded down a certain number of times. Office chairs that roll on wheels are also subject to government requirements for durability.

However, these stationary chairs were not subject to any government regulations. It is likely that most of the potential types of damage would provide some warning prior to a catastrophic event. In the case of the component that I investigated, if it broke, the seat would shift on the frame, probably as someone was sitting down. It might eventually twist and fall to the ground, but the person's buttocks, or at least their knees, would still be supported by the cross bars of the frame. It might be uncomfortable, but probably not life threatening. Of course, an active toddler might create some high strain rate load cycles.

In any case, the outcome of this lack of regulation allowed a new design by a new supplier to be immediately put into service without any durability testing. The field returns began to arrive after a very short time. I can't remember if the cause was "sharp thread roots" or "low strength", but obviously the two could, alone or synergistically, reduce the durability. The need for a fix was inconvenient, but probably less expensive than initiating a more sophisticated engineering development program.

To my mind, this is a business decision on the part of the manufacturer. As a Failure Analyst, I can offer ideas to make a more robust component, including use of a rolled thread, use of a higher strength material, or use of a larger diameter fastener. I can even suggest that they consider doing some basic stress analysis or a simple durability test. But moving to a higher level of design sophistication is unlikely to happen in that type of industry.

I call this type of engineering design practice "The Field Is the Test Lab"! Another name for this type of testing is "Guess-and-Hope".

Next in order of sophistication is the "Guess-and-Test" method of design. In my work, this is probably the most common. In most cases, the "Guess" is not a completely random one. It is based on experience with similar designs. Sometimes "Guess-and-Test" includes some analytical work. There may be a finite element analysis (FEA) to detect "hot spots" in the stress field, even if there is not a fine meshed non-linear analysis performed that realistically models the actual quantitative stress levels.

In my experience, most of these quick finite element analysis models are performed in order to determine what the specified yield strength should be. It is infrequent that the analyst even looks at the normal stresses, which are what drive a fatigue crack once it initiates. Usually, when the FEA data are provided, if I don't request the normal stresses, I'm given only Von Mises stresses. Often, the FEA is not even performed until the first design fails the durability test.

Guess-and-Test, increasingly supplemented with some sort of analytical work, is very common for brackets and other small parts in the automotive and related industries.

When I graduated from engineering school, I was young and naive! The Guess-and-Test engineering design criteria really shocked me when I first started doing materials engineering application work, rather than the process engineering that I thought I would be doing for my whole career. We had to take a statics course as part of the metallurgy and materials science curriculum. I thought that my life as an applications engineer would be easy. The design engineer comes to my desk and tells me that he needs a type of steel with a yield strength of 65 ksi. I go to my books and give him a choice of high strength low alloy grade 65, if it's a piece of thin flat stock that is to be used as is, or suggest that he might be able to get away with grade 55 if the part will be formed and the area of the 65 ksi is in the deformed area. I suggest a way to determine if we'll be able to "get away" with the lower strength material.

But this was a complete fantasy. The bottom line was that it was extremely rare that the designer had any clue as to what the quantitative requirements for the component were. One time, the designer showed me the print he had made up for his prototype. This was after the regular production had started and there were failures in the field. He informed me that he had specified "High Strength Low Alloy Steel, Grade 60 or Better".

I was really confused. "What does that mean, Grade 60 or Better?"

"Well," I was informed "I'll take Grade 60, 65, 70, etc. - anything that was at least 60 ksi minimum yield strength."

"But," I asked him, "Why would you test something that was stronger than you were planning to make it in production? How does that actually test your production design? Why didn't you make it Grade 60 or Worse?"

Of course, he didn't want his design to fail in testing. This person wasn't dull-witted (he had a master's degree in engineering from a fine university). This is just how the engineers of his time thought. Later, some of the automotive companies started creating material specification procedures that would not allow this type of thinking. However, there are probably still a lot of companies that don't have detailed requirements for prototype material specification.

A last note before leaving this case study of "Grade 60 or Better" is that Grade 70 and Grade 80 are not better than Grade 60. The best grade for the component is one that is adequate to meet the strength requirements and is readily available in the product form required, at an economical price, and that, if it requires a shape change, has enough ductility to facilitate the required shape change. When you do failure analysis of structural components, you will probably still run into people who think that higher strength is always better. Unfortunately, the world of materials engineering is not usually that straightforward. Sometimes increasing strength will make the material more susceptible to unexpected types of damage. The complicated reality is that materials selection and specification is an entire field of expertise unto itself.

While we are on the subject of prototype materials as related to the Guess-and-Test Design Methodology, a related concern is how the component is made. The world of rapid prototyping has exploded in recent times. Is a 3D printed part good to check fit and finish? Maybe. Is it good to check general appearance? Maybe. Is it structurally the same as what will be made in production? Maybe. Whoever is in charge of the prototype testing needs to understand the factors related to these issues.

DESIGN PRACTICE SPECTRUM

- **Guess-and-Hope (The field is the test lab!)**
- **Guess-and-Test**
- **Comprehensive Engineering Design**

Years ago, one of my clients had a small part made from flat spring steel. They tested the prototype, and it miserably failed the durability test requirements. They sent it to my lab, with a list of questions.

1) Was it a fatigue crack? Yes, it was a fatigue crack.

2) Did the hardness meet the spec? Yes, the material was exactly in the middle of the specified hardness range.

3) Did the heat treatment cause decarburization? No, there was no decarburization or surface pits, or any of the other common microstructure problems that can beset small springs.

4) What is the problem? We agreed that I would make a metallurgical cross-section. I saw a very strange feature along the edges of the part. It etched mostly white and I made a note of that when I submitted my report to the quality engineer. Eventually, we heard back that the part was a prototype made using an electric discharge machining (EDM) method. The EDM method leaves a brittle recast layer on the cut edges (etched white[85]). The part was never intended to be used for durability tests. They wanted something to show the customer.

This was an automotive part. Now the automotive industry requires pre-production parts to be made by the same processes used in regular production.

The change in prototype and pre-production part manufacturing requirements makes me think that these two stories that I tell were not the only incidences of problems related to differences between prototype and production part specifications and manufacturing methods.

The final major basic category of design practice that I want to review is what I call "Comprehensive Engineering Design". Figure 4.7 shows a flow chart of the fairly comprehensive engineering design methodology.

In my career as a Failure Analyst, I've been involved in two projects, which used all the levels of design optimization shown on this chart.

There were a couple of other projects where we used most of the levels shown, but the projects were for litigation and blame assignment rather than process or product optimization or improvement.

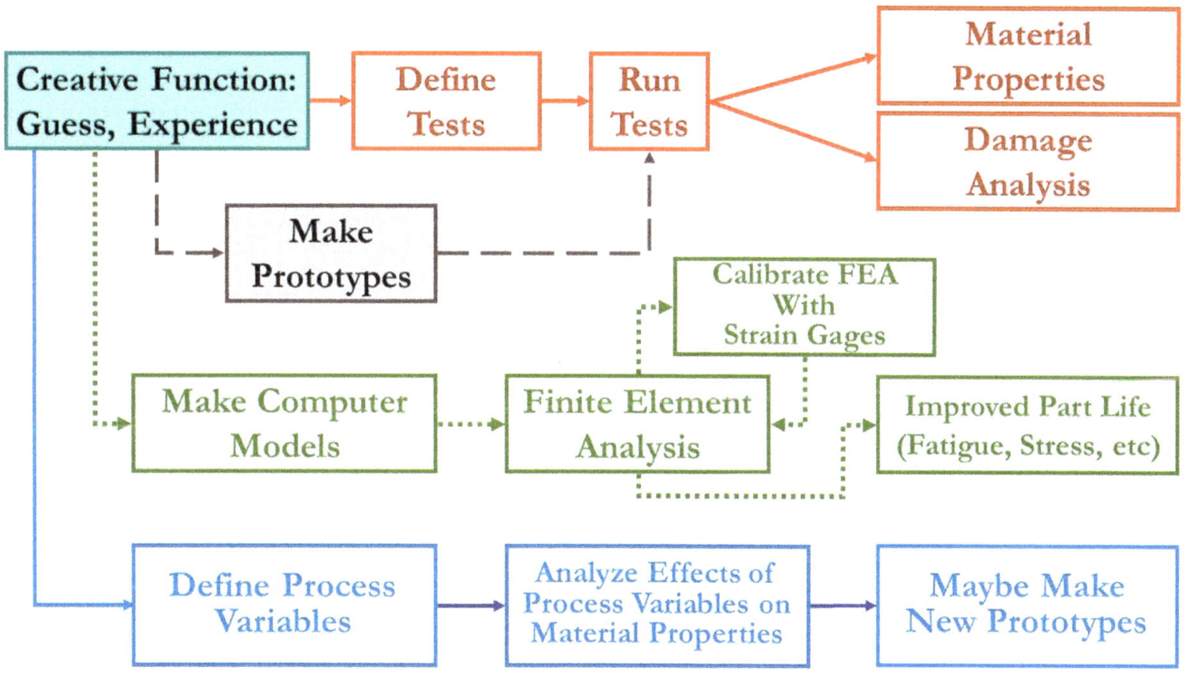

Figure 4.7: *Comprehensive engineering design optimization process*

[85] The white-etching surface layer on the prepared metallographic coupon is associated with the EDM recast layer. White-etching refers to the white appearance of the altered microstructure of a polished and etched steel sample.

Of course, there are methodologies between Guess-and-Test and Comprehensive. There are those that go beyond what is shown here as Comprehensive.

There are several technical issues that appear when an attempt is made to create a highly detailed engineering model of a real assembly subject to loads. The first one involves setting boundary conditions for the computer stress model (Finite Element Analysis or FEA).

There are a lot of computer models that are made by someone in an office who has never looked at the component or the assembly and has no idea of the functional requirements. In this case, the person making the model has little choice but to use default settings (make up a fantasy).

An alternative choice would be to ask the client who is requesting the model for information. This often seems to be a difficult or impossible task. The analyst is often not to blame. The company that requests the work, or those setting the budget for the analysis, may not have any idea of what it takes to create a realistic model.

There have been two FEA models that I reviewed which stand out for poor quality. Please note that I do not place 100% of the blame on the analyst for the result. If the people asking for the model don't know better, the analyst has little choice but to proceed in the best way they can.

> ***The first poor FEA model situation*** *involved a plastic part molded around a steel component. When the plastic cooled and shrank, the stresses were so high that the plastic part cracked. The analyst had actually used laboratory-determined thermal expansion data to predict the stresses in both the plastic and steel parts. Unfortunately, the people who determined the thermal expansion data must have been having a bad day. The values they came up with for the steel were higher than those that they measured for the plastic. They apparently never checked published literature to see if they were in the expected range of values.*
>
> *If they had bothered to ask their buddies in the metallography lab if the data made sense, they would have realized they had a major problem. Anyone who has used a hot mounting press in a metallography lab for a few projects knows that the plastic shrinkage must be managed. Plastic shrinkage is generally much greater than the steel's shrinkage. But this was a rush job. So, you get what you get.*
>
> *The FEA modeler plugged in the lab-determined test values and showed there were no stresses in the plastic at the crack location, so they concluded that something else was wrong. Of course, if they are over-molding polymer onto steel there would be no stresses generated by differential thermal expansion, because the inner steel part is being shown (incorrectly) to be shrinking more on cooling than the outer plastic part! Since there are always many things that can go wrong in any manufacturing process, this sent the parties on a wild goose chase.*

If you don't take the time to discuss how the FEA boundary conditions will be established, you are likely to end up with worthless results that bear no relationship to reality.

Garbage In, Garbage Out!

> ***The second poorly performed FEA model*** *began with a request to model the effect of an impact bending load at the end of a hollow vertical beam. The analyst decided to model the effect of a ball bearing being shot at a short distance from the end of the beam.*
>
> *They modeled the size and shape of the permanent crater made in the beam, rather than the deflection under load that was the designer's request. Again, it is hard to fault the analyst. If the person who wants the data doesn't take time to specify exactly what they want, it's unlikely to be determined.*
>
> *Apparently, nobody at the company that requested the work even noticed that the data were irrelevant. I suppose someone wanted an FEA file to show the boss that they looked at the stress levels. The problem was only noticed when I asked for the*

FEA to support my analysis of a durability test failure.

For most of the detailed FEA work that I've been involved with, the conversation with the analyst took several hours. In most of the cases where the end client was willing to gather strain gage data to compare with the analytical output, the measured values were within a few percent of the model. If you don't take the time to discuss how the boundary conditions will be established, you are likely to end up with worthless results that bear no relationship to reality. "Garbage in, garbage out."

It is also important to state here that the correct analysis uses the correct physics. Establishing incorrect boundary conditions are not the only thing that can go wrong. If the situation involves multiple types of physics (structural, thermal and vibration) then accurate modeling requires their use. If non-linearities are significant then using a linear model will have its limitations in accuracy. Also, if you are limited in time, then choosing the right physics, making the right linearity assumptions, and selecting the correct boundary conditions (preferably verified by a physical model) will deliver useful results.

Comprehensive engineering design methodology of the type shown in Figure 4.7 is naturally epistemological. Tensile test results may be compared with the stress levels shown by the FEA work.

For parts subject to fatigue, rough checks of finite element analysis may be compared with published SN (stress versus cycle life) data for similar materials.

Comprehensive engineering design methodology for structural components is best performed with materials at the low end of the allowed strength range. This will ensure that testing will realistically predict performance.

As a materials engineer, I understand loading geometry in a practical way. Therefore, I know it is far from simple to make a direct correlation between the boundary condition parameters an FEA specialist selects and, for example, a bending load at position y.

It would also be difficult for a FEA specialist to make a direct correlation between the boundary condition parameter selections and a torsional deflection in radians. This type of interdisciplinary work is among the most challenging aspects of failure analysis of structural components for me. So, consider this topic as an opportunity for expanding your understanding.

So, am I saying that the great Failure Analyst will always benefit from performing a comprehensive engineering design analysis? Well, that depends. The benefits of performing a comprehensive engineering design vary based on

- the situation in which the components are used,
- the engineering decision-makers and their desire for competitive advantage,
- the industry in which the client participates.

Sometimes regulations require FEA models. This is becoming more frequent in Europe. In cases where the government doesn't require it, the company may decide that it can save money to skip this analysis. Where physical tests are very expensive and time consuming, spending large sums of money on FEA models may be easier to justify economically as in the case of rocket design, aerospace vehicles, tall building design, or satellite missions and their survivability. When lives are at stake or multi-millions of dollars can be lost due to failure, comprehensive engineering design work will be a strong consideration.

When comprehensive engineering design work is part of a failure analysis, especially if there have been field failures involving safety issues, FEA and other modern tools provide a sound foundation for confidence in the fix.

As humans, we can rarely be sure we have 100% certainty. But using a process with multiple avenues for cross-checking the data will get us closer to having a justifiable confidence than by using the Guess-and-Test method. At the very least, we can claim to have been diligent in our efforts to address the issues. Diligence in engineering builds reputation. Sometimes that is a business advantage. Low cost has its own reputation and may provide a viable market position. But when it comes to safety or loss of huge sums of money, engineering leaders must consider that the consequence of failure could be personal liability (shouldering responsibility for lost lives and losing someone's fortune).

Comprehensive engineering design methodology is naturally epistemological.

Years ago, I heard a radio talk show panel discussing the problem of rail tanker accidents involving oil transportation. In my opinion, it was one of the better talk shows because the guests were well informed and their call screeners did a good job of keeping completely ignorant people off the air.

That day I was listening to the show with my critically-thinking-engineer hat on and was surprised that the host let the following statement go unchallenged. "Cost isn't an issue here. Safety is job one!" Certainly, the person saying this did not invent the statement. But the fact is this: cost is always an issue involving any real-world problem.

I want to be clear that I'm not promoting the type of accounting that attempts to quantify lives in dollars versus the cost of changing a design. I was alive when Lee Iacocca ran Ford Motor Company. Apparently, he was fond of saying "Safety doesn't sell" (1978).[86] *After Pintos started turning into fireballs, it was discovered that Ford executives had hard-heartedly calculated that it would be cheaper to pay off the estates of those people who were likely to be killed in gas tank fires than to fix the Pinto gas tank structural problem. Ford learned hard lessons: they took reputational hits, huge financial and legal hits, and after a damning investigation by the National Highway Traffic Safety Administration, Ford recalled 1.5 million Pintos and Mercury Bobcats.*

They are not the only company that had to learn hard lessons. Other companies have had to learn those same lessons over the years since then. There is a lot of daylight between "Cost is no issue!" and "bean-counting cruelty".

A critical thinker will question the broader context of the situation. If we spend all of our resources on safety, we'll never do anything else. It is impossible to have 100% safety in every human endeavor. What would it even mean to have 100% safety? Would we want that? Does that mean that everyone becomes immortal? I hope that isn't going to happen any time soon. As many have noted, improvements come when the prior generation finally stops preventing progress.

Here is a relevant example discussed in the media in the early 2000's. Should infants be placed in a crash resistant carrier when traveling on airplanes and should parents have to pay the extra money for a seat? Is it fair? Will it make a difference in a crash? Will it force parents to drive rather than fly? And is that fair since the death and injury rate per mile traveled is much greater for cars than for planes. Put it all together and "improved safety" on airplanes may actually cause a greater death and injury rate overall.

Going back to the safety of oil transport, I am all for improving the record. The few times I have investigated oil spills from trucks, I've been fairly unimpressed with the strictness of the government standards in the USA. Last time I checked (around 2005), the United States Code of Federal Regulations had a very minimal requirement for structural integrity. Furthermore, it applied only when the vehicle is traveling down the road in the usual manner. Except in certain specific situations, integrity of the tank in the case of a crash is not required.

In order to improve the safety record, would it be helpful to frame the discussion in terms of real costs and benefits?

Working on small projects may pale in importance compared with oil tank car accidents traveling near densely populated areas, poisoning the water supply, and causing massive explosions.

[86] https://www.tortmuseum.org/ford-pinto/ | American Museum of Tort Law

However, we remind ourselves that framing the situation in realistic terms will allow us to participate in realistic assessment of the level of risk.

4.11 Root Cause Analysis versus Failure Analysis

For those people who want to explore the nature of Root Cause Analysis, this section will be particularly fruitful in that it will provide the chance to build clarity in your thinking. Integrating these concepts into your understanding will improve your performance in the profession of engineering.

Many people confuse Root Cause Analysis with Failure Analysis. Not me - not this book - not us!

There's A Big Difference Between Root Cause Analysis and Failure Analysis

- **Root Cause Analysis** is often conflated with Failure Analysis. Many people believe that a root cause comes down to a single factor. The primary concerns of root cause analysis are the human and business factors, including company culture. A comprehensive understanding of the root cause must account for the complexity associated with multiple levels of causal factors.
- **Failure Analysis** must be performed at the physical cause level. The best failure analysis goes further and also touches on human and business factors. However, we absolutely must determine how the damage happened and the details of any unusual circumstances. The primary concern of failure analysis is to arrive at an epistemologically justified, data-based understanding of the sequence of damage, using physical evidence.

I was asked by some peers at ASMI to write a book about root cause analysis. Despite the general agreement for the need, I did not get traction for that project.

Much of the material in this book began as a start to a book for root cause analysis. I saw the need to focus first on thinking clearly and then upon methodology and the need for verifiable evidence.

.As I refined my understanding, and with no prospects for the root cause book, I decided to apply my lifetime of understanding thinking, methodology and evidence to failure analysis resulting in this book.

A common problem with root cause analysis is the implication it can be traced to a single main factor and thus creates avenues to falsely assign blame. The main problem that I have with the use of the term "Root Cause" is that it means so many different things to different people. I was a long-time participant in the "Yahoo Groups Root Cause Forum". At one point, Dr. William Corcoran (of NSRC and also a member of the forum) compiled a list of 100 definitions of "Root Cause Analysis". That is a lot of separate ideas for one term. One discussion concluded all failures were due to the creation of Adam and Eve. This was, of course, from a Western mythological viewpoint.

As Materials Engineers, we offer our clients and employers our knowledge of materials.

- Other people can measure the dimensions of the components.
- Other people can examine a mechanism's functions.
- Other people can review codes and quality systems.

But, as materials engineers, we understand that the manufacturing process creates the characteristics of a final product. We know the service environment may affect those characteristics. We understand the final geometry, as well as any heat treating or coatings, may affect our ability to measure the material characteristics affecting component performance.

If you want to drill down to a foundational truth, I will grant that all failures are a result of human error somewhere. The part itself can never fail. The part does what it does based on its structure and characteristics. The part responds to its environment. So, if there is a failure, it can always be traced back to a situation where a human failed to realize or account for a potential deviation from the design intent - or someone didn't install, use, inspect, or maintain the object properly, or it simply wore out before it was replaced. See Figure 2.3 (Origin of Causes Schematic).

When a failure happens, it's important to understand the design process may have been inadequate, missing the mark. The point is not to blame the design engineer. The organization has more influence on product integrity than individual engineers.

So, when my clients ask me to perform a "Root Cause Analysis", I tell them that I will do a "physical cause analysis" instead. If they really want to collect information on the much more involved human and cultural factors facilitating the physical failures, then I will do a root cause analysis for them. Usually, this distinction satisfies them. Sometimes they insist on a "Physical Root Cause" analysis, which requires only an understanding of the physical reasons for a failure, but allows them to check the box for a "root cause" analysis.

One strong point of proper root cause analysis is the idea of multiple, parallel levels of causes or factors. Experts working on complex failure analysis come to understand, sooner or later, that it is often impossible to assign a single cause to an event. It is rather a "constellation of factors" that resulted in a catastrophic deviation from normal operation.

A single factor's variation may be small and appear to be insignificant; but when combined with other factors, it is difficult to rule out an individual contribution to the "event". Author David Levy addresses this constellation of factors in his book.[87] In failures of physical systems, the psychological events are all the human decisions that go into designing, building, maintaining and operating a manufacturing company or a construction site, etcetera.

C. Robert Nelms, the founder of the Yahoo Root Cause Forum and the Principal Consultant at Failsafe Network, Inc., uses a three-tier level of Root Cause Analysis. He focuses on actionable understanding of human factors. One key to investigation is to have those who suffered the failure answer a question, "What is it about the way we do business that led to the failure?"

So now we have come full circle. A Failure Analysis is capable of identifying the physical reasons for a failure. A Root Cause Analysis helps identify human and cultural factors that lead to failure. As a Failure Analyst I am firstly concerned about the physical contributions to failure, but there is value in providing more – even when my client doesn't want to pay for a human factor root cause analysis. Many business leaders are not interested in redesigning the way the company does business. Providing differentiation in the definitions of Failure Analysis and Root Cause Analysis helps you to understand the motivations of your client.

4.12 Damage Categories, Mechanisms, and Modes

In this section, we briefly introduce my new integrated system for understanding the nested layers of terminology required to describe physical damage in structural components.

[87] David A. Levy, *Tools of Critical Thinking: Metathoughts for Psychology* | Chapter III, entitled "Multiple Levels of Description: The Simultaneity of Physical and Psychological Events".

4.12.1 Damage Categories

The primary damage categories, the types of things that can go wrong with physical parts, are 1) Deformation; 2) Fracture; 3) Wear; 4) Corrosion; and 5) Microstructure Damage. Here we distinguish these basic categories from damage mechanisms. We will acknowledge that much of the literature calls "damage mechanisms" failure mechanisms. We will address this further in Chapter 5.

4.12.2 Damage Mechanisms

Damage mechanisms as opposed to damage categories are complex processes that apply to particular alloy systems in particular environments and often are restricted to particular industries.

4.12.3 Fracture Mechanics – Mode

Many people use the term mechanism and mode interchangeably. Mode is specialized jargon in the field of fracture mechanics. Mode refers to a specific loading geometry. I recommend confining the use of Mode to its clearly defined Fracture Mechanics usage. I don't use it in Failure Analysis.

A. Mode I (one) is tensile loading of a part with a transverse crack.
B. Modes II (two) and III (three) are different orientations of shear loading.

There is no substitute for the word Mode in Fracture Mechanics. It is perfectly clear to those who work in this field.

4.13 Characterization and Process Condition

"What is it?"

This simple question is at the root of the concept of characterization.

> *When our children reach the age of grasping language, they gesture to what they want and we respond with a name. When we respond with a name, we are participating in the characterization processes. True, it's at the low end of the spectrum because we are providing little real information about the object, but we are providing a name that may be used to refer to it over and over again.*
>
> *When the same name is given for another object, we set the stage for the child to understand the idea of a category of similar items.*

People who work in materials testing labs may frequently be faced with similar questions. A client brings an object that was made by someone else and wants to know what it is "made of". Maybe they want to make their own version. How many dozens of times have I asked, "What do you want to know about it?"

How many times do they respond, "Just tell me what kind of steel it is!"

In some situations, this might work. Especially if the item has been machined from a piece of a standard bar stock, it might be possible to guess some of the strength levels. You determine the composition and match it to a common specification. You could supplement the composition analysis with a hardness test in order to gain a minimum amount of epistemological confidence in your determination.

But the essence of Materials Science and Engineering is learning to reveal the effects of the processing on the behavior of the finished product. See Figure 4.8. The Process determines the Structure, which determines the Properties.

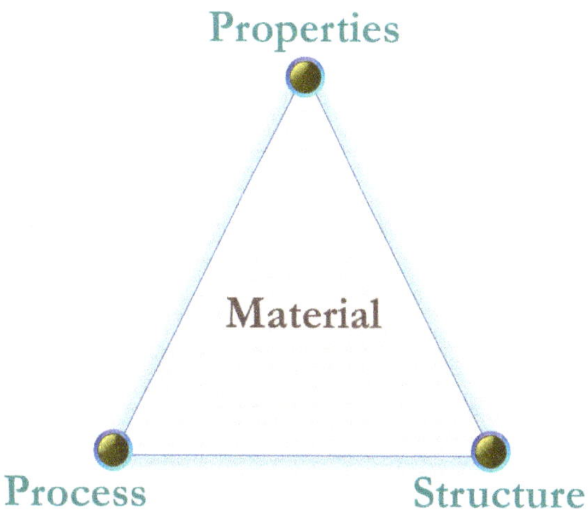

Figure 4.8: *Process-Structure-Properties Triangle*

- While machining doesn't usually change the properties of the material very much, if it is done with a dull tool, it can roughen the surface and result in a vastly diminished fatigue life.
- Any process that changes the shape of the product, either at room temperature or some other temperature, may have a large effect on the characteristics of the material.

Here is another point of view. From the manufactured-object's perspective, there is no single process. Rather, each minuscule volumetric element is experiencing its own set of processing parameters. Thus, an object may experience a subtle behavior change as a result of a complex machining operation.

For example, start with a 50 mm diameter piece of hot rolled bar stock. Cut it in half to make two equally long, full diameter parts. Machine-off a 1-mm layer (which leaves a bar of 48 mm diameter). On the other piece, machine-off a 4 mm layer (which leaves a bar of 42 mm diameter). It is likely that the average strength of the remaining material will be different from the original bar.

Medium carbon steel often has a low carbon layer on the surface. The lower carbon content, all other factors remaining equal, will tend to be associated with a lower strength. So, removing that layer may result in a higher overall strength.

However, steel tends to be stronger when it cools faster from the red heat used to produce a hot rolled product at the mill. So, the portions that had 4 mm removed may have lost their highest strength portions.

Now imagine you are starting with a completely unknown part. The part must have certain strength characteristics to perform its function. You must make an attempt to determine the strength at high stress locations. To do this, you must determine how the part is loaded.

If the small diameter portions are supported by a mating part, their strength level may be unimportant. But if they are not supported, these small diameter areas may experience the highest stress in service. Therefore, these are areas of interest. That's probably a good thing, because, in the case of a step-machined component, it would be much easier to get a tensile test coupon from the uninterrupted small diameter, than from the large diameter portions.

But wait! Most bar stock will be sold based on the strength at the mid-radius (half-way between the center and surface of a round bar) position.

So, how do you use the data from the test bar to answer the question: "What is it?" What should I tell the steel distribution center to sell? Do you ask for "Hot Rolled Carbon Steel, 50 mm diameter"? Or do you ask for, "Cold Finished Alloy Steel?" And, is it enough to specify Carbon or Alloy steel or do you need to specify a particular alloy composition? The question, "What is it?" is often a lot more complicated than most design engineers realize.

Is the answer to "What is it?" going to be used to specify material for a mass-produced product? A reliable design (as described in the section on Design Practices) covering comprehensive engineering design requires a thorough description of the material. And a reliable design requires forethought about processing effects.

You, the thinker, *must consider property variation as a function of position within the part.* ***You, the thinker,*** *must consider how process variations are experienced by each little layer.* ***You, the thinker,*** *must consider how a geometrical feature affects the structure and produces a behavioral tendency in varying circumstances.*

While characterization may be done for its own sake as a part of the design process, as I near completion of this book (2023), there are not many companies who routinely perform detailed materials engineering characterizations of their products. In fact, today, most companies are only willing to undertake the expense of doing this work after a painful failure.

Metallurgists and Materials Engineers have a term that summarizes the concept of property variation within a component. It's called "The Process Condition". Many industry standards for commodity metals and plastic bar stock are sold as the composition and description of how it was made. For example, if the last step of the sheet metal manufacturing was hot rolling, then it's called Hot Rolled Steel (or HRS). Likewise, if the last step was cold rolling, then it's called Cold Rolled Steel (or CRS).

NOTE: CRS is almost always HRS before it is cold rolled.

Thick, low carbon steel plate is sometimes sold as "Cold Rolled". I used to argue with tool and die apprentice metallurgy students that there is no rolling mill strong enough to squeeze a plate that thick at room temperature. Instead, it's hot rolled and then put through a cold mill to make the surface shiny. "Therefore," I argued, "the material should be called Cold Finished."

As a point of interest, this "Cold Finished" process has very little effect on the strength of the material when compared to true cold rolling, which can be performed on thin strip steel. During the true cold rolling process, the thickness is significantly reduced as the length of the strip increases; metallurgical changes accompany changes in the mechanical behavior.

Process conditions may be descriptive, but sometimes they are described in shop slang. You must know two things about this type of process condition description. One, you must know what the shop slang represents; and, two, that the words may not represent the true history of the material. I've included basic types of process conditions in Table C.

TABLE C: Types of Basic Process Conditions

Material/ Process	Metal	Polymer
As Solidified	Cast	Cast, Potted (e.g., two-part epoxies)
		Injection Molded
		Extruded, Spun (Fibers), Blow Molded (Thin-walled containers)
High Temperature Solid Shape Change	Hot Rolled (Plate, Sheet, Bar)	
	Forged (at least one dimension must be reduced)	
	Extruded (at least one dimension must be increased)	
Room Temperature Solid Shape Change	Drawn (Wire), Cold Rolled (Sheet and Bar), Cold Headed (Fasteners)	Stretched (Films)
		Spun (Fibers)

Characteristic structural features are created by many processes. For example, wrought metal parts, such as forgings and glass-filled injection molded polymer parts display flow lines that may be revealed in cross-sections. Figure 4.9 (next page) shows a cold forged steel test coupon made from a piece of round bar stock with flow lines.

A part made by another process such as casting will have totally different structural features.

Failure Analysts may often be called upon to confirm the process condition. In general, a medium technology forging will be much more structurally robust than a medium technology casting.

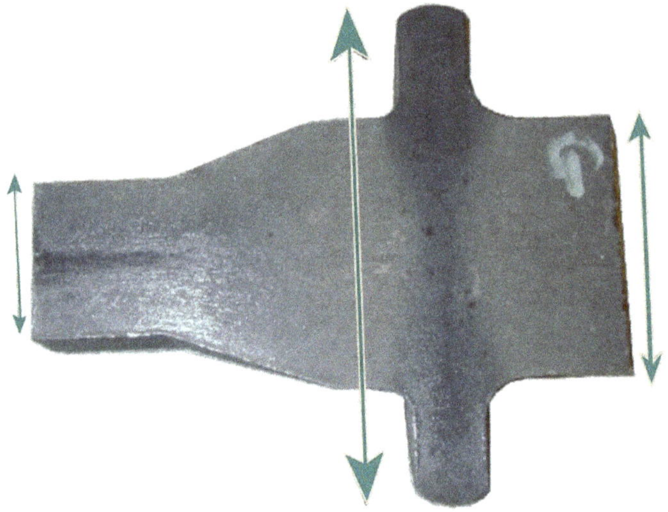

Figure 4.9: *Portion of a cold forged steel part sliced along a diameter*

The arrow at the far right shows the original bar diameter. The long arrow in the center shows the heavily worked, darker etching feature that has had its diameter enlarged by the forging process. The small arrow at the far left shows the portion that has been made smaller in diameter by the forging process. The cut surface was made to reveal the details of the metal flow by immersion in hot acid (macro-etching). We would expect different properties in the different zones of this component.

Figure 4.10 (below) shows a cross-section of a flattened ring-shaped part. The cut goes from the inside diameter (left) to the outside diameter (right). The visible cut was made along a true diameter (the cut would pass through the center of the ring on a radial line). The second cut was made parallel to it, which means it is not on a true diameter. It is along a non-diametral chord.

- Whenever you want to evaluate a component with cylindrical geometry along a longitudinal direction, it is often important to measure the size of features revealed by the cut.
- If the cut is along a true diameter, the measurements will be closely related to the actual sizes of the features.
- If the cut is not along a true diameter or radius, the measured values will be distorted or incorrect.

Also note in Figure 4.9 (above) that steel produced by a method that includes elongation in the solid state (or in the liquid state for polymers) has elongated features in the structure.

- It is not only steel; other materials produced by elongation in the solid state leave similar elongated features in the structure.
- Polymers produced by a method of elongation in the molten state also leave elongated features in the structure.

The original longitudinal (or rolling) direction of the bar stock used to make this forging was vertical as shown here in Figure 4.10 (below). The vertical lines have become curled as a result of the shape change. This provides the knowledgeable forging specialist information about how the shape change took place.

Figure 4.10: *Macro-etch coupon of a portion of a ring-shaped steel forging*

The arrow at the far left shows the inside diameter of the ring. The other three arrows show swirl features that mark aspects of the grain flow.

A common reason that Failure Analysts include macro-etch coupons in their investigations is to determine if there have been folds that were created on the surface of the part as it was deformed (forged). This fold is known as a lap; and can act as a damaging stress concentration if the stresses are high. See Figure 4.11 for an analogous example at the micro-scale.

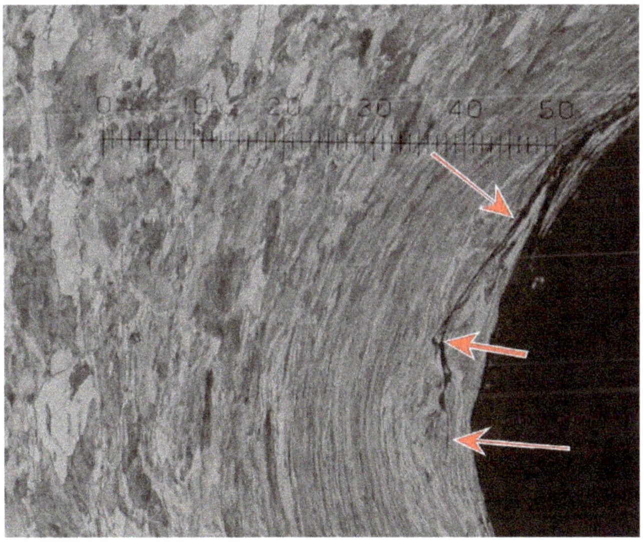

Figure 4.11: *Optical micrograph of lap in a bolt thread; micro-etched – magnification setting 200x; scale shows 250 microns (0.25 mm); smallest division is 5 microns.*

Lap discontinuities are often found in threads formed by rolling. This is a room temperature process. No heat is added besides what is created by the process itself.

Forgings have features that allow the material to reach closer to its natural high-strength range than other manufacturing processes. If a lap or fold ends up in a high stress location, the lap can act like an already initiated crack, thus reducing the durability of the component severely. Before we leave the subject of process conditions, please consider the concept of time at temperature. This is a key aspect of the process condition in general - and heat treating in particular.

Time at temperature is a very important concept in heat treating.

EXAMPLE: Imagine turning on a heat-treating furnace to 800 °C and allowing it to preheat until the furnace reaches the set point temperature as communicated by thermocouples at different locations. The load of parts to be heat treated is placed inside.

Obviously, if the parts were at room temperature, the entire furnace will be cooled down when the colder parts are loaded. Additionally, much of the heated air escapes. Furnace controls immediately work to warm the furnace back up, sometimes in steps.

The thermal conductivity of metallic parts is fairly high, but it is not infinite. Thus, it will take some amount of time for the parts to reach the desired 800 °C. That time will be longer for the interior than on the surface of the parts. If parts are large; if parts are loaded in many layers; or, if parts are in closely spaced layers, then some portions of the part or some parts, will heat up faster than others. This is why heat treating is partially art.

The heat treater must be familiar with the equipment they operate and consider the geometries presented.

They must select a thermal cycle program that allows all of the necessary material to reach the temperature required; and dwell long enough to allow the required diffusion to facilitate transformation to the desired microstructure - to produce the required characteristics. Overheating must also be avoided.

Time at temperature is a very important concept in heat treating.

4.14 Terminology Matters: What is a Defect?

Like some of the other terms we have considered in this chapter, the term "defect" has different meanings depending on who is using it.

Lawyers, quality engineers, building construction contractors, weld inspectors, radiographers, and consumers all use the term but may have different meanings attached to it. Similar to the word "failure", there are philosophical implications associated with the word defect. A Failure Analyst must understand the various uses in context.

4.14.1 Nondestructive Test / Evaluation Technicians

Nondestructive testing (NDT or NDE) technicians have their own terminology. Indications of discontinuities may be formally considered in their acceptance criteria specification documents to be defects. When there are interdisciplinary discussions going on, the failure analysis is well served by pointing out that the NDT people have their own definition of the word "defect".

4.14.2 Lawyers

The philosophical implications associated with the word defect are especially true on projects intended to provide support to attorneys with injured clients (due to a purported product defect).

- A defect is not something that we can necessarily see.
- It is not determined strictly as a result of sensory inputs to our brains.

In order to determine if something is a defect, we need to know what the object containing the (presumably objectionable) feature was supposed to do. We must know exactly how the feature facilitated the physical damage – or allowed the damage to happen sooner and/or more easily than expected.

4.14.3 Failure Investigators

For Failure Investigators and expert witnesses, we are expected to understand, apply, and communicate certain concepts. One of these concepts is: "We must know how the feature facilitated the physical damage – or allowed the damage to happen sooner and/or more easily than expected by a reasonable[88] user."

There is a chain of customers for information and the only person who can explain and manage this is the lead investigator. Each expert has their

[88] "Reasonable User" is a term of law in the USA. Even when a product is defective due to a design flaw, some courts will use one of two tests to find that the defendant has no liability. (1) Risk-Utility Test: the defendant is not liable for a design defect if evidence shows that the product's utility outweighs its inherent risk of harm. (2) Consumer Expectation: able consumer would find the product defective when using the product in a reasonable manner, or if a reasonable consumer would not find the product to be defective even when using it in a reasonable manner, then the defendant is not liable, even if the product's design flaw resulted in injury. | https://www.law.cornell.edu/wex/products_liability

working definition of defect. Yet, the Failure Analyst is the only person with a broad enough perspective to explain to a lawyer or to a jury the relevance of the feature and whether it is truly acting as a defect or merely an irrelevant discontinuity or feature. Indications are even spookier in that they are sometimes artifacts of the test method and represent nothing specific about the physical Subject Part. Careful discipline is required before allowing an NDT/NDE Technician to use the word defect in a forensic record.

4.14.4 Quality Engineers

For quality engineers, a defect is a characteristic of a component that does not meet a specific product requirement. In low and medium technology manufacturing industries, quality engineers are not usually expected to be able to predict the detailed effects of structural defects.

When a defect is strictly an appearance issue, such as a paint blemish in an automotive body panel, it is fairly straightforward to correctly conclude that the customer wants their new car to look absolutely perfect.

As soon as we start dealing with structural discontinuities that are invisible, it's a different story.

4.14.5 Here Are a Few Scenarios

- A slight thickness variation in a sheet metal stamping used on consumer products allows excessive deformation, but would not be visible to the average consumer. Is this a defect?
- What if the stamping is a housing for a power tool and allows the housing to be more easily dented? Is this a defect?
- What if the more easily-dented housing causes electrical interference, but the user doesn't realize the dents are more than normal wear-and-tear and the electrical interference is not a problem in using the product. Is this a defect?
- What if the dented housing causes electrical interference and the user is electrocuted as they work? Is this a defect?

How do we determine that the stamping deformed "too easily"? Maybe this is a bad example, as the wiring is supposed to be safe even in the case of a dented housing. Maybe all housings on such power tools are rigid plastic now, not metal. Maybe, maybe, maybe. The point is that it will be argued, and must be argued, in the United States of America under current law, based on someone's idea of reasonable.

EXAMPLE: If the damage category is fracture, we have a more straightforward situation. This is because with fracture, we are going to have one or more crack initiation locations. So, if the purported defect, let's say a void in a casting, is not at the crack initiation, we can end the discussion about the possibility of a defect having facilitated the fracture. It's not a defect in this case. It's a discontinuity. All materials have some type of discontinuity.

EXAMPLE: If the damage category is fracture and the purported defect is at the crack initiation, then we might be able to make a reasonable case that the fracture was facilitated by the discontinuity. But does that make it a defect?

If the crack initiation is where it would normally be due to "normal wear and tear", then we can't declare it a defect strictly based on its association with the crack initiation. We'd have to prove quantitatively, that the discontinuity's specific size and shape would reduce the strength or durability by a given amount. Then we might have "ammunition" for our argument that the product had a "defect".

It is often difficult and expensive to make a realistic computer model to prove such a premise. Testing with intentional defects would also certainly be more time consuming and expensive than simply performing a visual examination of the fracture surface and pointing out a suspicious looking feature.

Remember, a defect or even a defect that results in an injury does not always imply liability.

Let's take a quick look at alternative words that might be used in this context if we do not want to use the term defect. All material objects have features or characteristics. Characteristics that make the part what it is include the following:

- dimensions,
- alloy composition,
- paint thickness,
- colors,
- weight,
- reflectivity and more.

Features may also refer to typical discontinuities associated with the manufacturing processes used to make the object. For example, castings will generally have porosity.

- The porosity will not be a problem if the specific casting process was properly carried out. Non-damaging porosity is a normal characteristic of most casting processes.
- Porosity is a feature of most casting processes and if a Failure Analyst used radiography to inspect a casting, the radiographer may find an "indication" that there is porosity in the casting. The radiographer must then determine if this indication is permitted or not. Technically, it's not the radiographer's job to even state that there is porosity present in the casting. All the radiographer should be concluding is that there is an indication of the "porosity type".

4.14.6 Critical Thinking

Before leaving this topic, we should acknowledge that materials science students learn about the presence of crystalline "defects" in almost all materials.

- These may be submicroscopic such as a missing atom.
- They may be submicroscopic in the form of a row of missing atoms. This type of crystal defect is known as a dislocation.

  I describe dislocations and their consequences in Chapters 5 and 8.

But in no way are either missing atoms or dislocations to be considered manufacturing defects of any type. They are simply present by virtue of the way crystals are.

This is knowledge, but critical thinking involves a context. Humans are often careless with the frame of reference. So, how many materials science students stop and question themselves about using the word defect when referring to a plane of missing atoms? How many inexperienced Failure Analysts think it is appropriate to detail this in a failure analysis report?

Don't do it because it will just confuse everyone. Critical thinking tells us to take care to avoid using confusing words out of context. Dislocations are good. They allow us to make materials stronger than they otherwise would be, while still allowing us to form and machine them readily.

Missing atoms are good. They allow the use of heat treating to tailor material characteristics to requirements during manufacture and, separately, during use.

Consider your word choice carefully before using the word "defect"!

4.15 Manufacturing Considerations

Desirable microstructures of modern materials are obtained by processing materials according to proven methods. The benefits of modern quality science, aided in large part by application of statistical methods, have accrued to those material producers who have learned to tweak their processes to manufacture raw materials and products within a desirably optimized range of characteristics.

> EXAMPLE: *High strength low alloy (HSLA) steels are often sold to minimum yield and tensile strengths and minimum elongation values.*
>
> *This dual minimum requirement, for both strength and ductility, has created economic benefits for the*

producers; but it has created the occasional headache for many purchasers of small quantities of these materials. Small buyers can include anyone who isn't purchasing entire heats of steel (say 150 tons) on a monthly or weekly basis!

How does this tweak affect the small buyers? The optimization of tight controls on rolling temperature, thickness reduction schedule, and composition have enabled mills to optimize their material to meet both the highest strength class requirements and ductility requirements of the lowest strength class. Much of this optimized material is for flat rolled sheet and plate intended to be cold formed (stamped). Companies purchase the material anticipating a moderate strength level, only to find that it is much higher strength than they expected. Many small buyers experience the following consequences:

- *Expensive tooling changes.*
- *Higher tonnage presses to cause the desired shape changes.*
- *Higher frequency of splits or tears.*

As stated previously, this scheme benefits the producing mill because they can make one strength class and sell it as any of a whole range of strength class materials. While it does indeed meet the "letter of the law", it strongly deviates from the "spirit of the law". The experts who negotiated the original industry specifications for high strength low alloy steels developed them in the late 1800's.

For damage prevention, exert care in the specification of materials and consider how processing affects strength and other important material factors. It requires understanding and expertise to specify a material for a particular application. Be wise and seek expert counsel to avoid manufacturing and processing problems.

4.16 Mechanical Behavior – a Critical Concept in Failure Analysis

This chapter closes with the important concept and term "Mechanical Behavior". In short, mechanical behavior refers to how materials respond to forces. Materials deform as a result of forces interacting with them.

If the force is low, deformation (shape change) is not permanent and we call the deformation elastic. This shape change is like a rubber band that snaps back to its original size after being stretched.

At higher stress levels, the material may either break or permanently change shape.

> EXAMPLE: *Think of a piece of taffy. When it is warm, it deforms readily as you stretch it. After it is stretched and if you put it down, it does not return to its original shape because it has plastically deformed* [89]
>
> *But, if you put a piece of taffy in the freezer and try to stretch it out, it will break into tiny pieces. The taffy has broken in a brittle manner.*
>
> *In general, all other conditions staying the same, less force is required to break a part that can't deform permanently. If a material is capable of permanent shape change, it will require more force to break.*

IMPORTANT: Brittle and Ductile Behaviors are Condition and Process Dependent

- A part that is made from a material that can deform permanently before it fragments is called ductile.
- A part that has very limited ability to deform before fragmenting (breaking into pieces) is called brittle.

Processing errors and environmental damage occurring at room temperature may suppress a material's ability to permanently deform. When this happens in a material that is expected to deform permanently before breaking, we call it "embrittlement". Sometimes, low temperatures acting alone suppress ductility.

[89] People who have studied critical thinking may have noticed this is a tautology; more on this in Chapter 7.

Alternatively, other processing errors and high temperature damage may leave the material more susceptible to permanent shape change at lower forces than the designers anticipated.

Materials that deform leading up to fracture (as they are breaking) are known as ductile materials. Ductile materials may behave in a ductile manner (permanently change shape). But, as stated above, if something goes wrong, they may behave in a brittle manner (without accompanying permanent shape change).

Some materials are inherently brittle. They respond to low levels of stress by deforming elastically and going back to their original shape when the load is released. If their fracture strength is exceeded, they will separate in a brittle manner.

If the separation is a result of stress alone, rather than stress and a corrosive environment, the brittle crack may propagate at the speed of sound in the medium. Large structures (bridges, buildings, airplane parts, heavy machinery) can be subject to this type of brittle cracking. When it happens, it's almost always unexpected and generally a big problem.

> *EXAMPLE OF A VERY BRITTLE ROAD SIGN: So... I was driving home one winter night during an ice storm. The event I am about to describe is so traumatic that I rarely, if ever, drive during an ice storm now.*
>
> *It was zero degrees Celsius (that's always the temperature during an ice storm). The road was very slippery. I was going less than 20 km/hour. But I slid off the road and hit a road sign whose post was made from a channel of roll formed steel.*
>
> *I had managed to slow the car a bit but when I hit the sign, it made a clinking sound and snapped off at the base. The separation was the result of stress from my car and a cold environment; the brittle crack propagated at the speed of sound that I heard as the clink. That is the technical side of the story.*
>
> *Luckily for me, a tow truck company owner saw me slide off the road and came to help. He did not even charge me a fee. He got my car back on the road and I made it home.*
>
> *I thought the accident was the worst of my day, but when I arrived home, I received bad news that disrupted my intention to report the accident. My spouse informed me a childhood friend had died. His wife couldn't find my phone number and my friend was already buried and I had missed the funeral. Obviously, I was very upset.*
>
> *I reported the accident to my insurance company a day and a half later. The agent told me to call the police and tell them I broke the sign so they could write a report and the insurance company would pay to replace it.*
>
> *So, off I went to the police station and reported the accident.*
>
> *To my utter surprise, they decided I must have been consuming alcohol (I was not!) and wrote me a ticket for leaving the scene of an accident. That's the USA Injustice System in action. They write you up for something worse than you did so you can "plea bargain" down to what you actually did (hit a sign). The officer's writing was so terrible I misread the date and the court declared I failed to show and issued a warrant for my arrest!*
>
> *Thank goodness I decided to ask a lawyer to handle the court hearing. That's when I found out I had a warrant against me. The lessons ("meta-points") here are two-fold:*
>
> *1) The State of Michigan apparently has no effective specification for requiring ductile behavior of the steel they use for road signs, even at moderately low temperatures, despite the fact that parts of the state routinely achieve -30 °C degrees.*
>
> *2) Minor damage sometimes results in severe "consequential" damages. In this case, the poor steel specification resulted in my almost having to go to jail, and did result in \$1000$^{\underline{00}}$ of fees and expenses, and a LOT of anxiety!*
>
> *3) I learned one other major lesson from this traumatic event. I totally lost my courage to speak out and speak up for others. I felt and thought that I had to keep a low profile. My plea deal included*

> *wiping my record clean after three years. My fear pushed me to develop compassion and empathy for the vast majority of humans who routinely live life so much closer to the edge.*

Ductile materials, in general, are more forgiving of sudden (unplanned) loading than brittle materials. Environmental conditions may prevent materials from responding in a ductile manner despite their theoretical ability to do so.

Whether a fracturing material responds to a force in a ductile or brittle manner depends on three basic groups of factors.[90]

1. Rapid loading (high strain rate) can initiate brittle behavior.
2. Low temperatures will promote a tendency to brittle behavior.
3. Certain types of damaging environmental substances may also suppress ductility and promote brittle behavior.

Many products are subject to room temperature durability tests before they are released to the marketplace for use. If the product is then subjected to a low temperature "excursion" in service, it may respond surprisingly in a brittle manner.

SUMMARY: The readers of this book must come to have a deep understanding of the two categories of mechanical behavior: ductile and brittle. These are among the most important aspects of mastering engineering design and failure analysis of structures.

[90] Rapid loading and low temperature fractures are often indistinguishable based on evaluation of their appearance. In the absence of background data, it may be impossible to pick one over the other as the primary causal factor.

CHAPTER 5

Damage Classification

The Processes of Damage

5.1 Damage Classification Overview

Nothing lasts forever, not the tallest mountains nor the largest galaxies. Change is part of our dynamic universe. Living things are perhaps one small subset of the physical realm that, for a time at least, seem to defy the Second Law of Thermodynamics.[91]

The Second Law of Thermodynamics tells us that in natural conditions, entropy (disorder) will always tend to increase. Babies start out as a few cells and become more ordered as they mature into children and then adults. As we progress through time, the body accumulates damage. Our bodies are subject to stresses and strains. Inflammation impairs our range of motion and dulls the sparkling color of our youthful skin. Our blood vessels may get clogged and our internal organs lose function due to damage or disease. Eventually, the body dies.

> *My dear, very hard-headed mother passed away just before her 83rd birthday due to a fracture of her skull, as a result of a fall, which was due to damaged nerves in her feet that left her with no sensation of the floor. If we transferred this to the world of inanimate objects, we would say that there were consequential damages (broken skull) which resulted from the initial damage (nerve damage in feet due to complications of diabetes.) My mother loved and ate too much chocolate! Hard headedness is an acknowledged, if bemoaned, trait on both sides of my family.*

Healthier humans are often able to recover, to some extent, when damage occurs to the body. In

[91] Henry Schrödinger, in his effort to connect entropy with biological systems, postulated that order may come from disorder. He later concluded it does not. | Web Search: Life as a Manifestation of the Second Law of Thermodynamics by E. D. SCHNEIDER and J. J. KAY, Page 26; ScienceDirect.com Asset PII: 0895-7177(94)90188-0 |

the case of a mechanical injury such as a broken bone or a torn muscle and scrapes or cuts, humans and animals have some ability to reknit tissues. In the case of other imbalances (diabetes, electrical breakdown between neurons, bacterial or viral infections), medications or dietary and lifestyle changes can slow or reverse the damage. Deposits lining our blood vessels may be considered mechanical blockages resulting from chemical imbalance. Doctors sometimes address the consequences of this type of "damage" with mechanical means.

My dad had his carotid arteries scraped out. His doctor also prescribed medications that slow the deposition of the harmful materials in the first place.

See the deformation damage in a hollow cylindrical structure shown in Figure 5.1 (below).

5.1A *- Context* ***5.1B*** *– Close-up*

Figure 5.1: *Note the readily visible crumpling of the rust-toned sheet steel in this hollow cylindrical structure. This deformation damage was facilitated by the sheet-thinning through corrosion on the inner surfaces.*

Of course, we humans are more than our bodies, more than the physical aspects which support our lives in earthly society. Still, there are similarities. Illness and injury in living beings have their parallels in the basic types of damage to inanimate objects. In short, a damaged thing (be it living or not) has experienced a detrimental change.

For the purposes of this book, we are not so concerned with how we might repair the damage of a given thing. As Failure Analysts, we're most concerned with sequencing the events that resulted in the damage and communicating it. Now it's your turn.

EXERCISE 5-A: How would you categorize the damage due to a cancerous tumor? Consider both primary and consequential damage.

Sometimes the change due to damage is obvious at first glance and sometimes it's not. If we don't know what the object is used for, often we can't tell what's wrong with it.

That is why our investigations must start with visual inspection and background data collection. The details of how to do this are covered in multiple other sections of this book. We've already described the *looking-at* process in Chapter 3 and how to improve our skills at seeing what is in front of our eyes.

In this exercise, one of the major features we'll be *looking-for* is evidence that allows us to assign a damage category. This is so fundamental that it almost seems too obvious to comment about; but sometimes it is not obvious at all. When we clearly state the evidence (what is there for all to see), that's when we receive benefit from our damage category assignment. We are acting toward building a consensus about what happened to the object - or at least about the realm of possible scenarios regarding what happened.

Let's be clear here. Our goal is not to encourage mindless group think; our goal is to have an informed, intelligently reviewed set of evidence that begins to speak for itself, especially to those who have prepared themselves to hear and understand.

Early in our careers, we need to constantly remind ourselves: **damage is a natural consequence of the passage of time.** This is entropy in action. I constantly remind myself of that fact when I view manufactured items and civil constructions. I'm in good company. The Buddha noted, "All compound things are impermanent."

For some objects, the noticeable damage might take centuries to happen, so we personally might feel that we do not have to worry about it.[92] For other items, those subject to daily use in harsh environments, damage may start the moment that the object is placed in service.

Before we move on to the details of a damage classification system, let's take a moment to recall the difference between damage and failure. Make yourself some notes and then go back to the last chapter to see if you got it right!

EXERCISE 5-B: Damage is different from failure in the following ways (fill in the blank).

____

____

____

____

____

____

At the most basic level, all damage may be assigned a position on the spectrum of mechanical and environmental damage. Of course, the subject of the investigation has always experienced an environment. If it is part of a vehicle constructed on the moon for use in interplanetary or interstellar travel, the atmosphere might be vacuum plus streams of various and varying sub-atomic particles. The environment might include the chance of a collision with space junk. If there is a collision with space junk, we may have mechanical damage. And we need to remember that objects in outer space experience extreme temperatures! The point is, if we really want to understand damage, we need to consider the environment as well as both the intentional and unanticipated loads that the structure must bear.

Now let's assume the structure or machinery is to be used in Antarctica. Cold temperatures (environment) will surely influence the manner in which any mechanical damage happens even if it doesn't directly cause the damage.

[92] The damage to the earth's environment at its current severity level is due, in large part, to an attitude of not worrying about it. After all, how could a few little humans contaminate the ocean or the atmosphere?

EXERCISE 5-C: Was this damage to the handrail mechanical or environmental?

Years ago, I was asked to analyze an aluminum handrail (previously mentioned in Chapter 4, Section 4.7) that had been found with a classical macro-ductile fracture a few days after it was installed. The vertical members of the structure were cemented into the ground. There had been a cold snap. The horizontal members tried to contract in length and could not. As the temperature dropped, the stresses eventually exceeded the tensile strength. Was this damage mechanical or environmental?

Maybe a better way to think about damage categorization is whether it is a result of intentional or incidental stressors. The handrail designers in the example above had originally specified steel. Later, the owners of the property asked them to change to aluminum, which would not require painting. Nobody thought to check the coefficients of thermal expansion for the two metals.

Handrails, for most of us who casually use them, are considered "good" when they are rigid enough and "bad" when they are flimsy. But apparently this was not the first time such an event had happened. The structural code does give guidelines for lengths of handrails in our part of the state of Michigan, which does sometimes experience a wide spectrum of temperatures over the course of an average year. There were guidelines for steel, and a different set of guidelines for aluminum.

Did your classification consider more than one type of damage? Read on.

Other desirable handrail characteristics would be a surface finish that is agreeable to touch and, perhaps as the world population worries more about communicable diseases, antimicrobial. If that was the intended context and a specific handrail had been found to sicken a few hundred people, the handrail could be considered a failure even if it is itself completely undamaged. Now it's your turn. After all of the previous information, what damage category would you choose for this handrail-health calamity?

I hope you embraced thinking beyond a dichotomous method of damage categorization of mechanical or environmental.

☯☯☯☯☯☯☯☯☯☯☯☯☯☯☯☯☯☯☯☯☯☯☯☯☯☯☯☯☯☯☯☯☯☯☯☯

DAMAGE CLASSIFICATION IN 5 STEPS

Questions to Guide the Investigator

Step 1: Clarify Context – What did the part have to do? What was it made of?

Step 2: Compare-Contrast – What's wrong with it? What is the type of damage?

Step 3: Connect the Data Points – Create a possible history, a working hypothesis, for the sequence of damage.

Step 4: Contemplate Complexity – Are there other possible scenarios?

Step 5: Calibrate Your Confidence – You have learned what happened, and possibly have some ideas on how to prevent recurrence. You can explain your findings, and any remaining uncertainties, and solicit alternative ideas or agreement with your proposed scenario.

EXERCISE 5-D: Consider and Reflect on Damage Classifications in 5 Steps (Contemplate Complexity!). As you contemplate my words in the text box above, make notes on your reflections.

Every physical object, structure, assembly, etcetera., is subject to both environmental conditions and mechanical situations, which may interfere with their ability to perform the functions for which they were constructed. Often, it is impossible to assign damage to one single category. So, we begin our investigation by going down a common path, letting the visible damage classification and background data inform us.

- If it broke, we're going to begin by calling it a fracture, because doing so points us toward a set of tools that will help us understand what happened and maybe help us understand why it happened.
- If the thickness of the material appears to have been reduced by chemical action prior to the fracture, we're going to call it corrosion, because doing so points us toward a set of tools that will help us understand what happened, why it happened, and maybe what to do to prevent it happening again.

The main purpose of damage categorization is not necessarily to reveal an ultimate truth about what happened. The primary purpose of damage categorization is to take us to the next step down the path of understanding what transpired to facilitate that damage.[93] This allows us to find meaningful answers that fulfill the purpose of the investigation. This may be to assign responsibility for the payment of property damage or to prevent similar future damage, among other purposes.

5.2 Damage Categorization: Nested Layers of Terminology

The damage categories for structures and structural components come in five basic varieties:

- Deformation
- Fracture
- Wear
- Corrosion
- Microstructure Degradation

Deformation, Fracture and Wear are three different types of mechanical damage. Corrosion and Microstructure Degradation are two different types of environmental damage.

5.2.1 Damage Categories, Mechanisms and Modes

Let's bring our attention back (from Chapter 4) to the integrated system for understanding the nested layers of terminology required to describe physical damage in structural components.

5.2.2 Damage Categories versus Mechanisms

Here we distinguish the primary damage categories from damage mechanisms. We will acknowledge that much of the literature calls "damage mechanisms" failure mechanisms. They are conflating process damage with the end state of a non-functional structure. However, so you are very clear about the differences, we will use the term damage to refer to minor and major, or catastrophic damage alike.

[93] In some cases, usually driven by government or industry sponsored safety codes, it is necessary to assign a specific, known damage mechanism. In others, understanding how to prevent a recurrence is all that matters.

Damage mechanisms, as opposed to damage categories, are complex processes that apply to particular alloy systems in particular environments and are often restricted to particular industries. For example, chloride Stress Corrosion Cracking (SCC) of austenitic stainless steels in the chemical process industry differs from SCC in the boiler world.

SCC in the chemical industries is generally acknowledged to be a hydrogen embrittlement process, while SCC that degrades boiler components is related to crack tip oxidation.

5.2.3 Crack Propagation Type

To distinguish fatigue (progressive cracking due to cyclic loading, such as from vibrations) and monotonic overload (immediate crack) versus sustained loading (a single long-lasting load like creep or classical hydrogen embrittlement in steels), consider the term *crack propagation type.* This links loading conditions (based on their variation in time) with macro-scale crack appearance.

5.2.4 Crack Path Description

When referring to the micro-scale, I find the term *crack path* helpful to understand and convey the details of micro-structural interactions with the crack. This could include intergranular, cleavage, and other fundamental crack behavior preferences. *Crack path identification is usually a crucial aspect of determining the damage mechanism.*

5.2.5 Nested Fracture Description

See what we have accomplished! We now have a nested structure to link loading conditions (based on their variation in time) and industry specific damage mechanisms as subsets of the bigger damage categories. Later, we will add loading geometries to this system.

As you proceed through Chapter 5 and Chapter 8 more details about the integrated nested layers for the other damage categories will be revealed.

EXERCISE 5-E: On a piece of paper or in the box below, sketch the nested layers relevant to describing damage of physical structures. You may wish to update your sketch as you learn more.

5.3 Mechanical Damage Categories (Deformation, Fracture, Wear)

The simplest damage categories to understand are deformation, fracture, and wear. These three are ultimately due to the structure experiencing a stress level higher than it could tolerate. In other words, the stresses were higher than the strength. These are the basic categories of mechanical damage. They result in a change of shape, which may be localized or widespread. It may include temporary or permanent deformation. It may include fragmentation. Or the change of shape could be a combination of any of these dichotomous pairs:

- localized/generalized,
- temporary/permanent,
- deformation, fragmentation.

Often, a combination of these occur before someone notices damage.

IMPORTANT: The important thing to remember for now is that stress and strength are general terms. In order to use them effectively, we need to consider the way the relevant type of stress and strength were determined.

Figure 5.2 (next page) shows a zinc coated steel bolt that has sustained damage. As you can note, the language in the previous sentence doesn't reveal much useful information. "The stress exceeded the strength" is tautological. That means it doesn't contain real information; it only reinforces definitions.

When observing the bolt below, can you make a more useful statement, by further describing the type of stress and strength? (Maybe not on the first read!)

Figure 5.2: *This steel bolt is broken and the coating is worn (circumferential band) below the crack surface.*

SELECTED MECHANICAL PROPERTIES

Tensile Strength: Resistance to Fracture at fairly slow loading rates

Yield Strength: Resistance to permanent Shape Change (Plastic Deformation) at fairly slow loading rates

Creep Rate: Resistance to Slow Plastic Deformation (permanent shape change) at temperatures above 40% of the absolute melting temperature (degrees Kelvin).

Impact Strength: Measured in terms of energy absorbed during a rapid fracture, often at low temperatures, especially when evaluating steel toughness.

High-Cycle Fatigue Strength: Resistance to Fracture at loads that are below the nominal Yield Strength. This impedes plastic deformation. Locally though, the stress is above the tensile strength. High-cycle fatigue generally takes hundreds of thousands or millions of repeated load cycles before a crack initiates.

Modulus of Elasticity: Modulus is related to flexibility or stiffness. It is not a strength value, but a ratio of the stress to the strain. The numerical value for the modulus helps predict shape change. The range of stress in which the modulus of elasticity allows us to quantitatively predict shape change is from zero up to a value just below the yield strength. The modulus varies with the direction of atomic bonds within a crystal. If you look up a value for the modulus of a metal, it will usually refer to an average value of randomly oriented crystals.

Types of strength include tensile strength, yield strength, impact strength, creep strength, and fatigue strength. Each of these may carry requirements for further sub-categorization.

Wear is a complex process and not a singular or uniform process. We are making a deeper dive here because of that complexity. Wear may be viewed as a result of micro-scale processes involving stress. Factors influencing the specifics of a wear process may include contamination, lubricant degradation, corrosion, mechanical problems (such as misalignment or vibrations), excess friction, high temperatures and associated accelerated damage that comes along with high temperatures. All of these factors, and others, facilitate the wear process.

Pump components are subject to a whole range of additional specific wear damage processes, including cavitation, which is due to fluid pressure changes.

This book is about failure analysis of solid objects, so we won't be spending time on the "hows" of lubricant degradation. But you might imagine they are related to high stresses, high temperature, composition change, etcetera. In other words, the same factors that cause the damage in solid objects are those that operate to degrade lubricants.

5.4 Environmental Damage Categories (Corrosion and Microstructure Degradation)

Every part exists in an environment. Thus, this category affects every other type of damage. It may be an accelerant, a mitigator, or neutral. Every fracture, deformation, wear, or corrosion event happens at some temperature in the presence of a chemical environment. That environment may be the void of outer space; it may be the air you are now breathing. *Paying attention to the environment helps you understand the damage process in a holistic manner.*

Aqueous corrosion (happening in a moist, wet or water containing environment) is considered to be electrochemical in nature. The electrochemical reactions may be present at the micro-scale and may be reacting to microscopic differences in composition within a single component; or, the electrochemical reactions may be present at the macro-scale (such as in large components), which together create galvanic couples due to the presence of incompatible metals or alloys.

Corrosion has many different meanings. Some people lump any detrimental change due environmental factors into the category of "corrosion". But I find it useful to separate microstructure damage due to heat or other energy sources from chemically induced corrosion.

Microstructure damage may have little evidence of material loss or corrosion product buildup. Microstructure damage is generally not readily observable to the unaided human senses until subsequent (consequential) mechanical damage reveals that there is a problem. Many, but not all, corrosion processes immediately produce visible changes, including the "exemplar" corrosion process known as rust!

- "Plain" corrosion is environmental damage that involves chemical reactions between the Subject Component and environmental substances.
- Chemical interactions in the electron clouds of a material facilitate corrosion.
- Changes to a material resulting from energy impingement facilitate micro-structure degradation.
- Microstructure degradation may involve chemical reactions, but the elements involved are initially part of the material itself, not a foreign substance. As the damage progresses, the action of chemical interactions involving foreign atoms may accelerate the damage.

5.4.1 Clarify Your Thinking

Consider microstructure damage as coming primarily from excess energy rather than aggressive substances.

Microstructure Degradation involves physical changes to the integrity of the material due to energy impingement. This can be polymer chain scission due to ultraviolet light or loss of yield strength due to rounding or enlargement of second phase particles in an alloy steel.

This category also includes damage to microstructural integrity of nuclear reactor components subject to bombardment by high energy sub-atomic particles. Keep in mind, this last example is not a clear-cut case, because matter is put into the material, along with the energy.

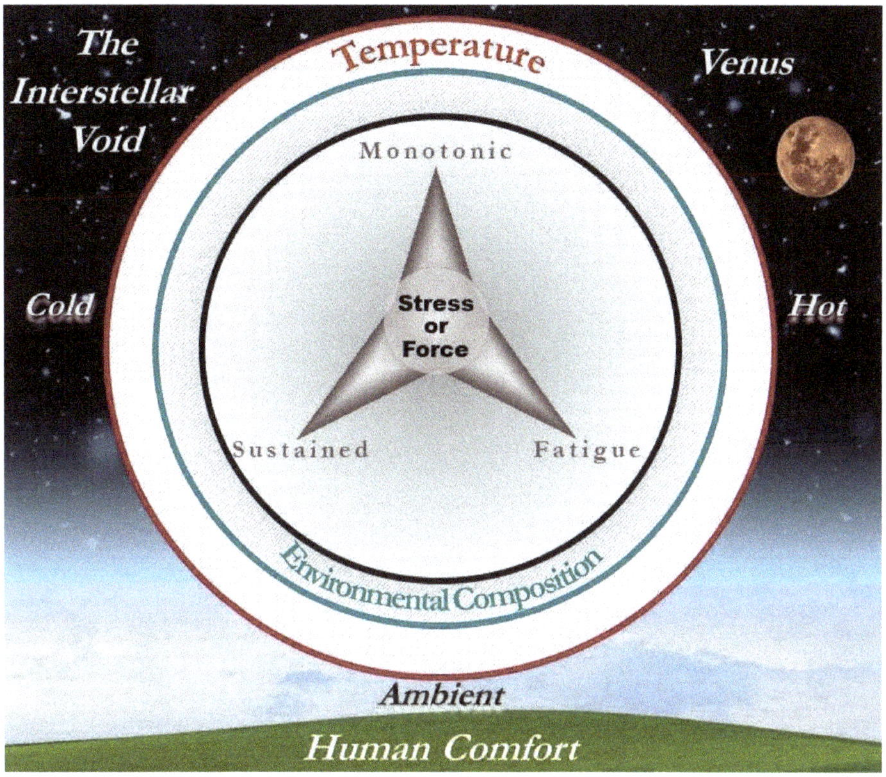

Figure 5.3: *Environmental and temperature composition are features of the physical world that facilitate damage.*[94]

Other examples of microstructure damage might be incipient melting at the grain boundaries of certain aluminum alloys (see the aluminum strap example on the next page in Figure 5.4), and creep in high temperature steam lines in power generation equipment.

Categories of damage exist mainly to give us a structure on which to build our understanding. To some degree, classifying damage to a structure always involves arbitrary distinctions.

Microstructure damage is complex. Heat, UV, and neutron radiation may all facilitate microstructure damage, separately or together.

5.5 Complex - The Final Damage Category ("Damage Mechanisms")

We will define Complex as damage due to service condition interactions where the environmental and mechanical factors synergistically degrade the functionality of a component.

The definition above is the essence of a damage mechanism. Some refer to this as a "failure mechanism" but again this would conflate the end state with a process.

The way the damage happens is affected by the environment. Usually, the damage results in a dramatic reduction in mechanical strength.

Stress Corrosion Cracking (SCC, in metals) and Stress Cracking (in polymers) could be considered

[94] For an alternative graphical viewpoint created by Aaron Tanzer, See *ASM Handbook, Vol. 11, 2002*, Section 4

Complex Damage. Other members of the Complex Damage group would include other "mechanisms" that we started familiarizing ourselves with in Chapter 4.

By Damage Mechanisms, I mean a particular material, perhaps in a particular process condition, whose integrity has been compromised by a particular substance to which it "is allergic". Crazing of polycarbonate in the presence of organic solvents and tensile stresses could be considered an example of complex damage, or alternatively as a damage mechanism. Season cracking of brasses in the presence of ammonia is another example. Many Damage Mechanisms are associated with specific industries or applications. The damage itself often has characteristic features. For example, chloride induced stress corrosion cracking in austenitic stainless steel usually has a branching crack growth habit. Many complex damage mechanisms are associated with particular microstructure features. Industry specific damage mechanisms may use identical terminology that means something very different in different industries, such as the example in Chapter 4 related to stress corrosion cracking. **REMEMBER** to keep separate the concepts of a damage mechanism (Liquid Metal Embrittlement) and a damage category (Wear).

The damage in Figure 5.4 (below) could be considered complex. The description in the figure depicts how information was revealed as the analyst changed scale during the investigation. We began with a visual inspection of an aluminum strip which had four intentional hollow cone features. The strip exhibited unexpected bumps. These bumps were noticed after heat treatment. The moderate magnification cross-section image (lower left) shows that the bumps are blisters.

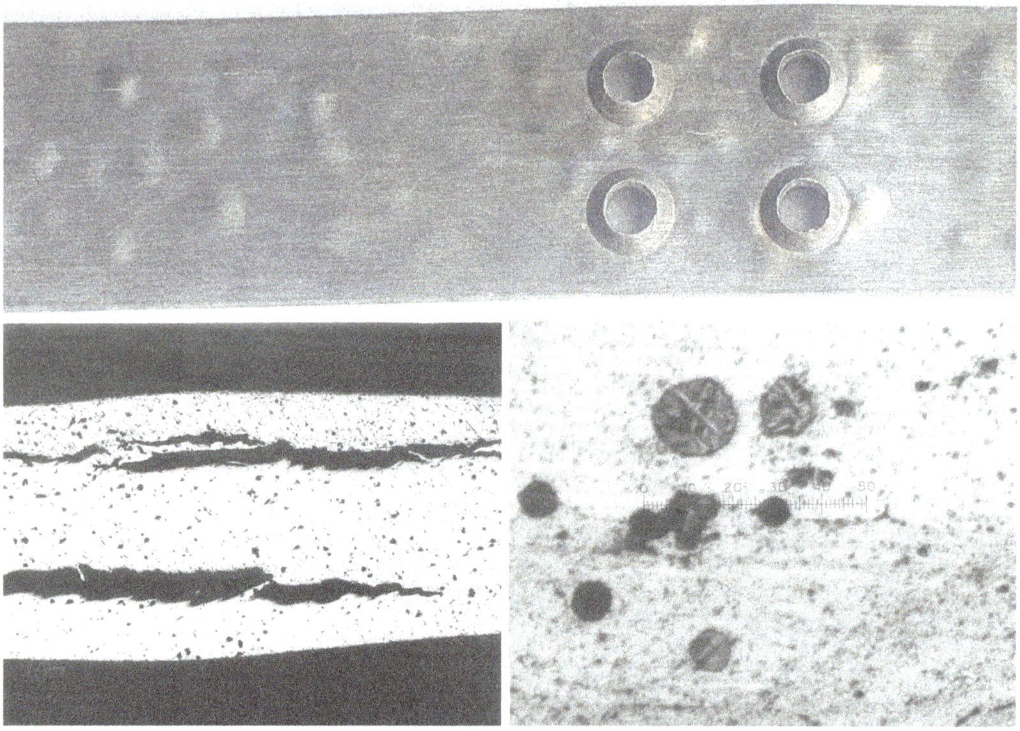

Figure 5.4: *The* ***upper image*** *shows a bumpy aluminum strip.*

Below left *shows a metallographic cross-section, which reveals internal cracks found to be associated with the bump locations. Note, also, the small dark dots.*

Below right *(higher magnification) reveals that the small dark dots visible at left have dendrites, not expected in wrought metal. This indicates microstructure damage due to excessive temperature exposure (local melting and re-solidification). Could the high temperature have caused the cracks as well?*

Continuing to look at Figure 5.4 we see the highest magnification image shown (lower right). Notice the needle-type features, called dendrites, within the larger dark circles. These are solidification features and imply that there was melting during heat treating. Dendrites will be discussed in Chapter 6.

Let's leave the complex categorization aside for now, and get back to corrosion.

5.6 Back to Corrosion and Microstructure Damage

Words are important! They reflect your clarity of thought, or lack thereof! Expert Failure Analysts know the difference between a deposit and a corrosion product.

5.6.1 Definition: Corrosion Product

The solid material (usually oxidized) that adheres to the remains of a part, or flakes away. A well-known type is red rust; it forms on carbon steel; it is a corrosion product. Some corrosion products are sulfur based, or are complex compounds of other elements.

5.6.2 Definition: Deposit

The term, "Deposit" is the collection of the foreign particles that are not the actual corrosion product but may be mixed with the corrosion product.

5.6.3 More On Corrosion

Many lay people think all corrosion is the same. When they talk about corrosion they are usually talking about rust, which implies macro-scale material removal and deposition of a crumbly reddish-brown substance. Also, corrosion has historically and primarily referred to metals. Aqueous corrosion, the kind most people are familiar with, generally facilitates removal of metallic atoms participating in a solid object and liberates them into a water-soluble ionic form, which is readily carried away from the part by fluid in motion. But, as noted above, the electrochemical activity may alternatively result in an insoluble oxidation product that adds mass to the part. Thus, a corroded part may weigh less or more than the original.

If we have the situation of an insoluble corrosion product adhering to one component that is tightly assembled to another, the volume increase may lead to deformation or fracture of adjacent components.

> *Prior to a major renovation, the Statue of Liberty had skin bulges. The skin bulges resulted from build-up of insoluble corrosion product between the inside surface of the outer copper alloy skin and the inner ferrous metal support structure. Cultivate your curiosity on this topic as an exercise.*

Corrosion products are, as a rule, less structurally sound than the metal they replace. Generally, they are more subject to wear. Furthermore, these deposits are often ceramic materials and may be abrasive. This may create a self-reinforcing feedback loop that accelerates material loss through a simultaneous abrasive wear process. Finally, in both corrosion and microstructure damage, as the original component's or material's strength properties are compromised, then deformation from the service loads or fracture may result. **NOTE:** we may have component strength being compromised as a result of either material loss, microstructure degradation, or both.

EXERCISE 5-F: Stop and run a movie in your mind. Imagine a clean original part gradually getting attacked by a corrosive agent. Imagine the structure's atoms washing away; think of the part thinning and losing mass.

Then rerun the movie with an insoluble corrosion product being created. Now imagine this result. Make notes on your reflections.

EXERCISE 5-G: Compare-Contrast - What is the difference between a material losing strength and a component losing strength? Do they always go hand-in- hand? Explain why below.

5.7 Thoughts about Classification Systems

The important concept to ponder at this point is that any classification system will be arbitrary to some degree. A classification system is a model of reality. A model can never be completely representative of the actual physical world.

> **Any classification system will be arbitrary to some degree.**

Should wear even be its own category? As we just discussed, wear is often due to simultaneous deformation and fracture. But here are two ways to think about reality.

First, consider the physical realm only. In this case, we can make simple true or false statements.

> EXAMPLE: *are you wearing a wrist watch at this moment? The answer should be "yes" or "no". It should be fairly simple to decide which is correct, which is truthful. I am, at the moment that I am writing this, wearing a wrist watch. If I had been carrying a wrist watch in my pocket, then the answer would be no, the same as if I were not in possession of any wrist watch at all. (Note that by the time I am working on the final editing, I have not been wearing a wrist watch for some years! But I am going to leave my example in the book.) …back to the watch that I had in my pocket. The answer would be no, I'm not wearing a wrist watch, unless... Unless wearing the pants which contain the watch makes you feel that the answer is yes. The point is that even fairly simple facts may be difficult to evaluate. One person may see the situation differently from another.*

But if you wish to consider the truth about conceptual things, such as a nation, health, democracy, then it's a whole different type of effort. We need a totally different criterion with which to judge. In such cases, we help humanity's cultivation of clarity if we make statements in light of whether they are useful to those trying to cultivate clarity, rather than that they are true or false.

> *For example, the statement, "I enjoy good health" may be a true statement for me to make, even though I have high blood pressure, arthritis, and asthma. I can still walk around, travel, work in my garden, drive to work, prepare my own food, etcetera. It is useful to me to believe that I have good health, compared to other situations that could prevail.*
>
> *There are also times when I need to remind myself that I suffer from some health conditions and need to make sure I have adequate rest. So, it's not a lie if I say that I enjoy good health,*

even if I know I have some health challenges.[95]

A classification system is a model of reality. A model can never be completely representative of the actual physical world.

Other conceptual things have similar problems. Is your nation a democracy? What exactly does that mean? Does it mean that the word democracy is in the formal name of the country? Such as the Democratic Republic of the Congo? Does it mean that all adults are able to vote? Does it mean that most adults are able to vote?

For that matter, what is a nation? Is it lines on a map? Is it a feeling of fellowship among citizens? Does it have to be a currently existing nation-state?

Author David Levy urges us to avoid the Reification Error[96], which involves conflating different categories of things. Knowledge of basic philosophical principles allows us to properly compare apples to oranges, credit unions to schools, religions to economic systems, green to salty, a football game to a chess match, and branches of literature to branches of science. These same philosophical principles urge us to use extreme care when comparing truth to computers and beauty to cars.

The first six pairs of entities are things, social organizations, sensory inputs, and elements of human culture. The last two compare ideals (truth and beauty) to electro-mechanical devices; another way of saying this is comparing concepts to things. When we mix categories of things in an unintentional manner, we're on the way to making the Reification Error.

Apples and oranges may be weighed, but it's much more difficult to measure the value of a school or the experience of the green color of its playground lawn. Apples and oranges may be measured, tasted, seen and counted; but the opportunities afforded a community by a credit union, the value of an education imparted to local youth, and the beauty of a colorful mural are harder to quantify.

How do damage categories fit with this discussion?

Fracture, deformation, wear, corrosion, and microstructure happen during service and may all be viewed as results of processes. They are more like the holistic experience of attending football games and less like the score at the end of the game.

In order to understand and communicate any of the things in the above list, humans need concepts. We use concepts to build our model of reality. Failure Analysts need a classification system to understand and communicate our model of reality. Our relevant professional model of reality deals with damage of physical components used in structures.

The very nature of a model of reality is based on concepts.

So, let's take a step back. Many people find it useful to have Wear as a separate category. Wear could be broken down to various combinations of deformation and fracture – as long as the category included "that results in material removal from the main body of the component". I'm comfortable with the usefulness of keeping Wear in its own category and am not recommending that we eliminate the category of Wear.

Within the category of wear, there are multiple sub-categories including: abrasive wear, adhesive wear, cavitation wear, Hertzian stress subsurface cracking leading to material loss, etcetera. Many industries find it useful to have a variety of

95 The Nobel Prize winning physicist Niels Bohr famously said, "The opposite of a great truth is another great truth." Great truths worth that name are truths that help guide our behavior and interactions with others. The purpose of great truths is inspiration, not factual description of physical reality.

96 *Tools of Critical Thinking: Metathoughts for Psychology* by David A. Levy | Page 9, Chapter II, entitled "Reification Error: Comparing Apples and Existentialism", page 8.

descriptive terms into which their subset of common wear situations may be categorized. Yet, there never was a moment when someone created a system of wear categorization that would be all-encompassing of every potential type of wear damage for all of eternity.

The same is true of corrosion.

There never was a moment when someone created a system of corrosion categorization that could be all-encompassing of every potential type of corrosion damage for all of eternity.

5.8 Going Deeper: Damage Categories

Corrosion damage is almost always a very complex result of chemical or electrochemical interactions between the material and the environment including the temperature. It is also influenced by the geometry of the structure in ways analogous to stress related damage.

> *For example, instead of a load transmitting cross-section being a key factor to the durability of the component, initially at least, it's the surface geometry. Crevices, roughness, differences in surface composition and anisotropy of the material being corroded all play a role in the local severity of the attack. Also, the relative amount of surface area of different materials in the same assembly plays a role, which can influence the way galvanic corrosion proceeds.*

Microstructure damage is a result of physical changes in the micro or molecular structure of the material. It could be due to thermal or other energy impingements on the part. In many applications, high temperature and oxygen contained in air exacerbate microstructure damage. Thus, microstructure damage often is accompanied by oxidation, which is an electrochemical process.

Oil drilling applications are subject to sulfidation, for similar reasons. This inevitably blurs the line between corrosion and microstructure damage, as we have emphasized on multiple occasions.

Perhaps this shows us that those who think that microstructure damage should be included in the corrosion classification are correct.

Some microstructure damage would never be able to proceed in the absence of stress. It's the combination of high temperature and stress that eventually results in a combination of deformation and possible fracture. And that's when we notice, "Oh the microstructure has deteriorated!"

Stainless steel in the presence of chlorides may crack at stress levels far below normal strength. This is called stress corrosion cracking, and is usually listed as a type of corrosion. But there is some evidence that the hydrogen from the corrosion process weakens the metallic material and allows it to crack. Others have found evidence that brittle oxides replace the metallic material and that is what cracks. Both scenarios are probably accurate in different environment–material–service combinations. Sometimes it's hard to figure out exactly what is going on. Usually, from a pragmatic viewpoint, our analysis (as described in Chapter 6) will provide hints for avoiding recurrence.

5.9 Damage Temperature and Loading Rate

A complication faced by those trying to master damage categorization is that there is no way to entirely separate the effects of the loading rate from those of the temperature. That is because the material's response to loading, whether deformation or fracture, is strongly dependent on diffusion happening at the nano-scale. Diffusion rates are strongly dependent on temperature.

> *Because of this temperature dependency of diffusion rates, sugar dissolves more readily in hot tea or coffee than it does in iced tea or coffee. That's why many restaurants offer iced tea and coffee beverages pre-sweetened.*

In short, there's really no way to understand the stress experience of any component without simultaneously characterizing the thermal environment in which it occurs.

In Figure 5.5 (below), you will see two elbows used in an industrial heat exchanger. Corrosion and wear are two of the most likely damage processes that allowed the leak. Heat exchangers experience high temperatures and wet environments, which may result in deformation as well as material loss and property degradation. The competent failure analyst will consider the steps needed to determine the most significant factors that facilitated the demise of the equipment.

I constantly practice creating a holistic representation of the damage process in my mind when working on any failure analysis. I like to imagine the component or structure in my mind interactively responding to changes in loading and temperature or the chemical species that might be attacking the material in a general manner or how it might be attacking in specific meso or micro positions within the envelope of the structure (such as crevice corrosion).

I take time to think about the potential effects of other sources of energy such as heat conveyed by convection, radiation, or UV light. It's a complex, changing picture.

Figure 5.5: *Two elbows used in an industrial heat exchanger*

When I do a failure analysis, I feel best about the data quality when I can actually run a movie in my mind that shows the component in its pristine, original condition, and imagine the loads flexing and stressing the component at the same time as the environment modifies the responses from those that might be expected when the load and environment are considered in isolation from each other.

Failure Analysts often use the term "failure mechanism". Let's leave aside the issue of whether it should be called "damage mechanism", what exactly does the term "failure mechanism" refer to? The word "mechanism" in failure analysis has a lot of meanings, so I tend to avoid using it. I don't want to confuse a situation. I want to clarify it. Therefore, as a best practice, I work to find other more specific and descriptive terms.

Some people call fatigue a "failure mechanism". A more specific and descriptive term would be to think of fatigue as a type of macro-brittle crack propagation that happens as a result of repeated loads.

Some people call stress corrosion cracking (SCC) a "failure mechanism". A more specific and descriptive term would be to think of SCC as a type of macro-brittle crack propagation that certain materials are subject to when exposed to specific environmental substances. SCC happens in the presence of sustained loads and tends to produce cracks with multiple branches.

Some people call dezincification a "failure mechanism." I think of dezincification as a type of complex corrosion process that affects certain copper-zinc alloys. It leaves them porous by dissolving away the zinc. The copper gets attacked too, but is sometimes redeposited on the part surface.

Do you see how using more specific and descriptive terms help you remember the damage, temperature and loading rate interactions and lead to making you a better failure investigator?

Now, let's take a more detailed look at each of the basic damage categories in turn. We'll start with the stress related ones, and we'll start from a simplified view, and we will momentarily set aside all of the simultaneous interactions we've just been discussing.

5.10 Understanding Fracture

Fracture means that atomic and/or molecular bonds have been broken. Fracture, cracking, or breaking results from separation of atoms along a nominally two-dimensional area *("Nominally two-dimensional" means more or less flat but it could have curves, steps, or bends)*. Fracture happens because the load exceeded the load carrying capability of the component. In other words, the stress exceeded the strength.

Figures 5.6 and 5.7A-C show visible results of monotonic and fatigue cracks, respectively. Chapter 8 will provide more details on how to recognize monotonic overload and fatigue cracks.

Note that the series of images in 5.7A-C depict only one side of the fractured object. The images focus on the fracture surface, the new surface that was created as a result of the fracture.

KEY SKILL: Learn to recognize *which* basic patterns on a part surface reveal that it is a result of a fracture event.

Fracture processes that proceed unchecked tend to create multiple fragments from what used to be a single component. As was just implied, the fracture process itself creates distinctive features on the crack surface as the crack propagates (at both visible and microscopic scales). In common structural materials, these features fall into a limited number of readily recognizable groups.

Learning to recognize these groups of features is the purpose of studying fractography, which is performed at multiple levels – macro, meso and micro.

- The macro features, first and foremost, are used to help us visualize loading.
- The meso features, first and foremost, reveal the crack growth direction.
- The micro features will be described later, but they often provide information about the integrity of the process controls maintained during manufacturing.

5.10.1 Exemplar Macro Appearances

Fatigue is defined as cracking due to repeated load cycles. Alternating light and dark bands are often present on fracture surfaces of parts that broke in fatigue. The different colors represent varying loads and/or varying environmental conditions. These alternating light and dark bands, sometimes hard to see, are among the most recognizable visual indicators of fracture.

Fracture happens because the stress exceeded the strength.

Figures 5.7A-C are macro-fractographs depicting fatigue fractures. These images were captured with the intent to depict what we will be calling meso-scale features when we get to Chapter 8.

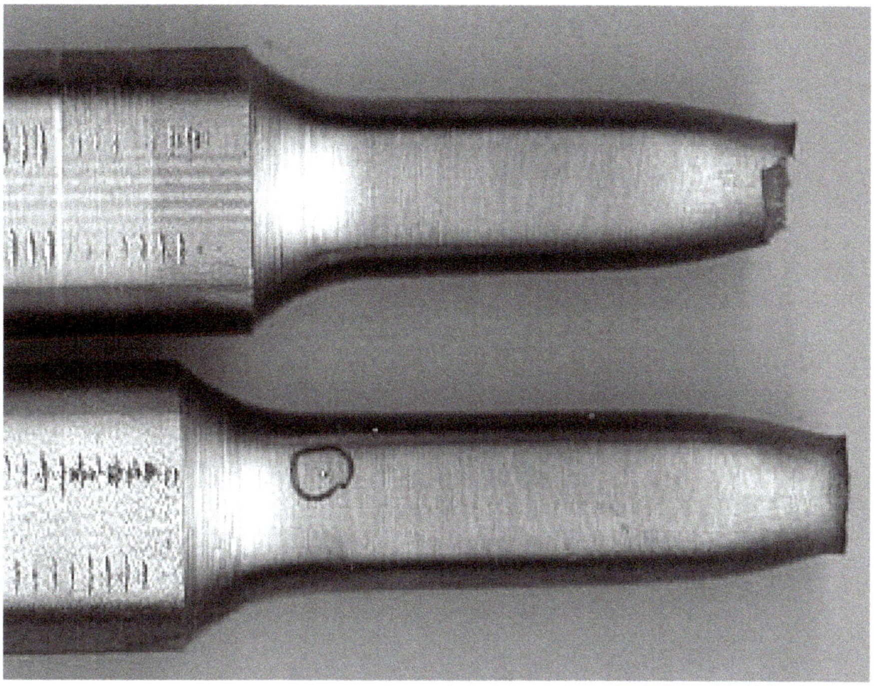

Figure 5.6: *Monotonic (single load cycle) fracture – the diameter of the gage length is 0.5 inches.*

Notice the pinched down areas (on the right) on the fragments which result from plastic (permanent) deformation as the tensile load was gradually increased until the part broke into two fragments. There is visible deformation (necking, or formation of a neck) immediately surrounding the crack, which is a strong indication of monotonic loading.

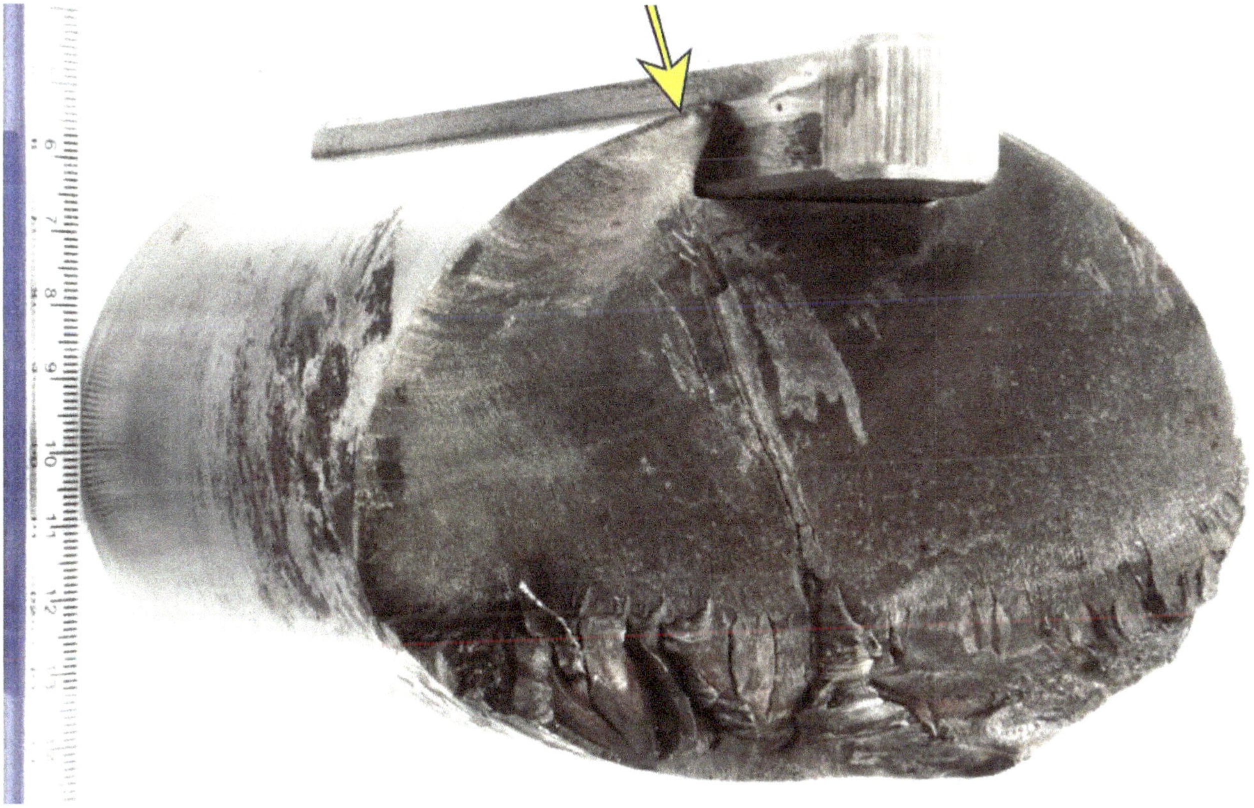

Figure 5.7A: *The yellow arrow shows initiation. Notice the fracture is smooth. This image is a macro-fractograph.*

Does this component look familiar?

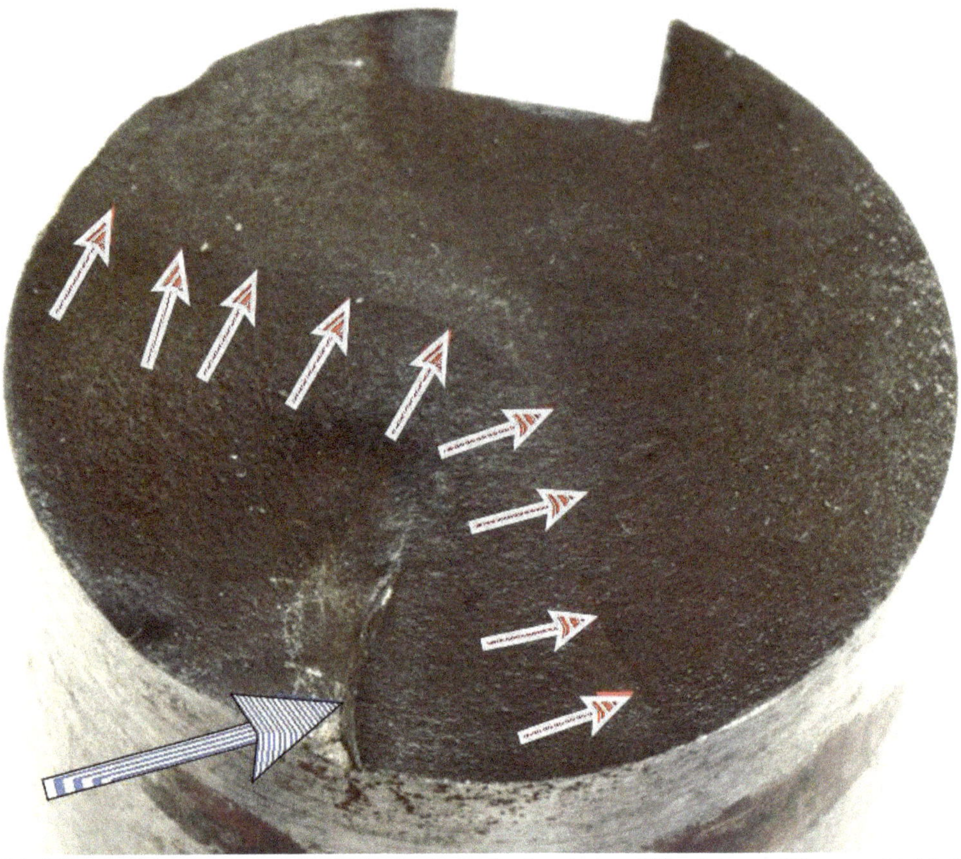

Figure 5.7B: *The crack is very smooth and alternating color bands are difficult to see. They are highlighted with small red arrows. The large blue arrow points to a step (ratchet mark), a classical indication of multiple initiation points. The diameter is approximately 35 mm.*

Figure 5.7C: *The crack started at the inner edge of the bottom of a cup. Curved dark bands are highlighted by arrows; these are beach marks.*

The crack was covered with aluminum from a die casting operation. The aluminum was chemically removed, but faint crack features are still visible. The micro-scale features are likely damaged and not visible, even with a scanning electron microscope.

5.10.2 Exemplar Micro Appearances

In order to understand fracture, we need to understand it at multiple levels. Figures 5.8A–B depict two of the main types of micro-scale true fracture features. The micro-void coalescence (MVC) features displayed in Figure 5.8A resulted from a crack process that happened in one single load cycle. We call this "monotonic loading". This is likely what the (hidden) center of the part shown in Figure 5.6 looks like with a scanning electron microscope.

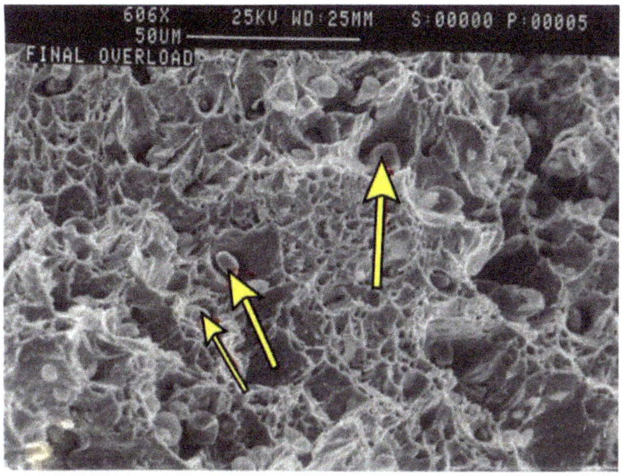

Figure 5.8A: *Micro-void coalescence (MVC) due to Monotonic (single load cycle) Tensile Load – this is the classical micro-scale ductile crack feature. Small spheroids at the arrows are the microscopic particles that encouraged the formation of the voids.*

These voids are a result of deformation prior to separation. They are not present in the original part. (Scanning Electron Microscope Micro-fractograph)

Figure 5.8B shows the classical appearance of individual crack arrest marks from a component broken by repeated load cycles, each one of which was insufficient to cause complete fragmentation. This appearance indicates fatigue loading, and could potentially represent something similar to what might be seen with the help of a scanning electron microscope when examining a piece of the fracture surfaces shown in Figures 5.7A or B. Each crack arrest mark shows how far the crack propagated during one load cycle. Note the original magnification setting of 2000x. Many fatigue cracks have features that are heavily influenced by the microstructure of the component. This complicates diagnosing fatigue during micro-scale examination.

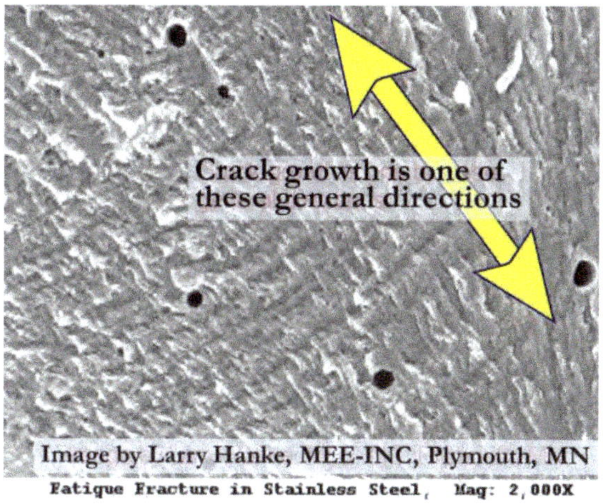

Figure 5.8B: *Fatigue crack in single-phase material – each fine line is a witness mark for the position of the crack tip during a single load cycle.*

The original magnification is 2000x.

Scanning Electron Microscope Micro-fractograph is provided courtesy Larry Hanke, Materials Engineering Inc.

We'll go into greater detail in Chapter 8 on Fractography, but for now we'll remind ourselves that we always need to be thinking about macro-, meso-, and micro-scale features; those that tell us about the loading, the crack propagation direction, and other details.

The analyst must also learn to distinguish the different types of progressive cracks. Is it a fatigue crack, due to variable loading, or is the crack a result of weakening due to environmental attack? Each of these types of cracks may look different, although sometimes a microscope will be required to see the relevant distinguishing features.

Environmental attack will often leave evidence in the form of insoluble corrosion products that may be detected with the eyes, and analyzed using the EDS microchemical analysis attachment on a scanning electron microscope. As you gain experience with examining fractures, you may get a pretty good idea as soon as you see the part, as to whether it was a single load cycle, a fatigue crack, or an environmentally assisted crack. And sometimes you will need to employ every fancy instrument available to figure it out.

5.10.3 Other Considerations

Although there are a limited number of "classical" fracture features, some materials have micro-scale features that are strongly influenced by their complex microstructures as well as the loading history. If these micro-scale features are coarse, they can change the macro appearance. This is most often true in castings, powder metal parts, additively manufactured components (Laser Powder Bed Fusion - LPBF) or fiber reinforced composites. Parts made by additive manufacturing (AM) may produce exceedingly complex patterns at multiple overlapping scales.

Furthermore, sometimes a component or a welded assembly will separate but when we examine the undesired free surfaces, there are no true fracture features. One example would be the case of a weldment with a lack-of-fusion discontinuity. This also happens in LPBF AM parts.

Sometimes, castings have oxide films or large voids present from the time of solidification, but these discontinuities are revealed only after a stress event.

Sometimes the extent of the discontinuity is sufficient to create a much higher stress level than anticipated by the design engineer. If 20% of the cross-section is unable to share the force transmission duties, the remaining material may experience a higher stress level than the design value. Depending on the shape of the discontinuity, its effective severity may be greater than its proportional area might suggest. In other situations, the stress level will only be slightly higher than the design intent. In the first case, the component may fragment unexpectedly the first time it's used, or soon after entering service. In the second case, if the component or assembly is subjected to repeated loads, a fatigue crack may initiate adjacent to the discontinuity. It may be years or decades later that the problem is discovered. Or, if the stresses are low, the discontinuity may never initiate a crack. Many designs are limited by rigidity, rather than strength. The stresses in such assemblies can be quite low.

EXERCISE 5-H: CULTIVATE CURIOSITY by sketching a component with a severely damaging discontinuity. Show how the forces are transmitted through the part.

Then sketch a component with a mild discontinuity, again showing how the forces are transmitted through the part.

Some examples of non-fracture features typically found on fracture surfaces are corrosion products, which are shown in Figures 5.9A and 5.9B (below).

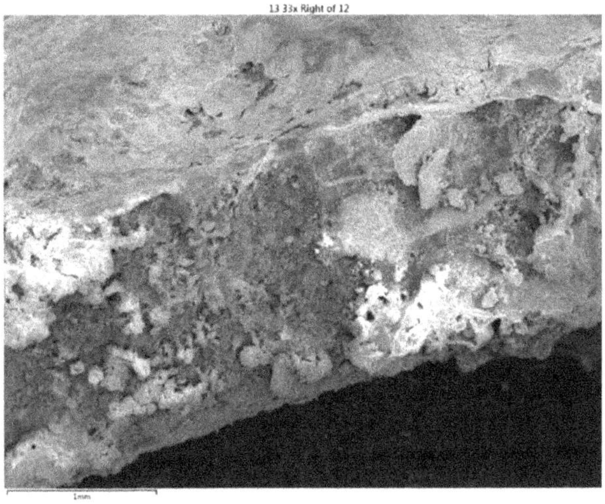

Figure 5.9A: *Corrosion product obscures the true fracture surface on this brass casting.*

Figure 5.9B: *"Coral" feature within a patch of corrosion product – this is not a true fracture feature.*

An example of solidification features in castings is shown in Figure 5.10. It's important to be able to recognize these non-fracture features. A shrinkage void may be revealed by a fracture event, when the surrounding material breaks. Corrosion product may have happened after the crack, or be a part of the damage process itself.

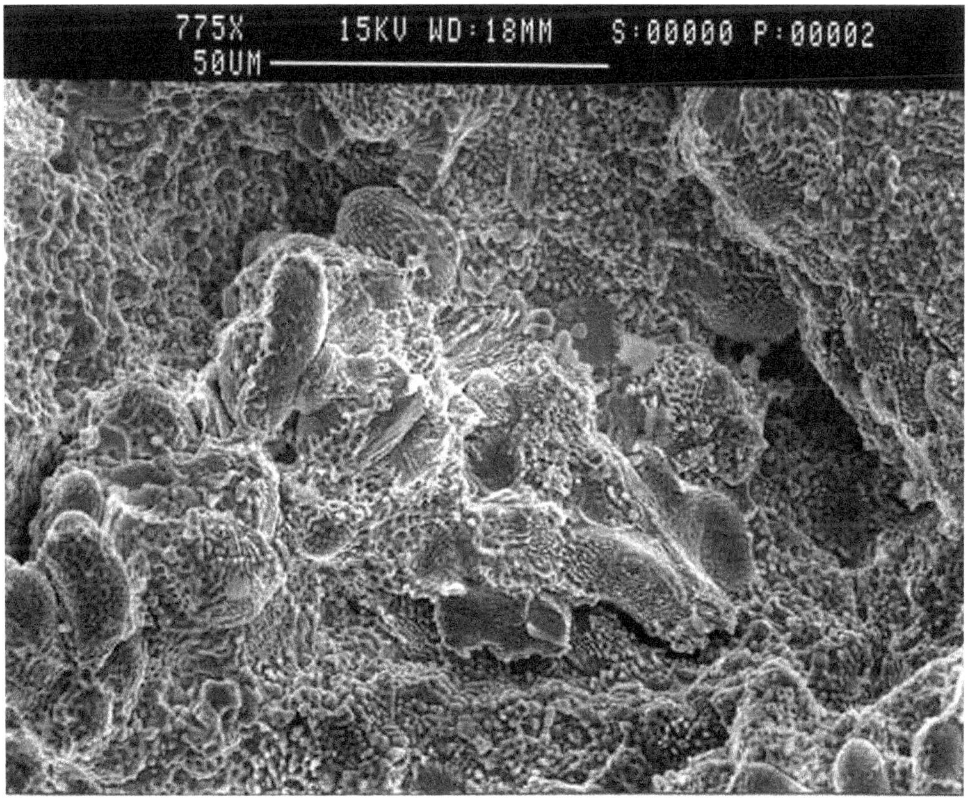

Figure 5.10: *This feature is due to a shrinkage void in a casting – not a true fracture feature.*

5.11 Understanding Deformation

Figure 5.11 shows a deformed and broken chain link. While the link was intact, the parallel portions were loaded in axial tension. Once one of these cross-sections fractured, the remaining ligament was loaded in bending. The material was ductile (able to permanently deform. But the crack behavior itself was brittle (no evidence of deformation, such as the pinching we saw in Figure 5.6, on either side of the crack).

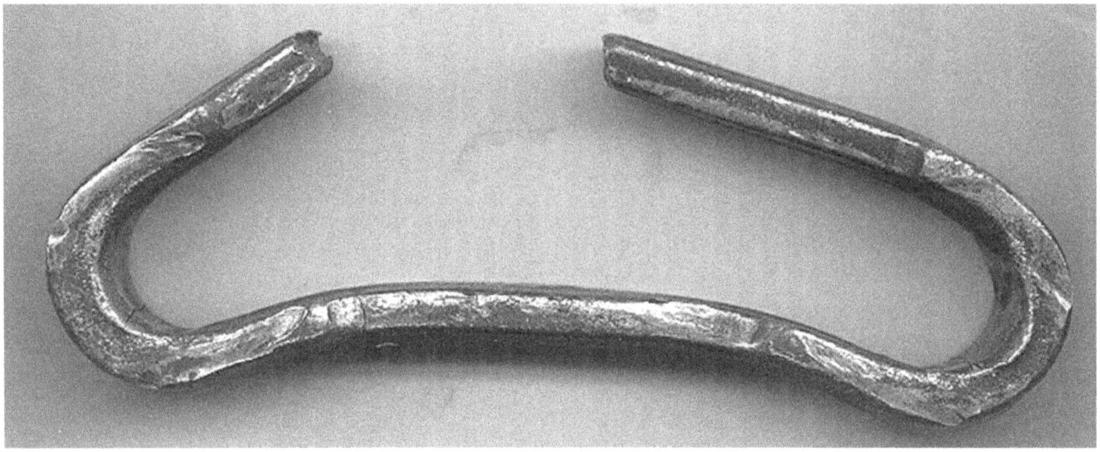

Figure 5.11: *This chain link fractured, which then allowed the deformation of the remaining load bearing portion.*

Deformation is influenced by temperature, stress and strain. Figure 5.12A shows the ambient and low temperature mechanical behavior of 1075 steel. It was probably hot rolled or annealed, although the information on process condition was not provided. The nearly vertical straight lines starting from zero strain is related to the elastic modulus as shown in the Figures 5.12A and 5.12B (next page) and is nominally constant for isotropic metals at a given temperature.

As the stress approaches the yield strength, the behavior no longer follows a straight line. The value at which the line bends away from the nearly vertical line to the right is called the proportional limit and indicates much more shape change for a given amount of stress change. This is related to the yield strength or stress level where permanent deformation starts. Also note, the proportional limit goes down as the temperature goes up. The change in yield strength response with temperature is seen on this graph to be much more pronounced than the change in elastic modulus. It's possible that this is a schematic, and the change in modulus was never considered by the person who prepared this graph. Finally, please note that the proportional limit goes up as the speed of testing goes up!

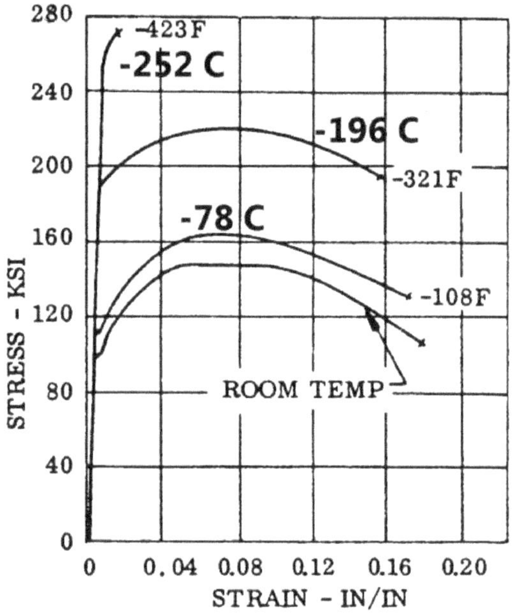

Figure 5.12A: *Effect of temperature on tensile properties – 1075 steel*
Note the negative signs indicating the test temperatures (test speed is not shown here) The image is provided courtesy of ASM (the reference: Boyer, Atlas of Stress Strain Curves, ASM International, page 198.

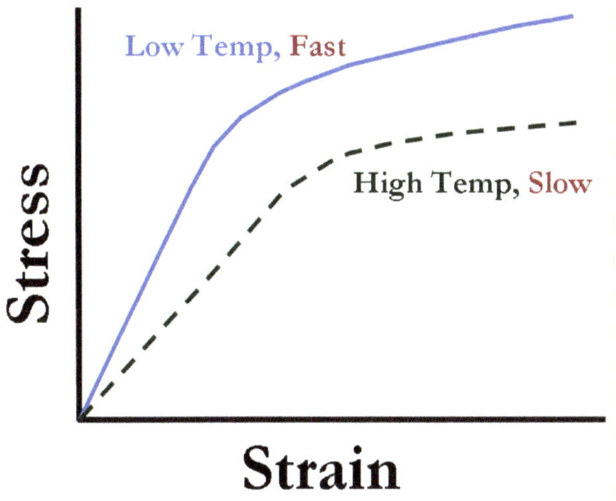

Figure 5.12B: *Low Temperature and Fast Loading Rate (Strain) result in a higher proportional limit (yield strength) and a higher modulus (greater stiffness). High Temperature and Slow Loading Rate (Strain) result in a lower proportional limit (yield strength) and a lower modulus.*

Deformation (like fracture) is a result of the stress exceeding the strength. However, in this case the strength that is exceeded is the yield strength. Atomic bonds are separated, but new ones take their place as soon as the separation occurs. The result is that there are just as many atomic bonds as there were prior to the deformation event, but some of them are now between different atoms. This type of deformation is permanent and it is called plastic deformation. If the load is removed, the part will not return to its original shape.

Plastic deformation can happen in metals as well as polymeric materials. Most ceramics cannot respond to stress with macro-scale plastic deformation. The yield strength is effectively higher than the tensile strength. There are some metals that have this response as well, such as grey cast iron. These materials are inherently brittle at the macro-scale.

In ductile materials, plastic deformation often precedes and accompanies fracture. It is important to learn to separate plastic deformation that precedes or accompanies fracture, such as the pinched ends of the fragments shown in Figure 5.6, from plastic deformation that happens as a result of a fracture, such as the chain link depicted in Figure 5.11.

The chain link experienced obvious deformation once the crack completely interrupted the stress path. The beginner might have to stop to think about the fact that the arc shape in the lower portion of the link must have happened after the crack in the upper portion interrupted the stress path. At that point, the stress in the lower portion was effectively more than doubled.

5.11.1 Elastic Deformation Failures

> *Occasionally, I've been asked to look at deformation damage that involves excessive or inadequate elastic deformation. In this case, when the load is removed, the part returns to its original shape. Usually, these situations involve some sort of spring device. The failure involves a spring constant that is too high or too low, a spring that is too stiff or not stiff enough.*
>
> *Once I investigated a mass-produced spring clip in the final stages of design validation. It lacked adequate retention force on a small but significant portion of the population of parts, but only after the seventh or eighth of ten required load cycles.*

The usual spring constant problem is a different situation from a multi-load retention force issue.

If a part returns to its original shape when the load is removed, then we know that, at least at a simplified level, the yield strength was not exceeded.

If the deflection was too great or too small, then it is the modulus of elasticity or geometric features rather than the yield strength that is the key characteristic that needs to be evaluated.

For heat treated carbon and alloy steels, the modulus of elasticity is essentially a constant. There's nothing we can do with our processing to change it. So, the most likely "cause" of the problem lies in some aspect of the dimensional characteristics of the component.

For glass reinforced polymers or other anisotropic materials,[97] the variation in stiffness is much more likely to be related to a material variation or a combination of variation in material characteristics and dimensions.

Did the stress exceed the strength or is this an issue primarily related to flexibility (modulus)?

In the case of the spring that met the requirements for the first seven or eight of ten required cycles, I asked the designer to review the finite element analysis for the component. He did so. The stress values were within a few percent of the 0.2% offset yield strength for the material.

I informed him, "It seems to me that you are really pushing your luck. No manufacturing process is free from variation."

"I'm dealing with very tight design constraints," he informed me. He knew he was pushing the limits, but he thought the design would work.

In this case, the heat treatment probably was a factor because the yield strength was only slightly exceeded on each test cycle and some parts were no longer able to meet the criterion.

"Remember," I pointed out to him, "the slope of the stress strain curve is far from linear as we approach the arbitrarily defined 0.2% offset yield strength. Very slight variations in dimensions could produce much more significant variations in deflection when the stress is in the flattened area near the yield strength."

Remember Figure 5.12?

Another thing to realize is that the 0.2% yield strength value that is published for a material and process condition, or microstructure and hardness range, is obtained at a specific strain rate. The strain rate conditions of the component durability test may vary widely from these standard test conditions.

Finally, we need to remember that the 0.2% offset yield strength is an arbitrary number that people agreed to use for standard material certification purposes. It does not reveal an objective universal truth about the nature of the material itself.

A good question to keep in mind for deformation failures is, "Did the stress exceed the strength or is this an issue primarily related to flexibility (the modulus of elasticity,[98] geometry, or both)?"

Again, we see that we must consider the details of the specific context when dealing with deformation problems. We need to determine whether it is some aspect of the strength that is the primary problem, or alternatively, whether it is primarily a rigidity concern.

We have discussed plastic deformation due to the stress exceeding the yield strength. We have also discussed permanent deformation near the elastic limit where geometry, temperature, strain, and modulus of elasticity matter more than yield strength. We are now ready to touch on buckling as it applies to tall, thin columns loaded in axial compression as well as long, thin tubular sections loaded in torsion (twisting). See Figures 5.13 and 5.14 A and B (on the next page).

Equations for columnar buckling may be readily calculated with fairly simple formulas (Euler's Formula) for simple shapes. Losing slenderness, the equations rapidly get complex. The equations for torsional buckling of cylindrical tubes are

[97] Anisotropic materials have different strengths or flexibility values when tested in different orientations because the material properties are not the same by orientation.

[98] The modulus of elasticity, or Young's Modulus, is related to the inherent strength of the atomic bonds; but in bulk materials, due to nano-scale crystalline imperfections, this strength is not available.

much more complex. The paper listed as the source of Figure 5.13[99] gives some advice for approximating the limits of stability for these types of geometries.

Figure 5.13: *This image shows torsional buckling.*

Figure 5.14A shows compression buckling of a long slender column in various end freedom conditions. The model was produced as part of a University project to demonstrate the different Euler buckling modes and the effect of boundary conditions on the critical buckling load.[100]

Figure 5.14B shows an example of compression buckling in nature (New Zealand). [101]

There are many other types of deformation. Many of them are important during manufacturing. How would we forge, stamp, or make coil springs if we could not intentionally control permanent deformation?

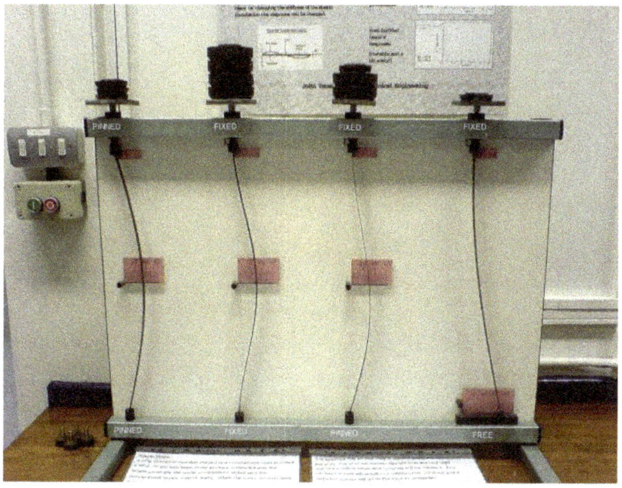

Figure 5.14A: *Compression buckling of a long slender column in various end freedom conditions*

Figure 5.14B: *Compression buckling in New Zealand*

Deformation failures might include unintentional spring-back in formed sheet metal stampings, inconsistent angles in torsion springs (see Figure 5.15), various changes of shape during either heat exposure during heat treating, or quenching / cooling after treating.

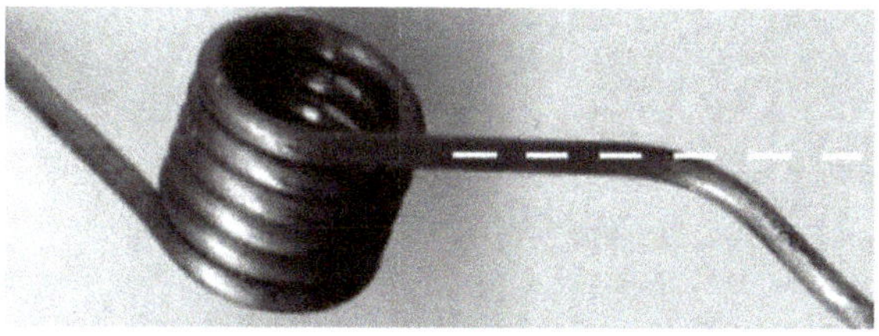

Figure 5.15: *In a torsion spring, the angle between the legs is critical, and can be difficult to control.*

[99] National Advisory Committee for Aeronautics Report No. 479, Stability of Thin-Walled Tubes Under Torsion, 1933, page 9, Figure 6 Detail. by L. H. Donnell | https://ntrs.nasa.gov/archive/nasa/casi.ntrs.nasa.gov/19930091553.pdf

[100] This image is provided courtesy of Grahams Child. It's called, "Buckledmodel.JPG" | licensed under creative commons http://en.wikipedia.org/wiki/Image:Buckledmodel.JPG

[101] This image is provided courtesy of 4nitsirk. The image is called, Mataikona Rocks (3) and is licensed under creative commons BY-SA 2.0. | https://openverse.org/image/45c06c32-08a1-4bd0-aa0e-8afc289be158?q=Compression%20buckling

Deformation failures include distortion due to unbalancing of residual stresses as a result of material being machined away, etcetera. Some plating processes leave high residual stresses, which could cause distortion in thin materials. But remember from previous lessons, it would be wrong to consider deformation automatically as a failure mode.

> *I always look forward to university professor presentations at the ASM International Failure Analysis Society symposiums; they present their research on mechanical properties of an alloy system. However, sometimes when they present, they say, "I loaded the tensile test coupon in a universal tester and increase the load 'until it failed'." That makes me laugh.*
>
> *It is an evocative reminder for me that Failure Analysts must be extremely careful with our language. Wouldn't it be more of a failure if it didn't break? Presumably they are going to keep increasing the load until it does break. We need to be conscious of the words that come out of our mouths.* ***Breaking*** *is not the same as* ***failure****.*
>
> *One such example in real life is when pop-top cans don't "pop". That's the failure: the fact that it DID NOT break.*

Figure 5.16 brings us back to more traditional failure analysis of service components. It shows a portion of a specialty conveyor chain link. Clearly visible are the asymmetrical indentations on an otherwise symmetrical component. These features were not shown on the engineering print.

There was a fracture elsewhere on this component. But these deformed areas most likely indicate something unusual and undesirable in the operation of the conveyor system. This deformation mark is an example of a "witness" mark.

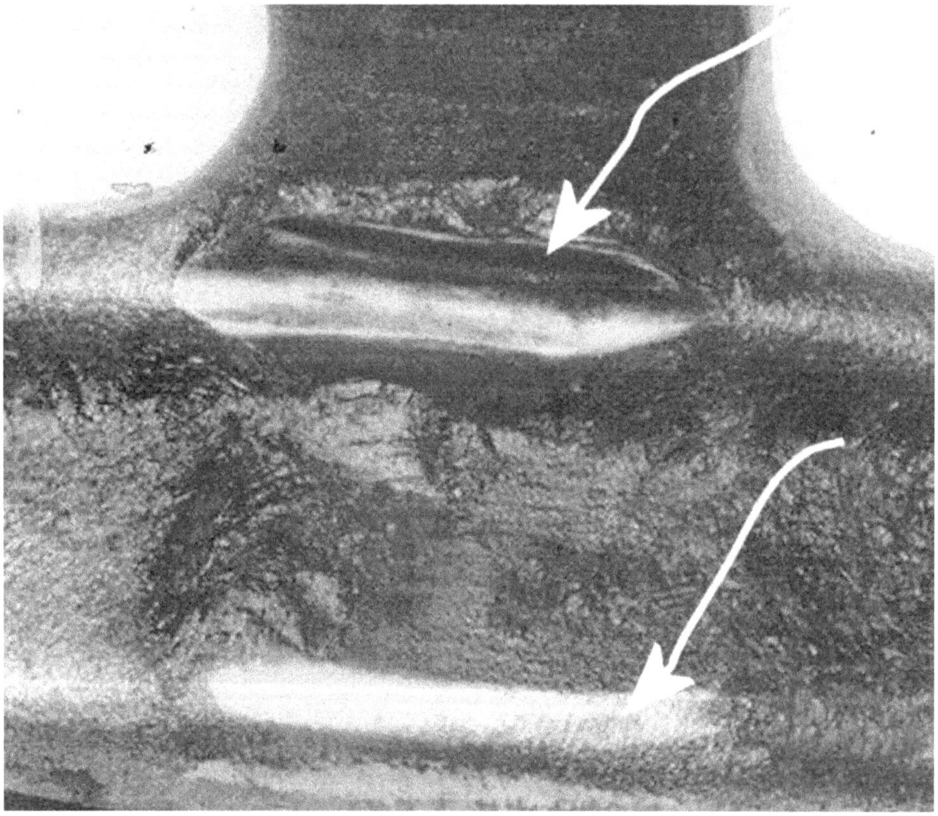

Figure 5.16: *Deformation mark on chain link*

5.12 Understanding Wear

The science used to study wear is called tribology. Wear and tribology live in a complex world at an intersection between science and engineering. In this book, we approach that intersection from the perspective of trying to understand how damage is proceeding.

When I first started performing failure analysis and someone handed me a worn-out component, I felt lost because over half of the load bearing cross-section was missing. And worst of all, what wore out was gone so there was nothing to analyze - not even any particles. The particles were gone because it was not a lubricated application. There was no grease to hold on to the liberated particles!

If you are lucky, you will have multiple components displaying a range of severity of damage. Or, perhaps, you will have multiple areas of a worn component that displays a range of severity.

If you are lucky, you may have some access to the debris particles created by the wear process. This may include bits of the damaged component(s) and it may also include abrasive particles that facilitated the damage. And of course, you will gather as much information as possible about the background situation.

In common everyday failure analysis, you probably won't have access to a tribology specialist. You must have some understanding of the basic ways in which wear takes place including: abrasion, adhesion, deformation leading up to fragmentation, contact fatigue, and fretting. Fretting is sometimes classified as a type of corrosion damage. If you work with pumps, you need to be able to recognize cavitation damage.

If your clients or customers expect you to propose a fix to the problem as part of your analysis, you will be able to find credible ideas to suggest in various failure analysis and tribological references.

We haven't gotten into the details of the full failure analysis process yet, but we must note the special requirements of evidence preservation for wear failure analysis. If the wear debris is available, such as when it is trapped in grease in a ball bearing assembly, a sample should be retained. If the component is to be cleaned for microscopic examination, care must be taken to save the grease. It is often a good idea to do this prior to sectioning. Why? Because the sectioning operation will likely contaminate the "natural" debris specimen; it could be confusing to analyze.

Disassembling a ball bearing assembly was not something that I understood how to do when I first worked in this field. If the ball bearing is all metallic, cleaning with solvent in an ultrasonic cleaner can be a good option.

Once I tried solvent cleaning for a bearing assembly that had some plastic components. The solvent dissolved the plastic and made a mess. It would have been more embarrassing if the client had not given me a dozen of the same model for my evaluation.

Finding a solvent that will dissolve bearing grease is also quite challenging. I've had to use acetone, naptha, alcohol and still ended up having to clean the balls, races, and cages tediously by hand with a cloth and cotton swabs and compressed air.

Filtering out the metallic or abrasive debris can be equally challenging when there are big blobs of mineral filled grease.

I'm sure that people who do bearing failure analysis every week for years have "tricks up their sleeves" to make this less tedious than it is for me. The important point here is, all this happens before you can perform the necessary examination with optical and scanning electron microscopes.

As with other types of damage, the wear analysis starts with a visual examination and document review. As just noted, there's often quite a bit of work involved in preparing the Subject

Components for the visual examination of both the components and the debris, when available.

When there is a lubricant involved, a decision must be made regarding whether it will also be analyzed. If it is determined that this is not necessary, we must realize that we are not performing a true failure analysis. We are simply doing some lab work to gather data. We may get lucky. Sometimes it is fairly obvious that the grease wasn't the problem. But if we haven't taken steps to conserve the lubricant, if there are questions at the end of the component evaluation, we'll have no way to double check our original assumptions.

The visual examination may tell the inexperienced Failure Analyst something about the way the wear is occurring, but it's more likely to simply reveal the severity of the wear. This enables the analyst to select individual components or test locations for more detailed evaluation with a microscope.

TIP: It's a lot easier to view a component from different angles with an optical microscope than when it's inside the sealed chamber of a scanning electron microscope. So don't skip the optical microscope evaluation. But for polished parts, the reflections may make it difficult to see the details of the damage that may be revealed with a scanning electron microscope.

Even more difficult than viewing the damage details with an optical microscope is the task of photo documenting the relevant features on shiny bearing assembly components. The reflections are difficult-to-impossible to entirely eliminate, and depending on the viewing audience, a lot of arrows and circles are going to be required to highlight the features of interest.

Our goal in all this is to determine how the damage is taking place. There are multiple possible scenarios; a few examples are listed below:

- Are there hard, sharp particles that are scraping material away?
- Are projections (bumps) being smeared (deformed) and then flaking off?
- Are two surfaces coming into contact so that small bumps or projections (asperities) are actually being micro-welded together and subsequently torn apart, leaving craters in one or both of the components?
- Is a localized oxidation of asperities creating an abrasive condition (fretting) which may lead to fatigue cracks in unexpected locations?

Once you have determined which possible scenario is allowing the damage, you may have the basic information you need to look for a solution.

Unless you have access to a nano-indentation hardness tester, it may be difficult to determine the hardness of the debris particles. An EDS microchemical analyzer (accessory equipment on many scanning electron microscopes) is helpful to identify the source of the particles. Examination of the shapes of metallic debris particles may distinguish those that are a result of subsurface cracking, versus smearing and flaking, and scraping or gouging action. Compare-contrast the worn component surface with the debris particles. Are both sets of data self-consistent? Do they tell the same story?

The following three case studies walk the Failure Analyst through different aspects of wear.

5.12.1 CASE STUDY: Steel Plate In a Conveyor (Machinery) Application

Client: Here is a worn steel plate that was used in a conveyer. What happened?

Discussion: The image in Figure 5.17A (next page) does not have a scale marker bar. It's from a project I worked on years ago and the particulars are gone from my records. However, the image was obtained with a normal film camera (Canon AE1 Program) with a 2x macro lens. What is depicted was visible to the eye. The application involved another component sliding over this plate.

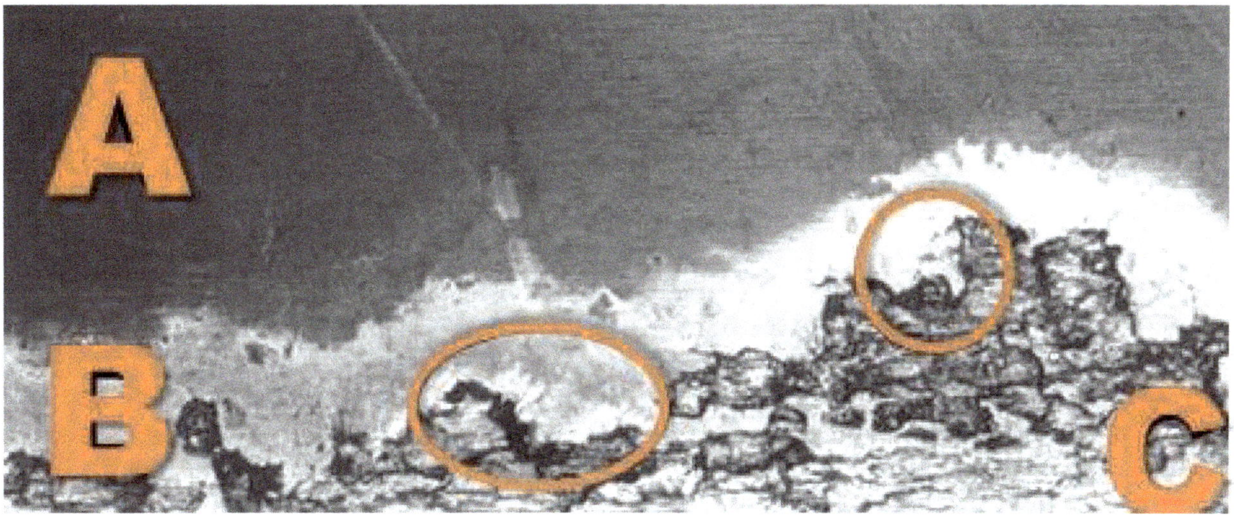

Figure 5.17A: *Wear on plate – industrial machine application (Area A is the as-manufactured surface of the plate. Area B is more highly reflective. Area C is the problem. Ellipses show smeared flaps at the boundary between B and C.)*

Perhaps the surface asperities have been smoothed out. An oxygen rich layer may have been polished off. Area C (the problem) is very rough. The circled areas give a hint that material is being smeared and the overall appearance of Area C indicates that material has been removed.

Figure 5.17B (next page) does have a scale marker bar. It was obtained with a scanning electron microscope.

- Here, the white-outlined arrows show several areas which have clearly been smeared out into flake shapes waiting to be broken away.
- The black-outlined arrow indicates a large crater that has presumably been created by the flaking mechanism.
- The small circled area with the white-outlined arrow shows a less severe area with the same basic scenario.

This gives some confidence that we're seeing one basic sequence of increasingly severe damage. Of course, it would be preferable to examine more than one area.

The area depicted by 5.17B (next page) is only approximately 0.2 mm by 0.15 mm. This clearly demonstrates one of the limitations of the scanning electron microscope (SEM). The magnification setting of the SEM was less than 500x, a very moderate magnification.

TIP: It is always important to make some effort to determine how representative your image data is when working with microscopic images.

This damage 5.17B (next page) was clearly a combination of deformation followed by fracture. We call it wear. I don't remember what fix I recommended, if any. But looking back on the data through the lens of experience, I don't think lubrication would have helped. A smoother surface to start or a higher hardness material might have been potential solutions.

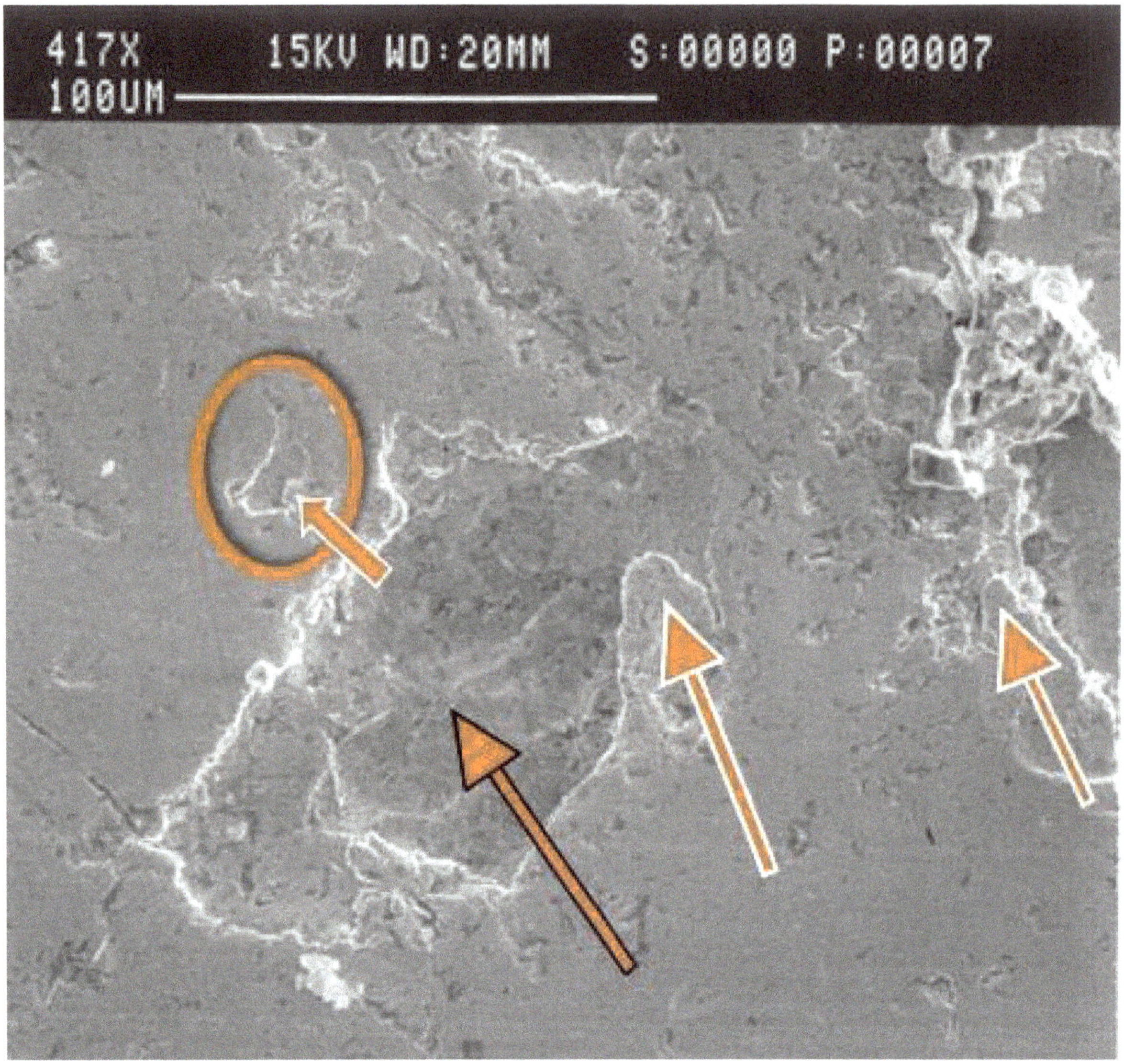

Figure 5.17B: *Wear on plate from industrial machine – image obtained using a scanning electron microscope*

5.12.2 CASE STUDY: Bearing Race Examination

Client: We'd like you to examine this bearing race and explain why it is worn.

Discussion: This damage scenario is similar to the previous case study and is depicted in Figure 5.18 (next page). However, there are more cracks visible where the flakes have not yet broken away. There appear to be more layers of smeared material. It is likely that the bearing race is much harder than the plate from the previous case study (5.12.1 CASE STUDY: Steel Plate in a Conveyor – Machinery - Application).

A common wear mechanism in bearings involves sub-surface cracks working their way out to the surface, but this is not obviously what is happening here. The material is also smeared.

The arrows show dark and bright areas that are image artifacts due to grease remaining on the surface. The grease is not conductive and renders the specimen less photogenic. Also, the electron buildup at the non-conductive areas interferes with clear details nearby.

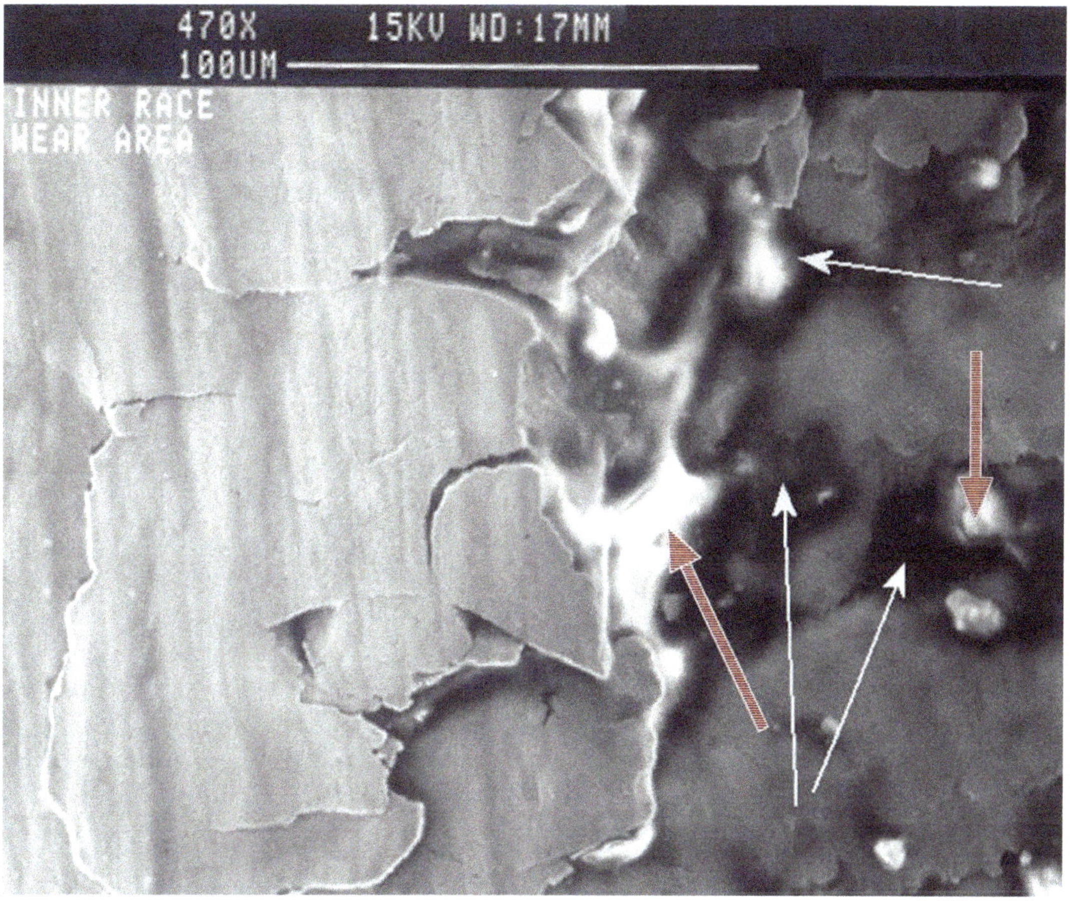

***Figure 5.18**: Bearing race wear detail – arrows show grease that remained on the surface of interest after cleaning in preparation for the SEM analysis.*

5.12.3 CASE STUDY: Fretting Damage and Associated Fatigue

Client: We'd like you to determine why this hard steel component broke.

Discussion: Figures 5.19A through 5.19C (next page) are increasing magnification views (details) of a tapered cylindrical part that had a poorly matching tapered cylindrical part fitted over it. The parts had been breaking at the edge of a worn patch. Instead of having a uniform contact surface, there were some areas that were not touching at all. Furthermore, an out-of-round condition on the mating part meant that it rocked back and forth relative to the stable position of the component shown.

Clearly the surface has some rough areas between smoother areas, as depicted in Figure 5.19A.

Note the magnification settings in Figures 5.19A-C. We have done what I always advise against: zooming right in to the micro-scale. In this case, it is to protect client confidentiality.

Figure 5.19B now renders visible some small particles. Again, we have evidence of charging interfering with the image quality. Cleaning the part would have also removed the particles that were clinging to the component and providing data about the wear scenario.

Figure 5.19A: *Fretting damage "micro-scale overview"*

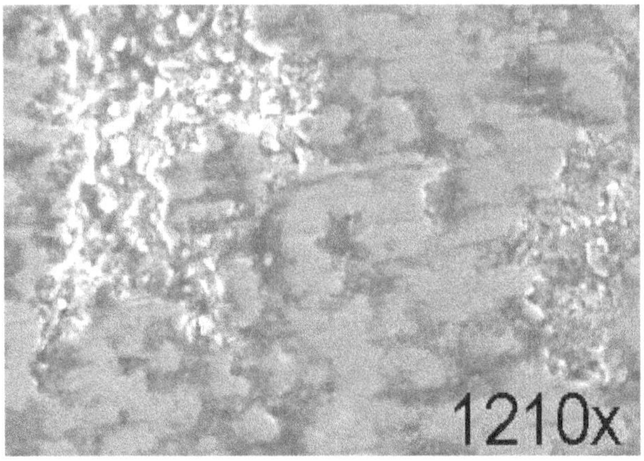

Figure 5.19B: *Fretting damage showing severely roughened surface*

Note that the cracks are only visible when the magnification is over 2000x as in Figure 5.19C (below). These were not the cracks that caused the component fragmentation. They were nearby. These "secondary cracks" allow us to imagine the sequence of damage that leads to the eventual fragmentation of the component. The "real problem" was wear, even though the inquiry was fracture. Fretting is often considered a "complex" damage "mechanism", because it combines oxidation, a corrosion process, and wear, mechanical removal of material, often leading to fracture, resulting from surface roughening. It's both complex and complicated!

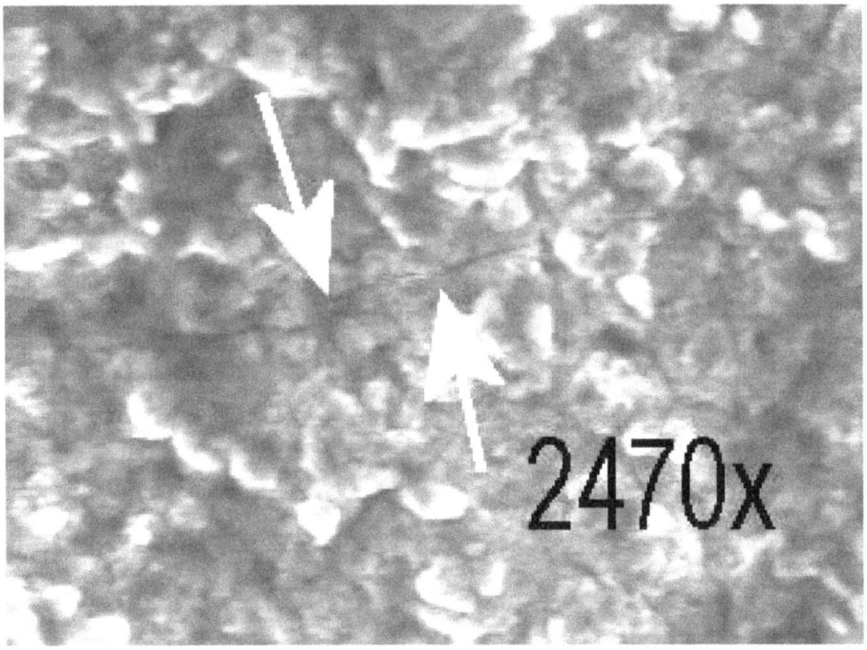

Figure 5.19C: *Note the crack is now barely visible in roughened area. This is not the crack that caused the fragmentation.*

5.12.4 Closing Thoughts on Wear

A few years ago, I was on a team tasked with making a decision about hiring a tribologist. The company had a lot of metallic and ceramic components that were subject to wear. The HR department had finally found someone they hoped was a suitable candidate. The team listened to his presentation on his latest research project.

In the corporate world, communication is as important as technical skills. We wanted to make sure that at least some of the other engineers working at the company would be able to communicate their requirements to him and understand his recommendations.

For whatever reason, I had in my mind the wear of clothing and textiles. I had come to understand that most normal "wear" did not take place while the garment was being "worn" by the human. Rather, the "wear" was a deformation and fracture mechanism, primarily happening during laundering and during hot air machine drying. The individual filaments unwind from the spun threads, protrude, and break off. That's what makes the lint we find in our washer or dryer lint filters. I wanted to get an idea of how flexible the job candidate was. While it isn't one of the most popular engineering disciplines, textile engineering exists and I had an acquaintance who was a textile engineer.

Our candidate was not very happy that I asked a question that had nothing to do with his tribology research. It caught him off guard. But it's valuable to be able to see that the process of fibers breaking off during laundering has similarities to metal particles flaking off of a machine component. It shows flexibility of mind. It shows that the concepts available to describe damage to a component are not just stuck in a narrow mental filing cabinet.

Who do you want to work on your team? Someone who is able to see that the "wear" of "wear and tear" in textile garments is mostly associated with "wash and dry"? Or someone who sticks to a narrowly defined description of the problem – perhaps misleading an investigation? Now textiles that are used to cover chairs, couches, and car seats are a different story because they are not typically laundered!

5.13 Understanding Corrosion and Environmental Damage

The study of Corrosion and Environmental Damage was disorganized and incomplete prior to the work of Mars G. Fontana. Under his leadership at Ohio State University, members of the metallurgy department systematized a large body of work. The first edition of *Corrosion Engineering,*[103] was published in 1967. I have the 1986 release.

In Section 1-3 of his book, Fontana rejects a definition of corrosion that applies only to metals. He considers that deterioration of paint or other polymeric material due to sunlight is just as relevant to corrosion engineering as is attack of the grain boundaries of a rail car wheel by copper (Liquid Metal Embrittlement). He includes rusting of exposed carbon steel in wet air and pitting attacks on austenitic stainless steels in chloride environments.

Fontana classifies eight forms of corrosion. He does not include detrimental changes to microstructure. The damage categorization system introduced in this book is partially consistent with Fontana's. I would include Liquid Metal Embrittlement (LME) in "Complex" or "Mechanisms", rather than corrosion, because we have both a foreign substance (copper contaminating steel) and an energy source (heat).

Conceptually, I think that spheroidization of carbides in alloy steel used in high temperature steam piping applications in power plants is similar to sunlight breaking molecular chains in transparent polymeric materials. But, as this type of damage progresses in high temperature steel piping (used in power plants) decarburization and oxidation accompany the change in the microstructure features. A more direct

---

103 A respected colleague insists that Fontana's organization of corrosion categories in *Corrosion Engineering* is inadequate for current failure analysis work and this colleague does a lot more corrosion work than I do. Personally, I have found Fontana's structure very helpful. But I issue a warning here: *this section of the book may have some limitations for people who specialize in corrosion and damage in severe environments. If you fit that description, you likely already have other resources on which to base your understanding of corrosion categories.* If you are new to corrosion and damage in severe environments work, be sure to consult other relevant works and people who work in the field.

comparison to the effect of sunlight on polymers (energy impingement) would be an aluminum alloy that was briefly exposed in service to a high-enough temperature. This could create incipient melting at grain boundary triple points.

Look at Figure 5.20 (below), depicting a portion of a propeller shaft from a pleasure boat. Would you call this crevice corrosion? Is it pitting corrosion? Is it galvanic corrosion? Is it all of the above?

Corrosion categories, like damage categories in general, are to some degree arbitrary. What a Failure Analyst needs to understand is that if the client is an insurance company, they want to decide if they are obligated to pay the boat owner for the repair. Sure, we could assign damage category names that may seem to help (but do not help), but what we really would benefit by asking is, "What happened? How did it happen?"

Figure 5.20: *This section of a stainless steel shaft was used on a pleasure boat. The left end was kept mostly dry. The right end was protected galvanically by a "Zinc".*[104] *Deep pitting corrosion was found at the unprotected area.*

As I have noted elsewhere, the definition of boundaries for any particular damage category is arbitrary to some degree. If we choose to devote our careers to corrosion, we might want to leave microstructure changes of metals to someone else.

Corrosion has a long history of being thought of as a chemical attack that removes portions of a metallic component by changing the metal atom into an ion; at that point it is then susceptible to returning to its preferred thermodynamic condition as part of an oxide or sulfide.

Throughout this book, I use the term "corrosion" to refer to that which occurs as a result of changes to the electron structure of a metallic material - changing the atom back to a state more in line with how it was found when it was part of

[104] This definition is repeated in Section 5.17: Boat propeller shafts often have sacrificial zinc anodes attached to them (these zinc anodes are called "Zincs" by people who spend a lot of time with boats). Zincs are intended to prevent corrosion.

the Earth itself.[105]

- Such corrosion, when it occurs in an aqueous environment, is electrochemical in nature.
- In a dry environment, it might be a purely chemical process, driven by simple thermodynamic considerations.

One of my peers told me that UV damage to polymers in the form of chain scission should not be included in microstructure damage. I get the point. It's not a type of damage that will be visible with materialographic methods. Should I create yet another category?

> *One of the mushroom identification books (yes, one of my hobbies is mycology) writes about this type of argument among fungus experts. Do individual experts tend towards "lumpers" or "splitters?"*
>
> *A lumper tends to group things based on partial similarity. A splitter tends to divide a new category at the slightest non-conformance to a single feature from the foundation group that inaugurated the category.*
>
> *I guess when it comes to polymers damage due to UV (in the form of chain scission), I'm joining the lumpers category. I'm always learning and as I learn more, I may change my mind. Maybe we will meet some day and I will have moved to the splitter camp!*
>
> *For example, I recently came across the outrageous decision of zoological specialists to put pandas in the carnivore family - but they eat only bamboo. They lumped pandas into carnivore category because of the shape of their intestinal tract. I may have just taken a step toward the splitters!*

105 I define this more natural earth state or more stable atomic configuration as one form of atomic comfort.

5.14 Eight Basic Types of Corrosion

I spend a lot more time working on mechanical damage than chemical damage, so I have many case studies of mechanical damage, and few for corrosion. My goal is to familiarize the reader with as many of the eight sub-categories of corrosion for which I have case studies. The student of corrosion who chooses to study the work of Mars Fontana further will be rewarded.

Given the Fontana scheme, where would we put the fully reversible softening of polyamides as a result of water absorption? I am not going to let the need to conform to a basic damage class keep me from doing the job of the failure analysis!

The basic corrosion categories I will discuss are:

1. Uniform Corrosion
2. Galvanic Corrosion
3. Crevice Corrosion
4. Pitting Corrosion
5. Intergranular Corrosion
6. Stress Corrosion Cracking

The categories I will not discuss (because I have no publishable case studies) are:

7. Selective Leeching
8. Erosion Corrosion

5.14.1 Uniform Corrosion

Uniform corrosion is the easiest to understand. It's the case where we have a material in contact with an environment that is aggressive toward it. Moist air in contact with bare steel will initiate the formation of an oxidation product that is usually some form of hydrated iron oxide. In this case, the entire surface of the steel part will be attacked.

> *I do not believe I have ever seen true uniform corrosion in my personal practice.*

What is important to understand about uniform corrosion is it's the easiest type to study quantitatively. If we can manage to set up a test cell where a coupon of the material can be relatively uniformly exposed to the environment of interest, we can track corrosion by simple weight changes. This allows standard corrosion rates in specific environments to be published and, thus, expected service life may be readily calculated. *If nothing changes!*

Has the severity of corrosion due to industrial atmosphere changed between 1945, 1970, and the present? How about an ocean side environment? Still, compared to all the other types of corrosion, this type is relatively predictable.

Corrosion may remove or add to the component weight, sometimes simultaneously. If a piece of steel rusts, oxygen and probably water are being incorporated into the test coupon. However, if the rust flakes off, it takes with it the added oxygen and the added water (from hydration) and some portion of the iron (and other alloying elements in the steel) that are no longer in the metallic state due to corrosion.

Uniform corrosion is a necessary category that allows us to build a foundation to understand the other types of corrosion processes. Note: some experts call this type of corrosion "general corrosion".

> *I was in a situation with a chemist (not a corrosion specialist) who performed a corrosion resistance test. The individual was reasonably astute. He made sure that welded portions of the assembly being tested were subjected to the aggressive service environment, as well as areas that were far from any weld effects.*
>
> *The change of weight over time was used to evaluate the severity of the corrosion. If a metallographic cross-section had been made after the weight change data had been collected, it would have been clear, despite the lack of weight loss for any of the coupons, that the weld heat affected zones were preferentially attacked.*
>
> *In actual service, the thinned down material adjacent to the welds allowed the normal service pressure to create hoop stresses in excess of the tensile strength of the material. His test design was good, and the testing he did was competent, but incomplete, resulting in a mistake in the data interpretation.*

Perhaps the biggest difficulty with data use from tests assuming uniform corrosion is from people who are not familiar with the details of how corrosion actually happens in real components. They may calculate a life expectancy based on an average thickness loss, while any real component is unlikely to be completely flat and completely free from variations in temperature and stress that may accelerate the damage in a localized area.

> *My practice has been mostly with automotive applications and, more recently, pleasure boats. I occasionally work with plant piping. As noted earlier, I don't believe I have ever seen a completely uniformly corroded part. Even flat panels tested in a perfectly proper ASTM B117 salt spray chamber tend to have stains and spots that are not uniformly distributed. Why might this happen? Let's consider some example*
>
> - *Structural steel is often covered by a high temperature oxide.*
> - *Cold rolled sheet steel usually has an oil film to prevent rust. Once the sheet is turned into a product, it is usually painted.*
>
> *Now, even the most straight-forward corrosion gets a bit complicated. If there is a bare spot in the high temperature oxide or an area missing oil, these locations will be attacked first. Alternatively, even if the coating started out adequately uniform, exposure to the environment is never exactly uniform. For example, if a stack of steel sheets is stored under a protective roof in a rainy climate and one vertical edge of the stack is directly under a hole in the roof, then that corner of the stack will probably start to rust first.*
>
> *Here is another example from years of practice. A client brought me a piece of carbon steel that had reportedly been used for over a decade as a liner in a wash tank for a coating process. A potential supplier convinced them to try out a new additive to the wash water. Three days later, there was a hole approximately 5 cm in diameter in the wall of the tank. The corrosion on the outer surface of the tank was virtually non-existent. The inner surface, which had been in*

contact with the new wash chemical, looked uniformly red: except for the hole. Upon inquiry, it was revealed that the heating element was directly behind the now perforated panel.

So, the uniform corrosion was not exactly uniform!

5.14.2 Galvanic Corrosion

Galvanic corrosion occurs when two materials are in contact with each other in an environment that is more aggressive to one member of the "galvanic couple". The contact need not be direct. Moisture (a conductive liquid) can carry electrons from one material to the other. In this case, the more susceptible material may be attacked. See Figure 5.21 (below).

Figure 5.21: *Galvanic corrosion attacks the carbon steel despite the paint – this is in a highly corrosive environment; it is clamped between two pieces of stainless steel.*

Intentional use of galvanic coupling is often used in my industries. Manufacturers put a layer of zinc coating on a steel component. The zinc corrodes much more readily than the underlying iron and any stray electrons will act to effectively remove the zinc and leave the iron alone. Even when a breach in the zinc coating occurs, the surrounding zinc continues to corrode preferentially and leave the iron intact.

Boat propeller shafts often have sacrificial zinc anodes attached to them. Zincs protect stainless steel shafts from corrosion.

Similar to zinc, aluminum is also more readily attacked in moist environments than steel. Some do-it-your-selfers (DIYs) find this out the hard way when they staple aluminum screen to a wooden frame using steel staples. In this situation, the aluminum screen corrodes readily and can be damaged to the point of failure - the tiny aluminum wires completely convert back to aluminum oxide. Aluminum doesn't rust and it doesn't turn red, but it does corrode! Sometimes, spectacularly, as you will see if you study the additional categories of corrosion in the major metallic alloy systems.

REMINDER: This is not a corrosion science or corrosion engineering book. It's a failure analysis book. Our purpose is to help you recognize the major categories of corrosion. Other authors will provide more information on the details of the corrosion process.

An important purpose of this book is to help you recognize major categories of corrosion.

One of the major misconceptions about galvanic corrosion is that there is an order of susceptibility to the damage type. Industrial charts list the galvanic series. Some people (mechanical and design engineers among them) think that all they have to do when vetting a new product design is make sure that there are no exposed galvanic couples that are far apart on the series chart. There are two major problems with this oversimplification.

First, you must be sure that the chart you are using is the relevant one for the environment in question. The most common environments in galvanic series charts are either salt water or plain water (whatever that is).[106] The order is different on each chart. If you are going to be in a more exotic environment, the order may deviate significantly from whichever of the more common charts you have consulted.

The second major problem with a galvanic series chart (used inadequately) is that the relative exposed surface area for each member of the galvanic couple has a large effect on the corrosion rate. For example, a large steel structure fastened with stainless steel bolts could be a robust design. The number of electrons lost by the iron in the plain steel structural members will be spread out over a large area.

It's possible we could have a similar situation to the steel staples used with the aluminum screen; the material being clamped by the stainless fasteners would corrode faster than the portions of the structural member farther away. But this might be an acceptable design, especially if the fasteners could be retightened as needed.

The opposite situation (carbon steel fasteners used on a large stainless structure) is likely a very bad idea if corrosion is a concern. And why else would stainless be specified? It's much more expensive! With so much less surface area than the large structural members, the fasteners could be prone to a rapid demise. I'm not saying this can never work; people get away with questionable designs all the time. However, it would be a good idea to diligently investigate before specifying such a material combination.

The equipment history for the assembly shown in Figure 5.21 was undocumented. But I'm fairly certain a corrosion specialist was not consulted for this assembly design. It is also difficult to believe that the original equipment manufacturer specified small pieces of carbon steel in direct contact with stainless steel to be placed in an aggressive industrial environment. Instead, this invitation to galvanic corrosion was most likely a fix put in place to combat an excessive corrosion problem. This fix probably included original carbon steel replaced by stainless. This is a fine example of the law of unintended consequences. In this case, the stainless itself performed well, but it accelerated the damage to the adjacent component.

THOUGHT EXPERIMENT – HOW COULD IT HAPPEN? Let's consider a coated steel product that has a missing area in the coating, a "holiday". The steel exposed to the environment represents a tiny surface area compared to the area covered by the unimpaired coating. Thus, we have created a galvanic cell.

Steel corrodes at that spot, preferentially protecting the rest of the material. See Figures 5.22A and 5.22B (next page).

[106] The word "water", in many Indo-European and Semitic languages, is etymologically related to the word "what". Over 5,000 years ago, when our modern languages were first taking their earliest recognizable forms, every new source of water was apparently considered an unknown.

This is one way to create pitting corrosion attack when the nominal composition of the material and the environment would have led us to expect uniform corrosion. This coating holiday may not be attributable directly due to the end product manufacturer's process. It could be the mill scale (oxide) present on hot rolled steel. Maybe there's a scratch or a dent that removed the oxide.

Even a microscopic galvanic cell, over time, can grow and subsequently cause extensive damage.

EXERCISE 5-I: CRITICAL THINKING – Confirming the Context

Follow along with my thoughts. Take a moment and think back to uniform corrosion. In a way, conceptually if not technically, classical uniform corrosion may be viewed as galvanic corrosion where atoms from the part are being sacrificed to unknown or unspecified items in the environment or "the whole earth". We have earlier defined critical thinking as confirming the context or the frame of reference. The frame of reference of the system here is the part subject to corrosion and "the rest of the world" that is acting like an electron sink.

Another aspect of the exercise is stray current corrosion. This is sometimes suspected when a totally unexpected corrosion event is discovered or when corrosion happens at a significantly greater rate than expected. It is often maddeningly difficult to prove or disprove.

As we noted earlier, all classical aqueous metallic corrosion is electrochemical in nature and involves galvanic considerations. There may or may not be anything different or odd looking about the results of a corrosion process that happens due to a stray current.

Figure 5.22A shows the outside of a corroded pipe. Figure 5.22B shows the heavy deposits on the inside. Note also the position of the hole in Figure 5.22B at the edge of the deposit. The deposit is heaviest at the bottom (in service) due to gravity.

The astute reader will consider that there are two factors potentially acting synergistically to corrode the same position, one from the outside and one on the inside. Although it is also possible that the hole created by the crevice at the edge of the gravity related deposit was followed by the stain on the outside surface.

NOTE: Thinning adjacent to a hole is a powerful clue to determining which surface corroded first.

Figure 5.22A: *Galvanized outside surface of a pipe with dark discoloration around a hole*

Figure 5.22B: *Inside surface of the same pipe with the same through-hole location depicted in Figure 5.22A*

5.14.3 Crevice Corrosion

My first true understanding of crevice corrosion came about by two events, closely spaced in time.

The first event happened in an industrial situation. Two pieces of aluminum plate had been clamped together with threaded fasteners. The aluminum plates reinforced a mechanical attachment to the wall of a tank. It was an aqueous environment. The corrosion was most severe along a line at the edge of the smaller reinforcing plate. The corrosion attacked both the smaller plate and the vessel wall to which it was attached.

The second case occurred as I brushed my teeth when I suddenly noticed grooves at the gum line along my back teeth. I immediately thought it may be crevice corrosion. I told this story at a local dinner meeting of our ASM International chapter. A member's wife disputed my assertions claiming the groove was wear, more likely from brushing my teeth too hard or in the wrong way, or with too stiff a brush.

It was a plausible scenario, yet I remained unconvinced by her diagnosis. I researched the subject and learned that both of these damage types (mechanical wear from brushing and crevice corrosion at the gum line) can occur in human teeth. Due to my tooth brushing frequency at that time, the grooves in my teeth were more convincingly explained by crevice corrosion than by wear.

Once again, classical crevice corrosion in metallic components is a result of electrochemical action in localized galvanic cells. The micro-environment in the crevice may be vastly different from the surrounding areas. In some cases, crevice corrosion can occur within hours. This is due to localized anti-entropic diffusion of aggressive ions into the crevice. Of course, this may not be the only reason for accelerated rates of corrosion.

Crevice corrosion is frequently found in piping systems used to heat buildings. Deposits from minerals precipitating from the flowing water can stick to the bottom surface of the pipe. This creates a protected area under the deposit and an area more prone to attack just outside of the deposit. We saw this exact pipe situation in Figure 5.22B.

Figure 5.23A (below) shows a cross-section of a shaft with classical geometrical crevice corrosion, and the smoothness of a fatigue crack. Figure 5.23B (of the same part depicted in Figure 5.23A) makes it clear to the knowledgeable observer that the crack started at an oxide filled pit. Is this crevice corrosion? Pitting corrosion? Was the rate of corrosion enhanced by the more corrosion resistant copper alloy (galvanic corrosion) sleeve?

Figure 5.23A: *Shows broken shaft with fatigue crack initiations (white arrow, far left) – crack is initiated in a crevice under a bronze sleeve (black arrow). Shaft diameter is approximately 10 cm.*

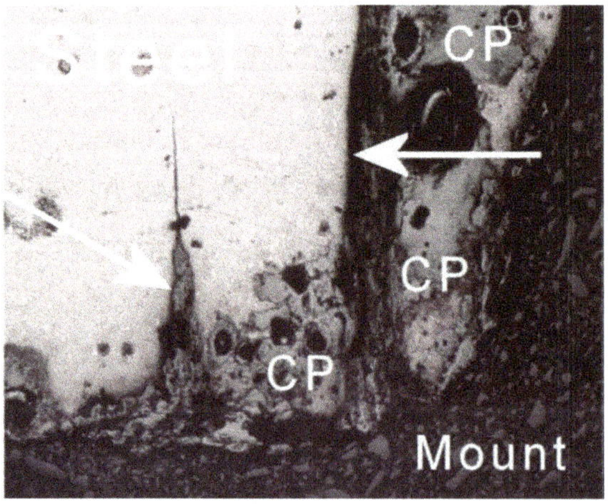

Figure 5.23B: *Shows a metallographic cross-section of the part from Figure 5.23A – the main crack profile is at the right arrow; and a secondary crack profile is at the left arrow. The original magnification was 50x; nital etch.*

5.14.4 Pitting Corrosion

Pitting corrosion has a similar mechanism to crevice corrosion, but it happens in the absence of any specific geometric crevice feature in the design. A bit of dirt can create a microscopic crevice at its edge. Thus, pitting corrosion is often more severe on the bottom side of a horizontal pipe run.

We just examined and discussed Figure 5.23 for crevice corrosion; we could just as easily consider it pitting corrosion - *if we did not have evidence* that the corrosion extended in front of and behind the plane of the page. Materials engineers need to learn to turn two-dimensional data into three-dimensional data.

Surface roughness is often enough to create a microscopic crevice in susceptible materials in particular environments. I am a generalist Failure Analyst and the most common pitting corrosion situations that I run into are in copper water lines and copper fuel lines, and 304 stainless steel components used in aqueous environments where chloride ions are present.

Figure 5.24 (below) details a pit partially covered with oxide on the inside surface of a copper pipe.

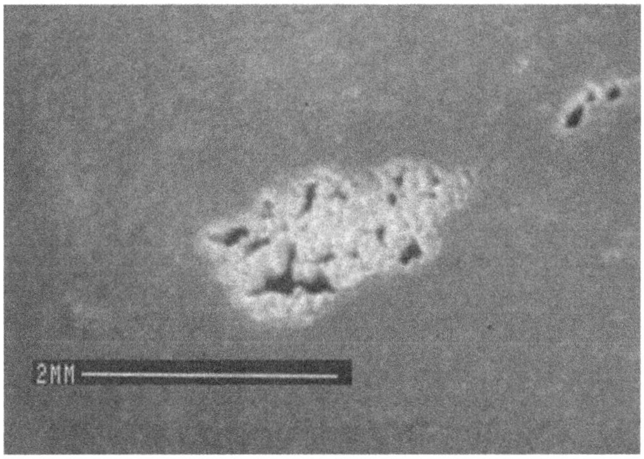

Figure 5.24: *Copper oxide "lace" covering a corrosion pit*

Pitting corrosion on 304 stainless steel can be the initiation site for subsequent stress corrosion cracking. That was the case shown in the severely spider-cracked plate material shown in Figure 5.25 (next page).

I followed the maintenance engineer to the table where the subject of the investigation was positioned. From twenty meters away, I told him, "It's stress corrosion. You have too high a level of chlorine in the environment."

His immediate response was, "Yes, there's chlorine in the environment, but it doesn't contact this component, which is in a protected housing."

This was a big lesson for me as a consultant. Some clients (most people!) are not happy when you can instantly see the solution. Maybe they feel bad that they didn't know the answer as quickly. In this case, they just had their boss cut a purchase order for me to visit the site and investigate; in five seconds before even approaching the part within arm's length, I announced the answer. Even though this was nearly twenty years ago and my eyes were better back them, I'll long appreciate this client for helping me understand the details of consultant client communication.

Communicating capably doesn't always mean in the simplest, most direct manner. I ended up performing some lab work to document the presence of the chlorides using standard metallurgical lab techniques. I was also asked to sample the material that was creating the "environment" and the client was surprised to find a hundredfold higher concentration of chlorides than they thought they had. They later informed me that their internal investigation revealed that someone had decided that if a little chlorine was good in this process, more was better. This is another example of the law of unintended consequences.

Stress corrosion cracking (SCC) is a combination damage category that needs an aggressive environment and a tensile stress, as well as a material and process condition susceptible to the aggressive substance. SCC is depicted beautifully in Figure 5.25.

Figure 5.25: *Stress corrosion cracking often starts as microscopic pits. Chloride attacked this 304 stainless steel.*

Since I am a Failure Analyst and not a corrosion engineer, I find it is essential to be familiar with the common "material-environment couples". I also find it essential to familiarize myself with basic processes by which the couples interact and facilitate corrosion. As a professional Failure Analyst and a business owner, I can recommend simple fixes in some of these cases.

For example, molybdenum containing 316 stainless steel resists chloride pitting attack better than 304 stainless. Minimizing pitting often maximizes the life of a structure. Even if the 304 has served a long life, clients often desire to extend the life of the replacement. Suggesting a more appropriate material for the application, such as 316 for chloride pitting resistance, is often appreciated.

5.14.5 Intergranular Corrosion

Intergranular corrosion tends to occur in specific materials in specific environments. Often it is related to a materials selection or processing error. Sometimes stress corrosion is intergranular, but often it isn't. This highlights the necessity to get the background data. In my system of categorization, intergranular stress corrosion would be a damage mechanism while pure stressless intergranular corrosion would truly be a subcategory of corrosion damage.

Intergranular corrosion may be a result of local composition differences in the grain boundaries, rendering them vulnerable to corrosion. Figure 5.26A (next page) depicts a scanning electron micrograph of intergranular features created by an aggressive acid environment. This corrosion could have been facilitated by improper heat treatment sensitizing grain boundaries (processing error).

Figure 5.26B shows a metallographic cross-section of the same Subject Specimen.[107]

Figure 5.26A: *Scanning electron micrograph shows classical intergranular (or Rock Candy) features.*

Figure 5.26B: *This cross-section (same Subject Part) shows a portion of the crack profile. Dark lines surrounding polygon-shaped grains are probably a result of metallographic mounting compound filling the cracks.*

Until the latter half of the 20th century,[108] 300 series (austenitic) stainless steels were susceptible to intergranular corrosion along weld seams due to precipitation of chromium carbides in the heat affected zones. Because some chromium is tied up as carbides, the amount of chromium left in solution is reduced within the iron-chromium-nickel matrix. *(If you are not familiar with microstructures, do not fret! We'll be covering basic information on microstructures, including grain sizes and grain boundaries, in Chapter 6.)*

BOTTOM LINE *(in this situation)*: carbide-lined grain boundaries are lower in available chromium, which is used to make a protective chromium and oxygen containing layer. This is an important part of how stainless steel is made stainless. So now, the astute reader will be correctly thinking that we have created a network of galvanic cells between the grain boundaries and the grain interiors. This type of intergranular corrosion is another example of traditional corrosion categories that fit into the electrochemical / galvanic corrosion type. Each category is intended to help us recognize what is happening, and what to do about it.

5.14.6 Stress Corrosion Cracking Corrosion

We have defined SCC as a damage mechanism, but Mars Fontana considered it a basic type of corrosion.

Figure 5.27: *The Panda[109] reminds us that categories are arbitrary. Pandas eat only bamboo but are categorized as carnivores because of the structure of their guts.*

Stress corrosion cracking occurs at stress levels as low as ten percent of the value expected to cause cracking in the absence of one of the specific aggressive chemicals to which the Subject Material has susceptibility. It is easier to show the relationship than write it. See Figure 5.28 (next page).

[107] Figures 5.25A and 5.25B are used by permission, courtesy of Larry Hanke, Materials Engineering Inc.

[108] At the end of the 20th century, most steel mills started making all 304 as 304L, a low carbon version that prevented chromium carbide precipitation.

[109] Photo by Kirandeep Singh Walia through Pexels @ https://www.pexels.com/@kirandeepsingh/

Note that the stress level at which stress corrosion cracking initiates and propagates is a function of the environment. Stress cracking (SC) is essentially the same phenomenon but the technical term, applied correctly, is used only for polymeric materials.

When I am wearing my "metallurgist hat", I find the term "stress cracking" a bit silly, since all true cracking is a result of stress.

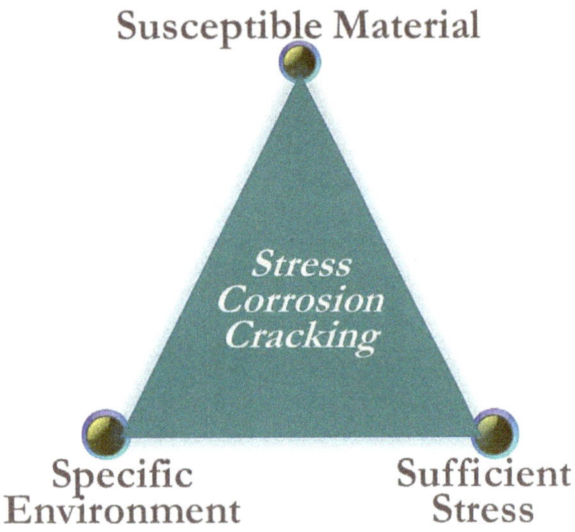

Figure 5.28: *A susceptible material in relation to a specific environment at a sufficient stress level will cause stress corrosion cracking.*

Figure 5.29 shows a typical branching crack pattern in a metallographically-prepared cross-section of stainless steel afflicted by SCC.

5.14.7 Selective Leaching and Erosion Corrosion

Because of business confidentiality, I have no case studies that I can share for:

1. Selective Leeching.
2. Erosion Corrosion.

But don't let that stop you from delving deeper. The student of corrosion who chooses to study the work of Mars Fontana will be rewarded. It is a great place to begin.

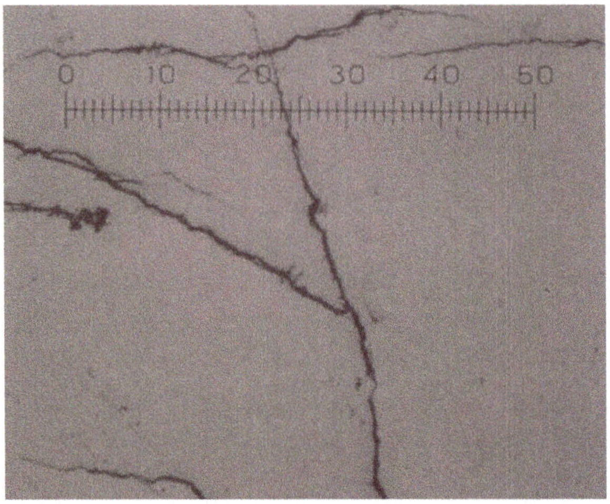

Figure 5.29: *Typical multi-directional branched cracks in stainless steel SCC damage (as polished) – the 1-50 marker bar is 1 mm.*

5.15 Microstructure Degradation

Microstructure damage is often invisible until it results in one of the other classical mechanical damage types. While investigating the classical mechanical damage, the expert Failure Analyst can find the microstructure damage.

For example, if you are called to a power plant to look at cracked steam pipes, you will already have in mind the potential for microstructure damage. The experienced Failure Analyst will know to look for microstructure damage as a result of the inquiry conversation with the client.

The desirable microstructures of modern materials are obtained by processing materials according to controlled methods. They often require finely balanced simultaneous inputs of heat (or cold!) and plastic (permanent) deformation. These optimized processes often rely on the presence and interaction of dislocations and/or second phase particles.

If you do not have a materials engineering background, you may not be familiar with the terms, "dislocation" and "second phase" particles.

WHAT IS A DISLOCATION? (PHYSICAL METALLURGY)

DEFINITION OF A DISLOCATION: A linear discontinuity within a single crystal or grain in a metal.

DEFINITION OF A CRYSTAL: An atomic structure with a regular, repeating order in three dimensions. Imagine a crystal, as shown in Figure 5.30 (below), where "Mother Nature" missed her count! There is an extra "half plane" (or missing half plane depending on your viewpoint) of atoms within the crystal. This has implications for how the material responds to loads and stresses.

A single dislocation within a single crystal will make plastic deformation easier (occur at a lower yield strength), but a high concentration of "tangled" dislocations could result in higher yield strength!

While individual dislocations are invisible even with the most powerful optical microscopes, new techniques in scanning electron microscopy may reveal them. They really exist!

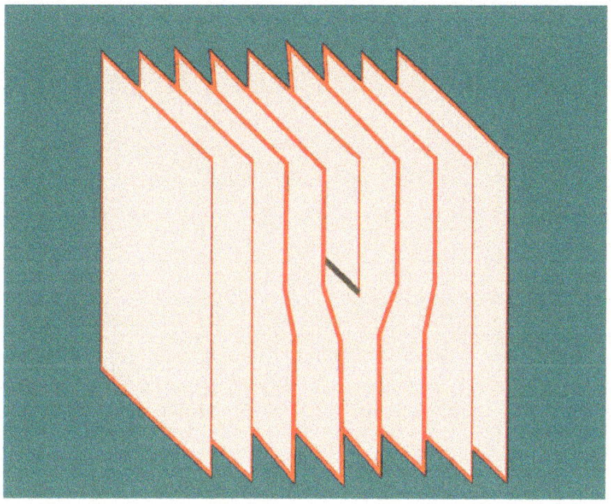

Figure 5.30: *Edge dislocation*[110]

DEFINITION OF A SECOND PHASE PARTICLE: A particle that is embedded in the matrix of the alloy and is of a composition that differs from that of the matrix (surrounding material), and is "engineered" to optimize material properties.

Just for fun, I'll explain a dislocation through another technical term and viewpoint: stacking faults. In Figure 5.31 (next page) we see three rows of corn taper down to two irregular rows on the left end of the top ear. A physical metallurgist could consider this a stacking fault. The lower ear has an extra (or missing) kernel; the kernel doesn't "know" which row it now belongs in.

Interaction of dislocations with other dislocations and second phase particles, and the presence of certain types of atomic bonds at the boundaries of second phase particles, results in a distribution of areas with locked-in strain energy. Creation and manipulation of these high strain energy areas is the foundation of many high strength conditions.

[110] Figure 5.30 is provided courtesy of the creative commons license. At the time of publication, the image can be found at the following url: https://commons.wikimedia.org/wiki/File:Dislocation_edge_d2.svg The original uploader was Wikityke at English Wikipedia. SVG made by User:Magasjukur2 [CC BY-SA 2.5 https://creativecommons.org/licenses/by-sa/2.5]

***Figure 5.31**: A stacking fault; by example of corn*

NOTE: If you are interested in how second phase particles create strengthening, a good research phrase is "concepts of coherent, semi-coherent, and incoherent boundary energy".

Strengthening obtained through dislocation tangles and control of size distribution of second phase particles is meta-stable. This means that the strengthening is generally reliable at room temperature, but at higher temperature when the atoms have enough energy to move around more frequently (through Brownian, or random, motion), they will often find more comfortable positions. When atoms find more comfortable positions, they lose their strengthening effects.

The main factors associated with greater atomic comfort are:

- a lower concentration of dislocations
- coarser (larger) second phase particles

We will do well to remember that change is always happening within engineered materials. Sometimes the rate of change is so small we don't need to be concerned about it. But in the physical world, atoms are always vibrating. In temperature ranges where humans evolved to live, atoms vibrate fast enough to occasionally change places within the solid state.

Higher temperature industrial processes enable much greater rates of atomic vibration and accompanying greater rates of diffusion (atoms exchanging positions with each other). If you take (or took) a physical metallurgy course, you will (or did!) learn that these particular microstructure changes (dislocation annihilation and second phase particle coarsening) tend to result in strength loss.

In general, greater comfort for the atoms means less strengthening. Other microstructure changes may result in a simultaneous loss of ductility. This would be the worst of both worlds. The component may eventually lose all as-built safety factors.

NOTE: This potential for microstructure changes while in service (and associated performance changes) is another reason we use the best practice of **Stop, Think, and Do Nothing!** before we conclude that the material used to make parts was out of specification. Chapter 9 details data interpretation and reporting.

> ***Stop, Think, And Do Nothing!*** **before concluding that the material used to make the "failed" part was "out of specification".**

Thus, when thermal energy continues to interact with material in service (especially over the long term such as years, decades, or longer), microstructure may evolve to a condition that is vastly different from its as-manufactured condition. It will likely not be as suited to the application as it was when it was new.

The nominally invisible microstructure damage may be detectable by various inspection methodologies, primarily metallography. Power plants often have requirements for nondestructive microstructure inspection of the surfaces of high temperature steel steam pipes. Inspectors look for changes in the pearlitic microstructure of the steel. These changes may include carbide platelets morphing into rows of spherical or spheroidal particles, dissolution of the small carbides, re-precipitation of spheroidal iron carbides in grain boundaries. Eventually, the carbon in the grain boundary carbides diffuses out towards the surface of the part, resulting in what is essentially a lower carbon steel than what was originally manufactured. Since carbon acts as a primary strengthener in many steels, we can readily imagine that the material is not as strong as it was on installation.

Figure 5.32 (below) shows a medium carbon steel tube microstructure from a power plant. It was in service for decades.

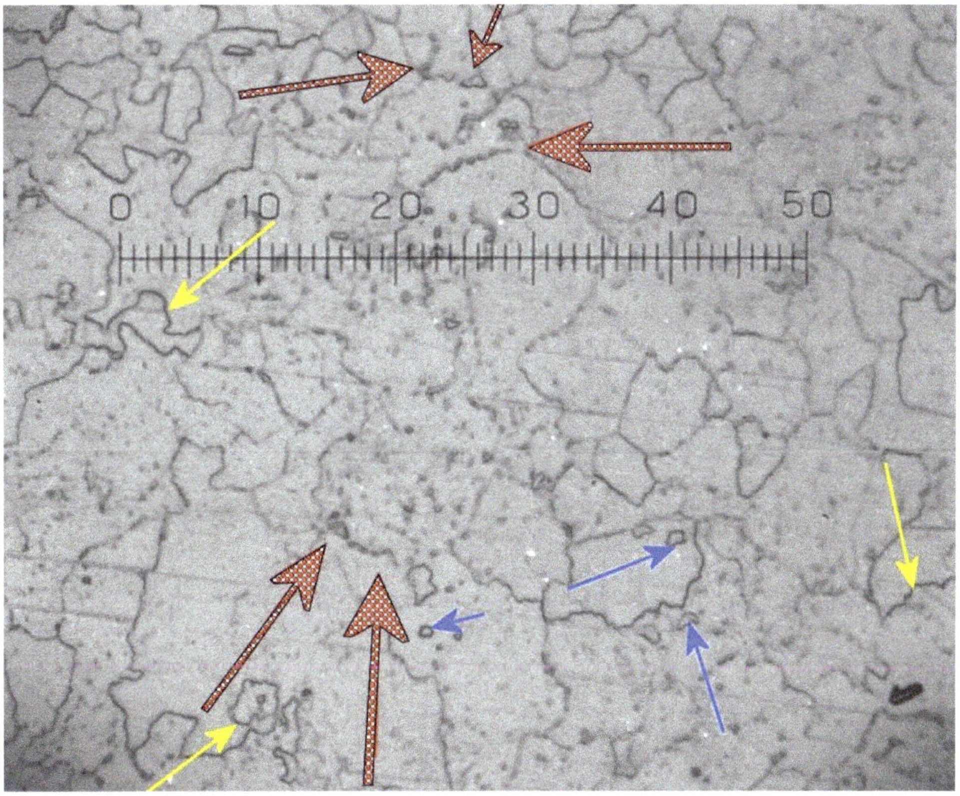

Figure 5.32*: Optical micrograph of a replica prepared from a metallographically polished and nital etched steel tube that was in high temperature service for many years – the 0-50 scale marker bar is 0.25 mm long.*

Note the small, rounded particles which are congregating along grain boundaries (red arrows, black borders), while others remain within the grain interiors (blue arrows). Note also that some of the grain boundaries have shapes that are far from the usual straight line or shallow arc (yellow arrows). The rounded particles are carbides.

CHAPTER 6

Practical Aspects of the Failure Analysis Process

The Practice of Failure Analysis

6.1 Overview

Quality failure analysis rests on the use of high-level human cognition. In Chapter 1 of this book, we articulated seven aspects of human mental processes that form the foundation of the way effective Failure Analysts use their mental skills. Let's review them here, in reference to the basic steps of a failure investigation.

6.1.1 Clarify Context

Define the frame of reference of the event, events, or service conditions that resulted in the failure.

The opinions of the stakeholders requesting or requiring the investigation are important to understand. This does not mean that we limit our suggestions for the analytical work to what they think will lead to the answer they want. But we must ensure that we have a way to address their concerns. If the stakeholders want suggestions for recurrence avoidance, we must understand the design and manufacturing processes, the procurement processes for the raw material or outsourced treatments (heat treating or plating, as examples), and service conditions. Low-key industrial failure analysis clients may choose to withhold much of this background data. That is up to them. It's a business decision. They are requesting an investigation of limited scope. In forensic analysis or government investigations, the background data will generally be revealed, sometimes with the help of orders from the court.

Being able to suggest an appropriate level of detail for the evaluation requires familiarity with the consequences of the failure. This is covered in greater detail in Chapter 7 on Project Scope Negotiation.

The tone that the lead investigator employs during discussions about background data collection and scope negotiation is of course related to the communication skills of the communicators. Persuading people that they'll be more likely to get the data they need if they provide the requested information is a skillset that requires a long time to develop.[111]

Communication skills (Step 7) are present in the background of every step.

6.1.2 Compare – Contrast

Compare–Contrast infuses every aspect of the investigation. Using standardized test methods whenever possible facilitates comparison of the Subject Part to others, as well as to published data and material specifications. The inability to use standardized test methods complicates interpretation of the data but is often a reality of this work. Perhaps the single most noteworthy item specific to failure analysis that is related to Compare–Contrast is selection of reference or control parts. This was briefly introduced in both Chapters 2 and 4.

6.1.3 Connect – Create

Connect–Create forms the foundation of creative thinking, "connecting the dots", getting to "the AHA! moment", where the data stops being data, and starts being useful information; perhaps sometimes even leading to breakthroughs in understanding or the emergence of new ideas. This may happen repeatedly over the course of a big investigation. Sometimes an insight will reveal that an earlier insight was incorrect.

6.1.4 Contemplate Complexity

Do we do this at the end of the project, after we have selected a preliminary idea for a fix, as was suggested earlier, or before choosing a fix? The experienced, competent and effective analyst cultivates a habit of evaluating the data as it becomes available, to ensure that all the relevant factors that influenced the production or service history of the component are revealed.

But some independent labs do not work this way. Analysts order a set of tests and evaluations; they may be mostly hands off until all the data is provided for preparation of the client report. This is not the preferable way to work, to my mind, as it reduces the chances to change the plan based on interim data.

A flexible process allows contemplation of the mutually interactive influences of factors revealed as the inspection proceeds. This provides a window into the functioning of a dynamic system. Seeing the whole system allows us to double check, triple check, and quadruple check our work, in an interesting way that sharpens the mind. Going down checklists, which must always oversimplify, dulls our intellect.

Those who perform failure analysis for litigation often end up in long negotiations about the test protocol. It can be difficult to pin down the description of what needs to be done. I once watched a group of people at a well-known lab do a leak test with contaminated solvent, when we were looking for evidence of contamination. They all had PhD's. They were not materials engineers with specialized knowledge of evidence preservation.

When you agree in advance to follow a specific procedure, and then realize that the product is not made the way you thought it was, it is highly desirable to have the flexibility to modify the testing to accommodate reality.

6.1.5 Calibrate Your Confidence

We need to make peace with the fact that certainty is limited to our eventual death and the necessity to pay taxes, as the old saying goes. Confidence is always limited, and in that knowledge, we can move forward, rather than becoming paralyzed by indecision.

Competent Failure Analysts make it a practice to include a description of sources of uncertainty that might cause significant doubt about the findings. If different interpretations are possible for some of the data, let the stakeholders know. As the expert who has performed the analysis, you are usually the one who is most suited to

[111] Contact Aliya Analytical for more information on developing this skill. | Or, request The Firebird Forum Newsletter of October, 2018, which may still be available through Richard Swanson on LinkedIn. It has 14 pages of advice on interviewing for engineering investigations.

pronounce an opinion about which interpretation is likely correct, and why you think so. This is a hallmark of a competent technical communicator. Calibrating your Confidence makes it easier to Communicate Clearly. You provide your reasoning, to the degree that the client is able to grasp it, rather than asking for blind trust.

6.1.6 Cultivate Curiosity

None of these thinking skill optimization techniques will work well if you have not learned a wide variety of things about the world. Continuing to gain knowledge about different subjects will help you keep your mind flexible. This practice will help you thin the mental filters that keep you from seeing what is present. Having a foundation of facts about the physical world gives you a greater chance of noticing anomalous (strange, uncharacteristic, inconsistent) data, which could otherwise lead you astray.

6.1.7 Communicate Clearly

The final report, whatever form it takes, in large disasters, will be studied by many people, and likely over a long time frame. If there is ever a recurrence, those doing the next investigation will certainly seek out the knowledge gleaned by previous investigators. The report, like the entire investigation, will benefit from being the result of a methodological process. This does not mean that the report has to be written in the dry and old-fashioned passive voice. It is ok to let the readers know when there was dissent about what tests to run, or how to run them. Letting future readers understand how the decisions were made (which tests were performed and how they were performed) is an important aspect of the communication work.

6.2 What Makes a Real Failure Analysis Investigation?

The three most essential differences between regular lab testing and the testing associated with failure analysis are:

1. the amount of mental work done to ensure that the physical tests will be informative (what is generally called "defining the scope" of the investigation),
2. the attention paid to the order in which the tests are performed (from less to more destructive), and
3. the ambiguous nature of the test results.

Evidence preservation, rather than immediate perceived utility of expected data production, must guide the order of the testing. Simply looking at the part may result in destruction of evidence, because in order to inspect the entire component, it must be moved from its original position and often, its original location. The position and orientation of the object is sometimes part of the evidence. Thus, evidence preservation marks the beginning of a materials-engineering-informed failure analysis.

As David Levy says, "To Observe is to Disturb."[112]

> *Sometimes, after the fact, you realize 'Whoops! Was the wide end of the crack on the left or the right side?" This happened to me this week, after 30 years of doing failure analysis. Fortunately, I had taken enough photos before disassembly to be able to see what the part originally looked like.*

Remember to prioritize preservation of both visible and invisible evidence while you plan and perform disassembly and / or sectioning.

Learning to respect the importance of evidence preservation is what will lead you to perform the tests in *the most advantageous order*. Note that I did not say, "the *right* order". While there is a general progression of standard tests used in failure analysis of structural metallic and polymeric materials, and their assemblies, it might vary a bit from project to project. The main variation is how much iteration is required in the first few

[112] David A. Levy, *Tools of Critical Thinking: Metathoughts for Psychology*, Second Edition (2009), [Chapter XXI, Page 142]; ISBN 978-1577666295, Publisher: Waveland Press Inc.

steps to document visual and macroscopic examinations; and it may include unique steps associated with revealing the damage.

And sometimes the most advantageous order is a business decision, rather than a technical decision. If parts are not immediately available for comparison to the Subject, you may decide to go ahead, and follow up with the comparison parts later.

Sometimes, the most challenging part of the job is revealing the crack surface. This is often the case in a crack that blunted or stopped before complete fragmentation. Think of a fatigue crack, which may render the object unfit for service, while leaving it in essentially one piece. Revealing the crack in those cases generally requires cutting the parts. Minimizing damage to the microstructure of the area near the cuts is a key part of preserving evidence. As you gain skill and experience, you may find yourself being able to extract a lot of information about the crack, even before you look at the crack surface itself. This is the power of understanding the loading geometry. Extra time thinking before cutting may save a lot of time in cutting. For large pieces of equipment, the old adage "Measure twice, cut once" is a good guide. Sometimes nondestructive testing can help us decide how to cut the part in order to reveal the crack surfaces. This will be described in more detail later in this chapter.

While corrosion and wear damage may also leave the component intact, these damage categories by nature affect a surface first. Sometimes you can perform a visual exam as the first step. Other times, disassembly will be required before the damage is visible.

Inspection with a borescope can be helpful to see what is going on inside piping. However, it may or may not be truly nondestructive. If there is a blockage, the borescope head may knock it out of place.

> *Years ago, I worked on an insurance claim where there was a question about whether a pipe was blocked. This was understood to be the key issue by all the parties present at the inspection. The pipe was removed, and it was handed to the expert for the owner of the home, which had been damaged by the failure. The expert was a professor at a local university. He held it up to the sky to see if it was clear. I - younger then than now (!) and younger than the other "expert" - watched in dismay as he then lowered it to the ground and banged it to loosen the blockage. This was, of course, before any of the other experts had been able to see the condition of the pipe interior. After banging it and watching some dirt come out, he smiled and said, "No blockage." This was beneficial to the case of the home owner. If the pipe had been blocked, the consequential damage would have been considered the owner's own fault. This is one of my prime stories for illustrating that highly, but narrowly, educated people are not necessarily the best Failure Analysts.*

BOTTOM LINE: Do the less destructive activities before those that are more destructive. Disassembly is generally destructive to some degree.

David Levy quotes R. D. Laing[113], as follows: "The scientific method is based on tampering with what would be happening if we were doing nothing to it."

Base your visual inspection and disassembly / crack revelation procedure on the most readily available practical manner of preserving both visible and hidden evidence.

What is hidden evidence? Part of hidden evidence are hidden cracks, which might require a nondestructive testing method to reveal them. The appearance of the surface on unopened cracks is another type of hidden evidence.

The initial stages of the investigation must focus on preservation of both visible and hidden evidence.

---

113 R. D. Laing, *The Facts of Life* (1976); ISBN: 0-394-49666-3, Publisher: Pantheon Books, a division of Random House, NY

Magnetic particle inspection, useful for finding hidden cracks at or near surfaces of magnetic materials, is considered a type of "Nondestructive Testing". However, it requires magnetization of the component. This may later interfere with image quality in the scanning electron microscope, especially if the component is not completely demagnetized between MT (Magnetic Particle Testing) and SEM (Scanning Electron Microscopy).

Microstructure is another type of hidden evidence, at least until it is revealed. Sometimes crack surface features will reveal something about the microstructure features. Hardness and other mechanical properties are certainly not just hidden, but invisible. Electrical conductivity, which may be relevant in some situations, is also invisible.

Some of these characteristics are very sensitive to even small amounts of heat or deformation. Compared to aluminum and other lower melting point and soft alloys, steel is usually quite forgiving in this aspect. But care must still be taken to keep heat and deformation away from critical areas.

This chapter will provide a mental structure to facilitate evidence preservation during both "nondestructive" and destructive testing. We might do well to remember what real estate salespeople say determines property values: Location! Location! Location! In our case as Failure Analysts, you won't go far wrong to remind yourself that failure analysis is about Evidence preservation! Evidence preservation! Evidence preservation!

When the testing is complete, you will have data. Hopefully you will have good data. Good data, by definition, is not necessarily data that you like, because it fits your preconceived notions of what "caused the failure". Good data, by definition, is data that you have a meaningful record of the frame of reference in which you collected it. Good data, by definition, is data for which you can explain, in detail sufficient to support interpretation, how it was obtained.

This will generally include a detailed map, or other record, of the locations of any micro-scale images, locations and orientations of test coupons extracted to facilitate mechanical, micro-structural, compositional, and other tests.

At the end of the failure analysis, sometimes you are still "scratching your head", wondering what really "caused the failure". As we have noted, what you are preferably asking, "What factors facilitated the damage?" But even as we clarify our own mental conversation, people in other work functions will still ask us why it failed. So, either way, even the most competent Failure Analyst will sometimes have to say, "I don't know why this XYZ Gizmo broke / corroded / wore out / stopped working." But if you have good data, you will rarely be at a total loss to suggest improvements in the design, or help the attorneys with information that allows them to decide who bears the responsibility for the problem.

Good data, by definition, is data with a meaningful record of the frame of reference in which you collected it.

Even with good data as defined above, sometimes you will not know what was different about the subject that made it prone to an unexpected demise. Usually, in the case of manufactured components, as opposed to civil structures, this is a result of inadequate "reference parts" for comparison (See Chapter 4). In civil structures, because most are built to well-recognized standards, this is sometimes the result of unrecognized and complex system interactions.

Another thing to remember is that sometimes you will be wrong. I found out years after I worked on one project that I was wrong about the cause of the fractures. But my advice to fix the problem still worked! Sometimes we get lucky.

Another lesson from the critical thinking toolkit is: the cure doesn't prove the cause, which is a quote by David Levy. [114]

FAILURE ANALYSIS IS ABOUT:

Evidence preservation!
Evidence preservation!
Evidence preservation!

THIS POINT IS CRITICAL

6.3 Basic Structure of a Failure Investigation

The aspiring competent Failure Analyst will always remember that the first step of an investigation or project is to do nothing. I call this "Step Zero".

Yes, that's right. Do nothing. Just sit or stand and do nothing.

This is difficult, especially for people who are steeped in Western culture. Asian culture has resulted in a higher tolerance for stillness and quiet.

If the silent "Step Zero" parts of the investigation are not done with the care that they need in the given situation, the value of the entire investigation may be lost.

The next step is to clear your mind. If you don't have a personal practice to do this, simply counting silently from twenty to one will be a good start.

> *Sometimes, this is not possible. Years ago, I arrived at the scene of a structural collapse, and saw one of the company's employees with a can of orange spray paint in his hand, ready to mark the pieces to be cut out. Yet only one side of the "interested parties" had been provided the opportunity to see the evidence. I was younger then than now! I probably yelled "Hey! Stop!" I was so excited; I probably didn't even yell "Please!" Thinking back on this, my failure analysis skills have improved since a big life event resulted in a quantum leap in my store of patience.*

The failure analysis process continues from "Step Zero" with four simultaneous activities that I will call "Step 1". Without these four activities, we do not have a true failure analysis underway. Step 1, always nominally nondestructive to evidence, includes:

A. Collection of Background Information and Selection of Reference Components;
B. Evidence Preservation, sometimes including Dimensional Inspection;
C. Visual Inspection, perhaps followed by Radiographic Inspection, or Review of Images from Remote Failures. Eddy Current and Ultrasonic Inspection are essentially nondestructive for some (but not all) part and crack or discontinuity geometries; and
D. Optical Microscopy, Image Stacking Photography, Dimensional Inspection.

These are not necessarily four sequential sets of activities, but may be iterative and interactive. That's why I think it will be helpful to ingrain them in your mind as a unit, called "Step 1". Radiographic inspection, including Computed Tomography, is becoming more widely available, and is particularly useful for complex assemblies and cast parts. Parts made by powdered metal and additive manufacturing techniques may also benefit from radiographic methods.

As you try to preserve the evidence, you will be doing a part of the visual inspection. As you preserve evidence, you will be thinking about what materials or components or records would help to interpret the data. For broken parts, if the size and shape of what you are examining is amenable to inspection of the fracture surface, macro-fractography would be part of Step 1. So would optical microscopy and perhaps even scanning electron microscopy, if no cutting is required to accomplish the work.

[114] David A. Levy, *Tools for Critical Thinking: Metathoughts for Psychology*, Second Edition, Chapter XVI, pp 104-109

Almost all failure analysis work will include some testing activities that are usually destructive in nature, such as composition and strength characterization. This central step, Step 2, is almost always destructive of the evidence in its "pristine state". This is what many people think of as "the real" failure analysis work.

> **While "NDT" methods are not destructive of the component, they may be destructive of the evidence!**

EXERCISE 6-A: If you have actual experience doing failure analysis work, please take a moment to reflect on how YOU think about the different steps along the path of a typical failure analysis project.

____

____

____

____

____

____

____

____

____

____

The silent parts of the analysis are usually the most important. And I repeat, if the silent and often undocumented Step Zero and Step 1 parts of the investigation are not done with the care they need in the given situation, the value of the entire investigation may be lost.

The first part of Step 2 of the failure analysis process generally includes optical microscopy, especially if this has not been part of Step 1. When the part is geometrically unsuited for direct optical microscopy and must be cut to facilitate microscopy, this will be the case.

Step 2 also includes those types of non-destructive testing that, from the Failure Analyst's viewpoint, are not really nondestructive! We must remember that the field of Nondestructive Testing was primarily developed to help determine fitness for service. It was named by the people who developed it for this purpose. Often NDT is performed before final manufacturing steps, although it is also widely used for in-service inspection. This latter would be the case for civil structures, aircraft, power plants, pipelines, and other applications where safety is a concern. Ultrasonic inspection may require immersion in water, or application of a couplant gel[115]. If the part has a rough surface, it may need to be cleaned, perhaps even ground, in order to inspect it. That will remove the potential contamination or corrosion product you may need to characterize, and it will definitely change the surface finish.

> **MENTAL STRUCTURE TO FACILITATE EVIDENCE PRESERVATION**
>
> ***Step Zero: Mental Preparation***
>
> ***Step 1: Evidence Preservation***
> - **Documentation and planning**
> - **Anything that may be done "nondestructively"**
>
> ***Step 2: Destructive Testing***
> - **Revelation of invisible characteristics**
> - **Any work requiring sectioning**
> - **Ongoing evidence preservation**
>
> ***Step 3: Evaluation of Data***
> - **Data Evaluation**
> - **Report preparation**
> - **Ongoing evidence preservation**

Dye penetrant and magnetic particle inspection may leave foreign residue (unrelated to service

115 Water or a couplant gel conducts ultrasound and facilitates ultrasound inspections.

conditions) on the Subject Part and crack surfaces. As previously noted, the process of magnetizing the part to perform the magnetic particle inspection may reduce the quality of any future scanning electron microscope images. This may be dealt with if you have a de-magnetizer. If the magnetic characteristics of the part are related to the way the damage happened, magnetizing the part will cause confusion and potential loss of information.

Especially if corrosion is suspected as a contributing factor in the damage to be evaluated, the resultant surface contamination from dye penetrant or ultrasonic couplant gel will change the surface composition. While it is true that you can often get a small specimen of these suspected contaminants (the dye and developer for dye penetrant, the magnetic particle solution for the magnetic particle inspection, etcetera), it can still add a few layers of confusion when you are trying to interpret your EDS[116] microchemical analysis data.

"NDT" methods are not destructive of the component, but they may be destructive of the evidence!

If the optical microscopy was performed as part of Step 1, the next test method on the list, if it is to be used, will generally be scanning electron microscopy, or some sort of focus stacking enhanced light microscopy image collection. To repeat, if the part is small, this also may be done as part of the visual examination of Step 1. But if sectioning is required to allow the scanning electron microscope to be used, then SEM will be part of Step 2, as the sectioning definitely puts us into the destructive part of the investigation.

We are studying materials-engineering-informed failure investigation!

The next part of a materials-engineering-informed failure investigation will be an attempt to characterize both the microstructure and the behavior of the Subject Material. This will almost always require sectioning, although there are hardness tests that may be performed on the surface of a component that are relatively less destructive. Likewise, a part surface may be polished to allow microstructure documentation, without gross "destruction".

If it is desired to determine the actual yield strength, tensile strength, fracture toughness, core microstructure, bulk composition, then there is no way around the gross destruction of the Subject.

The failure analysis process ends with two sometimes simultaneous or overlapping activities. Let's call this part of the work Step 3. These are:

- Evaluation of the data and self-consistency checks
- Preparation of report or other type of communication with project stakeholders

Often, one of the desired outcomes of the failure analysis will be a recommended fix. In the case of forensic work for litigation, usually it is not the Failure Analyst's job to be involved with the fix, unless the analyst is working for a company who had responsibility for the design and / or manufacture of the Subject of the investigation. For insurance work, even if a determination of fault is requested, the analyst is not the one who determines whether the insurance company will cover or pay (two different things!) for the loss.

In between the essentially mental activities of Step Zero and Step 3, the bookends of the investigation, we have a series of other activities that are usually destructive of the physical evidence. To prevent unnecessary loss of information, the competent analyst makes it a practice to keep good records of exactly what was done, how the cuts were made, the orientations

[116] Energy dispersive spectroscopy (EDS) is used for qualitative and semi-quantitative analysis of materials. Typically, SEM instrumentation is equipped with an EDS system to allow for the chemical analysis of features being observed in SEM monitoring.

and positions of sections removed for mechanical testing and microstructure evaluation, etcetera.

But before we go on, look at what we have just achieved: Failure Analysis in Four Steps.

6.4 Exploring the Contents of the Failure Analysis Toolbox

We'll start with a high-level overview of the activities that are part of a failure investigation. Recall that the goal of much failure analysis work is to understand a chain of events that culminated in the damage / failure that we are investigating. How do we do this?

I find it helpful to see my work as a process where I use "The Failure Analysis Toolbox". Failure analysis work benefits from different activities, based on the severity of the consequences of the failure, and the resources available. Some of the "tools" in the box are physical tools, such as cameras, microscopes, ultrasonic testers, dye penetrant kits, band saws, lathes, burning torches, and hardness testers. Some of the "tools" are aspects of our human nature, including our eyes, brain, patience, communication skills, and education.

Every tool exists because it is useful to someone at some point in time. On the other hand, every tool has limitations, and creates distortions in the picture that we are building in our minds. This is as true of our eyes and brains, as it is for a dye penetrant kit or a scanning electron microscope. These distortions are a natural result of how our brains evolved. A big part of compounding competence in any intellectual endeavor is learning to recognize when the distortions will interfere with the utility of our interpretations.

Our eyes see some things well, such as color differences and basic shapes. In other circumstances, our eyes see poorly. An overlapping list includes color differences as well as relative sizes and patterns. Think back on Chapter 3. Scanning electron microscope images show small features clearly, but electron images can't reveal color. Features that are clearly visible to the eye sometimes completely disappear when the subject is viewed as an electron image.

Sadly, I have lost track of the number of times I have had to take something OUT of the scanning electron microscope because the "obvious" feature (when examined with photons) that I wanted to examine became indistinguishable from surrounding features when examined using an electron beam. I am embarrassed to admit this, because after 35 years of doing failure analysis work, I really should know by now to mark the areas of interest before I put the part into the SEM! A black Sharpie®[117] marker is very useful for this purpose. I have utilized them to create more visible arrows, circles or lines than the fancy silver ink pen I bought for this express SEM use.

Tensile tests are good at ranking strengths of materials, in the conditions defined by the test method. However, it is the rare situation where the loading rate and temperature of the test are exact reproductions of the service conditions. We will do well to remember this, as we plan our investigation, and as we wrap it up and try to communicate our findings. Prepare to discuss not only the specific values, but the order of the strengths of either multiple materials, or the same "nominal material" but in different process conditions. Strengths may change when we change the test conditions in what sometimes appear to be minor ways.

As we open our mental toolbox, we find a little card, with the words Step Zero, printed in large font. Take a moment now, and find a card, and print, preferably using a dark, wide point marker, these two words: Step Zero. Or take a photo of the text box below, print it out, perhaps multiple

REMEMBER STEP ZERO!

times, and affix it to surfaces that you will notice. I used this method 10 years ago to remind myself

[117] Sharpie is a trademark of Sanford, LP and is registered in the United States and other countries.

to slow down. I found and printed pictures of snails and turtles, and taped them to my walls and the side of my computer and to doors at eye height. Over time, it worked. Change always takes time! If you are naturally too cautious of a person, maybe you would want to choose photos of lions or sharks instead!

Making the Step Zero stickers, rather than simply reading about it, will impress it upon your neural network, and increase the likelihood that you will remember this important step whenever you are participating in a failure investigation. If you are at the point in your career where you are assembling a "go kit" for field work, make this your first acquisition. Think about taping it to the lid of your kit.

6.5 Overview of the Failure Analysis Process - STEP ZERO and 1

First, we do nothing. We clear the mind of distractions; and then, while looking at the "evidence", we start to perform our visual examination. Chapter 3 went into a lot of detail about the general aspects of the visual examination. Most of the discussion of the visual examination in Chapter 3 was geared to helping you learn to *Look-At* the Subject of your investigation. But that is only half of the job! The other half is learning to *Look-For* the specific things that an expert damage/failure investigator knows how to see.

6.5.1 Visual Exam

The basic technical purposes of the visual exam are as follows:

- Determine the primary damage category (deformation, fracture, wear, corrosion, or microstructure degradation). Chapter 5 has more detail on this aspect of the visual exam.
- Get a general sense of the condition of the Subject Part, Assembly or Structure. In other words, does it look like new? Is it well maintained or does it show evidence of contamination, dirt, or mechanical damage? Sometimes it is difficult to visually distinguish chemical attack from mechanical damage. That is why the full evaluation must sometimes be completed before answers may be confidently provided. We prime our minds to receive the answer when the data becomes available by asking questions repeatedly; starting early in the investigation, "How did this damage happen?"
- If the part is deformed and broken, we will look for clues to the order of the damage. For parts that are deformed and/ or broken and corroded, or deformed and/ or broken and worn, we will again look for clues to the order of the damage. Note that fresh steel fracture surfaces from single load fracture events may rapidly flash-rust to a red color. This doesn't often happen with fatigue cracks, as the fatigue process is generally accompanied by oxidation. The greyish oxide protects the underlying steel from rust.[118] Sometimes it is difficult to determine this from the visual examination alone.
- If the primary damage category is fracture, we need to attempt to determine where the crack initiated, as that is where much of the microscopic part of the evaluation will be focused. Initiation identification is key to carrying out the destructive testing in an effective manner. Often, the initiation area is damaged. If it's damaged, it may be difficult to recognize. Finding the final separation features will often help us to improve our theories about the initiation location. We should routinely try to determine both the initiation and the final separation locations. That provides epistemological confidence for our findings. This is a key aspect of Calibrating

[118] Rust is different from iron oxide, which occurs as FeO, Fe_3O_4, and Fe_2O_3. Rust actually has water molecules as part of its structure. There are several different types of rust. Some are red and some are black. Once, a client sent me a part for analysis of the red rust. By the time it arrived at my lab, the red stains had turned black.

our Confidence, one of the primary steps of Compounding Competence in Failure Analysis.

- For corrosion damage, we need to determine whether the damage is general (meaning widespread), or alternatively, is localized.
    - Or, is it generalized, but obviously more severe in one area?
    - If it is localized, is there a type of geometric feature that is present at the most severely corroded areas?
- Now is a good time to ask how we know if something is corroded! For metals, it is not always readily apparent that the material has true corrosion product, which is a result of oxidation (or other chemical change) of the material, or whether it has a deposit of foreign material that has collected on the part. Early in the investigation, it's generally safer to call it a deposit. Sometimes there will be bare metal, with obvious thinning. There are cases that can be difficult to confidently identify as corrosion or mechanical damage, particularly wear. They could have both types of damage simultaneously.

Material property degradation may not be visible until it allows subsequent deformation, fracture, or wear. Property degradation, or microstructure damage as we have previously called it, is generally a result of exposure to high temperature (or bombardment by nuclear particles, in the case of nuclear reactors, or the sun for polymers) over an extended time. This may allow the microstructure to evolve into forms that have lower strength. The composition may also change, as oxygen from the atmosphere allows the metal to return to its thermodynamically preferred oxide form.[119] Polymers may also oxidize. The oxidized material will not have the characteristics we've worked so hard to impart to the optimized, engineered products.

For broken Subject Parts, there is a whole other aspect to the visual exam which is called "Macro-Fractography", which will be briefly covered in this chapter, and will be covered in greater detail in Chapter 8.

The conclusions that we reach during this *Looking-For* part of the visual exam, have consequences for every remaining step of the analysis.

To quote David Levy again, "...entirely false beliefs can generate their own reality."[120] One of the experiments that he cites showed rats to behave more intelligently when their handlers had been told that the rats were smart, and less intelligently when their handlers had been told that the rats were dull.

Please contemplate the complexity of the layers of mental filters that humans have.

Check out Francis Bacon's Four Idols.[121]

- The Tribe
- The Market
- The Cave
- The Theatre

---

[119] Most commonly used structural metals are found in nature primarily as oxides or sulfides. Iron, aluminum, copper, titanium, nickel and chromium all must be chemically treated to separate the metallic element from the oxygen or sulfur that is bound to it in their ores. Thermodynamics is the science that explains why metals prefer to exist as compounds of oxygen or sulfur. Once we purify metals for use, they are always ready to recombine with the oxygen in the air as soon as the conditions permit!

[120] Tools for Critical Thinking: Metathoughts for Psychology, Second Edition, Chapter XXII, page 148. The run up to the 2020 presidential election in the USA provided a painful example of a significant fraction of the population creating and inhabiting an alternate reality. Time will tell how long it persists. My neighbor who mows my lawn and plows my snow has fallen prey to this trouble. And he of course is convinced I have fallen prey! He is still mowing my lawn and plowing my snow, fortunately!

[121] Oxford Dictionary Quick Reference; *Idols of the Mind* | https://www.oxfordreference.com/display/10.1093/oi/authority.20110803095956892;jsessionid=A3CAA6F2BF7DA9C922D6C00EF4194733

If we are confident that corrosion was not a factor in the failure / damage, we won't preserve the corrosion evidence, and then obviously we'll have more trouble noticing it, even if it is present. And if it is present and we notice it later, we may be less able to accurately determine the specific type of corrosion.

Our photographic record of the visual exam will often be consulted during the investigation. We may use it to improve our confidence in our growing understanding of the situation, or, when necessary, to justify our need to change our opinion of how the damage happened. That is why it's so important to do a careful, thoughtful job of the visual inspection, and associated photographic documentation of the visual inspection. If you capture a feature in an image, it doesn't prove you recognized the importance of that feature, but it is present in your records so that you might notice it later, and someone else might see it and point-it-out to you.

> *As I told my son when he was learning to drive a car, just because I saw him turn his head, that doesn't mean I know he saw and processed the data coming to his retinas!*

The visual exam is not totally separate from specimen selection. If you didn't realize that you needed to do a certain mechanical test, and you didn't earmark material to do it, you may not have enough material left to perform the test. Mechanical test coupons generally require the most preplanning, as they usually require the largest chunk of material.

6.5.2 Iterative Thinking Process: Define-Plan-Do

Most of the stages of the investigation, including the visual exam, are iterative and interactive.

Figure 6.1 shows a schematic of the overlapping, iterative, or interactive nature of each group of steps during the failure analysis process. In order to clearly define the specific questions to be answered by the investigation, we need to be engaged in both collection of background information and visual inspection, or review of images from remote failures. Defining the purposes of the investigation will require communication with stakeholder representatives. The Failure Analyst will likely be the one most capable of getting specific here. Whenever possible, take the time to ensure that all stakeholders are satisfied with the purposes of the investigation as-defined (Patience!). Again, if we don't list what we want to determine, the chances of having the needed data later may be low. Especially for small components, or small damage areas, if we don't have a plan to perform a test initially, there may not be enough material left to go back and do it later.

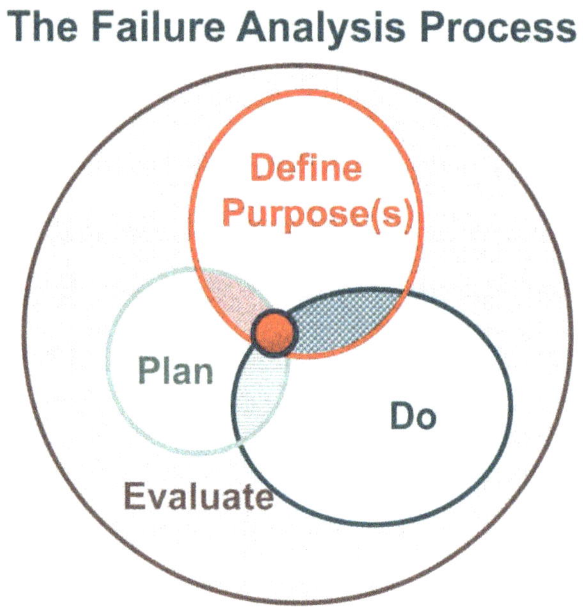

Figure 6.1: *An abstraction of the main set of mental activities required for a failure analysis*

Obviously, there is no hard boundary between defining the purpose(s) of the investigation and planning the investigation. However, by planning the investigation, I am referring to specific selection of test methods and specimen locations. In order to know whether there is enough material to perform all the tests that we would like to run, we need to know what size test coupons will be extracted from the Subject Component or Assembly. If it is an assembly, we will need to have a plan for disassembly. Many times, some degree of disassembly is needed in order to assess the damage. Or, alternatively, a radiographic inspection may be performed.

6.6 CASE STUDY: In Scope Definition

During the time frame of this book's manuscript preparation, a client brought an electronic assembly that was housed in a sealed plastic box.

In this case, we had to plan and perform the disassembly before we could perform the visual inspection of the individual components.

The client had carefully used a Dremel® tool to score the housing, and then used an *XACTO®* blade to make the final cut which opened the box, so that we could perform the inspection.[122] What about evidence preservation? How was evidence preservation achieved in this project?

Essentially, we performed the disassembly under a microscope and took photos before and after removing each component.

When a repeat failure occurred a few weeks later, a different technician was assigned to open the housing. That person was in a hurry. They did not ask the first person how they had opened the housing, and decided to use a Dremel® to cut through. When we finally saw the internal components, they were sprinkled with small particulate debris, most of which was found by EDS to likely be bits of the plastic housing and abrasive particles. Some of the plastic got very hot, and was found melted onto the internal components of the assembly. This did not simplify the interpretation, and a significant amount of the allotted project time had to be used to try to determine what the debris particles were. It would have been easier without the added debris problem amplified by the poorly planned housing opening procedure.

This project illustrates the intersection of the three overlapping "Define-Plan-Do" activities of the failure analysis. The client wanted to know why the assembly was no longer fulfilling its function. Since I am not an expert in electronic assemblies, I asked them to say more. They refined their goal to inspecting the individual internal components for damage, and damage categorization.

We have now defined specific goals and can ask relevant questions! A project definition of "find the root cause" is not helpful in the same way as these specific goals and the resulting questions.

Their product knowledge led them to speculate that mechanical damage was a possibility, as was thermal damage from an electric arc event.

Corrosion from contamination was not thought to be likely, due to the sealed housing, which appeared to be in good condition on initial inspection. However, contamination events do sometimes happen during manufacturing.

When we got to the examination and found the debris, we were forced to determine its origin, because contamination had not been on the original list of what we were looking for.

Fortunately, it was visually clear that the relevant damage was mainly mechanical. Several of the sub-components were broken. The particular broken fragments appeared clean. The damage was similar to that found on the first set of parts, where there was no debris issue.

EXERCISE 6-B: How would you classify an electric arc event if required to fit it into the classical four damage categories of wear, deformation, fracture and corrosion? What about in the five category system of wear, deformation, fracture, corrosion and microstructure degradation? Classify on this basis.

____

____

____

____

[122] Dremel ® is a registered trademark of Robert Bosch Tool Corporation | X-ACTO ® is a registered trademark of Newell Brands

6.7 CASE STUDY: In Evidence Preservation (Second Example)

I would like to share another evidence preservation lesson I learned after many years in the failure analysis business. A client submitted a tiny component for documentation of damage using the scanning electron microscope. I removed it from its envelope, and after examining it with the optical stereo microscope, I must have been distracted. A phone call? I set the part down and then could not find it. When I got to the SEM, and opened the envelope where it had been, it was gone. Sadly, this was not a very good example of evidence preservation. I spent two hours walking back and forth between the office where the optical microscope was, and the lab where the SEM was. I studied every tiny bit of dirt or shiny particle on the sidewalk. It was night. This was not fun. I gave up and went home. In the morning, I repeated the search, to no success. A colleague came over to use some of my equipment. I told him my sad tale. He offered to help find it. We repeated the walk to the lab. Finally, we went back to the office, and I was getting up the courage to tell my client that we had lost the part. Suddenly, my colleague spotted it, lying on my counter top. I was relieved that the part had not been exposed to a salted sidewalk. This was a tested part that had gone through a very long and expensive test. Had I found it outside, in the winter, in Michigan, lying on the salted sidewalk, the investigation would have been compromised.

Now, whenever I have small parts, I make sure that they are double or triple wrapped before they move from one laboratory area to another. This has become a formal part of my informal quality system.

For the above-mentioned electronic assembly, each small component was wrapped in a piece of lens paper, which was folded, and then inserted into a small manilla coin envelope. Then all of the coin envelopes were stored in a small plastic box. This is also a part of evidence preservation. Most of my career has been spent looking at larger components that would be pretty hard to lose.[123] With small parts, for me, and new lines of business, in general, we have to be alert to new challenges in evidence preservation, as in all aspects of the work.

Chapter 7 presents a methodology for negotiating the scope of the work, so considering evidence preservation and its challenges in terms of time and resources will make for an accurate scope of work.

6.8 Visual Exam (Part 2): Looking-For, the Power of Macro-Fractography

If the damage we are characterizing is fracture, we need to include macro-fractography as part of the visual inspection. At first thought, we might assume that the word macro refers to something that is large. But, just as in economics, we have macro and micro versions that address aspects of the use of resources that are very different. The terms macro- and micro-fractography refer to more than the size of the feature we are looking at, and it refers to more than whether or not we are using a microscope.

In the Nineteenth Century, when these terms were first used, the people doing the work were looking at broken railroad axles and wheels and heavy farm equipment. Macro-fractography was generally performed without the need for microscopes. But now, there are a lot of very small mechanical devices. The bio-medical device industry is a primary developer of very small components. Some are so small that a scanning electron microscope is required to see the part in enough detail to be able to visualize the damage. So, the term macro-fractography has been refined

[123] Although a large package delivery company lost a 200 kg chunk of steel removed from a large casting. We had spent about 20 labor-hours and I can't remember how many canisters of oxygen to feed the burning torch, in order to remove this 200 kg segment. Perhaps the box in which it was shipped was not sturdy enough. That's what the delivery service said.

to refer to looking at fracture features in order to determine basic facts about the loading geometry and mechanical behavior of the part. Micro-fractography now refers to how the crack interacts with the microstructure.

Fractography at all scales can reveal the presence of manufacturing discontinuities that facilitated the crack initiation. It depends on the size of the component and that of the discontinuity. Fractography must always proceed from the macro-scale toward the micro-scale. Chapter 8 explores fractography in great detail.

Fractography uses visual cues to systematize and simplify categorization of features of broken parts. The visual cortex is always in use during an examination and the eyes are usually in use. This is true whether or not microscopes are deployed.

- Macro-fractography categorizes fragment shapes.
- Meso-fractography notes surface roughness and coloration patterns.

Once macro and meso fractography is done, the analyst is informed, able, and empowered to determine:

1. the loading geometry;
2. the material's ductile or brittle response to the loads that resulted in the crack; and
3. the crack initiation and final separation locations, and thereby, the direction of crack propagation.

MICRO-FRACTOGRAPHY

Micro-fractography helps us understand how the microstructure of a component influenced the progress of the crack.

- Micro-fractography reveals whether the crack runs in the grain boundaries or through the grains.
- Micro-fractography reveals whether the crack runs along the boundaries of second phase particles or through them.
- Micro-fractography reveals the presence of highly localized plastic deformation that may be associated with, or pre-date, the crack process.

Micro-fractographic features also allow us to notice whether the damage is due to a specific damage mechanism. For example, stress corrosion cracking, where chloride ions in the environment of an austenitic stainless steel component may lead to (primarily) transgranular, branched cracking. Micro-fractographic data can allow us to confirm or discard a particular damage mechanism associated with a known constellation of features.

Micro-fractography can reveal microstructure features that evolved during service. This is usually going to be a concern only in the case of service environments that modify the microstructure, such as high (occasionally low) temperature, extremely long times at ambient temperature, or the presence of corrosive substances. Comparing and contrasting the micro-fractographic features at the origin and final separation ligament can also sometimes be used to confirm loading geometry.

Usually, a scanning electron microscope is required to perform micro-fractography.

Once the three steps (mentioned on the previous page) are finished, the analyst explores whether:

A. the component was used in a manner consistent with its designed purpose, and
B. the material behavior appears consistent with expectations based on how it was supposed to have been manufactured.

Applying macro-fractography also may reveal whether:

C. the presence of certain types of manufacturing discontinuities or defects exist.

Meso-fractography entails pattern matching of fracture surface textures and colors. Analysts do this in order to categorize the type of crack as being due to a monotonic-sustained load, a monotonic-dynamic load, or a cyclic load. All of these types of cracks may or may not show evidence of corrosion.

Additional features that are sometimes revealed by the application of meso-fractography include a nominal safety factor for cyclic (fatigue) loading, severity of stress concentrations, and information for (or against) proper assembly, or maintenance issues.

How to interpret fracture features will be elaborated in Chapter 8.

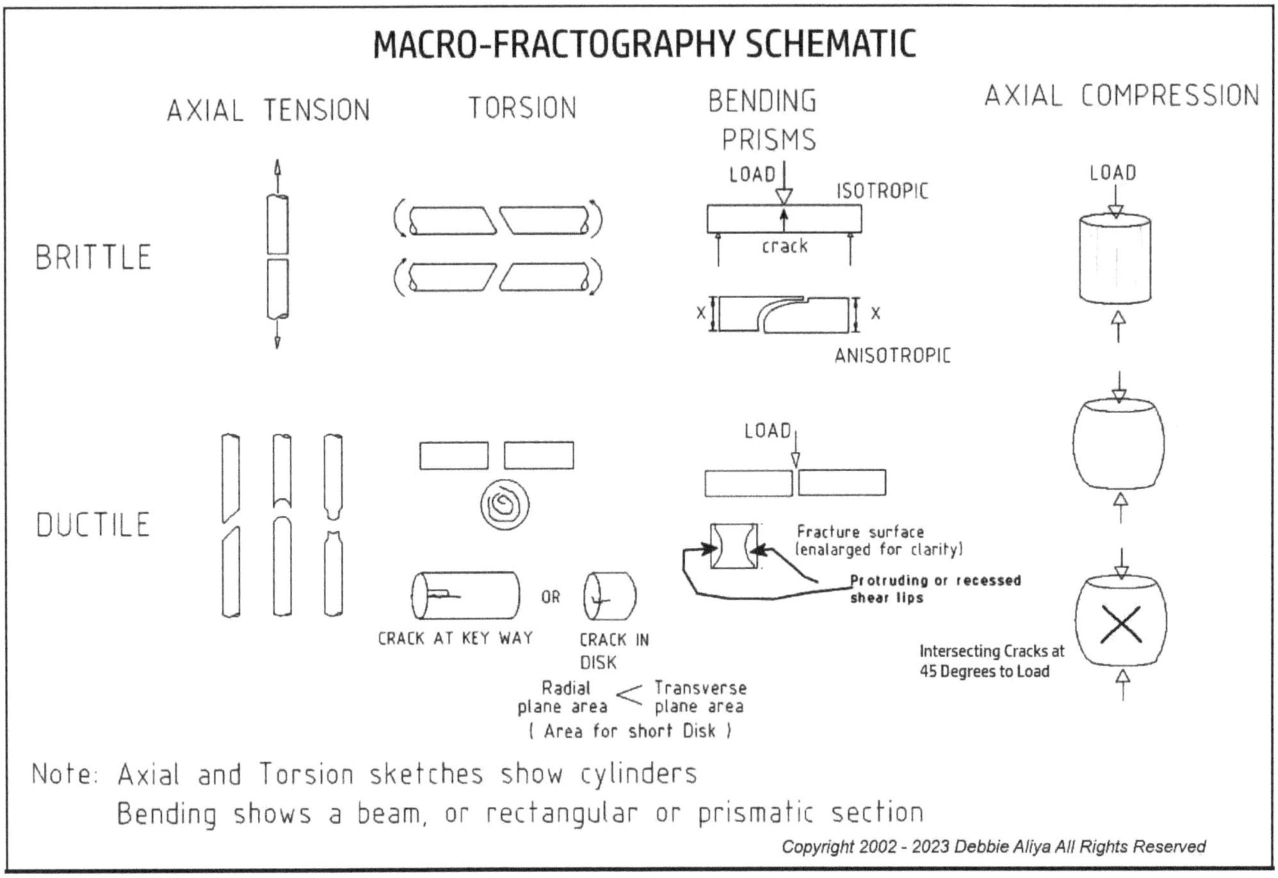

Figure 6.2: *Schematic of fragment shapes and orientations matched to the loading geometry for simplified shapes – this is a schematic of the resulting orientation and shape of fragments in simplified component shapes that have been loaded to the point of fracture. This structure helps categorize macro-ductile and macro-brittle cracks of varying loading geometries. This is the essence of macro-fractography!*

Figure 6.2 (above) shows a schematic of the resulting orientation and shape of the cracks in simplified shapes that have been loaded to the point of fracture. The examples of simplified shapes include cylindrical and prismatic shapes (often rectangular, but could include triangles and other polygons in one of three dimensions), subject to simple loading conditions. This includes tensile (stretching), compressive (squeezing), bending, and torsional (twisting) loads. After you become familiar with the fragment appearance of simply-shaped parts, continued practice will reveal the secrets of more complicated geometries.

NOTE: Macro-fractographically, brittle bending in isotropic materials, and brittle tension are similar to ductile torsion! We'll learn other features to distinguish these three later.

Figures 6.3 (below) and 6.4 (next page) show details of brittle and ductile axial tensile cracks in cylindrical components. Sometimes, macro-ductile and macro-brittle behaviors are almost as easy to distinguish from each other as many think they are. If we are sure that the loading geometry was axial tension, and sure that we started with a uniform cross-section along the length of the part, then the parts depicted in Figures 6.3 and 6.4 can be positively identified as having behaved in a macro-brittle and macro-ductile manner, respectively.

Figure 6.3 shows Macro-Brittle Behavior in axial tension; and it was diagnosed based on no visible change of diameter at the crack position as well as crack orientation (transverse, or perpendicular to the direction of pull) .

Figure 6.3: *Fractured threaded fastener head*

A threaded fastener head was friction welded to the end of a piece of bar stock of diameter equal to the fastener head. The upper (larger) diameter end was gripped and the smaller diameter threaded end was threaded into a fixture so that a tensile axial load could be applied. The crack, surprisingly, was in the larger diameter section.

Note that if we look closely at the fracture features of a macro-ductile, constant cross-section, axial tensile specimen, as shown in Figure 6.4; "the plot thickens!" The fracture features are macro-ductile at the edges (oriented at a "slant angle") and macro-brittle (perpendicular to the axis) in the center. Note that the entire fracture surface area of both specimens shown in Figures 6.4 and 6.5 would likely be found to be MICRO ductile, if examined in a scanning electron microscope. We would probably find evidence of ductile dimples - micro-void coalescence (MVC). The micro-voids would likely be elongated at the slanted outer edges of the crack and equiaxed in the flat center.

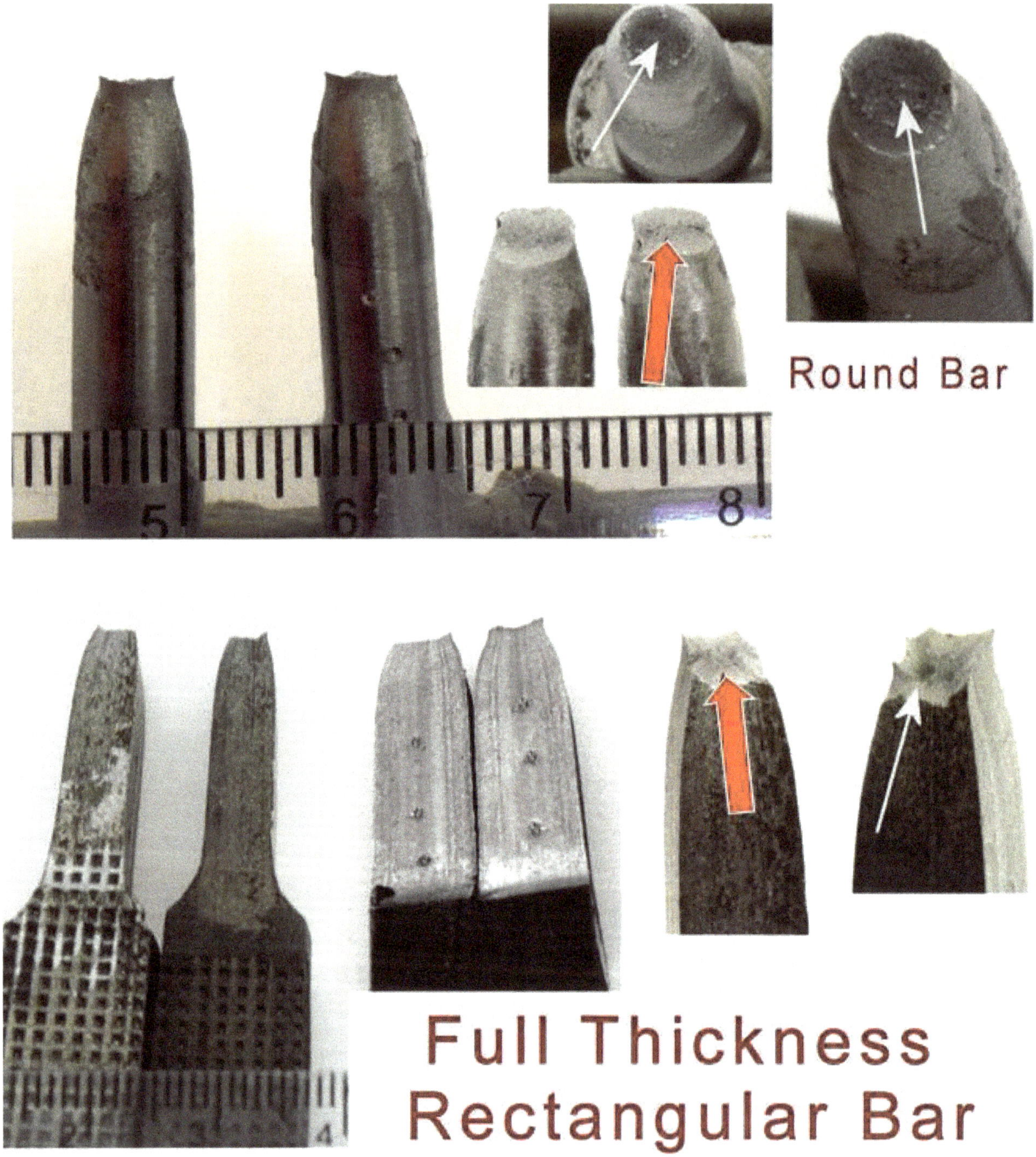

Figure 6.4: *The "Old School" method helps us recognize macro-ductile pinching / deformation. The upper image shows a cylindrical tensile bar with severe pinching. This is the classical sign of component ductility. The thick red arrows and white arrows show macro-brittle portions of the break that would be expected to have micro-ductile features. The lower images show rectangular tensile bars from the same component as the round (above).*

Note that we must show enough angles that the topological characteristics (deviations from flatness) are revealed.

Figure 6.5: *Threaded rod broken in axial tension (flat bottom and slanted portions of the crack)*

It has classical pinching on both sides of the crack, but the crack itself is forced away from the perpendicular by the helical stress concentration (thread roots). The angle of the crack is a crucial analytical feature. We will cover this in great detail in Chapter 8. For now, simply note that it is important to take pictures from all angles. When viewed at a favorable angle, we can make out both the flat cup bottom (arrows) and the slanted portions of the crack.

Years ago, I was enjoying some adult beverages with colleagues after a technical conference. The man sitting next to me told me that the center of a plastically deformed pinched area of the classical ductile cup and cone fracture was macro-brittle. That seemed very strange to me. I soon managed to ask a colleague, "Does Bill know what he was talking about?"

My colleague assured me, "Yes, Bill is the real deal."

It took quite some mental re-arrangements and extensive cultivation of clarity to be able to understand the value of using a self-consistent system to assign macro-scale fracture features, even in the face of contradicting "what everyone else knows". The careful contemplator of complexity soon realizes that every system developed to categorize real physical objects will have some arbitrary designations. It is now ok with me that I call the bottom of the "cup and cone" fracture macro-brittle, even as I know that if I look at it with a scanning electron microscope. I will find that it has micro-voids (the classical micro-ductile feature). Macro- and micro-scale features must be evaluated independently on their own merits. Within a short time of this discussion, Bill became my macro-fractography mentor.

Figure 6.6 (below) shows the classical helical crack resulting from twisting a brittle (tapered) cylinder. In this case, the material is a carrot.

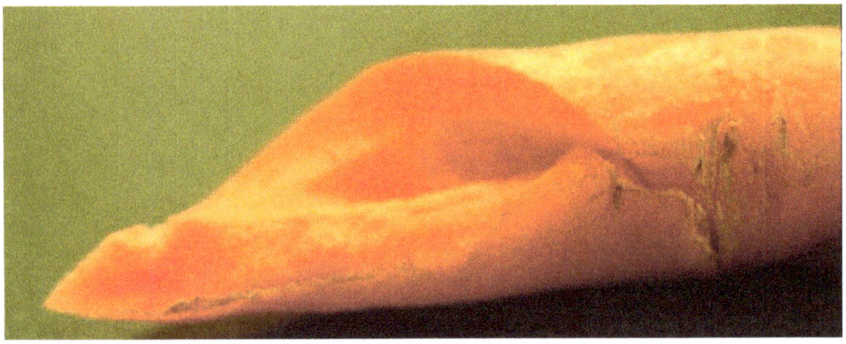

Figure 6.6: *Carrot fractured in torsion*

Note brittle torsion (upper left of crack). Note ductile torsion at transverse scar (near right side of image) and short radial (lengthwise) ductile torsion in core.

Figure 6.7 (below) shows an edge view created by the direct shear loading geometry, which is not shown in Figure 6.2 (Schematic of shapes, cracks and loading), but will be discussed in Chapter 8. The poor control of the associated burr (tooling maintenance problem; the actual burr feature is not visible here) has allowed a crack to initiate in this heavily cold worked material. Note the commonly found deformed or sheared portion and a ductile fracture portion, referred to (correctly) in the sheet metal stamping industry as "shear" and "fracture" respectively. The "fracture" area usually shows micro-scale ductile features (MVC, see micro-fractography section later.)

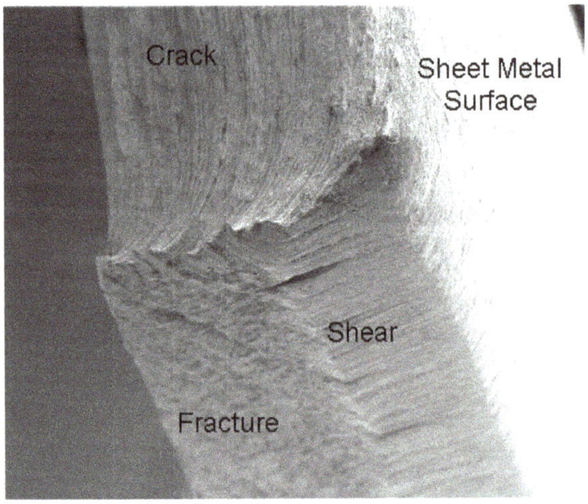

Figure 6.7: *Stamped sheet metal part with crack*

Bottom shows edge created intentionally by use of tooling in a sheet metal stamping operation. SEM image - 50x original magnification. Curvature of crack (curl) due to cold working. Note additional tiny perpendicular tear above the word "Shear". The crack itself likely initiated because of the burr, which is not visible in this image.

When performing macro-fractography, it's important to keep in mind the fact that some materials may respond in either a ductile or a brittle manner, given factors such as how rapidly the loading took place, whether there were corrosive agents in the environment, and whether the crack happened all at once, or in a progressive manner.

Fatigue cracks are one example of a progressive macro-brittle type of crack. By definition, fatigue cracks happen below the nominal yield strength of the component. However, at the local level, which may be a single grain when the crack first starts to propagate, the stress must exceed the tensile stress, or there would be no crack. Often this situation results from an accumulation of invisible damage that eventually creates a feature with a high degree of stress concentration. Accumulating this amount of damage may take thousands or millions of load cycles. But eventually a true crack may form. And then, for every load cycle that is severe enough, the crack will get a little longer and wider – or deeper and wider. In materials that have the same strength characteristics in all orientations, in some loading geometries, the crack will generally grow at the same rate in all directions. This means that the inner edge of the crack will take on a classical crescent shape, if, as is often the case, it starts at the surface of the part. As the crack grows, its surface may become oxidized. As the humidity or temperature change, the color of the oxide that forms may vary. This means that many fatigue crack surfaces will take on a striped appearance. See Figure 6.8 (below). .

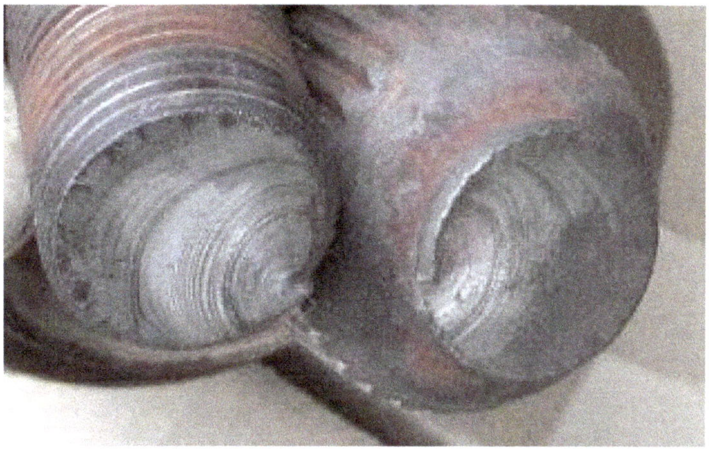

Figure 6.8: *Mating fragments with typical "beach marked" fatigue crack*

Note that the curvature of the distinct beach marks covering most of the crack surface is opposite the classical curvature. In other words, the arcs appear to be spreading from what other details inform us to be the final separation. The final separation is at 4 o'clock on the fragment shown at left. Do not worry! We will explore this in greater detail in Chapter 8.

6.8.1 Iterative Visual Exam For Fracture

This provides one example of how the iterative interpretation of visual evidence may be used. If we see a striped appearance, such as shown in Figure 6.8, we may temporarily assume it is a fatigue crack. That means it is macro-brittle, since by definition fatigue cracks happen below the nominal yield strength. If you see obvious signs of gross deformation that appear to be associated with the crack event, as opposed to happening as a result of (and following) the crack event, it's not a fatigue crack. So, if you see a flat, transverse, striped crack, it's macro-brittle, then you know the loading was unlikely[124] to be torsion, because macro-brittle torsion cracks must have a helical shape. Ruling out a loading geometry gets you closer to knowing what loading did facilitate, or at least could have facilitated the crack. Chapter 8 will cover the important subject of macro-fractography in greater detail.

6.8.2 A Few More Things To Look-For

Witness marks, or damage from something hitting or rubbing the part, can give us hints that there were loading events that might not have been part of the designer's awareness. See Figure 6.9 (next page).

While macro-ductile torsional loading does not generally result in pinching adjacent to the crack, if the part had longitudinal grinding marks, you may be able to see that they are twisted. See Figure 6.10 (next page).

[124] But not impossible - that will be discussed in the macro-fractography section of Chapter 8

Figure 6.9: *Witness (wear) marks consistent with crack initiation*

The worn away gold colored coating shows motion back and forth around the plane indicated by the red arrow. The view of the crack surface, with the red arrow position noted, allows us to imagine the motion that initiated the crack. Take a moment to think about this if it isn't obvious to you. Can you see the faint crescent shape (also called "thumb-nail" or "lentil") at 4 o'clock on the fracture surface? The crack started at this feature.

Figure 6.10: *Grinding marks on Subject Part*

Note spiral twisting of longitudinal grinding pattern. This is macro-ductile, despite the lack of pinching. There is clearly significant permanent deformation.

> ### Things to *Look-For*
> 1) General appearance
> 2) Primary damage category
> 3) Potential contributing damage categories
> 4) Witness marks
> 5) Fractographic features if there is fracture
> 6) Discontinuities

Note the rows of parallel (and faintly perpendicular) features on Figure 6.11 (next page). These are a result of solidification features known as dendrites. In general, when we see solidification features, including dendrites (such as seen in the overheated aluminum strap of Chapter 4) or voids or other features that are neither post fracture damage (rubbing, etcetera) nor dirt or debris, we may temporarily assume that there was a manufacturing discontinuity. If there is evidence that the fracture started at the edge of such a feature, we may start to consider the possibility that this manufacturing discontinuity influenced the timing and ease of the fracture event. Considering the possibility that a discontinuity contributed to the timing and ease of fracture (or corrosion, etcetera) is a long way from concluding that the component was defective, or even that the discontinuity caused the failure.

We must always keep in mind three mutually interacting key factors:

1. It broke or bent or lost material to wear because the stress exceeded the strength.
2. It corroded because the environment was more severe than the material could

withstand (even in the absence of a discontinuity).

3. We must always consider the process condition (as-molded, hot rolled, case hardened, etcetera) as part of what defines the Material. We need to remember that the Material never refers to the elemental or molecular make-up alone.

Figure 6.11: *Solidification features found on a portion of a fracture surface (not true fracture features)*

These solidification features (dendrites) indicate that a discontinuity was present at the time the casting was made.

6.8.3 Nondestructive Testing

Why should we perform Nondestructive Testing (NDT) as part of a Failure Investigation?

Nondestructive testing or nondestructive evaluation (NDE) can be useful at several points during a failure analysis. One example is for specimen selection, especially when many similarly damaged parts are available for the investigation. You may wish to know how the crack propagated into the core of parts that are not completely fragmented. Is the crack flat, or does it follow multiple planes, or is it curved? You may wish to learn the extent of a crack so that you may plan your cuts to reveal the fracture surfaces. You, or your client, may wish to perform NDE on the apparently intact portion of a damaged structure, to determine if the damage was more extensive than visual inspection reveals. You may wish to select an intact area for mechanical tests. If the Subject is a machinery component that caused financial pain when it "failed", the company may wish to inspect the "sisters" of that same component present in other machines, or its replacement.

What are the methods available? NDT methods fall into two main categories: Surface and Volumetric.

NDT Category: Surface Methods - Visual Testing (VT) and (dye) Penetrant Testing (PT) are strictly surface methods. The discontinuity must be open at the part surface in order to be detected. Magnetic particle inspection (MT) is a near surface method. This means that surface discontinuities are most easily detected, but sometimes, slightly subsurface discontinuities are detected. By the term "Surface Discontinuities", I mean those that are open at the surface.

Just because the discontinuity reaches the surface, it doesn't mean it will be detected. It might be too small, or unfavorably oriented. For magnetic particle testing, the part we're inspecting must be magnetic. Detectability with any NDE method is strongly dependent on the size and geometry of the feature. Larger features are not necessarily more readily detected. Surface finish strongly influences detectability.

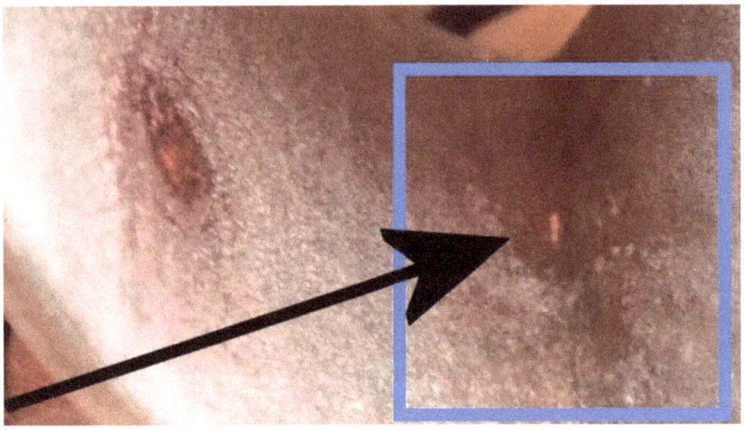

Figure 6.12A: *Dye Penetrant Testing revealing a small indication on a stainless steel casting*

Illustrated here is regular penetrant testing. There is a related method where a fluorescent dye material is used, and the examination is performed using black light. Fluorescent penetrant has the potential to reveal smaller discontinuities.

With VT, PT (Figure 6.12A), and MT (Figures 6.12B-D on the next few pages), the indication that you are seeing is directly related to the size and shape of the discontinuity that produces it. However, it is not exactly, "What you see is what you get." These methods do not provide any information about the depth of the discontinuity. But the discontinuity shows up on the part where it is, ***if*** it shows up and ***if*** it's not an artifact. That's a lot of "ifs".

Volumetric methods require much more skillful interpretation, especially phased array ultrasonic. Computed tomography (CT) is perhaps the most intuitive in the relationship between the detected indication and the physical discontinuity, although CT data may also be misinterpreted.

Some parts are inspected with surface methods multiple times during production, as additional layers are machined away. The surface methods are generally easier to perform than the volumetric methods. Multiple inspections allow manufacturers to avoid spending money on parts with known unacceptable discontinuities. Of course, some scrapped parts may eventually be "cleaned up" (meaning all of the discontinuities are removed). To some degree, the specifications for acceptable and rejectable indications are arbitrary for parts that don't have detailed quantitative design data.

APPLICATIONS NOTE: VT may be used to inspect all parts that are large enough to see. This is also nominally the case for PT, although surface finish features may limit inspectability of all NDT methods. As previously noted, MT may only be used on materials that have a strong magnetic response (generally steel, or ferritic or martensitic stainless steel).

Visual testing is more or less what it sounds like. The inspector examines the parts, specifically looking for one or more specified features. Visual testing as an NDT method is a totally different procedure from what Failure Analysts do. Visual inspection for failure analysis is open ended. VT as part of an NDE inspection is strictly defined as looking for the features shown on the negotiated list of indications.

Furthermore, a VT procedure may include cleaning. The test methods also require a certain level and uniformity of lighting. Inspectors performing NDE work, including VT, must have vision tests at regular intervals. Of course, having 20/20 vision does not prove that the individual was paying attention when performing the specific test in question!

Penetrant testing (PT) usually requires cleaning, then soaking the parts in a liquid penetrant of controlled viscosity and surface tension. Then, the liquid is removed from the surface (rinsed). A developer is used to coat the part surface. This draws any penetrant that has filled cracks, voids, etcetera, toward the part surface. The penetrant and developer are made from contrasting colors, to promote ease of detection of discontinuities. Parts that pass this test must usually be recleaned to remove the developer.

With MT, the parts to be tested must be saturated with a magnetic field. Magnetic fields are directional, so many part geometries require at least two different rounds of testing, usually with mutually perpendicular magnetic fields. This might be circumferential and axial, for shafts. Thus, MT requires a method of creating a magnetic field in the part. MT also requires particles that will be attracted to areas which leak magnetic flux due to discontinuities. Some MT particle materials are used dry. These are available in different colors to contrast with the part appearance. The images in Figures 6.12B-D show the results of a liquid fluorescent magnetic particle test as it appears when illuminated by black light.

Note also that all of the surface method indications are visible to the eye. For methods that use black light, you may need to accustom your eyes to the dark for a few minutes before reliably being able to see the indications. But for the standard methods, you don't need a microscope to see the features of interest, and you don't need a computer to interpret the data.

> *I often stand next to the NDT technician and have him*[125] *show me the "indications" as they are revealed. Often, I photograph the results, after they mark the area with the indication, using a wax pencil or paint stick. Once the test is complete, the indications generally (or at least often) return to invisibility.*

Note use of the word "indication". An indication is just that. It indicates that there MIGHT be a discontinuity at the indication location. Many indications have characteristic shapes. Part of becoming a certified NDT technician is proving that you can distinguish a crack-type indication from a pit or a lap type indication. Many NDT technicians get so used to speaking to each other, that they forget, if it was ever clear to them, that they do not know, and cannot tell, what the indication is due to, without additional information. Sometimes a Visual Test will provide the "epistemological confirmation" that it is really a crack, rather than a lap. Other times, metallographic cross-sections might be required to confirm the nature of the discontinuity. Obviously in that case, the part must be destroyed to get the confirmation.

It is always challenging to perform NDE to support a failure analysis because there is rarely an applicable standard procedure to carry out the work.

Before we move on from magnetic particle inspection (MT), let's talk about quality control of NDE methods. Every method has a set of requirements that the technicians and laboratory or testing service must follow in order for the data revealed to be confidence inspiring. A key point of interest is having demonstrated and

[125] As of 2023, I am still waiting to meet a female NDT technician; although in 2009, I did meet a woman professor specializing in the physics and cognitive psychology aspects of NDT.

documented that we can find the smallest indications that we need to find.

Figure 6.12B: *Liquid fluorescent magnetic particle test as it appears when illuminated by black light*

A yoke style magnetizer is being used at left (upper inset), for a disk-shaped part with through holes parallel to the axial direction. The red tint on the inside surfaces of the small holes is visible dye penetrant residue. At right, a ring-shaped magnetizer is used to prepare a cylindrical shaft for fluorescent MT. The colors are due to the use of black light to highlight any indications revealed by the fluorescent particles present in the bath.

Before we move on from magnetic particle inspection, let's talk about quality control of NDE methods. Every method has a set of requirements that the technicians and laboratory or testing service must follow in order for the data revealed to be confidence inspiring. A key point of interest is having demonstrated and documented that we can find the smallest indications that we need to find.

With NDE used for failure analysis work, it's often not done "to code" because the "failed object" is not going to be certified for safe use. But that doesn't mean that the technician should not follow the usual protocol. Figure 6.12D shows the use of a pie gage, the way that magnetic particle test (MT or MPT) procedures are validated. This is a small circle ("pie") with fine lines of different sizes. I value when the technician demonstrates to me that they can see the lines on the gage when they magnetize it alongside the part that I am having them inspect.

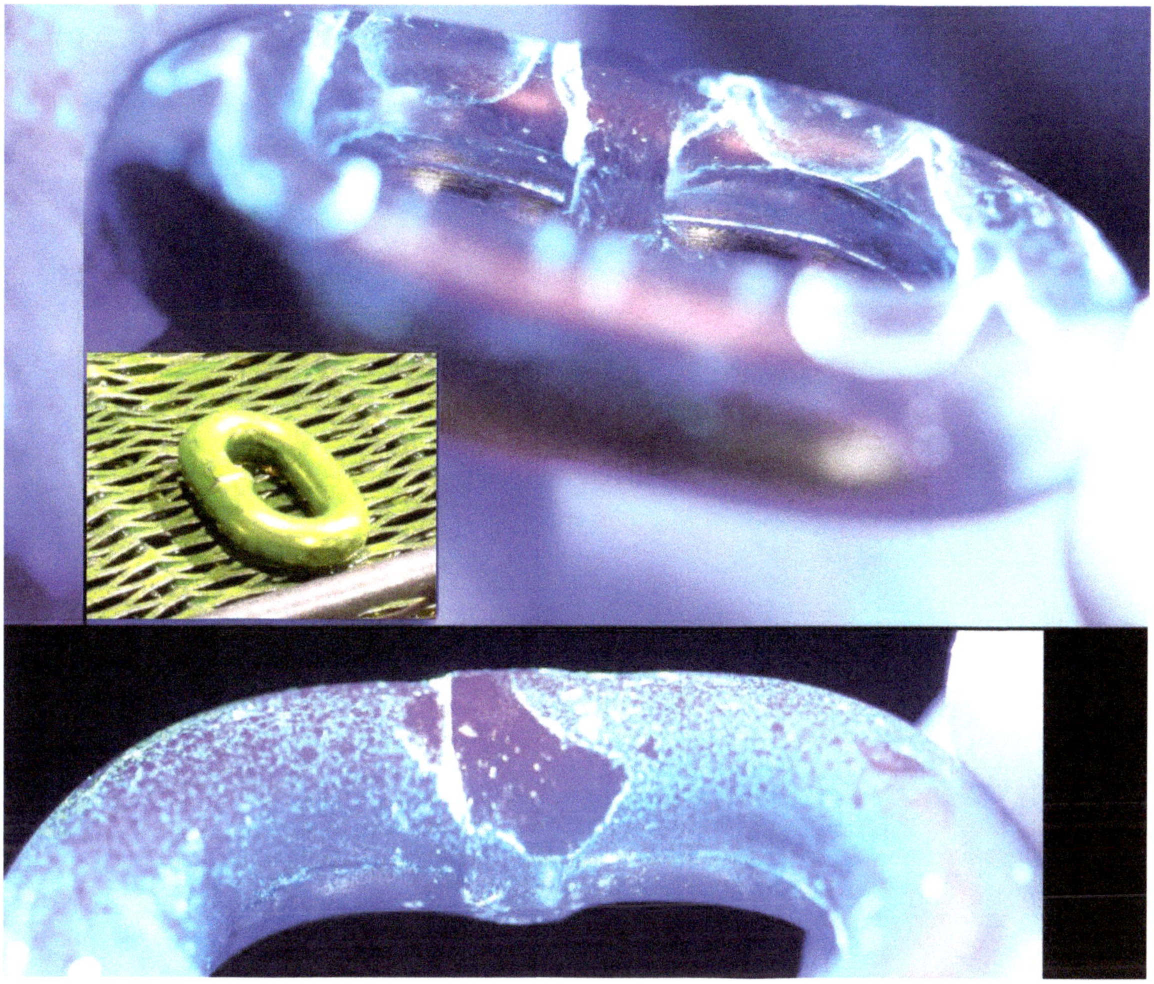

Figure 6.12C: *Liquid fluorescent magnetic particle test as it appears when illuminated by black light*

Small inset shows daylight (shop light) image of a chain link sitting on a mesh basket that was used to bathe the link in the fluorescent magnetic particle solution. Upper and lower views show appearance in black light (fluorescent) after the part surface was rinsed. Note the confusing scallop features (in the upper image) from where the link was in contact with the basket during immersion.

As previously noted, visual testing (VT) when performed by an NDT technician is very different from the type of visual inspection that was described in Chapter 3 of this book. VT for NDT in commercial applications is almost always reduced to looking for a series of features on a check list. The VT can be performed to make sure something is present (a drilled hole, a screw, whatever), or absent (a scratch, a pit, a stain, some dirt, a crack, etcetera). Because of the commercial pressure to perform the inspection in an economically viable way, there is no time for *Looking-At.*

People may think that it's simple to perform Dye Penetrant, but it requires skill to interpret the results and knowledge to recognize when the surface is clean enough to run the test or if it is too rough to achieve reliable results. The "visual test" may be the most challenging as it is the most open ended.

Figure 6.12D: *Liquid fluorescent magnetic particle test as it appears when illuminated by black light*

Pie gage on top of small curved piece of plate, held between legs of a magnetizing yoke. There is a faint bright line visible in the center sample is following the axis of the tube that the test piece was cut from. It is immediately to the left of piece of dark dirt. The bright line is due to magnetic leakage from a groove intended to demonstrate that the test set-up is correct.

The difficulty of doing VT may be illustrated by the work of the airport baggage inspectors through radiographic imaging of our belongings. Government auditors have routinely found that these "VT Technicians" miss firearms, knives, and other obvious dangers while focusing on the 400 ml bottle of sunscreen or any odd-looking object. A year after the "911" events in the USA in 2001, I was going to Florida from Michigan to teach a fracture analysis course.

I had a broken-off stub of a shaft with a sharp corner. They let me take it, after I showed them the notebook for the fracture analysis class. The "boss" decided I was legitimate. But I would not even try that now. In 2009, I had a lot of talking to do to be able to take my ball foot micrometer to California! These inspectors have a difficult, tiring and thankless job to do.

6.8.4 Failure of the NDT

Failure Analysts are sometimes asked to help with failure analysis of an NDT inspection. Why didn't the inspector see this, relatively speaking, gigantic (1mm x 5mm) crack opening during the dye penetrant test? See Figure 6.13 (below).

Well, the inspectors had been given exemplar parts with cracks that were tightly closed. They were looking for a small linear feature, not a (relatively speaking) big, gaping crack. The failure to detect a large crack made the purchaser of the parts unhappy. They should have specified a VT to supplement the PT. In fact, at least two methods of NDE are almost always required to have a reasonably high probability of finding the most common types of discontinuities in manufactured products.

Figure 6.13: *Missed crack during inspection after dye penetrant testing*

A clearly visible (approximately actual size as shown if printed) open crack that was missed during dye penetrant inspection. We proved it was there during the test, by demonstrating (with EDS microchemical analysis) the presence of penetrant and penetrant developer residue on the crack surfaces.

By the way, the inspector of the object shown in Figure 6.13 had recently passed his vision test!

One important point to remember about these surface methods is that once the test is done, and the parts are sorted into "accept" and "reject" bins, there is not necessarily a way to go back and see what the inspector saw. Photos of the indications are generally only obtained when the methods are done to support failure analysis, forensic work, or other special situations.

NDT Category: Volumetric Methods - Volumetric methods of testing "interrogate" the volume of the material, not just the surfaces. Ultrasonic testing (UT), is one of these methods, however, with the other primary volumetric methods, radiography (X-Ray, or RT, and especially CT, or Computed Tomography), we do retain a visual record of the inspection. We have either the film, or, more often now, a digital file obtained through use of an electronic cartridge.

The effectiveness of all volumetric methods is strongly linked to geometry of the test piece. For radiography, very heavy and very thin sections on the same component will require different exposure times, often necessitating double, or triple testing. But ultimately, RT results in a series of images containing features that resemble features of the part itself. Bright spots or dark spots appear in a position within the envelope of the image of the part that is related to their position in the object.

Figure 6.14A (below) depicts a curved piece of sheet metal with a lap weld along the bottom edge. Black oxide present on hot rolled steel has flaked away from the weld and several other somewhat rounded areas, which are labeled 1-7. The "regular photograph" is shown at left. A radiograph of an overlapping area is shown at the right. In the X-ray image, dark areas are thinner than the main area, as more x-rays reach the film. Bright areas are thicker, as they prevent some of the x-rays from reaching the film, so it stays clearer. The weld bead is thus thicker than the rest of the material. Note the narrow, dark band below the bright weld bead (see horizontal red crayon highlight along bottom of right image). This was a narrow band of severe corrosion along the edge of the weld. This resulted in a groove that followed the heat affected zone of the weld.

The numbers 14-27 at the top of the right-hand image show up in white because lead numbers were attached to a tape and then the tape was attached to the surface of the specimen. These reduced the concentration of x-rays available to make the film turn dark, and served to help describe the image.

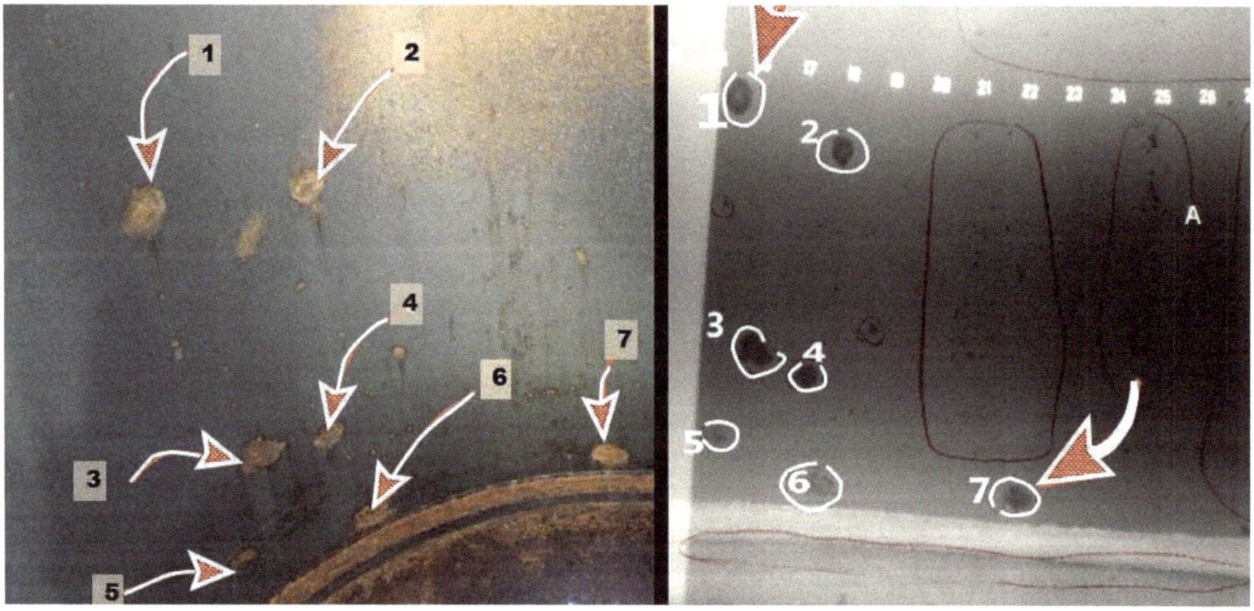

Figure 6.14A: *Curved piece of sheet metal with weld along bottom edge*

CT, or computed tomography, is a newer radiographic method that results in a three-dimensional model of the part. It shows the features in three dimensions as the viewer rotates and zooms in and out on the image. CT does not suffer from the same types of limitations that RT does. There is not an issue with the shapes of the features of the part, including any discontinuities, being distorted. For thin, flat parts, unlike that shown in Figure 6.14A, many of the geometry issues are not problematic. Still, the distances between the pits look a little different between the photograph, which is reducing a curved surface to a flat one, and the radiograph, which in this case, actually helps us to see actual distances, because the film was curved to match the part and then flattened before this photo was taken.

Note that radiographs should almost always have a penetrameter image. This is a piece of metal with small holes that is included in the "x-ray shot". The thickness of the material with the holes is matched to that of the item being radiographed. The test method or code that the radiographer follows will generally specify a type of penetrameter based on the thickness of the material that the x-rays have to traverse. Note that for a part that is not flat and lacks uniform thickness, the x-ray path may be much greater than the thickness of the material. The x-ray source is usually something like a point. So, if the angle that the x-ray is traveling is not perpendicular to the part surface, the x-ray will have to travel further before it leaves the part and heads on to the film or image capture device. These geometric facts will result in varying shades of grey on the radiograph.

The smaller the penetrameter hole that is visible in the radiograph, the better the quality of the radiograph. What does that mean, specifically? That means that you have proven you have lined everything up and calculated the exposure so that you have a chance to see small features in the volume of the part. If you can't see the penetrameter holes, you don't know what else you might be missing.

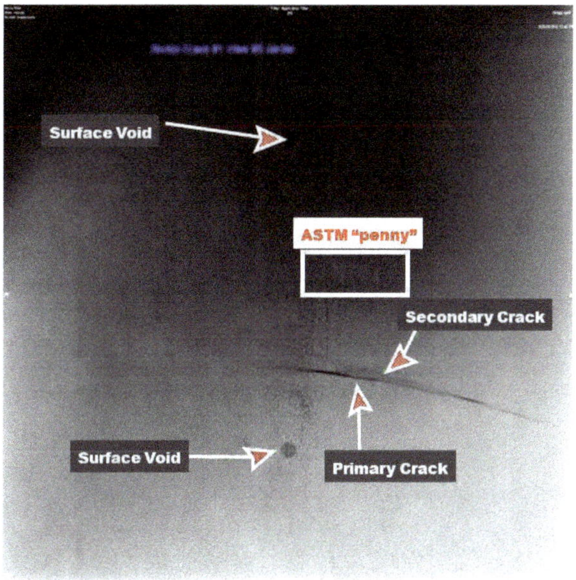

Figure 6.14B: *Faintly visible ASTM Type C wire penetrameter*

This type of penetrameter is known in American radiographic technician slang as a "penny". It is harder to see than the crack and several other discontinuities.

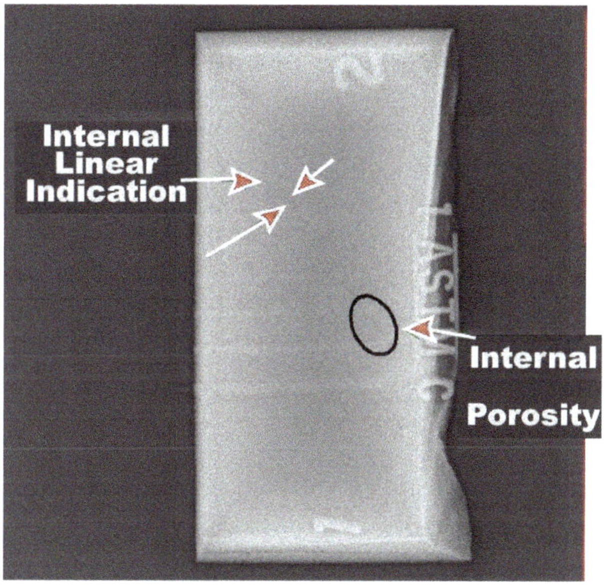

Figure 6.14C: *Visible ASTM Type C penetrameter*

Here, an ASTM Type C penetrameter is readily visible. The technician noted indications that are very faint to my eyes.

I prefer for the radiographer to use a "hole" penetrameter to the other main type, a wire penetrameter, because I am usually looking for voids of various types. Philosophically, if I am looking for a void, I want to know that I can see a void. Knowledgeable radiographers have told me that it doesn't matter if they use the hole type or the wire type. The wires are simply a series of larger to smaller diameter wires that are attached to the part. If the small diameter wires are visible, that indicates that features that size in the casting should also be visible.

While Figure 6.14A does not have a penetrameter, Figure 6.14B (above left) shows a wire penetrameter used during radiography of a heavy casting. It was made to a specification published by ASTM (American Society for Testing and Materials). The letters are barely visible to me, and the wires are invisible. But the trained technician told me that he could see at least a few of the wires. This radiograph was performed so that I could find a sound volume of metal for tensile and Charpy tests that I wanted to perform. The casting from which this chunk was cut was known to have a lot of porosity and shrinkage voids. I wanted to keep those out of the test bars.

KEY POINT: Be sure to speak with the radiographer about the selection of the penetrameter! And when the radiograph is examined, have them mark (a red wax crayon is useful on film, software allows marking on digital files) the general area and the smallest hole or wire that they can see.

Let's get back to Ultrasonic Testing now. This is the other primary volumetric method. There is a variation of ultrasonic testing called phased array UT (PA UT), which can overcome some, but not all, of the limitations due to complex geometries. The more advanced PA UT actually provides a three-dimensional image of the part with the sound-reflecting indications shown in proper position, orientation and size. This type of PA UT provides an image similar to that with which we are familiar from seeing in medical applications, such as babies in mother's bellies! But the task is much more difficult as the parts that people might want to test this way are vastly more different from each other than even a very skinny, heavy boned woman, from a very fat, small boned woman, and their offspring.

There is another type of phased array ultrasonic testing that provides graphs that are not visually intuitively related to the shape of the parts. But, like radiography, produce data that may readily be saved for others to examine.

Ultrasonic testing requires that an ultrasonic beam of energy be introduced into the part. The lack of a uniform cross-section, which will interrupt the beam before it gets to the far end, may be compensated for by use of angle beam techniques, but a much higher level of skill is required to interpret the data.

Figure 6.15 (next page) shows a cylindrical steel tube being inspected with an angle beam technique. The technician moves the probe along the part surface, and looks at a screen for the signal of the beam being reflected back to the probe by a far surface. The green peaks seen at the far left of the screen will generally move back and forth, and get taller or shorter, as the technician moves the probe along the part surface. The peak positions are related to either the position of the far surface that the beam is bouncing back from (the "back reflection") or a discontinuity. If we know the distance to the back surface, then we can separate an unplanned for, undesired surface due to a crack, a forging fold, shrinkage porosity, oxide films, etcetera, from the known surface, by its presence at a distance different from the known surface position. We clearly have moved away from any obvious relationship between the visual cue on the screen and the appearance of the item being inspected when using UT.

I have been involved in recommending ultrasonic NDE for large screws on stamping presses, after one was found to have a fatigue crack that had taken at least three years to propagate.[126] This is a good scenario for an NDT technician. They can schedule annual inspections, and have a good idea that they will be able to find a crack prior to it causing complete, catastrophic separation.

In another case, use of a compact tension fracture mechanics test[127] allowed me to determine the duration of stable crack propagation time for a roll in a steel mill, before it experienced catastrophic separation, interrupting production. Again, this allowed recommendation of a nondestructive inspection frequency. Tie rods on an extrusion press were another example of a single, unexpected, catastrophic, production-interrupting crack propagation event that led to the company specifying immediate NDE on the remaining tie rods on that machine, and other similar presses.

---

126 Based on visual inspection, some assumptions, and detailed review of tooling change records. This is another application of macro-fractography.

127 The compact tension fracture mechanics test is performed by applying equal and opposite forces through the two holes in the specimen, to propagate an initial sharp crack.

All three of these jobs were done with ultrasonic testing (UT). When conditions for UT are favorable, it's a quick method, and does not involve use of hazardous materials or processes (which radiography does). Favorable conditions include a simple shape, such as a uniform cross-section shaft, without cross drilled holes, grooves or diameter changes, with smooth, clean, ends. Sometimes, if all of the criteria are not favorable, the entire volume cannot be inspected, but useful information may still be obtained.

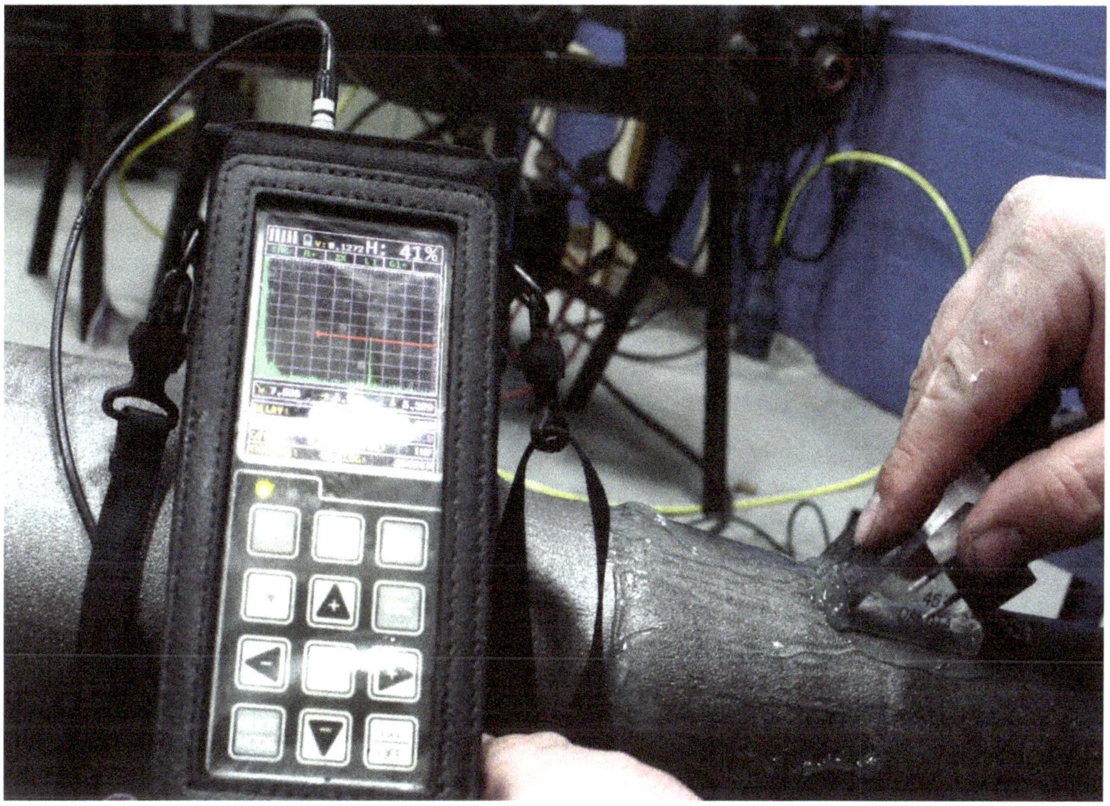

Figure 6.15: *Angle beam ultrasonic test of cylindrical tubular component*

6.8.5 Applications And Limitations of NDE Methods

Radiography is not usually particularly good for finding tight cracks, but if the geometry of the part is reasonably simple, ultrasonic testing may reveal the size, orientation, and position of cracks. Magnetic particle inspection is good for magnetic materials including carbon and alloy steels, and ferritic or martensitic stainless steels (and potentially even heavily cold worked austenitic grades, which have been at least partially transformed to martensite). But the discontinuities must be close to the surface of the part in order to be detectable.

Notice that I wrote "detectable". This is different from "detected". Just because a discontinuity of a particular type is theoretically detectable by a particular test method, does not mean that it will be detected in all cases. Many NDE methods need a human (or machine) with visual perception capability. The inspector might be having a bad day, and miss an indication. Use of an appropriate quality control tool, such as a pie gage or penetrameter, is crucial to understanding what the inspector might or might not have been able to see. Did the inspector see nothing because the part was good? Or because the inspector did a bad job? Always ask the inspectors what they think they would have been able to see. Always tell the inspectors as much about what you are trying to find as you can before they run the test. That will increase the chances that the method will reveal the features you are looking for, if they are present in a detectable orientation and location. That said, it must be noted that most NDT inspectors, especially those who lack the training required to obtain an internationally recognized Level III inspector's certificate, may not be able to carry on this type of conversation. But at least you will know this is the case.

It must also be understood that some part shapes are just too complex to get any meaningful NDE data from evaluations. Other methods of quality control must be found to ensure their structural integrity. Circumferential fillet welds that join a ring to a tube are essentially impossible to inspect with any standard NDE method. These are common welding geometries, and some other method of ensuring the structural integrity of the weld is needed.

If you are doing NDE to support a failure analysis, you will have to explain to the NDE service that your goals are different from a standard inspection. When it is possible, I try to work with technicians who have attained Level III standing in ASNT, the American Society for Nondestructive Testing, or an equivalent national organization from a different country. I have found it extremely frustrating to try to work with less qualified people to support failure analysis. They rarely have the background in knowledge of the physics behind the method to be able to explain their findings to me in a way that I can explain them to my clients. Even with Level III technicians, I try to find those who have attained certification in multiple methods. These individuals must know a lot about the methods they are using and the components they are testing, as well as some basics facts about the manufacturing methods they are made with. They even learn a little bit of metallurgy and materials science.

They understand that the cleanliness of the parts prior to being tested, or the lack thereof, will interfere with the probability of detection. Experienced technicians know that surface roughness can create "false indications" that look similar to cracks or laps, when being inspected by magnetic particle (MT) or dye penetrant testing (PT), useful for non-magnetic materials. They will be more likely to tell you when your part is simply not fit to be inspected with the method you have requested.

It may be very difficult to discern any indications that are potentially a result of an actual crack or other type of discontinuity when the surface finish is too rough. Thread roots in fasteners naturally collect the indicator fluid, whether it is magnetic particles or visible dye. That makes it more difficult to find cracks in thread roots. If a crack is filled with oxide or grease, the visible or fluorescent dye, or magnetic particle fluid, may not be able to do its job. There are many "little things" that an NDE technician needs to know to support failure analysis work. It's worth it to find and pay for the most qualified one you can, or give up on using this technique until you can find a qualified, competent person.

6.8.6 Other NDE Methods

The methods we've been discussing up to now include only systems to detect missing material within the envelope of a part, internal surfaces, or foreign material of a vastly different density. For example, tungsten inclusions from TIG welds in steel will show up bright in standard radiography, as it has greater density and blocks more x-rays than the steel matrix.

There are other methods that can be used to evaluate the electrical conductivity and magnetic response, which can then be associated with different microstructures. Eddy current (electromagnetic testing, ET) is one such method that can be used to sort for different responses to heat treating, for example, or mixed materials in lots of small commodity parts. See Figure 6.16A. Eddy current, once set up and calibrated to the parts in question, can be performed very rapidly, such as a second or less, in automated systems.

Eddy current testing may also be used to detect surface flaws, such as laps in deep drawn parts, and cracks in thread roots. This may be a good option, as we recall that dye penetrant and magnetic particle inspection of thread roots can be complicated by the excess concentration of the test solution in these locations. Eddy current, like UT, is safe, and requires no special precautionary methods, such as those needed for radiography. Also, similar to RT, it's nominally a dry method. UT does use a liquid or gel couplant to transfer the energy beam from the probe to the part, but it's not applied to the crack surface. Eddy current, on the other hand, is truly dry. However, like UT, we move very far from having any obvious, visual cues as to the relationship between the data produced and the shape or features of the component being evaluated.

Yes, eddy current testing is dry, quick, and safe. But it is inherently difficult to interpret. Each inspection job must essentially create its own set of standards. If you are trying to sort for cracked parts, it might be fairly straightforward to prove that the settings for accept/reject are appropriate, by performing another test method to find the cracks, such as MT or PT. Those slower methods are often used to confirm the ET set up.

But for microstructure sorting, some of the parts, generally those at the extreme ends of the elliptical fields shown in Figure 6.16B, must be sacrificed to actually examine the microstructures. Otherwise, the reasons for the differences in electromagnetic response will be completely unknown. There is no direct linear relationship, for example, that shows ET response as a function of hardness of 1050 steel bolts of size M8, Grade 12.9.

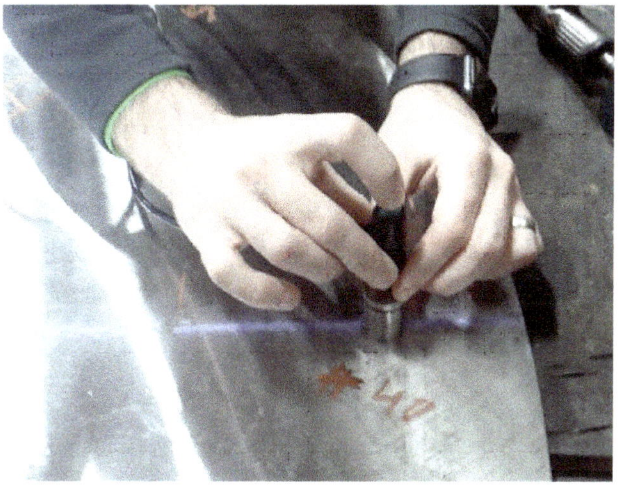

Figure 6.16A: *Eddy current probe on flat part being evaluated for microstructural uniformity*

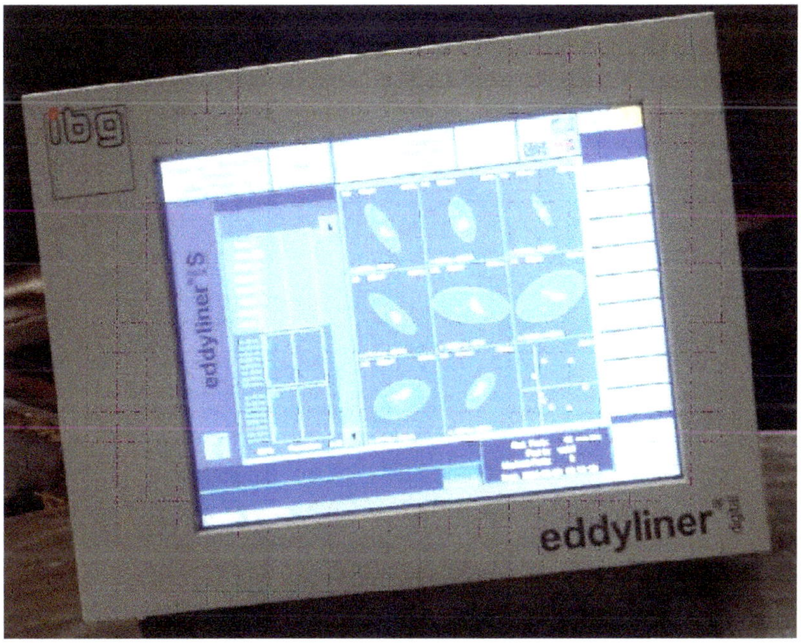

Figure 6.16B: *Eddy current display showing multiple test locations on the part in Figure 6.16A*

Up to now, we have broadly explored many critical introductory aspects of performing a failure analysis that must precede destruction of the Subject of the investigation. Take a moment and review Figure 6.17 as a reminder of the basics of a failure investigation.

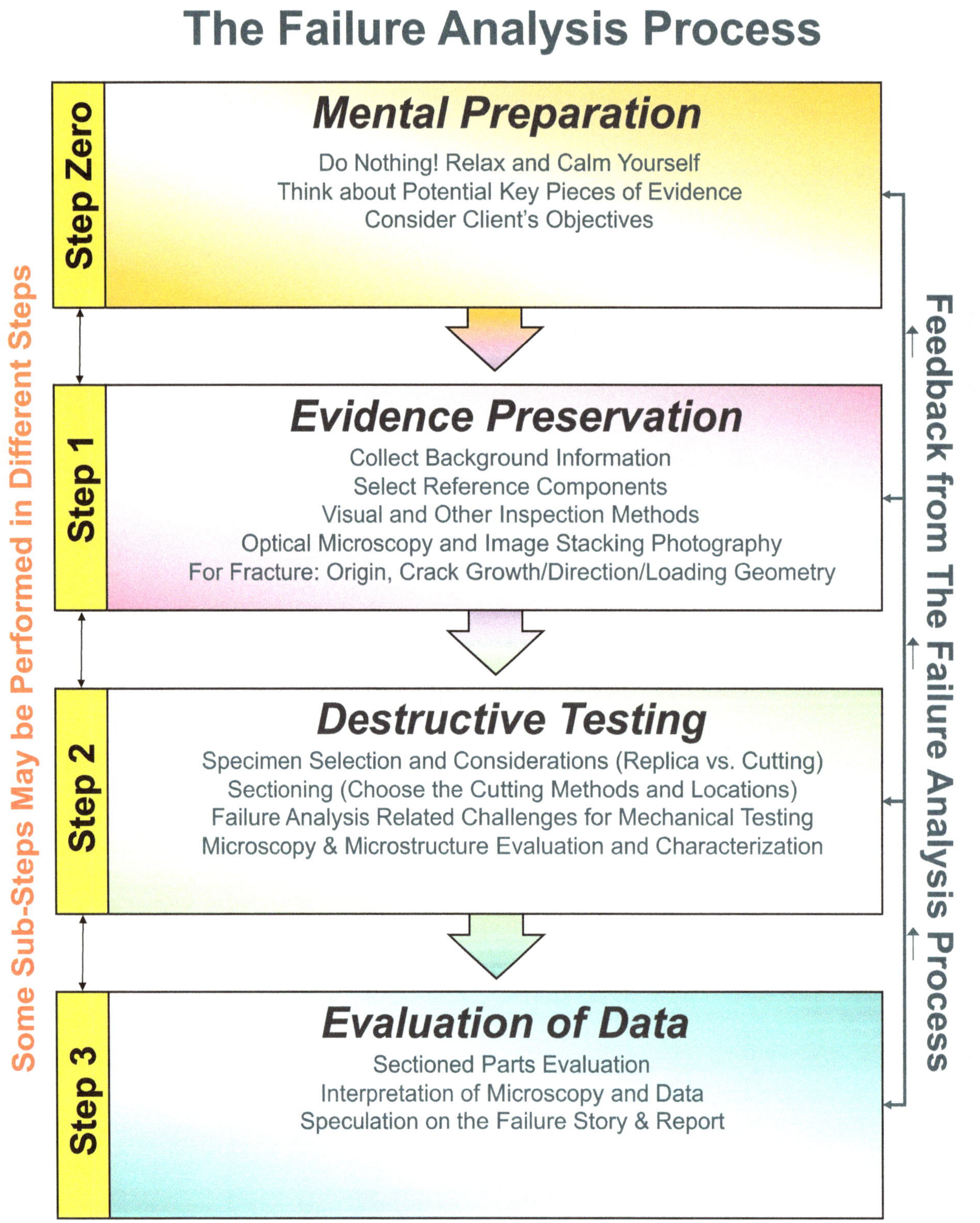

Figure 6.17: *The Basics of a Failure Investigation are an interactive process. When combining critical thinking and a materials-engineering-informed investigation, your chances of discovering the "story" of how a part failed is greatly improved.*

6.9 Review Key Points from STEP Zero and STEP 1

In the Step Zero and Step 1 portions of the investigation, our careful consideration revealed that almost every aspect of a failure analysis may be, at least in some situations, destructive to evidence, even when not destructive to the part. A key concept that we studied in detail is the difference between destruction of a part and destruction of evidence.

As participants in a failure analysis and readers of this book, we need to cultivate clarity for ourselves, as well as for others on the team who may not have considered the issue of evidence destruction in as much detail as we may have!

> **Failure Analysis evidence is easier to unintentionally destroy than the damaged parts themselves.**

"To observe is to disturb" may refer to physical characteristics and may refer to our thought processes. Knowledge builds on itself, and if some of what we know is wrong, it will be more difficult to see what is in front of our eyes, or to hear the importance of a fact that someone is telling us. Reminding ourselves to Compare-Contrast before every new step will help build our ability to slow down long enough to ask useful questions. "What, exactly, are we trying to achieve?" is often a good question. "To whom will we have to present our findings?" is a good question.

We will ensure our ability to continuously learn when we pair broad questions, such as the purpose of the analysis, with detailed questions, such as "What's that debris particle?" or "What's that shiny spot?" Effective failure analysis requires skill sets as well as knowledge. Skill sets include knowledge that must be practiced to ensure mastery. Most of the tasks of failure analysis fall in that category.

> **We must understand the Design Intent before we decide how to cut the part.**

Interviewing stakeholders (who suffered a failure) is a skill set. There are meso-scale questions such as "Is this all the physical evidence?" and "Where is the mating component?" We ask these meso-scale questions between the overarching macro view and the detailed questions specific to the particular investigation.

6.10 Overview of the Failure Analysis Process - STEP 2 (Destructive Testing)

At some point we will need to move on from pure evidence preservation and background data collection and get to the subsequent aspects of the work. Thus, we move into Step 2. The Step 2 portion of the failure analysis is almost always destructive. How you think about this is critical.

I've broken Step 2 into the following topics designed to prepare you to think like a Failure Analyst. Investigations may unfold differently:

- How to select specimens for destructive testing
- When to use the alternative replica method
- Tensile test coupon location selection considerations
- Choosing the cutting methods
- Choosing the cut locations
- Failure analysis related challenges for mechanical testing
- Microscopy in the failure analysis process
- Microstructure evaluation and characterization

> **The Step 2 portion of the failure analysis is almost always destructive.**

6.11 How to Select Specimens for Destructive Testing

The first task that we routinely confront is to decide which portions of the Subject Component will be used for which tests. This, like many of the earlier tasks we have studied, is usually an iterative process. Remember, Step Zero. Relax, then think.

For wear and corrosion, we first need to have determined – during the visual and/or nondestructive inspections – the least and most damaged areas of the component. We also need to have already determined whether the damage is something that might have been anticipated or that is totally unexpected. This second question can only be determined in light of knowledge of how the product is used, or what we might call the "Design Intent". Often, people who are familiar with the product, even if they do not have strong technical backgrounds, are valuable sources of opinion about whether a failure or damage event should be considered unusual.

> *There are those who have a misunderstanding about the nature of objectivity. These people want the failure analysis team to work "blind", in other words, in ignorance of the stakeholders' thoughts about what happened, or even absent background information, in an attempt to get what they consider to be pure data. But the world is too big. There is too much to know. An independent lab or consultant will usually lack the background required to visualize all the potential scenarios that could shed light on what happened.*
>
> *Performing the work blind leaves the investigators open to drawing ridiculous conclusions that could seriously damage professional credibility! The requesters must be encouraged to have at least a basic level of trust in those who are carrying out the analysis for them. It is much easier to determine whether the evidence is consistent with various potential damage scenarios, or not, than it is to create a scenario from thin air.*

In a perfect world, or at least a perfect failure analysis situation, we have current engineering drawings (as of the manufacturing date) for both the assembly and its individual components. We have access to someone who is familiar with the service conditions that the Subject experienced. If access to engineering drawings is not available, we will need to inquire about people's memories or opinions about the material used or research what is typical in that application; whether or not it is of uniform strength, and, if present, the use of coatings. If we are lucky enough to have a knowledgeable resource who can tell us if "that shiny spot" is usually present, on either new, or normally used components, we will be in a stronger position to be able to perform an effective and informative analysis.

People who are familiar with a product are often valuable sources of opinion about whether a "failure" or "damage" is unusual.

If the damage category is fracture, a primary consideration in specimen selection is preserving the crack initiation for microscopy. Saving the suspected final separation area is also helpful. Often, if the crack initiation area is not so damaged that it is essentially pointless, it will be worth examining the initiation area with a scanning electron microscope (SEM). The higher magnification and depth of field afforded by the SEM is often the key to identifying a particular type of damage.

The initiation is critical to identify, because it must be studied to see if there was anything unusual about that location. In other words, was there a discontinuity that allowed the strength of the component to be lower than anticipated, even if the material was otherwise properly processed? This is part of why fracture analysis is easier than wear or corrosion analysis.

It is critical to identify the fracture initiation because it must be studied to determine if there was anything unusual about that location such as a discontinuity that lowered the strength at that point.

Often, with wear or corrosion, the original area that was damaged is gone! There's no way to tell if it specifically had any anomalies. In those cases, we are forced to draw inferences based on the characteristics that we reveal in the remaining areas of the Subject Part.

In the case of corrosion, if there is heavy corrosion product or deposit on the surface, we may seek a less damaged area, where it is possible to observe the underlying material, as well as the residue from the environment. As in the case of fracture, careful consideration must be given to how the area to be examined is selected. Often, with corrosion in particular, multiple locations for compare-contrast will be needed to gain an understanding of how the damage is originating and progressing. These two aspects of corrosion damage often represent two entirely different processes. Pitting corrosion, for example, may initiate the damage. Stress corrosion may be the damage mechanism that leads to end-of-life.

For wear damage, removal of loose wear debris may be fairly non-destructive. Grease will need to be removed from bearings to reveal the underlying damaged surface. For small ball bearing assembly inner races, this may be done without cutting, but that may be difficult for outer races, as the viewing angle may not be favorable unless the ring is cut.

Now that we have introduced the idea of specimen selection and factors that influence it, let's move on to considerations of actual cut procedures as it relates to selection. In general, if the specimen is not very large, it is preferable to examine the Subject, or specimen cut from it, directly. Unless the part is small enough to fit into the electron microscope whole, in an orientation that is in line with the image detectors, as we just noted for bearing races, it will need to be cut. If it needs to be cut, you need to consider how to cut it. In other words, you need to know how to extract specimens.

Specimens will need to be extracted if you are going to perform strength testing to get at the basic issue of why the stress exceeded the strength. Microstructure evaluation usually requires specimen extraction. Core hardness requirements may be performed with micro-indentation methods on prepared microstructure specimens, but may require their own cuts in some cases. It is important to extract the specimens as close to the damage as possible so the data is relevant.

If corrosion appears to be the major or even a contributing factor to the damage, or if there is a deposit or debris that needs to be characterized, then it is best to cut the part dry.[128] If the part is extremely hard, that will be difficult. So, there are many considerations to be weighed prior to deciding how the part will be sectioned, especially if you have a complex situation.

Figure 6.18 (next page) shows one side of a broken weldment, as I marked it up for sectioning. The weld is not visible in this view. The part in the photo was made from flat rolled sheet steel with a punched hole, through which a pin was inserted and welded on the backside (not visible). I don't recall what happened to the mating piece or why I needed to cut the part before performing the SEM work - this piece was less than 25 mm wide. Maybe the client was "in a rush!"

In addition to planning the cutting process in view of preserving the crack surface for the scanning electron microscope examination of the crack initiation, it is often helpful to examine the final separation area. First of all, it is fairly common that the initiation area is rubbed, dirty, or otherwise damaged, rendering confirmation of the initiation location difficult. Confirming the final separation event, which is often ductile at both the macro and micro levels (i.e., has parabolic micro-voids[129]), can shed light on the loading geometry. This is true, even though micro-fractographic examination is usually thought of primarily in terms of shedding light

---

[128] Alternatively, bearing debris, such as grease with particulates or corrosion product or deposits) may be removed and set aside, so that a metallurgical wet saw may be used.

[129] The details of the basic types of micro-fractographic features will follow in the micro-fractography section of Chapter 8.

on the interaction of the crack with the microstructure.

In other words, we have just contradicted an earlier section of the book, or have more accurately modified the frame of reference in which we may proclaim: micro-fractography is performed to gain insight about the microstructure or its interaction with the environment. The competent analyst knows how to substitute another tool from the kit when the "obvious choice" isn't applicable or when general epistemological confidence is needed.

Specimen Selection requires careful consideration.

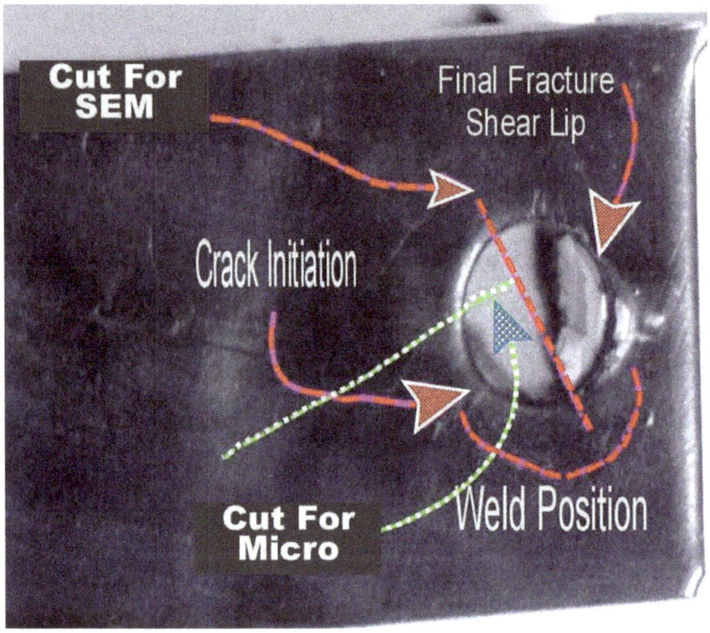

Figure 6.18: *Cutting broken weldment with inserted and welded pin for SEM (micro-fractography)*

Note the dark crack initiation thumbnail and the indicated final separation 180 degrees opposite (shear lip). The dark nearly radial "cliff" is possibly a result of weakness due to grain orientation effects. The crack can be seen to have re-initiated in several locations along the "upper" edge of the cliff. The cut for the "micro" (mount) provided grinding allowance for final exposure of the examination plane near the primary crack origin. Leaving more than half of the diameter was helpful to hold the metal in the plastic mounting compound. The remaining (right) piece of the pin could have been used for bulk quantitative composition analysis after the SEM exam was complete.

6.12 When to Use the Alternative Replica Method

For many components, the total fracture surface is small enough that the suspect initiation and final separation will be visible in a single electron microscope session. If this is not the case, and there are difficulties in cutting, especially with large, hard, corroded parts, the scanning electron microscope exam may be conducted on a polymer replica. The use of replicas has been previously described, in the visual examination chapter, for the purpose of capturing corrosion product or deposits. But they are also useful to replicate the microscopic details of the fracture features. The main drawbacks to the use of replicas are:

- Contamination of the surface of the part with the replicating compound;
- Defects in the replica, such as air pockets or entrapped dirt or dust, which may be confusing to the analyst;
- An inability to meaningfully use the backscatter electron imaging mode to rapidly screen for composition differences related to microstructure features on the crack surface;
- An inability to get any EDS (energy-dispersive spectroscopy) composition data, except for entrapped corrosion product or debris;
- Very rough surfaces and crevices, which may pose problems such as trapping portions of the replica when the rest of the replica is peeled away from the part.

These disadvantages aside, replication can be a good option for large components and it is less destructive to leave a trace of replication compound on the crack surface than to cut the crack surface. However, I have been involved in litigation matters where the parties have not agreed to allow replication. An example of replication is depicted below. Figure 6.19A shows the replication site after a building fire caused concern that the steel structure had been overheated to the point of softening the high strength steel. Obviously, nondestructive testing was required. Figure 6.19B shows the microstructure of one of the earliest types of High Strength Low Alloy structural steels with a specification in the ASTM system. This building was about 80 years old in 2010 when I performed this inspection. The microstructure revealed is called a bainitic structure.

Figure 6.19A: *Replica method applied to Truss 1*

A building fire caused concern that the steel structure had been overheated to the point of softening the high strength steel of construction. The charred wood is clearly not fit for continued use. The silicone replica is at the arrow head. The steel turned out to be fine.

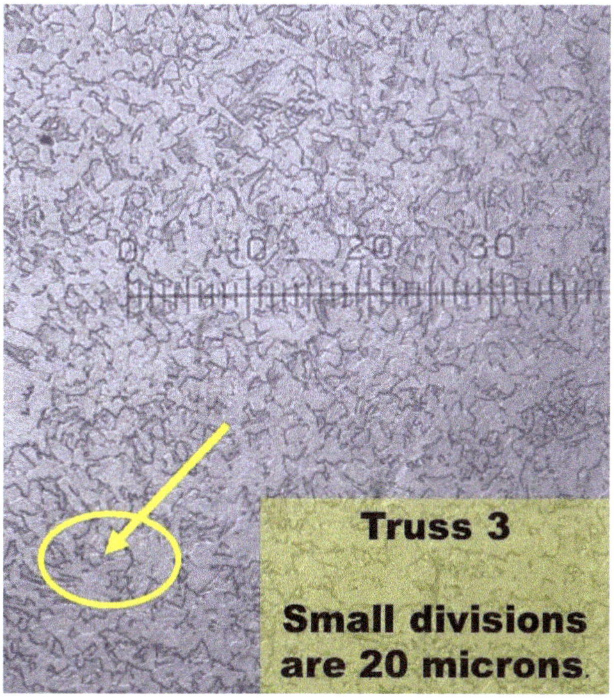

Figure 6.19B: *Replica method applied to Truss 3*

The replica had one protruding feature (yellow oval) which would have been a pit in the original. However, there's no reason for there to be a square pit. Likely, there was a cubic second phase particle, and the iron holding it in place dissolved away during the etching, and the particle fell out.

Getting a perfect etch for a replica is an art!

6.13 Tensile Test Coupon Location Selection Considerations

Let's focus our attention on Specimen selection and location considerations for destructive testing of tensile test coupons. Selecting tensile coupons will stretch your capacity for handling more complex considerations to make effective decisions with fractured components. This is a Step Zero moment before you cut. Relax. Think.

The tensile tests, or other standard mechanical tests, often require the largest portion of the Subject Component. So, we need to start thinking about how we are going to get the material that we need before we make any other cuts. Most bar stock of moderate diameter requires the certification tensile test to be obtained at the mid-radius of the bar. For most of the loading geometries that are present in real parts of real machines, the highest stress is at the surface. Thus, I will often "compare-contrast" the strength of a tensile test as close as I can get to the surface with one at the specified mid-radius position.

This issue of selecting the position of mechanical test coupons applies to all materials. Heating and cooling rates are important factors that influence the microstructure and thus the behavior of the components. Especially with large sections, the surface and core could experience vastly differing temperature change rates during manufacturing and possibly service. Tensile tests, impact tests (Charpy, for example), compact tension tests (if fracture mechanics[130] is to be a part of the analysis), and microstructure specimens, may all benefit from multiple test locations to support failure analysis. Why did the impact stress exceed the impact strength at a position just below the surface and after a very shallow fatigue crack had initiated? You must check the strength as close to the relevant position as possible to get an answer.

Did the material have characteristics in line with expectations based on the specification? You may need to check a different location! Compare-Contrast!

[130] Fracture mechanics is the only quantitative way to predict how damaging (or likely to influence failure) a particular discontinuity is in a particular application. Fracture mechanics was developed to support the aviation industry. Other industries make use of this discipline as well.

6.14 Real Life Problematic Examples of Tensile Test Coupon Selection

If you are doing a good job of selecting the test bar location that is as close to the damage as possible, then you will encounter challenges. Fractures, discontinuities, missing material from corrosion and other issues complicate extracting real life test coupons and therefore the data you will be able to acquire.

This is very complex activity requiring full attention to all of the details that you have gleaned from the background investigation, the visual inspection, and your general knowledge of engineering principles. Allocation of enough material for mechanical testing is essential.

If you are not scratching your head and double checking your thought process, you may be in for a big surprise – and not a happy one – when you try to interpret the data.

While a full length solid test coupon is desired, the following sub-sections provide an introduction to the roadblocks and complexity that frequently arise during mechanical test coupon selection and suggestions to handle them.

6.14.1 Problems: Tensile Testing Tubes

Using the entire tube as exemplified in Figure 6.20 (next page), has many advantages over cutting "flat" test coupons. The elongation values will generally be higher in full section tubes. Most metal tube specifications are based on the results of a full diameter test. The load bearing cross-sectional area calculations are easier for a full circle than a straight sided arc. In order to overcome the lack of a reduced cross-section that forces the break in the gage length, tubes are plugged at the ends, leaving the gage area in the center of the length unsupported. Generally, the tube will break in the unsupported area.

Figure 6.20 shows three pairs of aluminum-plugged stainless steel tube tensile bars. Behavior consistency within each pair was much greater than between the pairs. One pair was as-manufactured; one pair was subject to a hot durability test; and one pair was subject to a cold durability test. Each pair was made from a different part number (different versions of the manufactured part). The end plugs were installed so the unsupported length is equal to the minimum gage length for a test coupon and the plugs smoothed on their inner edges to allow pinching of the tubes to happen without interference under the test tension.

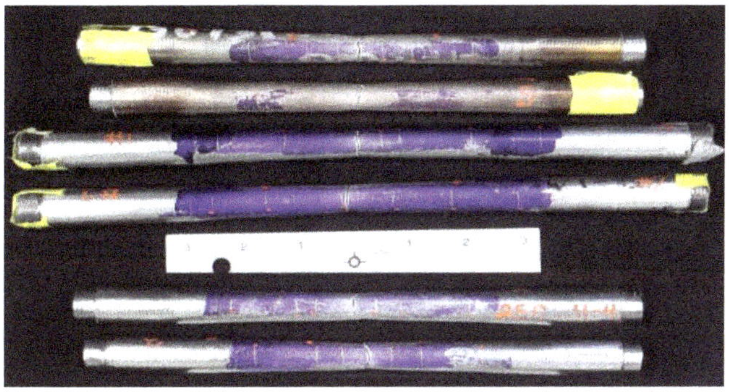

Figure 6.20: *Tensile test coupons: stainless steel tubes with aluminum plugged ends – scale shown is in inches. Gage length for the center pair was 2 inches, and for the top and bottom pairs was 1 inch.*

Test locations were kept controlled to the degree possible. Note that each pair of tubes was initially cut to a different length. The red dots were placed by the test lab to help them determine if the break was within the center half of the original two inch gage length. From this view, it's difficult to see which coupons had the greatest elongation. The original report indicates 56% for the center pair, and 75% for the top and bottom. How much of that difference is due to differences of microstructure and how much is due to gage length difference is debatable.

Sometimes, however, using the full tube section is impossible. Such an example is shown in Figure 6.21. The scale used in the image is cm. I could not use full tube sections because there was a large hole in the part from which the test coupon was to be made.

With some creativity we decided to cut the tube in half eliminating the hole and then custom machined the other half for tensile testing. A special equation, a nightmare to apply, is required to account for the difference in cross-sectional area due to the curvature of the tube wall.

In Figure 6.22 (next page) we see the preferred full diameter tube tensile test method. Plugs are inserted to support the ends where the test bar is gripped by the machine. These plugs were creatively welded into place to provide more gripping area for the tensile test machine fixtures.

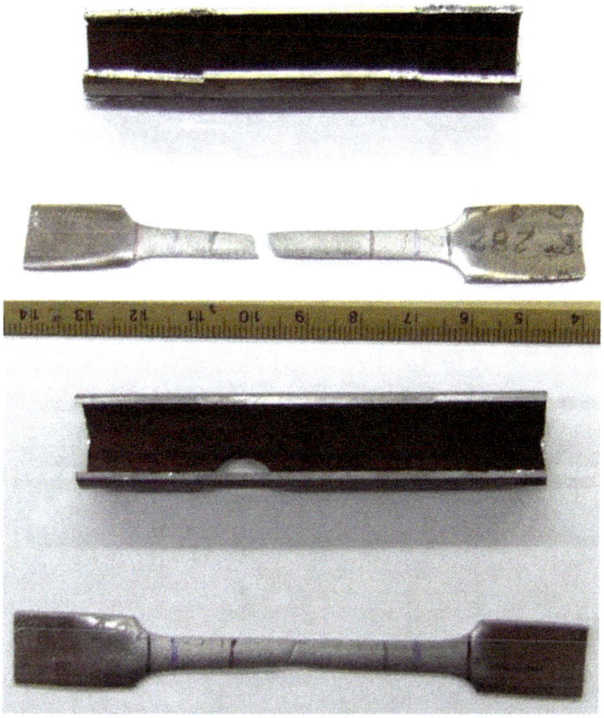

Figure 6.21: *Tensile test tube was too short and had a hole in it.*

Half of the tube was used to machine a "dog bone" coupon and then the tab ends were flattened. This avoided deforming the gage length. Note the high degree of stretching of the bars compared to the original length of the tubes (from which they were cut).

The yellow circles in Figure 6.22 highlight the multiple, overlapping sets of gage marks (tiny pricks) that are allowed when the test coupon might break away from the center of the gage length. Having pre-positioned multiple sets of gage marks, at least one pair is likely to have the fracture centered in its length.

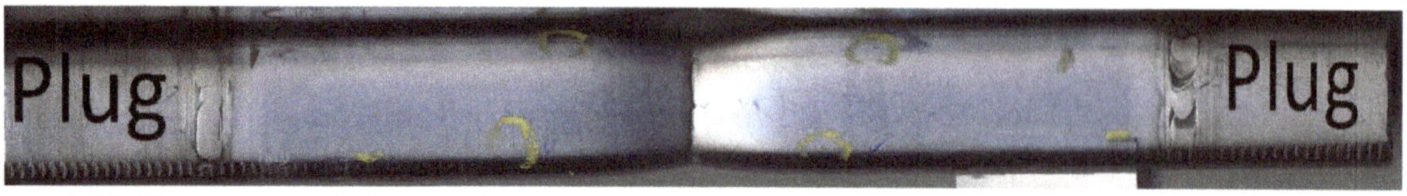

Figure 6.22: *Test coupon of high strength material made longer by welding inserted plugs – yellow circles show multiple sets of gage marks*

6.14.2 Problems: Tensile Testing Odd Shapes

Figure 6.23 shows a tensile bar made from a steel stamping.

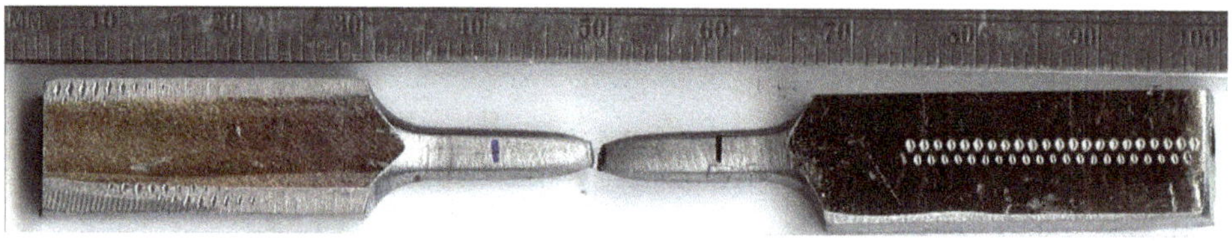

Figure 6.23: *Tensile bar made from a steel stamping*

The lab (not the writer of the book!) preferred to machine this flat stock into a round test coupon. But this was a problem because it eliminates any chance for the surface layers to contribute their strength or weakness to the measured value.

Compare the image in Figure 6.20 (previous page) to the test bars in Figure 6.24 (below), both of which show groups of coupons from a single manufactured product line. Each pair of test bars in Figure 6.24 was cut from a single injection molded thermoplastic part. Plastic tensile tests, especially those cut from parts rather than molded test bars, tend to have fairly large behavior variation from part to part. This occurs even within a single lot.

The plastic tensile test specifications usually require five replicates. In failure analysis, that's often simply not available.

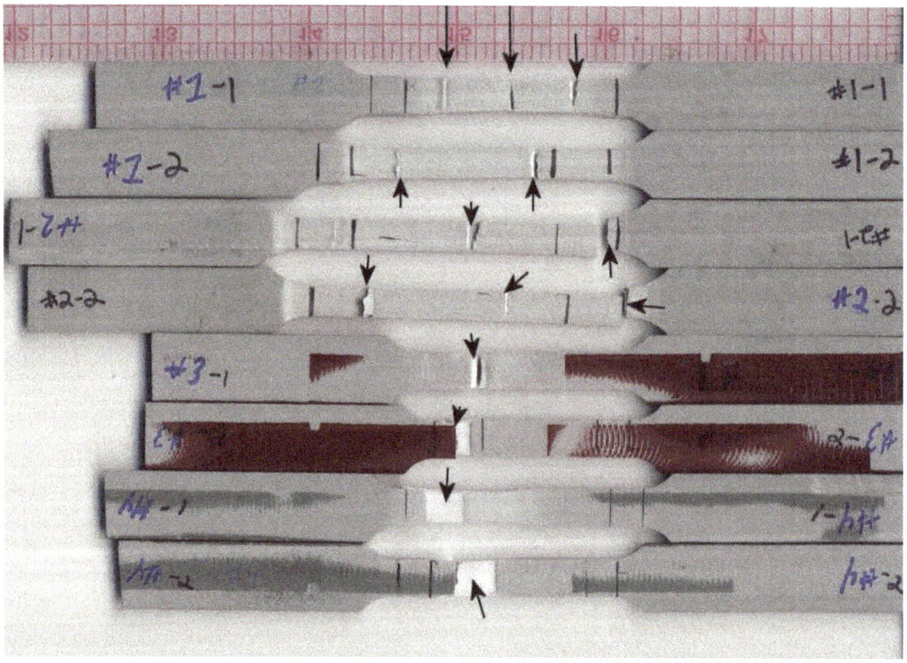

Figure 6.24: *Thermoplastic test bars cut from injection molded parts*

Note that some of the broken test bars have multiple areas with severe thinning (dark arrows) – some of which were far from the ultimate separation position. Some breaks were far from the center of the gage length.

Figure 6.25 shows a tensile bar that inadvertently contained a portion of the crack in the gage length. The strength values were far below specified. So was the elongation value. The crack was not within the specified center half of the length of the narrow area of the gage length. The lab reported the test results without noting a discontinuity in the gage area - even though ASTM specifies that coupons are to be discarded and another one prepared if a defect is found to have influenced the crack position.

That is why I always request labs to return my broken test bars. ALWAYS. Even the best test labs might not notice features of interest to the Failure Analyst.

Figure 6.25: *Tensile test bar with discontinuity (fatigue crack) in specimen*

This tensile bar was made from a part with multiple cracks in it and inadvertently contained a portion of the crack in the gage length. The crack was not within the specified center half length of the narrow area of the gage length. The lab reported the test results without noting the discontinuity in the gage area. The discontinuity is marked as an initiation thumbnail. In fact, it is probably just a portion of the end of the local fatigue zone.

The importance of getting mechanical property data as close as possible to the "failure location" cannot be overstated. It is often challenging to do so because of size limitations. Many laboratories will object to doing this work because it cannot be done according to the standard test methods. They fear an auditor will think they are incompetent because it looks like they are doing the testing wrong. It behooves the Failure Analyst to develop a long term relationship with someone in the machine shop who can shepherd these projects originating from unusual circumstances.

6.15 Choosing the Cutting Methods

Now you have been exposed to tensile coupon preparation challenges along with location selection. It is time to go back to address how to choose a cutting method. As already noted, the first choice to be made is whether to cut dry or whether a coolant is acceptable. In general, when the visual examination shows no evidence of a corrosion product or deposit, it is acceptable to use a wet method.

A common, preferred method is a metallurgical abrasive cut-off saw. Equipped with the right abrasive blade, this will leave the least amount of sub-surface microstructural damage. Of course, it must be operated by a careful, attentive, properly instructed technician.

An even better way to minimize damage layer thickness is using a precision diamond saw. So why doesn't everyone use them? Because, these saws are appropriate only for very small parts and they are quite slow when compared with an abrasive cut-off saw.

In both cases, the saws are generally operated with covers closed; so, creative fixturing is often required for odd shaped parts.

Even if the part wasn't odd shaped when it was manufactured, cracks often result in shapes that are difficult to fixture. Even if you can imagine a clamping arrangement that isn't difficult to fixture, the need to protect the crack surface during cutting often results in difficulties.

For example, a longitudinal cut on a small diameter round shaft with a transverse crack is a challenge. You may have to pick from options that most machinists would consider a terrible clamping practice. Technicians performing failure analysis work must be instilled with a proper work attitude, which values protection of the evidence, especially fracture surfaces, over efficient and even theoretically more stable cutting set-ups.

If you have a very large and challenging part, getting the crack exposed may be the most challenging part of the project.

Band saws are a good choice for non-hardened materials of large or odd shapes or for parts that must be kept dry.

NOTE: *if there is no coolant, the parts can get quite hot. The microstructure of most steel that can be cut with a band saw is generally not going to be modified in any significant way by the heat generated during band saw cutting.*

This is not the case for aluminum and other lower melting point materials. These must be cut with much greater caution. Some people recommend only hand operated hack saws for such materials.

6.16 Choosing the Cut Locations

Before making any cuts, it is wise to have a general idea of how all required specimens will be extracted (removed) from the component. Figure 6.18 is a Subject Part where specimens were extracted for cross-section microscopy and SEM.

If you plan to perform mechanical testing, such as tensile tests, Charpy impact tests, or fracture mechanics tests, take the time to:

- examine the part,
- measure the part,
- lay out or mark the surface of the part in order to prove to yourself that you have enough material to do all the desired tests;
- alternately, take a photo and mark it with proposed test coupon blank locations and sizes.

Although this may seem obvious, it can be a creative effort in a complex situation with limited Subject Material. Reengage with Step Zero.

EXERCISE 6-C: Find a damaged part, and sketch a layout for test coupon extraction on a separate piece of paper. Mark it to show what positions in the extracted coupon need to have traceability maintained. Make up a naming system that includes each piece to be cut out. Specify a cutting method, or multiple methods, as required.

6.17 Failure Analysis Related Challenges for Mechanical Testing

Over the years, most of my supplier test labs have ruined at least one irreplaceable test coupon. The most recent event occurred when the package containing the test coupons was lost by the shipping company. The shipment was not insured for the amount invested (approximately $\$4000^{\underline{00}}$). We had been planning to perform micro-fractography on the crack surfaces. They gave us $\$100^{\underline{00}}$ for their mistake.

Irreplaceable parts may be ruined despite your best efforts, multiple conversations, and emphasis that non-standard work is involved. Failures in the failure analysis process happen.

The problem that commonly arises is that the part is too small to provide a standard test coupon. We discussed this in the series of examples of tensile testing in 6.14. Even if the part started out large enough, the crack or corrosion or other damage process often reduce the original volume available for testing. American test methods ASTM A370 and ASTM E8 commonly use 2 inch or 50 mm long gage lengths. Many laboratories prefer or require a coupon length four times as long as the gage length. Both of these ASTM methods and other standard tensile test methods from Europe and Asia allow smaller test coupons of standardized geometries.

In general, the gage length of the specimen is four times the width or diameter. It is area that is actually providing the numerical test results and is shorter than the overall length. But what about tubes that are tested full diameter? What about wire rod that is too small diameter to machine with standard shop equipment? In these cases, the product form (such as tube or wire) specifications will generally take precedence over the general test methods.

Any deviation from whatever is the standard geometry (e.g., normal or product form specific) may provide results that differ numerically from expectations. Ductility measurement values are the most subject to coupon geometry related variability. It is often extremely difficult to determine whether the original material used to make the Subject Component "met the spec" at the time of manufacture.

In the case of service environments that expose components to high temperature and corrosion, interpretation of mechanical test data is as much of an art as it is a science. Judgment must be exercised; experience is valuable; published data is useful; and knowledge of the actual service conditions is important. If any of these are not available, the Failure Analyst must be aware that such knowledge is absent.

Small formed or heat-treated parts made from standard metal sheet, plate, bar, and wire each present their own challenges to the Failure Analyst trying to choose a specimen that will represent the characteristics of the portion of the Subject Part where the damage originated. It's helpful to test the Reference Parts the same way as the Subject.

Castings and powder metal parts, injection molded parts, and items made by the modern additive manufacturing processes also each present their own challenges.

If a void shows up in a powder metal part or a casting and it ends up in the gage length, the material may appear to be out of conformance. But many original material specification conformance decisions are made based on separately prepared test coupons, which may bear little relationship to the values found for coupons cut from the Subject of the investigation.

It is important to work towards getting good test values to represent the strength of a part that was damaged by a mechanical process (fracture, deformation, or wear). Hardness values may be substituted but they provide less insight about yield or high strain rate response. Often, a difficult to interpret tensile test is still the best choice to help us understand why stress exceeded the strength.

For mechanical damage, stay focused on how to get the best answer to the question, "What was the strength?" A quantitative answer to this question, including estimates of its uncertainty, will help us understand how the stress exceeded the strength.

KEY POINT: Uncertainty related to the test method error (load cell and micrometer are examples) is usually minuscule in comparison to that associated with strength variation within the component. In addition, strain rates and temperatures of the failure-related-stress-event and the laboratory test are unlikely to be closely matched.

EXERCISE 6-D: Look up the data for tensile strength, impact strength and hardness. Find at least some material that has data available for multiple test temperatures.[131] Are you surprised at the variability? Tell your colleagues / classmates what you have found.

[131] There are many websites and publications with material property data under different conditions. I trust you to find information.

6.18 Microscopy in the Failure Analysis Process

Now that we've made plans to obtain all the specimens that we need, it's time to start the microscopic inspection.

If the damage category is fracture and if you have been unable to identify the crack initiation during visual examination, then you must include a plan to attempt to determine the crack origin (or origins) using microscopic methods.

The crack initiations of small coil springs made from small diameter wire are often invisible to the unaided eye; but they may be clearly visible with an optical microscope.

Smaller components made from wire, such as many medical devices, may require higher magnification. We may need to use a scanning electron microscope to determine where the crack initiated and to see wear spots that might be associated with the initiations. Wear spots may not be at the crack initiation, but might provide information that helps explain an unusual failure such as unplanned coil contact or other evidence of incorrect installation.

As we have noted, there's no hard boundary between visual inspection and microscopic inspection.

Optical and electron microscopes are both used to reveal small features. As we have already noted, every magnification device must be appreciated for its capabilities. While magnification devices or instruments help us see what is simply too small to detect with the unaided eye, every type of magnification device introduces some sort of distortion. We discussed this back in Section 6.4.

Figure 6.26 (next page) shows a trinocular microscope, which makes it convenient to view and capture digital images at the same time.

The binocular microscope body is shown at **A**, while the camera (**B**) is held on an arm. **C** indicates the computer. Also at the **C** location, the vertical white arrow shows a small inverted plastic cup that I use to keep dust off of small parts and mounted specimens. The lower arrow at **C** shows a small "blob" of clay, which is handy to help position and secure whatever you are looking at. Location **D** indicates a paper cylinder that is my crude and economical way to make a light diffuser. It helps to minimize excessively bright reflections.

Location **E** shows the location of my 0.5x objective lens (below it). This lens helps me view larger parts at magnifications as low as 4x. The working distance (focal length) is longer for a low magnification objective lens. This can help increase the options for angles of incident light. The 1x lens, attached to the microscope, is to the right of the top prong of the letter **E**. This is the standard lens used on most optical microscopes.

And finally, the small cylinder at **F** allows me to replace the silver-colored adaptor that is under the camera at **B** and reduce the magnification. With the original silver-colored adapter in place, it reduces the magnification of the image to 60% of what it would be with the one shown in place under the camera.

Why is that important? Because this enables me to take photos at magnifications as low as 6x. Overall, for viewing, this set up allows 4x to 80x, and image capture from around 6x to around 150x.

The small piece of steel and other untidy collection pieces helps me hold odd-shaped parts in place for viewing.

The first thing you might notice in a light microscope image is that the colors are different from the ones that are visible with direct viewing. New optical microscopes may have color correcting software. Color changes and having to use bright light for both viewing and photography may hide features that were readily visible to the eye. That is why a visual inspection should always start with the eye - unless the part is truly too small to see that way. Only after the visual inspection is complete do we go on to view the part with an optical microscope. Then we move to a scanning electron microscope (SEM).

IMPORTANT: It is always advisable to examine the Subject with an optical microscope prior to using the scanning electron microscope (SEM).

Figures 6.27 and 6.28 (next page) show the scanning electron microscope and the sputter coated used for sample prep for the SEM that I use in my lab.

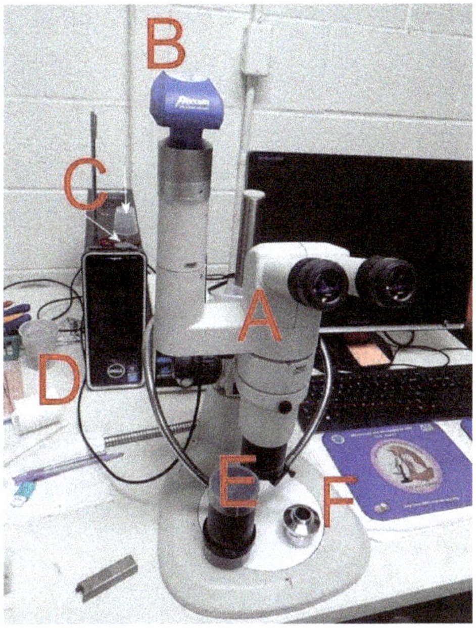

***Figure 6.26:** A typical set up for my Nikon SMZ1000 trinocular microscope*

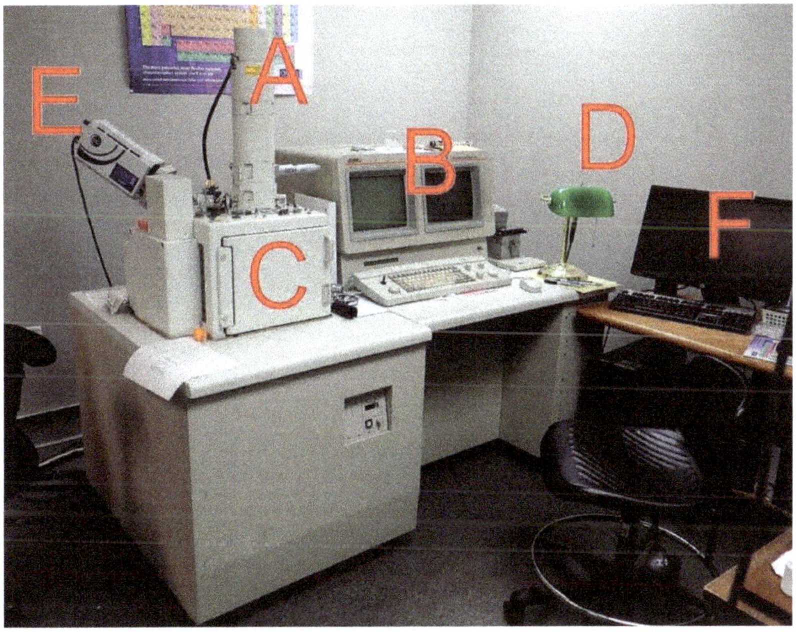

***Figure 6.27:** I still use this JEOL 5800KV SEM (circa 1998) in my lab.*

In Figure 6.27, Location **A** shows the column, at the top of the SEM, where the electrons are generated. **B** indicates the image monitor, at left, while the right CRT allows display of the controls for the stage and imaging set up. Newer SEMs are PC controlled, and would lack the entire section shown at **B**. **C** shows the door to the chamber. This microscope has a maximum working distance of 48 mm. A flat specimen can be viewed at magnifications as low as 18x at that working distance.

To the right of Figure 6.27, **D** shows the location of a small light I use so I can see the keyboard when saving data. The new PC controlled SEMs can be operated in normal lighting. Location **E** shows the EDS microchemical analyzer. This unit is one of the "new" Silicon Drift Detectors, which does not require liquid nitrogen cooling. Location **F** is the computer screen. The entire computer system has been set up to capture both the SEM images and the EDS data.

For corroded metallic specimens that have reduced electrical conductivity, I often choose to coat the surface of interest with palladium, using a sputter coater, rather than aggressively cleaning the corrosion product. This improves the image quality. Rather than use the Low Vacuum mode on my SEM that requires greater skill, I use the sputter coater. Evaluation of polymeric materials is also facilitated with sputter coating. Some microscopists use gold or carbon rather than palladium. I like palladium because it does not cause interferences with the EDS data that I often require.

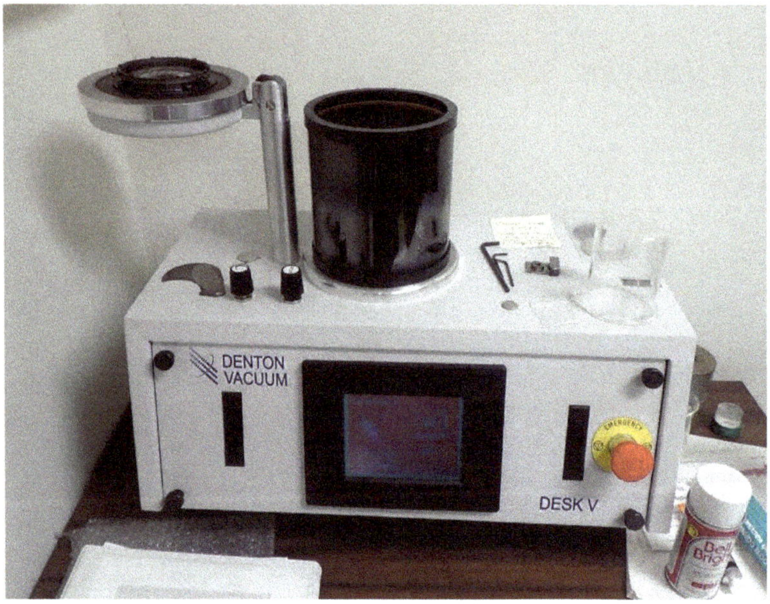

Figure 6.28: *Sputter coater in my lab*

The sputter coater is my go to resource to facilitate better SEM image quality for any poorly conductive specimens, including metallographic mounts. See Figure 6.28 (above). I can coat fracture surfaces with corrosion products, as well as organic or ceramic components. Note the upside-down glass beaker sitting on the sputter coater at the front-right corner; I use it to protect parts from dust as they await coating.

While a SEM has more depth of field at high magnification, which allows for better focus, you can only view the specimen in an enclosed chamber, because electron optics depend on a vacuum. You cannot just reach in and turn the part around to look at the hidden side, as is possible with a light microscope. Robotics assisted SEMs may reduce this inconvenience in the future.

Extremely high magnification work is best done with the specimen positioned very close to the image detectors. Short working distances mean that the lowest magnification is quite high. What does that mean in a practical sense? It's very easy to get lost when working at high magnification because you can't see the edge and core simultaneously.

Despite these limitations, the scanning electron microscope has so many advantages for the Failure Analyst. It is often a key part of the investigation. Especially when equipped with the commonly available EDS microchemical analyzer. This analyzer identifies the elements in contaminants, corrosion products, and microstructural features. The SEM is a powerful tool. In fact, SEM/EDS technology has greatly facilitated the science and art of failure analysis. It became widely available in the second half of the twentieth century. Despite the drawbacks of the

SEM in terms of accessibility to the specimen, it often provides key pieces of data to increase the analyst's understanding of the situation.

There is another major drawback to the SEM. All images are in greyscale. Electron wavelengths are not within human's visible color range. Thus, it is often helpful to mark the test locations before putting the specimen in the microscope. Areas that look obviously different due to color variations are often difficult or impossible to see in greyscale.

6.18.1 SEM Detector Types (2)

1. **Secondary electron images** (SEI) result from near surface interactions between the beam of electrons and the sample's "electron probability clouds". SEI images primarily result from topological features; note that:

- Protrusions appear bright,
- Pits appear dark.

Non-conductive contaminant may appear either dark or bright. Being able to recognize such "dirt" or other artifacts unrelated to the story we are looking to uncover is a key indicator of the skill of the microscopist.

2. **Backscatter electron images** (BEC) result from interactions between the beam of electrons and the atoms' nuclei in the sample. Backscatter electron images provide some information about composition, based on atomic weight.

Both macro- and meso-fractography sometimes require the use of a SEM. This is true for very small parts. The general appearance of the Subject and the orientation of the fracture can be documented and confirmed.

It's no longer possible, in the age of micro- and nano-scale structures, to separate macro and micro examination based strictly on the criterion of the magnification level being employed. In general, macro- and meso-fractography examinations focus on geometry and general appearance, while micro- has to do with features related to the material and process history of the item being examined. For micro-devices, these may all be completely interwoven.

For example, for very small broken parts, finding the crack initiation is often impossible without the help of a SEM. Although this work is being performed in a microscope, this is correctly called macro- or meso-fractography! Although we are working with a microscope, we call it macro or meso-fractography (See Chapter 8) because we are looking at general appearance, fragment shape, and trying to find the initiation (also called "origin").

IMPORTANT POINT: Examining the crack surface at the micro-scale, at the origin and at a reference position away from the origin (such as the final separation), is one of the most useful applications of the SEM. This helps confirm the conclusions drawn, if any, from the macro- or meso-fractographic inspection.

6.19 SEM for Micro-Fractography

In light of our knowledge of materials science, proper acquisition and interpretation of the SEM images will help us understand the crack's interaction with the material's microstructure. I like to call this activity "determination of the crack path". I use the term "crack path" when referring to features showing whether the crack follows or avoids grain boundaries, second phase particles, and inclusions. It may also show whether a layer of deformation is associated with the crack surface. SEM micro-fractography may be performed in one of two main ways.

First of all, you can simply try to match the appearance of the crack path to images in published references, such as the *Metals Handbook Volume 12,*[132] which is dedicated to fractography. You can also consult an article on Metallography

[132] *ASM Handbook Volume 12: Fractography*, (1987) Publisher: ASM International, Pages: 517, ISBN: 978-0-87170-018-6, Hardcover, classified as Failure Analysis Fracture with over 1,900 illustrations and 41 tables. There is also a new edition.

for Failure Analysis by Frauke Hogue.[133] Published references are very useful except with new materials. New materials can be difficult because they may not yet be fully documented or characterized. You may need to make your own reference fractographs in known fracture conditions (tensile bars, Charpy bars, etc.). Finding information on older and no longer popular materials can also be a problem.

> *Recently, I had to get out my 1985 edition of the Metals Handbook Volume 9 Metallography[134] because the 2004 version did not have a single microstructure image for bronze alloy 836. I was trying to determine if the second phase particles I found on the crack surface were typical of the alloy in question. The older book had one image and a second image of a closely related alloy. Even this was helpful for my project.*

The second way of interpreting SEM (micro-fractographic) features is only useful for those who already understand basic physical metallurgy (or polymer physics).

If you already have an idea and expectation for the microstructural features related to the material in question, this second method will work for you. Constituent shapes revealed by the crack help you understand both whether the component was manufactured properly, and how the crack is interacting with the different constituents. This is the essence of where a materials science background will provide benefits.

- For example, one might ask, is the crack favoring the grain boundaries in the aluminum rich phase and continuing right through the grains in the zinc rich phase?

NOTE: EDS attachments available on many scanning electron microscopes can also be used to identify groups of like shaped features by composition. The scanning electron microscope and EDS microchemical analyzer can often shed light on physical metallurgy questions, and occasionally on polymer physics questions.

- Is the crack propagating from small particles?
- For a glass fiber filled polymer, are the glass fibers protruding from the crack surface, or lying parallel to it?

Your ability to interpret micro-scale data will suffer if you don't develop an understanding of physical metallurgy or polymer physics (for metallic or polymeric components respectively). Without this knowledge, your ability to render useful opinions and provide good advice on fixing manufacturing process problems will be limited.

Using method one, you could match an appearance of a crack feature to a reference image, but you may not be able to ascertain whether the microstructure shows evidence of proper or improper processing. And if it was improper, what exactly was improper about it.

Understanding these details lets materials engineers shine in failure analysis work. These skills are less critical for technicians assisting engineers.

MENTAL FILTERS

Engineers use specific observations gleaned through the mental filters of theory.

It behooves the Failure Analyst to remember that the images impressed on the mind are highly processed as a result of our pre-existing beliefs and knowledge.

Another useful failure analysis application for SEM is identifying corrosion products and contaminants. Although the EDS provides elemental composition only rather than molecular

133 "Microfractography and Metallography for Failure Analysis", by Frauke Hogue; 2021 Edition of Volume 11 of the ASM International Metals Handbook, Publisher: ASM Int'l

134 *ASM Handbook Volume 9: Metallography and Microstructures* (#06044G), Published by ASM International, ISBN: 0-87170-706-3

or ionic states, you can often infer information about compounds.

This again shows that the more you know about science, the better you will be able to use your specific failure analysis skills.

6.20 The Value of Case Studies

Let's look at some examples of scanning electron microscope examinations. All of the following three case studies required cutting to perform the work. This is Step 2 work. These examples will reveal that the visual examination is a fluid process that crosses all scale boundaries.

Like compound interest, your competence will grow as you vary the situations in which you work.

6.20.1 CASE STUDY: Chain Link

See Figure 6.29A (below). From the color, it appears to be plated with zinc and sealed with a dichromate treatment. It broke during a test at a lab. The damage category is *fracture*. I used yellow and black paint to mark the cut locations. Using the same color paint on both sides of a crack makes it easier to keep track of the pieces after they are cut.

Figure 6.27B (below) may be the same link (or another link I examined during this same project). It appears to include the crack initiation.

This will only be clear to the experienced eye. Don't worry. We'll go into this subject in greater detail later, which will help you develop a more experienced eye if you do not currently have one.

Figure 6.29A: *Broken chain link – this chain link image was taken with a digital camera.*

Figure 6.29B: *Broken chain link (SEM image) – this image appears to include the crack initiation.*

In Figure 6.29C (next page), we are now at a high enough magnification (the setting is 400x original magnification) to see details that are invisible with an optical microscope.

But it is not really obvious what is going on until we get to 2000x, which is shown in Figure 6.29D (next page).

Look at Figure 6.29D; although we don't have any classical fracture features, it looks like there might be intergranular features with some deposit or residue obscuring the details of the crack.

So, this might indicate a quench crack, that opened prior to the final manufacturing step and retained quenchant and/or chemicals from the electroplating process.

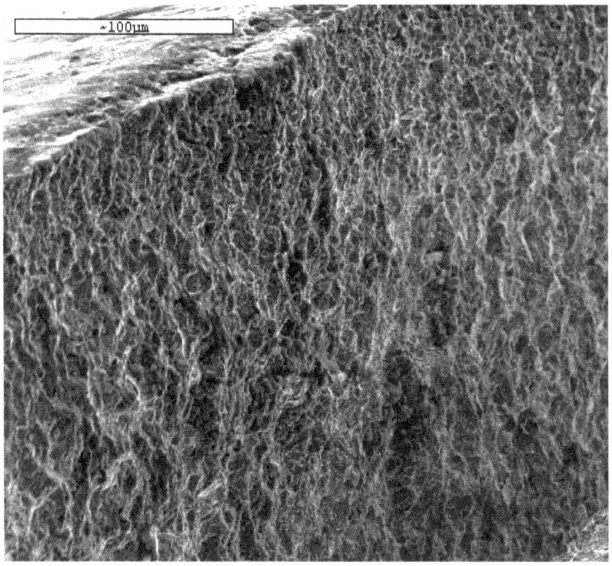

Figure 6.29C: *Broken chain link – 400x original magnification setting*

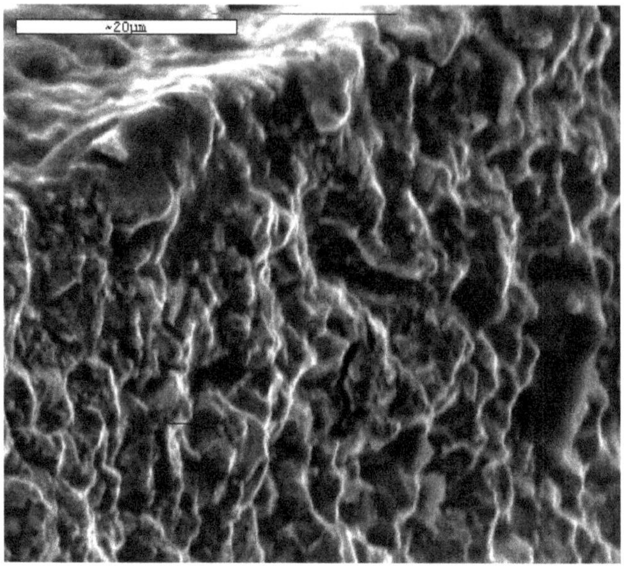

Figure 6.29D: *Broken chain link – 2000x original magnification setting*

Figure 6.29E (below) shows the location 180 degrees across the fracture surface from the area shown in Figure 6.29B (previous page). The sharp change of crack orientation is associated with the shear lip, a feature that is formed during the final separation of the fragments.

Figure 6.29F (below) shows a detail at the inside corner of 6.29E. Here we can clearly see classical micro-scale fracture features. It appears to be a mixture of cleavage (rapid, brittle fracture) and micro-voids. Micro-voids, the features associated with ductile fracture, are usually found at the final separation for room temperature crack events in steel components.

To capture Figure 6.29G (next page), I zoomed in to an adjacent area of the shear lip.

Figure 6.29E: *Broken chain link shear lip (180 degrees across from the area shown in Figure 6.29B)*

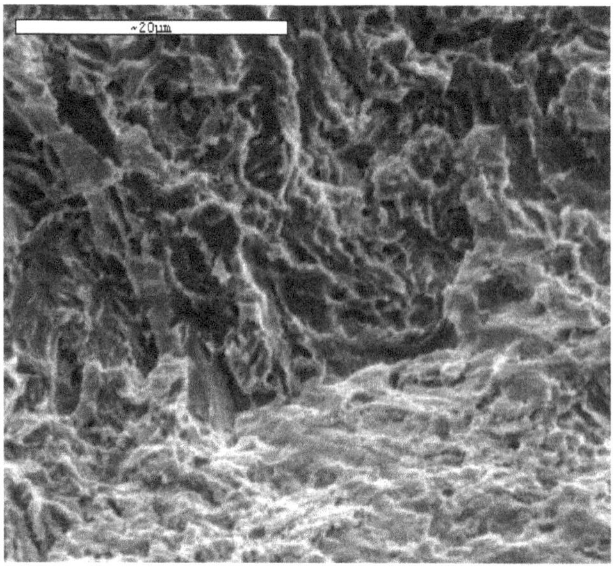

Figure 6.29F: *Detail of 6.98E (left) – image shows end of primary crack and beginning of shear lip.*

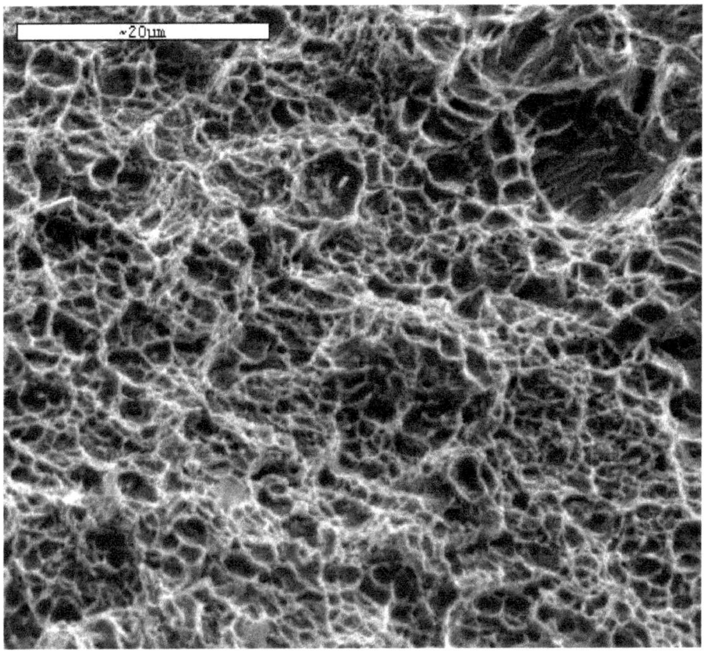

Figure 6.29G: *Micro-voids at final separation (shear lip)*

We have just documented intergranular fracture on one side of a crack, and micro-voids across the way. These micro-scale features were found on macro features that indicate initiation and final separation, respectively.

Congratulations! We have justifiably increased our confidence in our identification of the initiation and final fracture areas performed during the macro-scale inspection. We have calibrated our confidence!

Zinc plating, when improperly applied to hardened steel, can facilitate hydrogen embrittlement. When that is the case, the fracture features are usually cleaner. This usually happens after the plating process. That's why I think this might be a quench crack.

This case study is a great example of why I believe the most appropriate reasoning style to apply to failure analysis work is abductive rather than deductive or inductive.[135] Abductive reasoning allows us to be realistic about the lack of perfect data and enables us to do the best we can with what we have.

6.20.2 CASE STUDY: Another Chain Link

Figures 6.30A and 6.30B (next page) are two views of the same fracture surface. They depict the importance of varying the angle of incident light to highlight different features. This chain link is broken. Thus, the damage category is fracture.

Figure 6.30A has white-out areas, due to an unfavorable angle of incident light. However, the fan pattern indicating the crack initiation position at 6 o'clock is clearly visible. A side view (Figure 6.30B) shows what appear to be two shear lips at those white area locations!

135 Deductive reason goes from general theory to specific conclusions, while inductive reasoning goes from the specific to the general. Engineers use specific observations gleaned through the filters of theory. It's an interactive, iterative reasoning methodology rather than a one-way street, as the logicians describe it. Abductive reasoning is described in Chapter 4.

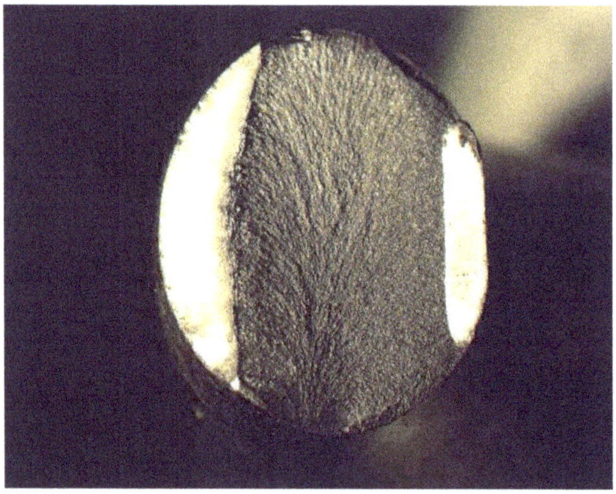

Figure 6.30A: *Fragmented chain link shows a fan pattern – the crack initiation position is at 6 o'clock (digital camera)*

Figure 6.30B: *Fragmented chain link side view – the side view shows two shear lips at the areas that were white in 6.30A. (digital camera)*

NOTE: See how the eye can find the matching feature in the previous image of the sequence above! This is excellent practice!

EXERCISE 6-E: Experience for yourself "context" by taking a series of photos from the macro- down to the micro-scale that clearly documents a material feature.

Figure 6.30C: *Detail of 6.30A – view of initiation fan (digital camera)*

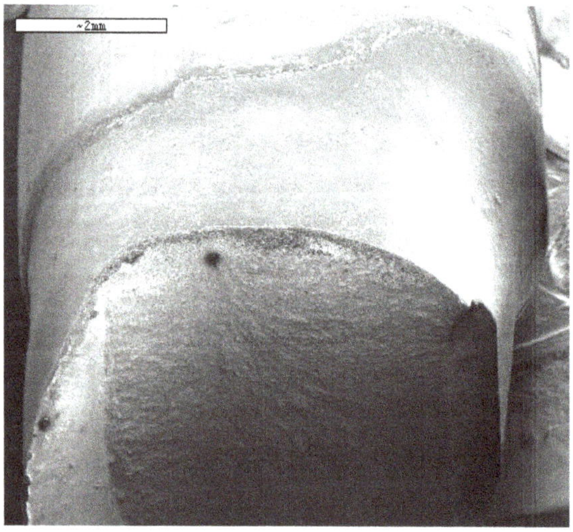

Figure 6.30D: *Mate to fragment shown in 6.29A-C – note the black spots (SEM image) on the fracture surface are "dirt"*

Figure 6.30C (above) shows a clearer view of the initiation fan. Figure 6.30D shows the mate to the fragment shown in Figure 6.30A-C. Note that Figure 6.30D shows how the SEM sometimes accomplishes much with a single view, even when being used essentially as a macro-camera.

Figure 6.30E (next page) zooms-in to the initiation and Figure 6.30F zooms-in to show a nominally vertical (slightly slanted) ridge pattern that we know is parallel to the crack direction based on the fan pattern direction.

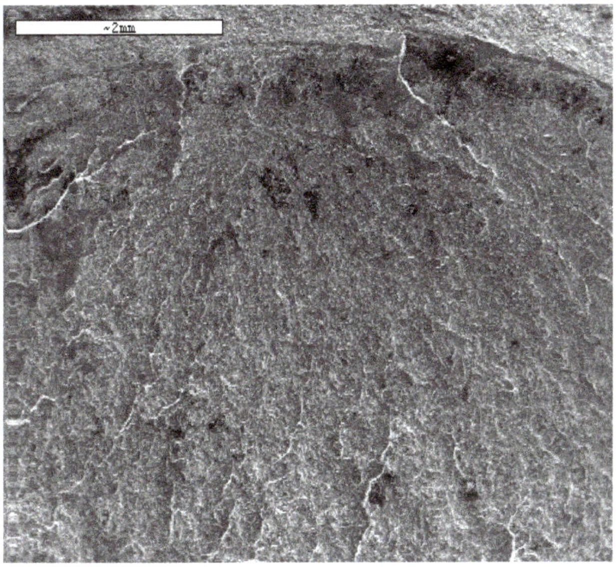

Figure 6.30E: *Detail of Figure 6.30D*

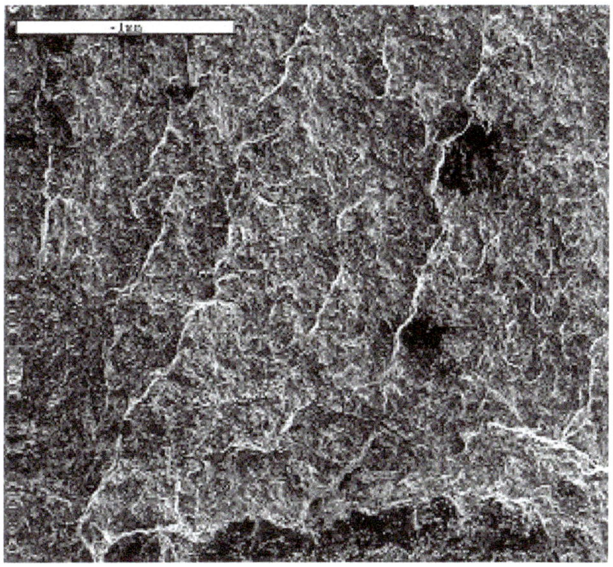

Figure 6.30F: *Detail of area below 6.30E – nominally vertical ridge pattern that is parallel to the crack direction*

In Figure 6.30G (below) we see a fairly classical looking cleavage fracture. These features are characteristic of a crack that grows at the speed of sound in the material, propagating faster than plastic deformation can respond. Figures 6.30G and 6.30H are both SEM images.

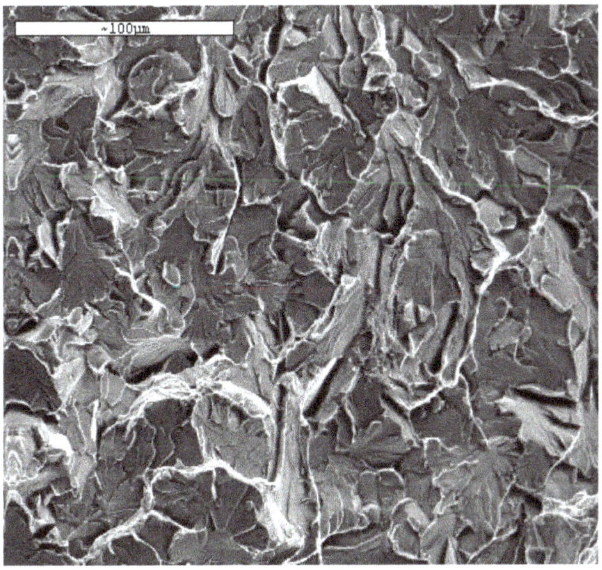

Figure 6.30G: *Detail of ridge pattern of 6.30F – note cleavage fracture, which are true micro-scale features*

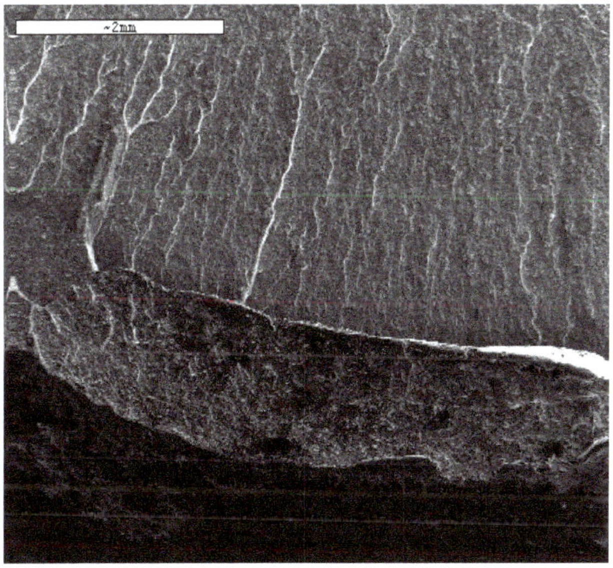

Figure 6.30H: *Fragmented chain link shows a "mini-shear-lip". It was found 180 degrees from the initiation fan.*

6.20.3 CASE STUDY: Injection Molded Polymer Housing (Cracked Stud Boss)

This case study explores an off-white injection molded polymer housing depicted in Figure 6.31A. There is a "boss" (reinforced hole feature) whose purpose is to position and hold a threaded steel stud. The damage category is fracture. The orientation of the crack is what we would expect from a macro-scale brittle response to a hoop stress.

Figure 6.31A: *Cracked boss on plastic injected molded part*

A detail of an off-white injection molded polymer housing (optical image from a digital camera)

Figure 6.31B: *A close-up of a major crack initiation on the plastic injected molded part*

Revealed crack surface from another assembly provided for the project (optical microscope)

Hoop stresses are found in the steel bands used to wrap wooden kegs, or in steel "tires" used to wrap spoked wooden wheels. Hoop stresses are also present in cylindrical walls of pressure vessels.

The classical fan-shaped initiation feature is marked at the left in Figure 6.31B (above). Two other initiation features were noted during the inspection, but not photographed "in all of their glory". They are highlighted with orange arrows to the right of the Figure 6.31B.

Apparently, the scanning electron micrograph images I obtained during this project were in the early days of my career as I first started capturing digital SEM data. There was no way to record a scale marker bar. I recorded 20x as the magnification capture setting and put it in my original file name. See Figures 6.31C-E.

Notice the extremely flat and featureless area at the center of the concentric arcs in Figure 6.31C (next page). This feature is called a "mirror" and is indicative of brittle fracture in glassy materials. Some types of polymers are "glassy" in structure.

The features captured in Figure 6.31C are not greatly more informative than they are in Figure 6.31B (above). However, with Figure 6.31D (next page) we are entering new territory.

In Figure 6.31D we are now able to see very fine features that give us confirming epistemological confidence in our diagnosis of what the polymer people call "stress cracking".

This is analogous to the metallurgy term "Stress Corrosion Cracking". Both of these descriptive crack propagation terms indicate that an **aggressive substance** in the presence of a **susceptible material** and a **sustained tensile stress** combined to allow the crack to propagate. Someone not aware of this could mistake these concentric circles for fatigue. See Chapter 8.

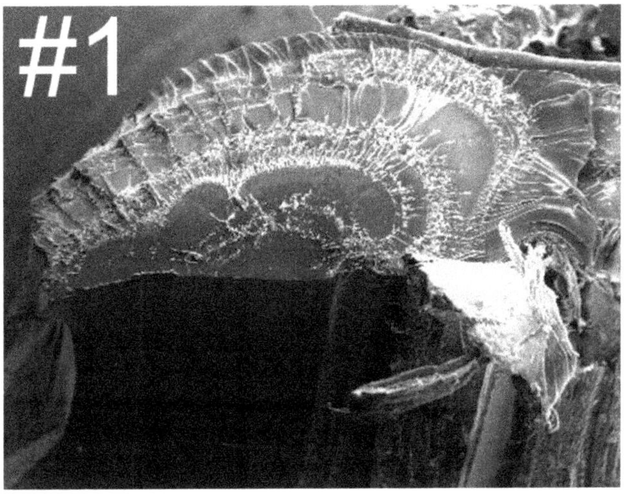

Figure 6.31C: *Major crack initiation on plastic injected molded part – shows a mirror area (SEM image; 20x magnification)*

Figure 6.31D: *Glassy polymer Stress Cracking features are visible (SEM image; 190x original magnification).*

Notice the beautiful flower shaped crystalline features. The EDS indicates these crystals were made of zinc that might have had the stud plating as its source. Is it possible that the increased volume of the corrosion product versus the original plating might have contributed to the zinc corrosion and created hoop stresses? This is a fine question to have at this point in the investigation.

Finding the mirror area at the initiation allows determination of brittle behavior at the mirror. The tear ridges in Figure 6.31E indicate that despite the mirror's brittle behavior, the material has some micro-scale ductility. Why? The tear ridge features indicate that the stress intensity increased to a point to drive the crack faster than the aggressive agent could diffuse to the crack tip; thus, eliminating the possibility for further embrittlement.

6.21 Microstructure Evaluation and Characterization

If your failure investigation has not yet required sectioning, you may still be in Step 1. But now you are usually moving into Step 2 because you are beginning microstructure evaluation.

Microstructure evaluation will always be destructive to some degree, especially when making a cross-section rather than a surface replica (See Chapters 3 and 5).

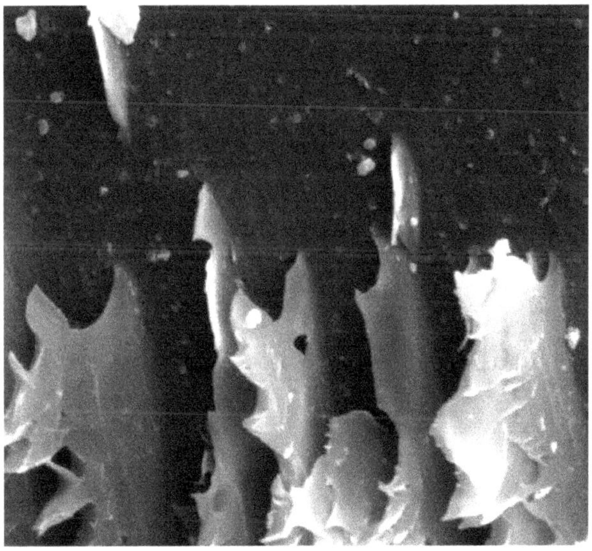

Figure 6.31E: *Tear ridges (now heavily cold worked) that indicate the material has ductility (SEM image; 1750x original magnification)*

Although the SEM may have revealed some microstructure features, the classical microstructure appearance will only be produced with a cross-section. A true microstructure specimen will be required to compare to published microstructure reference images.

In Chapter 3, we introduced the idea of microstructure evaluation in order to help non-materials engineers understand microscope images. We also committed to going into more detail. That commitment is fulfilled in this section. Here is a more methodological approach to understanding the value of microstructure work. This includes:

1. documenting the characteristics of the microstructures to demonstrate that it is nominally as expected - or not.
2. revealing specific features related to various categories of materials and manufacturing methods.
3. revealing the presence or absence of discontinuities.

The first key point has to do with both the material type and the processes used to manufacture a material and/or a component. This will influence the interpretation decisions you make going forward.

The second key point of microstructure evaluation for failure analysis support work is to gain information about the processing history of the raw material used to make the part.

6.22 Obtaining Microstructure Images

In earlier chapters, we talked about how microstructures may be interpreted to people without a materials science background by showing the differences between two patterns and by explaining that those structure differences are associated with differences in material behavior. But we did not explain in detail how those images were obtained. Here, we explain.

Notice that Figure 6.32A (below) shows two mounting presses side-by-side in the Aliya Analytical lab.

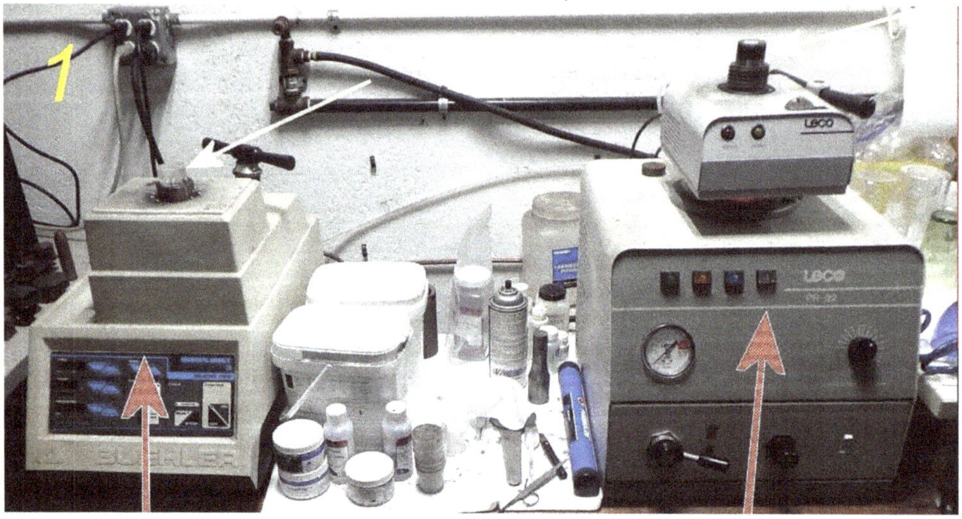

Figure 6.32A: *Two mounting presses are shown side-by-side in the Aliya Analytical lab.*

At left, the upper arrow shows the resulting clear mount made from an acrylic compound. At right, the upper arrow shows the resulting black mount made from an epoxy compound.

Figure 6.32B (next page) shows a part that has been cut to make a microstructure specimen. It's resting on the mounting press, a common tool used to embed (or mold) the microstructure specimen in hard plastic - a key step in revealing the microstructure.

Figure 6.32C (next page) shows a detail of two metallographic mounts, polished and etched: a transverse cross-section of a welded steel tube (left) and two transverse sections of steel rods (right).

Figure 6.32B: *Shows a piece of steel cut for metallography*

The steel part is resting on top of the mounting press platen. The platen will be lowered, and the mold cavity filled with plastic mounting compound. The mold cavity will be sealed and pressurized, resulting in a "mount".

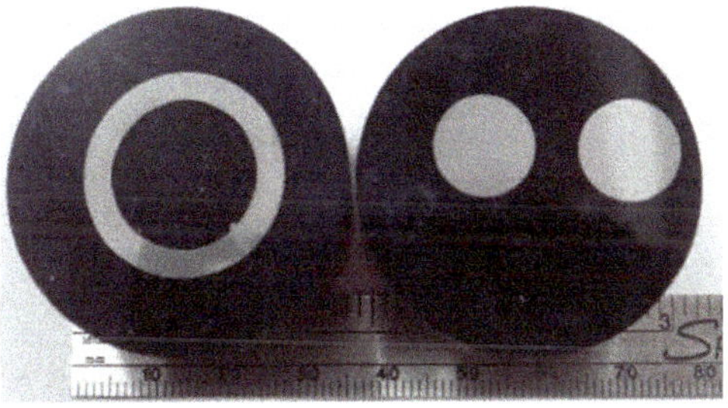

Figure 6.32C: *Two metallographic mounts, polished and etched*

Figure 6.32D shows the metallographic grinder/ polisher used to create the low distortion surface. Semi-automatic metallographic polishers press with an appropriate force, chosen by the technician, using a sequence of cloths treated with increasingly fine abrasives. This results in a mirror finish. We then examine the polished parts, using an optical metallograph, (Figure 6.31E). Polished specimens may be etched to reveal features. There are entire libraries of books about how to examine and document microstructure features using cross-sections. Older references which were developed for metals use the term metallography, while newer references concerned with metals, ceramics, polymers and composites use the term materialography. See the *ASM Handbook Metallography and Microstructures (Volume 9-2004, Pages 229 – 241)* to explore my detailed article on specimen selection and extraction for failure analysis and certification testing. In this section, we'll cover only an extremely brief introduction to how specimens are prepared for the microstructure, or materialographic, inspection.

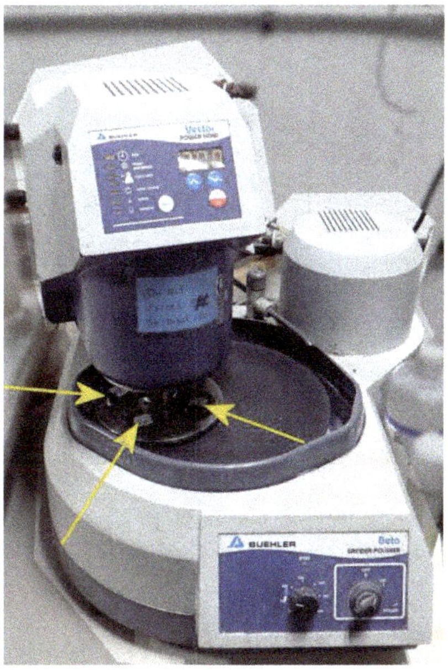

Figure 6.32D: *A semi-automatic materialographic polisher*

This unit holds up to four mounts (3 are shown by yellow arrows).

In Figure 6.32E (next page), **A** is the location of the binocular assembly. The stage control knobs are hidden along the right side of the microscope.

B is the location of the 1.5 inch (38 mm) diameter micro-mounts awaiting inspection under glass beakers.

C is the location of a handy micro cloth, sold for cleaning of spectacles, which I use to remove oil that can ooze out of cracks or voids in the specimens.

This microscope has six objective lenses. I would really have liked seven, but the cost was more than I could manage. The six that I have are 2.5x, 5x, 10x, 20x, 50x and 100x, allowing capture of images with the 10x multiplier, to range from 25x to 1000x. The software allows stitching of multiple adjacent fields, so that effectively even lower magnification views (of larger areas) are attainable.

The small blue camera attached to the microscope (to the right of the red **B**) is quite high-resolution. The Zeiss proprietary image format allows a single capture photo size of 14 MB.

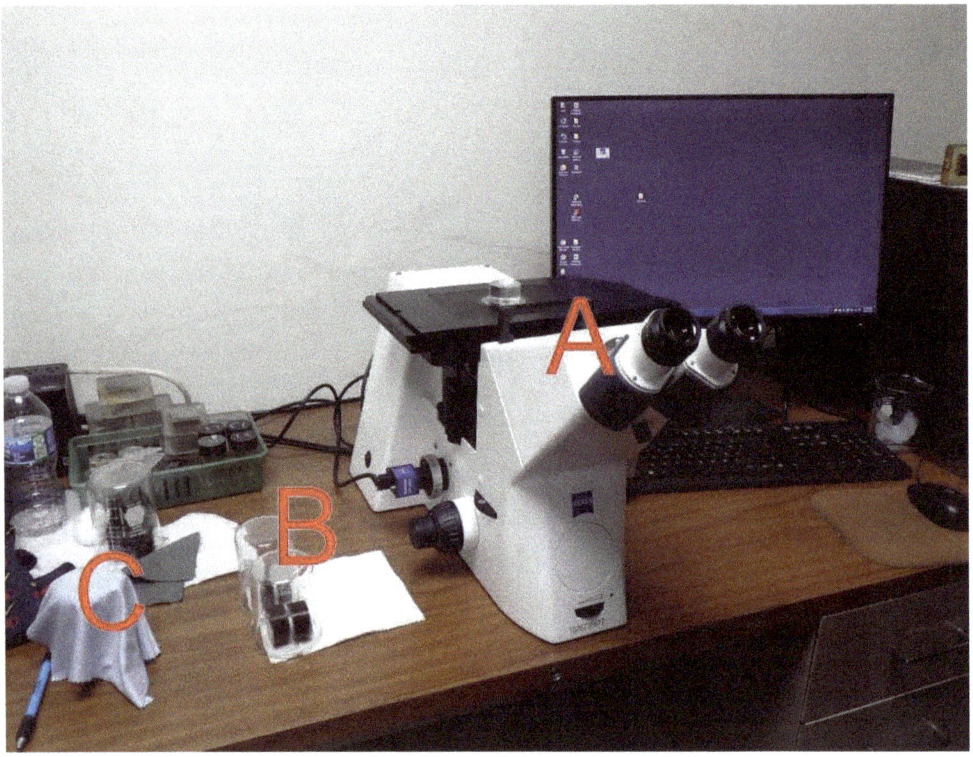

Figure 6.32E: *Zeiss Axiovert Metallograph for examination of cross-sections*

6.23 How Microstructure is Developed

Now, for the theory; this involves the explanation of how microstructure is developed within materials and components as they are being manufactured. The microstructure revealed in any given part, at any given location, examined along any given plane, is a result of a compromise between two competing forces of nature.

- Driving force of thermodynamic equilibrium.
- Restraining force of kinetic resistance.

The thermodynamic driving force and kinetic resistance are key concepts in materials science.

These two competing forces, acting in parallel with the manufacturing processes, create the resultant microstructure and associated material properties.

6.23.1 The Driving Force of Thermodynamic Equilibrium

Any pure material [136] at a given temperature and pressure will "want" to be present in a particular condition. It could be solid, liquid, gas or plasma.

If an ice cube is stored in a freezer for a week at minus 20°C and is then removed from the freezer and placed in a glass on a dining room table, at a normal ambient temperature for human dining activity (+20°C), then the outer layers of the ice cube will immediately start to experience stress. The outer layers of the cube experience stress because the water molecules *want* to be present as liquid water. At exactly zero Celsius, the water

[136] A pure material is comprised of a single type of atom or a molecule; water is a pure material in the company of its nominally identical sisters and brothers with no contamination.

molecules don't care much whether they are free to float around in a liquid, or are constrained by the crystal structure of solid ice.

Whenever a material is present in a state (solid, liquid, gaseous) that is different from the state that is natural for that condition, then we have a thermodynamic driving force that *urges* the atoms to change state.

However, it is not so easy for the material to make this change. Heat transfer is not instantaneous. Even if the surface layer is above 0°C, the layer immediately below it might remain colder. Hence, there exists the restraining effect of kinetics, or diffusion limited processes. Furthermore, a **single molecule or atom cannot change from a solid to a liquid.** There must be a group of atoms or molecules that together form a nucleus (or seed particle) in the new state or configuration. Therefore, there is always a delay when a material makes a state change resulting from a thermodynamic driving force.

Changing from solid to liquid because of heating is usually faster than changing from liquid to solid because of cooling. This is especially true for crystalline substances. Note that essentially all common structural metals are crystalline in the solid state. Polymers however, may be crystalline, amorphous (glassy, non-crystalline), or semi-crystalline.

When a portion of liquid metal is solidifying, the individual atoms must find their place in the ordered solid, one atomic unit at a time. Polymer chains have a more complex solidification process. They often require the long molecules to fold themselves in a certain manner, one segment at a time

State changes (from liquid to solid, solid to liquid, or even from one solid to another) are often diffusion limited. In other words, it takes time for the atoms or molecules to move around and form another structure or even to give up their structure and become amorphous (as solids do when they become liquids). The temperature-driven changes don't start to happen until the thermal energy is conducted past or through the surface or boundary of interest.

Most changes of state require diffusion. While liquids tend to be homogeneous (uniform), the solids that form from mixtures often segregate into areas of drastically different composition upon cooling. Changes of crystalline arrangement can also take place completely in the solid state. This could be atoms moving from one type of second phase particle within a matrix of a different arrangement to another type of second phase particle. It could be atoms moving to a different second phase particle of the same type. And like the Magic 15 puzzle[137] shown in Figure 6.33 (below), the atoms can only move one little jump at a time.

NOTE: There are some common solid state transitions that happen all at once, with the atoms keeping their nearest neighbors. The martensite that forms from austenite in carbon steels is an example of a solid state *diffusionless* transformation.

Figure 6.33: *Magic 15 puzzle – the puzzle's objective: arrange tiles in numerical order from top left to bottom right. The tiles can be rearranged only by moving one tile at a time into an empty spot.*

Diffusion rates increase as temperature increases. The associated thermodynamic driving force may increase or decrease as a result of the same temperature change.

BOTTOM LINE: All this is to say that microstructure interpretation gets pretty complicated, pretty fast.

[137] 15-puzzle.svg: en:User:Booyabazooka - This was derived from: 15-puzzle.svg and the image is in the Public Domain; title is *File:15-puzzle magical.svg*; (created 14 Oct 2013).

6.23.2 The Driving Force of Boundary Energy

Boundary energy is another concept supporting our understanding of a whole range of microstructure features. Every boundary has energy associated with it. A boundary is always more energetic than the material it surrounds. That is an interesting and important concept.

Atoms at the boundary are "not quite sure" which side of the boundary they belong to. This results in a less ordered and higher energy structure at boundaries. The most common boundaries are:

- "phase boundaries" associated with the local composition differences that form on solidification of mixtures of atoms, and
- "grain boundaries" between areas of a solid structure with different crystal orientations.

Everything else being equal, a microstructure with a large area of grain boundaries is more energetic than one with a small relative amount of boundary area.

- A microstructure with a large area of grain boundaries is a fine (small) grain material!
- A microstructure with a small relative amount of boundary area is a coarse grain material!

Another way to think about boundary energy is at the macro-scale in society. Families live in apartments or houses and must pay for that boundary to have privacy.

Nations have boundaries and must pay for signs to mark them and have soldiers ready to defend them (border guards to patrol and control those going in and out).

So, we live in a world that has nested layers of boundaries. Each layer serves different purposes, but all require energy to build and maintain.

Just ask Donald Trump!

Materials that have a matrix with many very small particles embedded within them (hardened heat-treated steel, for example) have more boundary energy than the same volume of material containing the same volume fraction of particles but as fewer, larger particles (annealed steel, for example).

This is important to understand because small grains and small second phase particles are often the foundation of a stronger, intentionally process-optimized material. But if the atoms are stuck in high energy positions, they will move to a lower energy state as soon as they have an opportunity to do so (for example, in an elevated service temperature). This is why many materials soften in high temperature service. Materials may be metaphorically anthropomorphized and understood as "lazy". They'd rather be in a low energy state! When heated for extended periods, they get enough energy from the environment to have the atoms diffuse away from the high energy positions.

This is "creep", or as we discussed in Chapter 5 (concerning Damage Categorization), over time, microstructures degrade due to energy (heat) input.

Boundary energy comprises much of the thermodynamic driving force, and when given a chance, results in microstructure change.

> **The twin concepts of *thermodynamic driving force* promoting microstructure change and *kinetic limitations* restricting change are among the most important and significant aspects of materials science. They are the most relevant aspects for interpreting microstructure changes in service, especially for elevated temperature service.**

6.24 Process Features

As we noted, one of the primary reasons for examining microstructures in Failure Analysis is to determine whether the Subject Part was processed properly. To the experienced materials engineer, a quick look at the microstructure can reveal whether the component was made in the as-cast or wrought condition, or made from a powder process. We will look at these in their historical order: Cast, wrought, powder.

6.24.1 As-Cast Features

Many engineered components start out in liquid form.

- Castings and injection molded parts are two examples.
- Even forgings usually start out as a cast ingot, if made by a traditional process.
- Many powder metal materials are atomized (or made into individual particles) from a molten state.
- Laser Powder Bed Fusion Additive Manufacturing often starts with particles that are direct-solidified and then the particles are remelted as they are merged into a component. Poorly optimized LPBF processes may have rows of unincorporated powder particles, which can create a plane of weakness.
- Metal powders are the foundation of some thermal and cold spray technologies.
- Traditionally made hot and cold rolled steel, aluminum and copper start out as molten metal that is poured into molds for solidification prior to the rolling processes.

Although a copper wire may have little remaining evidence of the details of its experience as a liquid, a thin sheet of stainless steel often does retain visible evidence of micro-segregation that happened during the original solidification.

Figure 6.34A (next page) shows a casting example where the discontinuity is clearly seen to be present (not propergating) between rounded features. When the primary (predominant) micro-structural feature is rounded, it is a good working assumption that you are looking at a casting.

Figure 6.34B (next page) is another area from this part; but as we view it at higher magnification, we see the "crack" is not really a crack at this location. It is completely curved. This means it is a discontinuity that was present when the casting solidified. If we could separate the two pieces and look at them (whatever magnification we used) we would not find any intergranular, cleavage or micro-void features. Instead, we would find non-fracture features.

NOTE: This image does not include any information about how the cross-section was oriented and positioned with respect to the crack origin.

Now that I have showed you the importance of *looking-for* rounded features, I have some disconcerting news. Sometimes rounded features don't look rounded. If they are dendrites, you may see something like Figure 6.34C (next page). A dendrite is a type of crystal that starts to grow in a cold area and extends itself outward as a column.

You will notice there are patches of rounded features mixed with long features. These are all the same type of structure. Some happened to be cut along their length instead of in a transverse manner. The long branches often have smaller branches associated with them. These are known as secondary dendrite arms.

IMPORTANT: If you are not familiar with this type of structure, please take a moment to imprint it in your mind. This is especially important if you plan on working with metallic components as part of your failure analysis work.

I could have shown Figure 6.34D (next page) in the previous section as it is a scanning electron micrograph. However, I'm showing it here because it is a topological view of some dendrites.

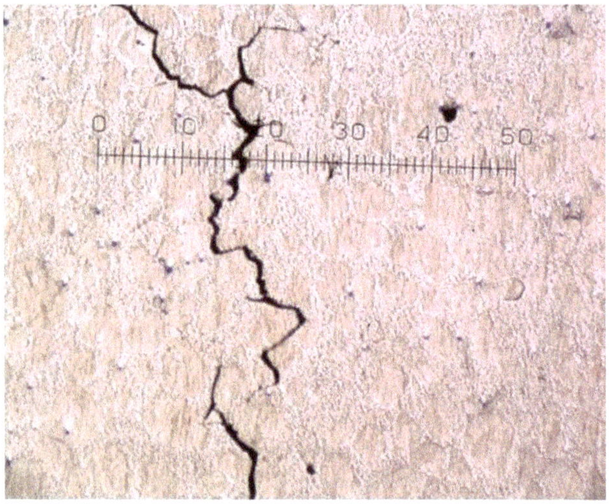

Figure 6.34A: *A casting example*

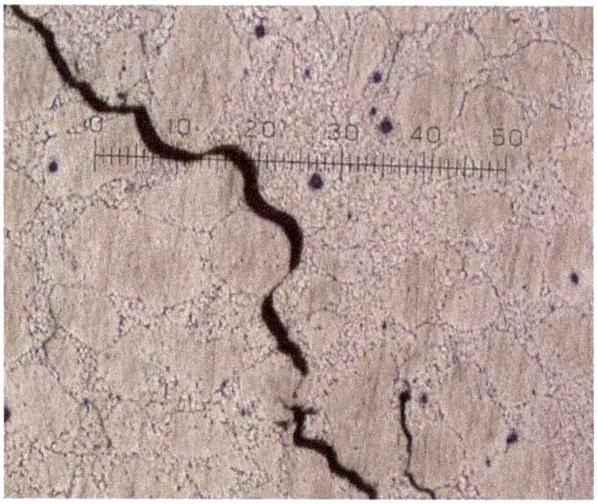

Figure 6.34B: *Close-up of same casting example*

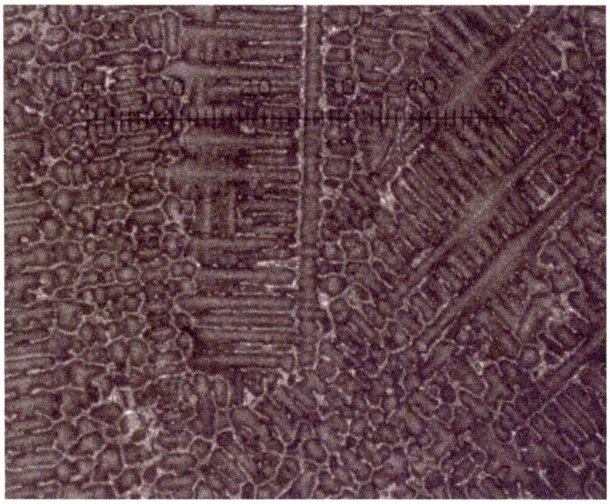

Figure 6.34C: *You first notice the dendrite's columnar feature.*

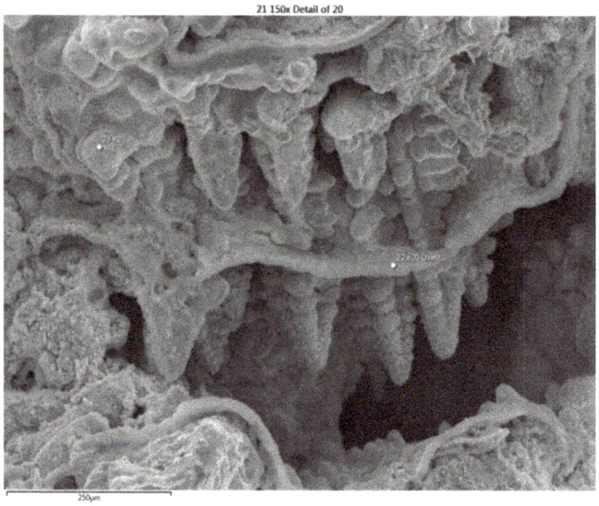

Figure 6.34D: *Topological view of dendrites (SEM)*

EXERCISE 6-F: Take the time to imagine how this three-dimensional structure depicted in Figure 6.34D would look if it were prepared as a two-dimensional microstructure specimen. Would it matter in which direction the cut was made?

This is a very important concept for materials engineering. In fact, it is one of the most important concepts in materials engineering.

Often, we must make use of revealed patterns that represent only two of the three dimensions that it takes to describe any volume of material. And the fourth dimension, time, is equally important to keep in our frame of reference.

Time means that when we see a structure such as those shown in Figure 6.34 image series A-F, we have an understanding that the material was once liquid; and the manner in which the liquid solidified left its mark on the structure and properties of the final solid item.

Figure 6.34E (below) is also an SEM photo and a portion of a "crack surface" taken from a casting. This is a fairly high magnification image and we don't see any long dendrites with distinctive secondary arms such as depicted in Figure 6.34D (previous page). This image was from a die casting, which solidifies very rapidly. As we learn to interpret micrographs, we will become more able to see the nested layers of structure.

The fine features, tiny pockets and bumps, are telling one story, while the larger features, the rounded ovals in the lower left corner, along a diagonal (up and to the right), are telling another. Both features are typical for castings.

In other words, we are looking at a structure that formed on solidification. The shape it has is the shape it had when it first solidified (not counting machining, where material might have been removed without changing any structures of the remaining material). But it is possible that some of the fine features could change were the component subject to heat treating.

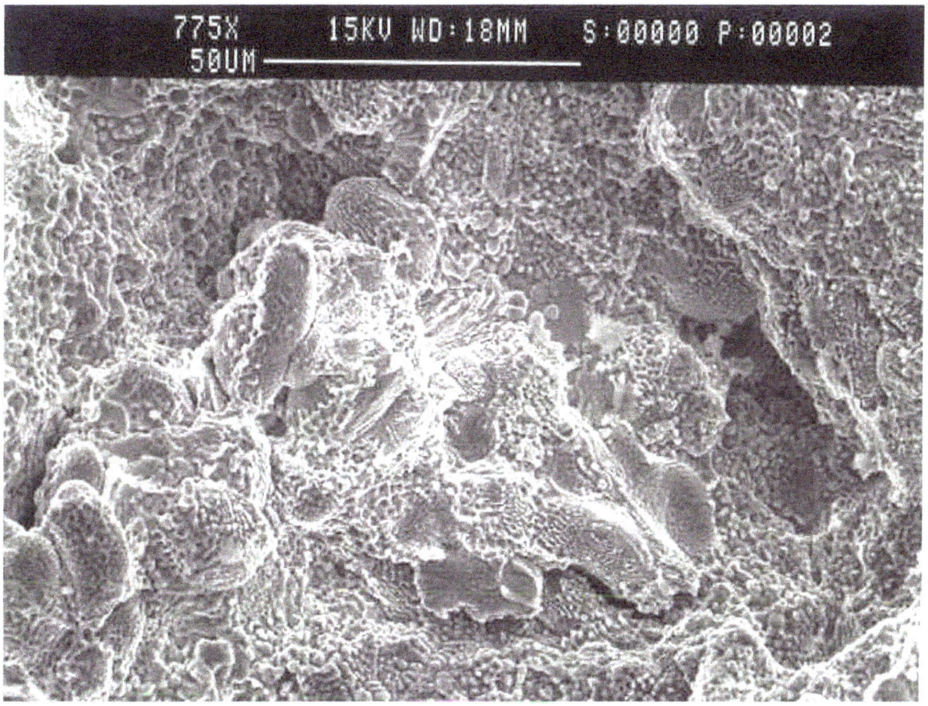

Figure 6.34E: *A SEM photo of a casting's crack surface*

In closing this section, I have a few more important notes to make:

IMPORTANT: Not all rounded fracture features found in metallographic analysis of cast materials indicate a problem.

Figure 6.34F shows ductile iron, otherwise known as nodular iron. More correctly it is called "Cast iron with nodular graphite".

The spherical (or spheroidal) areas, which show up as near circles in two dimensions, are formed when the material is still mostly liquid. But this is the desired shape of this graphite constituent in this material.

IMPORTANT: As a Failure Analyst, you must learn the desired characteristic shapes (structures) for each material that you intend to work on.

Because the strength of the boundary between the graphite and the iron matrix is very low, even in the highest quality ductile iron, the crack will often go around the graphite nodule, rather than through it. This will reveal a nominally hemispherical void, which may actually look like a void, or it might be filled with the graphite nodule.

EXERCISE 6-G: Draw a line through the Figure 6.34F (below); your line should represent an overall crack direction. Cross at least one of the graphite nodules. Then see if you can figure out the details of the local crack direction as it might be influenced by the microstructure features you see. Consider the description below to support you.

The image (6.34F) depicts almost all pearlite, except for the graphite (dark circles); I've marked one in the center right as "A". There are four noticeable and two very small areas that are essentially mono-color and stripeless. These are additional areas of ferrite. In the light and medium grey areas, you can see the alternating layers of light and dark, slightly curved and angled stripes of the pearlite. Sometimes the platelet spacing is too fine to be visible, even at 1000x. This image was originally obtained at 50x, so the spacing was quite coarse!

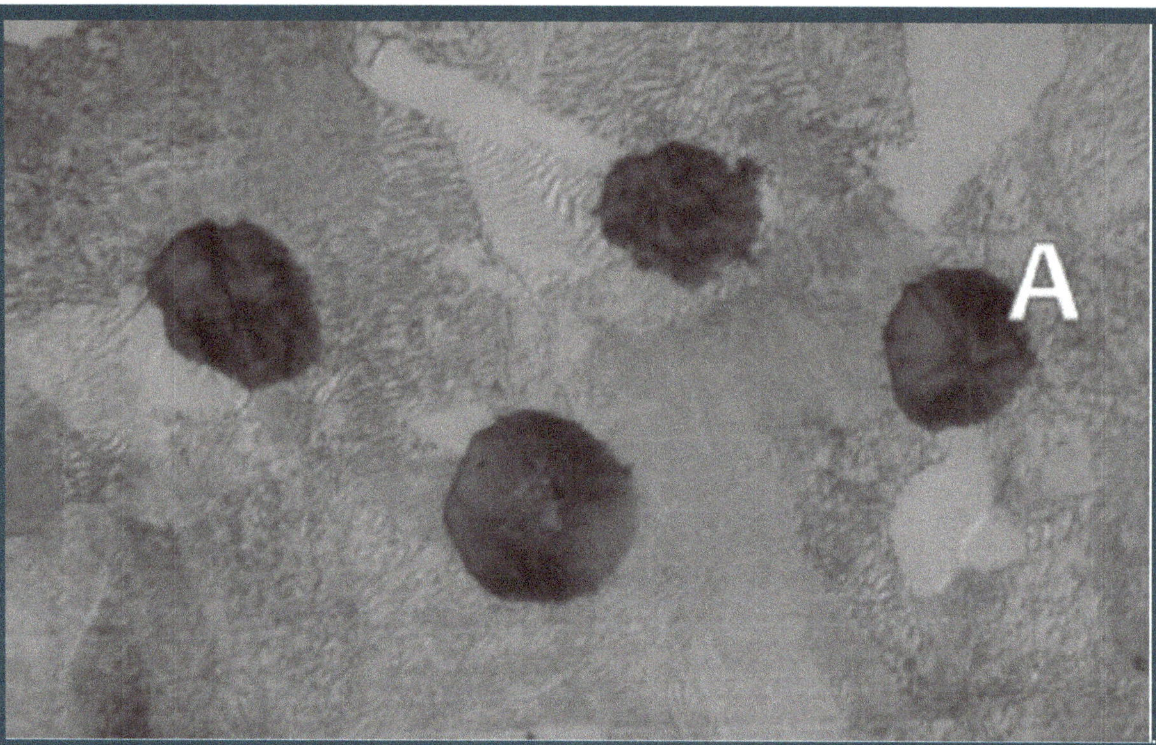

6.34F: *Cast iron with nodular graphite*

Make your notes in a digital document or a notebook. **NOTE:** For those without ferrous metallurgy background, It will likely require some research or discussion in an online forum.

MORE NOTES ABOUT PEARLITE: Pearlite is the structure that forms on slow cooling from red heat in steels with about 0.8% carbon. Steels which vary from that carbon content may still form pearlite on slow cooling, but it will be present alongside other configurations of ferrite (iron with virtually no carbon) and cementite (iron carbide).

6.24.2 Features Created By Solid State Shape Change (Deformation)

The majority of common metals used in modern times are used in the wrought (worked or deformed) condition. That means the shape has been changed in the solid state. We compare-contrast wrought materials to cast materials (which get their shape directly from the shape of the vessel into which the molten liquid is poured for cooling). Wrought and cast are the two main historical structural processing types. Someday, this may change. Additive manufacturing is becoming more widespread every day.

Additive manufacturing may be viewed as a micro-scale sequential casting process. However, as of now (2023), products with the highest strength requirements using metallic additive manufacturing process must be isostatically hot-pressed (HIP). Basically, this is a compaction process; it squeezes the material together to eliminate voids. The individual particles change shape during this process, blurring the boundary between cast and wrought. Certainly, the shape change during isostatic hot pressing of metallic additive manufactured parts is much less than in traditional wrought materials.

The major traditional processes used to make wrought material or components are hot rolling, cold rolling, forging, extrusion and wire drawing (see Chapter 4, Section 4.13, Characterization and Process Condition, Table C for effects on shape due to process). It is important to recognize the effects of solid state shape change on the microstructure. If a metal is worked at low temperature, it will generally get harder and often stronger as it is subjected to more cold work (shape change). Also, if a metal is worked at low temperature, it will generally get harder and often stronger as it is subjected to reduction in cross-sectional area perpendicular to the direction of elongation.

The term reduction comes from mill rolling to create bar stock from larger cross-section billets or from creating plate or sheet from slabs (see Figure 6.35 on the next page). Even if you are not familiar with manufacturing processes, you might get an idea that a slab is a parallelepiped or prismatic section or three-dimensional shape (pulled from a rectangle in a direction perpendicular to itself).

Imagine a slab of cheese - a large, thick, uniformly shaped rectangular piece of cheese. To make plates or sheets of cheese that are reduced along the thickness dimension (the smallest of the three directions, by definition) you typically squeeze the thickness rather than pull the length. The width usually stays the same. If you try this with cured (store bought) hard cheese, such as cheddar cheese, the cheese will break. Cheese is brittle.

Billets are usually square or circular in cross-section and quite long in comparison with their cross-section. They have to be squeezed from all radial directions to create bar stock. The finished bar stock is often quite a bit smaller in cross-section, and the total length of the bars made from a billet is much greater than the original billet length. So again, we have reduction of cross-section in the direction transverse (perpendicular) to the direction which is elongated in the solid state.

Think of the length of a bead of toothpaste that used all of the toothpaste in the tube. It is m much longer than the tube in which it was sold.

TO SUMMARIZE: the production process of the most commonly used metallic materials includes the material being subjected to permanent shape change, otherwise known as plastic deformation. Plastic deformation brings microscopic change along with the obvious shape change.

Some people think of solid metallic materials as stiff toothpaste. They think that if they push hard enough on a piece of metal, they can force it into any shape they want. This is not true.

Most common (and even most uncommon) metals are crystalline, which means at ambient temperatures on Earth their atoms take on a regular, repeating arrangement in three dimensions. Crystals have different characteristics in different directions. In some directions, the planes of atoms can slide over each other relatively easily, and do so at a low force. But when pushed in a different direction, the atoms may get stuck.

Deformation is more complicated than it seems!

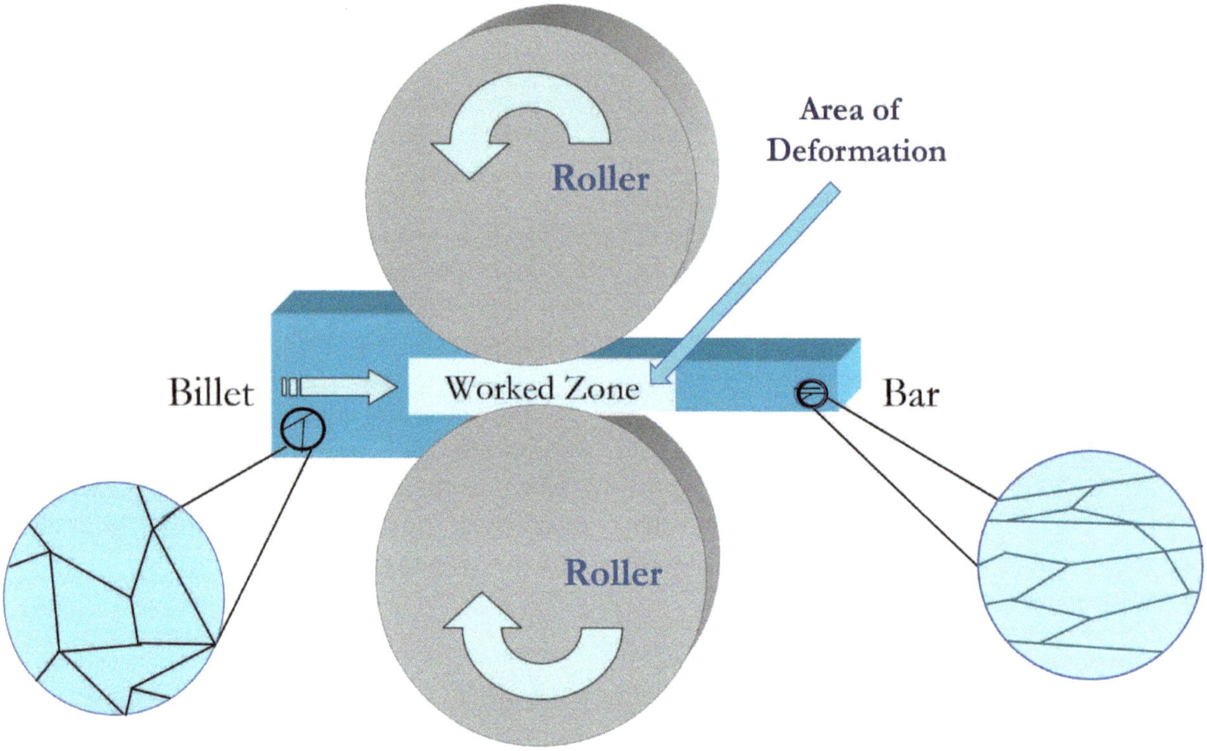

Figure 6.35: *Mill rolling creates bar stock from billets or sheet from plate.*

REMEMBER: Sometimes the atomic planes sliding past each other encounter additional friction because tiny particles (precipitates) exist along the interface between the two planes. These particles are often an intentional feature of the alloy in its solid state and are known as second phase particles, because the crystal structure is different from the matrix (or surrounding phase – material). They improve specific material properties and are governed by thermodynamic equilibrium and boundary energy in the material. These particles were discussed in Sections and Subsections 6.23.

Understanding this in its full atomic-scale detail is fairly complicated. If you have studied physical metallurgy, you will hopefully have realized that I've been talking about the effects of slip and dislocations, in addition to second phase particles. If you have not studied physical metallurgy, the important point to remember is that there are some changes in structure that are fairly easy to understand with the application of optical metallography (grain shape changes). However, there are other changes that are not quite so easy to understand by the lay person. But we will give it a try in the next paragraph!

Figure 6.36 (next page) shows an ear of corn that has one and then two extra rows of kernels for a short distance, which resembles a dislocation in a metal. It's a partial row of atoms rather than kernels. In three dimensions it's a partial plane of atoms. The atoms "become confused".

Figure 6.36: *Corn/Maize dislocations*

The confusion about which plane of atoms a particular atom belongs to makes it easier for an atom to change allegiance (Please forgive my attribution of human emotions to mere atoms!). When atoms permanently change their bonds to their nearest neighbors, shape change occurs.

If etched, some types of cold worked materials reveal microstructures with dark, needle-shaped areas within light colored areas. Lightly cold worked austenitic stainless steel, for example, may reveal local dark etching material volumes. These areas have high concentrations of dislocations. They respond to an etchant solution more readily because of the extra energy associated with the localized deformation. In Figure 6.37A (next page) notice the groups of closely-spaced, parallel, very narrow lines, which are darker than the adjacent material. These are areas of concentrated slip bands. In Rectangle 1 of Figure 6.37A also note that these features are not randomly oriented. Within a single grain, they are either all parallel to each other, or at approximately 60 or 120 degrees to each other. This is a result of the invisible crystallographic orientation, which limits the deformation to those planes that find themselves positioned in certain, specific, orientations.

How do we know that the grains themselves are oriented in particular directions? Aside from their showing us their non-randomly oriented slip bands, the atomic alignment can be detected with the x-ray diffraction test method. X-ray diffraction has some uses in failure analysis, but it is beyond the scope of this book. EBSD (electron back-scatter diffraction) is a newer test method that is a scanning electron microscope attachment that can also shed light on atomic alignment, more commonly called crystal orientation, and is also beyond the scope of this book.

Figure 6.37A depicts 304 stainless steel, which is an austenitic (face centered cubic) grade of stainless steel. These grains are not heavily deformed. They are still nominally equiaxed.

If a metal is worked (shape changed) at high temperature, then it is basically annealed and softened in comparable amounts to the strengthening effects caused by the plastic deformation. This is the definition of hot working.

Another interesting bit of information we can discern from this image is that it was likely hot worked, then cold deformed, and then annealed (or reheated to soften it and allow new grains to replace the deformed ones), and then slightly cold deformed. That is a lot of information.

How do we know this? The annealing twins shown (Figure 6.37A) in rectangles 2 and 3 reveal this, but only if you have a background in physical metallurgy to recognize the annealing patterns.

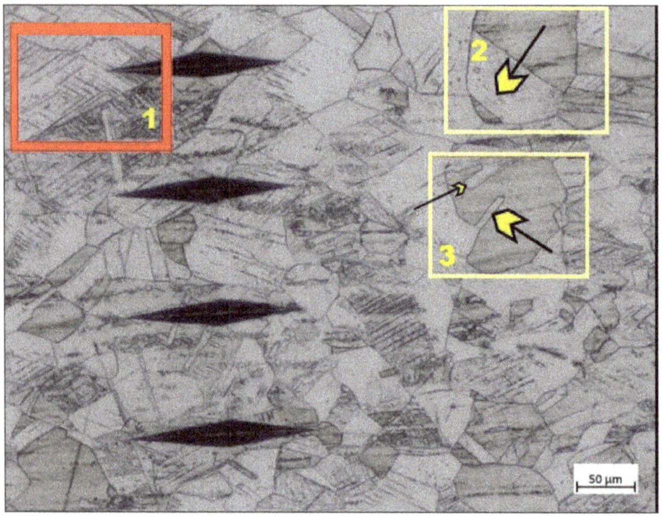

Figure 6.37A: *304 stainless steel with grains that are not heavily deformed*

Within rectangles 2 and 3, the light colored (light etching) areas are crystallographic mirror images of the adjacent dark colored areas.

Slip bands (rectangle 1) versus annealing twins (rectangles 2 and 3) are shown in an austenitic stainless steel prepared as an etched metallographic cross-section.

Getting back to our main narrative, what exactly do we mean by hot and cold in the context of working metals? Well, we have already provided the definition.

- If the material, upon cooling to room temperature, has about the same hardness characteristics that it had before the working process, it was hot worked.
- If the material has substantially harder characteristics, it was cold worked. Cold working generally raises the yield strength and the tensile strength as well as the hardness.

Yield and tensile strength are modified at different rates as the cold reduction proceeds.

Each alloy composition will have its own appropriate range of cold working or cold rolling processes. Cold rolling (specifically referring to flat stock rather than bars, for example, or forgings) usually means that no heat is added to the stock before the reduction. Any temperature change experienced by the material is a result of the work being put into the material by the rolling process, and turning to heat, mostly via internal friction between the atomic planes, as we have already described.

Hot rolling will also have a specific range of temperatures. There are a few commercial processes that make use of "warm work". It gives significantly more strength than hot working, but retains more ductility than cold work.

Figure 6.37B (next page) shows a heavily cold worked piece of medium carbon steel, made into spring wire, and then into a coil spring, which then broke during a durability test. The image is of a longitudinal section parallel to the length of the wire.

We can no longer see the individual grain boundaries that were so obvious in Figure 6.37A. However, we do see a sharp 90-degree corner – it shows the crack profile. This is likely near the crack initiation. From the slight curvature of the wire surface profiles we can see the crack is a macro-ductile crack if the spring was experiencing torsion (coil springs generally do experience torsion).

Usually fatigue (which is likely how this crack propagated in its durability test), follows a macro-brittle plane – by definition – as fatigue is a crack that propagates below the nominal yield strength.[138] But in torsion, the normal and shear stresses (the ones that drive brittle and ductile cracks, respectively) are equal to each other. In

[138] The term "nominal" refers to a category used to group together items based on shared properties. For example, structural steels are often given a grade based on tensile properties - a minimum yield strength.

torsional fatigue loading, we do often see theoretically ductile-macro crack orientations. The reason for this in a long, skinny cylindrical part is that the circular cross-section is much smaller than a plane oriented along a rectangular plane parallel to the diameter. There is much less material to resist the stress.

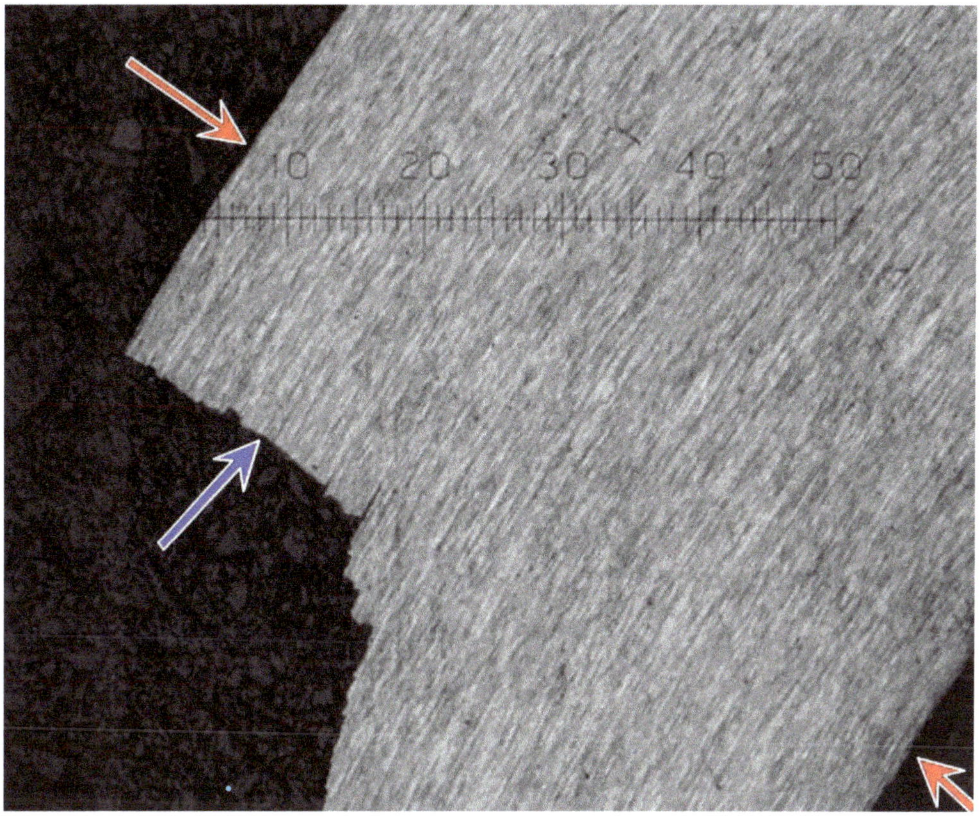

Figure 6.37B: *Cold worked steel spring wire made into a coil spring (Coil ID at right, Coil OD at left) – metallographic cross-section is etched. The dark background is the plastic laboratory added mounting material. The red arrows show the original wire surface. Original magnification setting ~50x*

We may also note the portion of the crack (blue arrow) that is perpendicular to the cold working, or elongated, grain direction. Have you ever tried to break a piece of wood perpendicular to the grain? Good luck! Again, this is possible here because of the small relative size of the circular cross-section compared to any other path the crack could take.

But even here, once the crack is too deep for the spring to carry the load, the crack turns to follow a more easily separated longitudinal grain boundary direction. This is true even when the remaining crack area is much greater than it would have been if the crack had kept going in its original direction. So again, we have a lot of information in a single image - Figure 6.37B (above).

Figure 6.37C (next page) shows a transverse cross-sectional view of the outer curved surface of a piece of hot rolled (or cold rolled and annealed) carbon steel bar stock. Since it is not stainless, we don't see any annealing twins. The grains are nominally equiaxed. But since this section was not cut parallel to the rolling direction, then it's possible that we are missing evidence of elongated grains that would have otherwise been visible. Each grain is uniformly colored and the grain boundaries are clearly visible.

Figure 6.37C: *Apparently equiaxed grains*

This is a high magnification view of a transverse cross-section of a cylindrical part.

The same is true for Figure 6.37D of a 316 stainless steel spring. It has nominally equiaxed grains; no strong evidence of working after the last temperature exposure; and grain structure is uniform. How do we know that these structures are wrought? We see the clues: uniformity (or homogeneity) and absence of rounded features left from the casting process. These are our clues.

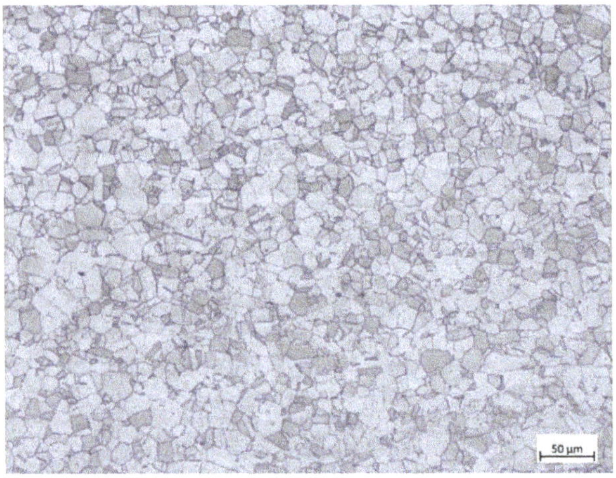

Figure 6.37D: *Cold worked and annealed 316 stainless steel spring wire – the metallographic cross-section is etched. Original magnification setting ~50x*

Note that even though we see no evidence of directionality in this image, it doesn't mean that the characteristics are exactly the same in all directions. There are features that we cannot readily see in an optical micrograph. One of these is crystallographic orientation. While some alloys may be etched to reveal relative uniformity or homogeneity of crystal orientation, this requires a lot of metallographic expertise and care.

Figure 6.37E shows a high temperature alloy etched with a solution that provides contrast based on the development of an oxide film whose thickness is dependent on the crystal orientation. Similar colors have similar crystallographic orientations. Most of the grains are shades of blue, but a few are red. Most of the red areas are annealing twins (sub-grains whose orientation is at a fixed angle to that of its mother grain, such as seen in rectangles 2 and 3 in Figure 6.37A).

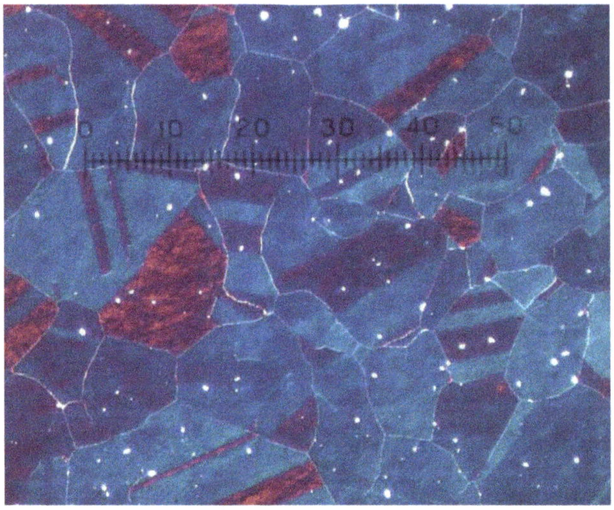

Figure 6.37E: *High temperature alloy IS etched to reveal hints of the uniformity of the crystallographic orientation in the component.*

Crystallographic texture influences many properties. People who form sheet metal into complicated shapes may use a device called a *MODUL-r*[139]. It measures the relative speed of sound in the longitudinal and long transverse

---

139 MODUL-r is a testing system that helps stamping houses determine the suitability of sheet steel for deep drawing applications. | https://www.aimil.com/RESOURCES/RESOURCEFILE/63_Drawability_Testing_System.pdf

directions of a sheet. The relative speed of sound may be correlated with the much more difficult to measure direct grain orientation, which would require x-ray diffraction.

Figure 6.37F (below) shows a moderate amount of grain elongation in the direction running horizontally in the image. The elongated diamonds are Knoop 500g micro-indentation hardness-test witness marks.

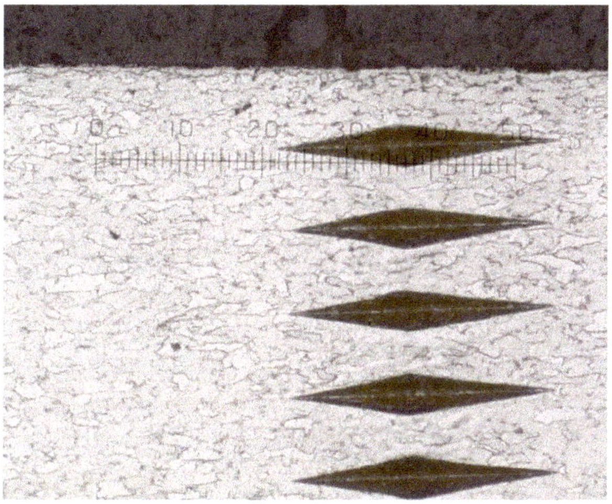

Figure 6.37F: *Grains are moderately elongated in the direction shown as horizontal on the page.*

Many wrought structures have evidence of micro-segregation. While casting also has micro-segregation, it is distinctively different looking from the long, straight bands common in wrought metals. Figure 6.37G depicts a medium carbon steel. The long dark bands have higher carbon, in the form of iron carbides, which can fairly readily dissolve and homogenize at red heat. However, the manganese that is segregated with the carbon can't diffuse easily. The manganese atoms are almost the same size as those of dominant element in steel, which is iron. Thus, even as the material is heated, reheated, worked and reworked, and all evidence of dendritic structure is gone, we still have features that are essentially a result of conditions present in the original solidification process.

The banding seen in Figure 6.37G is a result of the invisible manganese segregation, which happened during the original solidification of this steel. It may sometimes be documented with the use of a line scan in an SEM equipped with an EDS microchemical analyzer.

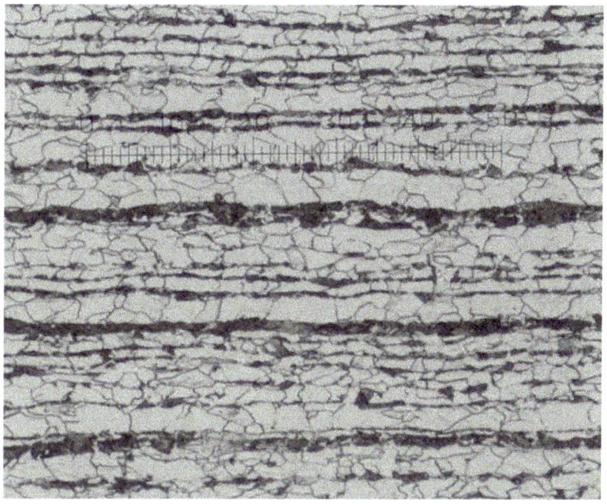

Figure 6.37G: *Banding in medium carbon steel*

Powder Metal Microstructures - We will now examine a powder metal (PM) part. See Figure 6.38A. PM is neither as-cast nor wrought. It is "its own thing". Powder is poured into a mold and compressed, then heated. Voids are left between the incompletely compacted particles. The shapes of the individual rounded particles are from their original solidification. Could you have mistaken this for a casting?

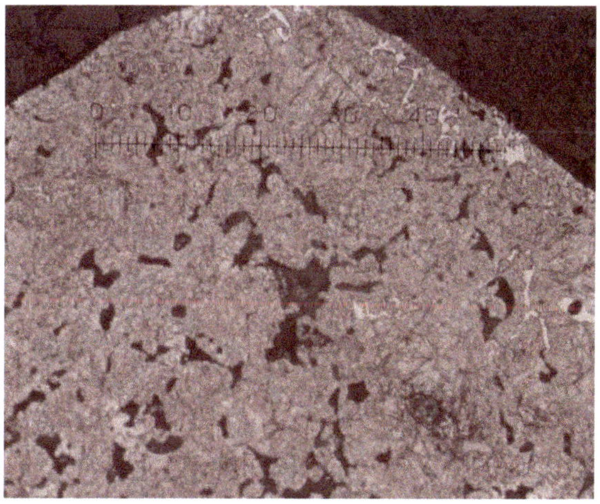

Figure 6.38A: *Cross-section is made from powder metal process. Magnification is probably ~50-200x.*

The component depicted in Figure 6.38B (next page) is also a PM component. It is made from irregularly shaped iron particles, not spheroidal ones such as used to make the part of Figure 6.38A. This raw powder metal was iron mixed with copper and graphite. The compacted "green" fragile shape is transformed into a tougher single part by heat treating above the melting point of copper.

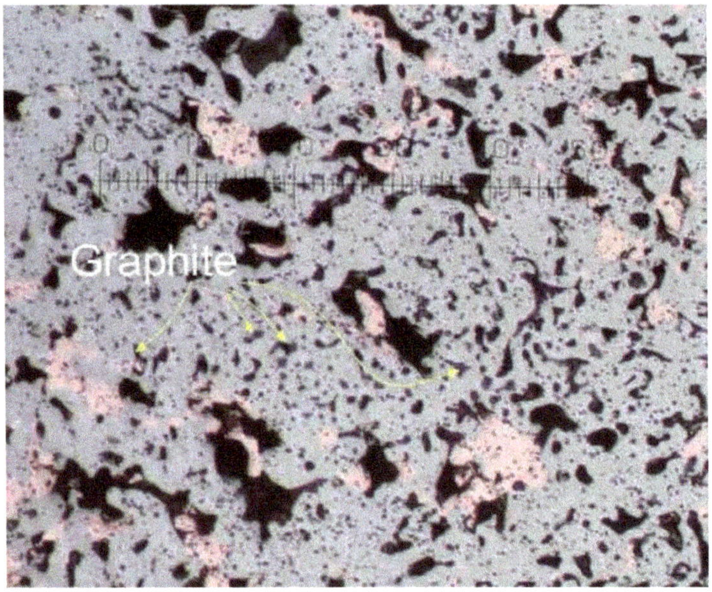

Figure 6.38B: *Powder metal image is produced from un-etched or as-polished metallographic cross-section.*

Cross-section shows the voids, graphite, and copper within the iron matrix. The graphite and voids are difficult to tell apart in this view. Magnification is not recorded (probably ~200-500x).

This is essentially a micro-scale brazing process. There is no remaining direct evidence of the original shape of the copper particles. Notice that incomplete compaction is not always considered detrimental. Self-lubricating components use the voids as an oil reservoir. Not all powder metals have a liquid phase. Some gain strength by solid state diffusion - or sharing of atoms that are in close contact with each other from two different particles. Both of these processes are likely happening here.

6.24.3 Structures Created By Heat Treating: More Structure Change In the Solid State

Heat treating seems like magic to people who do not have the scientific foundation to understand it. How can you heat a piece of metal, cool it down, and then leave it reliably softer or harder, stronger or weaker, or more or less malleable, as desired by the heat treater, or their customer?

Heat treating has been performed for much longer than it has been understood. Luckily for those of us living in this modern age, we do understand how to encourage the crystallographic transformations that are required to make soft steel hard and to make hard steel resilient. Hard, resilient steels are required for sword blades. They are achieved by heating to red heat, cooling quickly, and reheating to a moderate temperature below red heat.

This is much more comfortable and fits much better into modern ethics than the original sword finishing method, which involved quenching into a human abdomen, and was always fatal for the "quench tank" who was considered to be a sacrifice to the deity who would ensure the structural integrity of the sword.

Heat treating is performed on components that are made directly in the near final shape from castings and powder metal processes as well as from wrought products such as those machined from bar stock, forgings, or sheet metal stampings. Each process step affects the microstructure, although some processes affect it more than others. Proper machining may leave little evidence. Perhaps a very thin layer of deformed grains that looks something like Figure 6.37B (previous page) in the previous section on deformation related microstructures.

GOING DEEPER: The Historical Work of The Alchemists

Heat treating has origins, as does refining of metals from their ores, in the work of the alchemists. Centuries ago, alchemists laid the knowledge foundation for modern chemistry. Many alchemists were motivated by fantasies of getting rich. They simply needed to wrestle the secret of turning lead into gold from the bosom of Mother Nature! Sadly for them, these greedy people had no direct payoff for their efforts. It was not until the advent of nuclear physics that it became possible in vague principle to change the number of neutrons and protons within the core of an atom. Now we know that gold is made in supernovae, iron being the end of the line of element evolution in regular stars.

The materialistic dream of the alchemists still eludes us. On the other hand, many alchemists performed their work as an exercise in spiritual transformation. Boiling and re-boiling certain solutions required a huge amount of patience. They used elaborate glassware, which would readily break if the thermal stresses were excessive. These alchemists achieved increased patience and the ability to focus their minds on small details, an important personal quality for the Failure Analyst!

In Figure 6.39A (below left) we see a piece of medium carbon, hot rolled steel bar stock that has been subjected to induction hardening. Induction hardening is a strengthening heat-treating process that concentrates the heat on the surface, just where the strength is usually needed.

In Figure 6.39B (below right), at the boundary between the hardened (below left) and as-hot-rolled (right) positions, both heat-treat-influenced and as-hot-rolled structures are visible. But what is this image telling us?

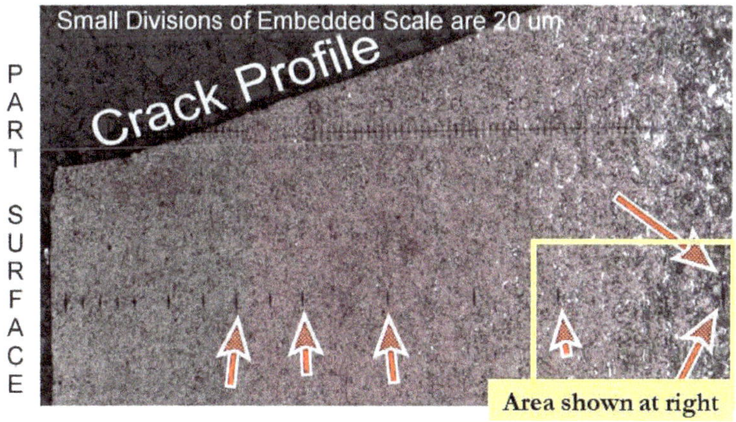

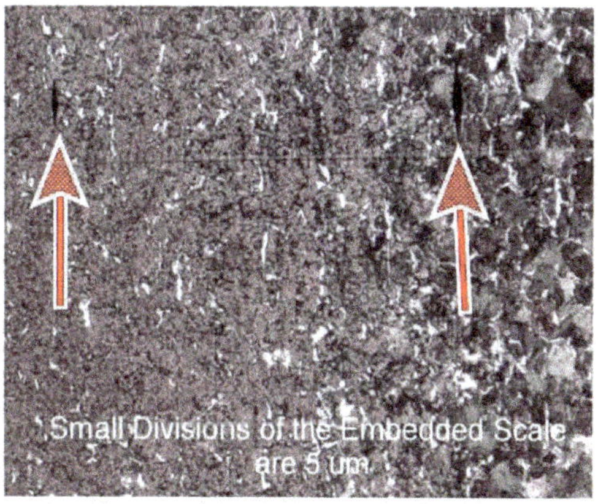

Figure 6.39A: *Shows the longitudinal section of a broken induction hardened medium carbon steel bar – hardened case is the mostly featureless area, except for lab made 500 gram Knoop profile indentations. Note the very small size of the indents at far left (hardened layer).*

Figure 6.39B: *Detail of 6.39A – Knoop 500g hardness indents are shown at arrows. One indent is on each side of the boundary between the induction hardened and as-hot rolled zones. Smaller indents mean harder material. Original magnification is 200x.*

This part cracked in fatigue due to torsional loading and we can see that the crack is at an approximately 30-degree angle to the part surface. This will be further explained in Chapter 8.

The science and art of microstructure interpretation allows room for improvement in even the most experienced practitioners. Yet, there are a few things that even an inexperienced person may readily notice.

1. Within each image, the white network on the right is more noticeable than it is on the left.
2. The matrix on the right is mottled and patchy, while the left side of the images (closer to the part surface) is more uniform in color.
3. The two diamond shapes at the red arrows in the Figure 6.39B are substantially different sizes. These are hardness test witness marks, made by an indentation hardness test method. Assuming they were made with the same hardness test conditions (these are Knoop 500-gram micro-indentation hardness tests), then the material on the left (smaller indent) is much harder (resisted indentation more effectively) than that on the right (larger indent). This hardness data is consistent with heat treatment described above.

One of the main purposes of hardening by heat treating is to refine the size of the features within the structure. Especially when there are hard particles embedded in a softer matrix, which is the case in steel, there is a local stress concentration around the hard particle. Local stress concentrations make it easier for cracks to initiate. When we homogenize the composition at the micro-scale level, we eliminate the small number of large hard particles and make a much larger number of much smaller, less damaging hard particles. Not only are these small particles less damaging, but if we perform the heat treating properly, these small particles can actually act to strengthen the matrix instead of weakening it. Many such homogenization processes change the shape of the particles to further reduce the stress concentration.

Particles can change from needle or plate forms to spheres or spheroids. The details about how these tiny particles enhance the strength and toughness of the alloys without a change in overall composition is a subject in physical metallurgy that we have already mentioned.

With heat treating, we can homogenize (make uniform) an inhomogeneous material. This allows us to eliminate the weakest areas. In some alloys, even the originally stronger areas end up being stronger than they were before the homogenization. Now that's alchemy! We have thereby made it more difficult for a crack to propagate.

We can see this on the left side of Figure 6.39A. Notice that most of the white network is gone near the surface. The white areas on the far right of both A and B are known to ferrous metal heat treaters as undissolved ferrite. They are almost pure iron. Almost pure iron is very soft. Once it starts cracking, the crack propagates easily. In other words, it is soft and brittle, the worst of all worlds for the steel metallurgist!

Because Figure 6.39B was obtained at the deep edge of where the induction heat penetrated, it is not surprising to see evidence, as we move from left to right, of less and less heat having affected the patchy (pearlite) original hot rolled microstructure. But as we will learn in the macro-fractography section in Chapter 8, the highest stress is often at the surface. So, if we can harden the surface of the part, we can often increase the effective strength of the entire part.

Induction hardening is a rapid manner of heat treating. It is much faster than putting the part in a furnace and waiting for the parts to heat to cherry red by convection of hot air or radiation (in the case of a vacuum furnace). Time is money, as business people like to remind us. So having a rapid method to strengthen the part where it is needed most is an excellent tool for the engineer.

Figures 6.39C and 6.39D (both on the next page) show a metallographic cross-section of an induction hardened part made from a medium carbon steel bar. Figure 6.39C shows the overview and Figure 6.39D shows a higher magnification view of the transition zone from case to core.

The network of white etching ferrite in the grain boundaries is only visible in the higher magnification view, which is from a different location around the circumference of the cross-section.

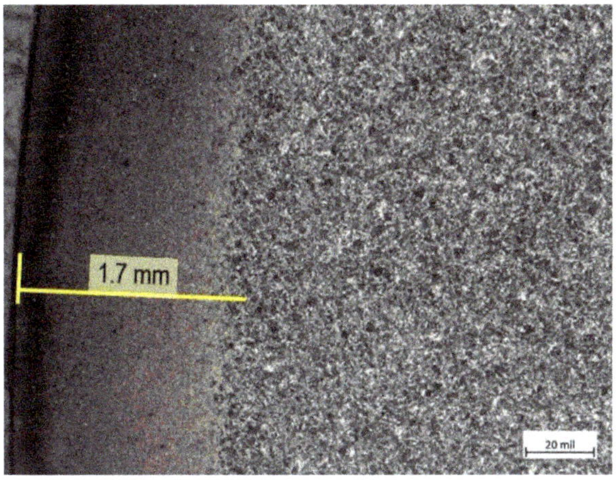

Figure 6.39C: *Transverse section of medium carbon steel bar, induction hardened – original magnification is 25x (nital etch).*

The case depth is shown as being 1.7 mm deep. It is quite homogeneous compared to the as-rolled bar stock microstructure that remains in the deeper layers.

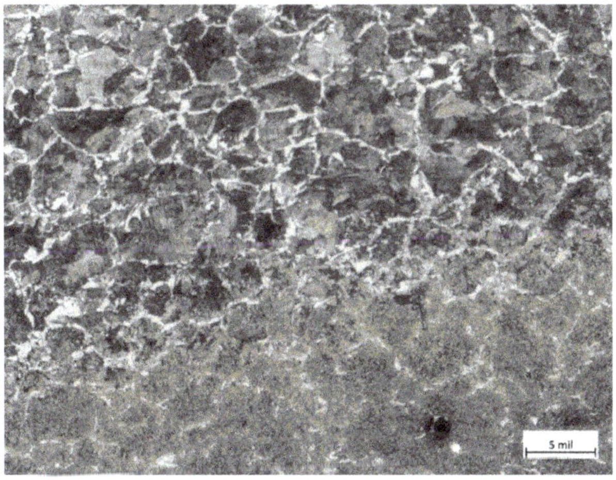

Figure 6.39D: *This is a detail of the same cross-section shown in Figure 6.39C, but at a different location – boundary of heat penetration during induction hardening (original magnification is 100x, nital etch).*

Figures 6.40A and B (next page) show an example of an undesirable superficial surface layer; but most of the part has reasonable quality structural characteristics. The part's surface is at the left. This material started out as hot rolled, medium carbon steel bar stock.

The lower magnification view (100x original setting) of Figure 6.40A reveals features that indicate micro-segregation similar to Figure 6.37G, but they are much less severe. The material of Figures 6.40A and 6.40B (both are on the next page) has been heat treated by quench and temper hardening. This process leaves the material at the high temperature where diffusion is facilitated (necessary condition for homogenization) for a longer time than induction hardening. Despite the trace of residual micro-segregation, the microstructure overall looks like it was achieved with an appropriate hardening and homogenizing process.

Based on what you see in Figure 6.40A, can you determine the rolling direction of the material?

Now, look at the row of Knoop 500-gram hardness indents. The two largest are at the far ends of the profile (both left and right) as shown in 6.39A. A larger indent shows evidence of softer material. As we have previously noted, the highest stress in most parts is at the surface. So having a soft spot near the surface was probably not the original design intent. So, what can we learn here?

Note the very thin, intermittent white layer at the left edge of the specimen near the long orange vertical arrow in Figure 6.40A. Immediately adjacent to the narrow white band is a much wider dark band. Then we have the mildly micro-segregated area reaching all the way to the far side of the image.

Figure 6.40B tells the story to those who have knowledge of ferrous metallurgy microstructures. The microstructure provides evidence of thermo-mechanical process history. We've already nailed down the presence of the micro-segregation and the rolling direction, but what is the white band to the left of the dark band?

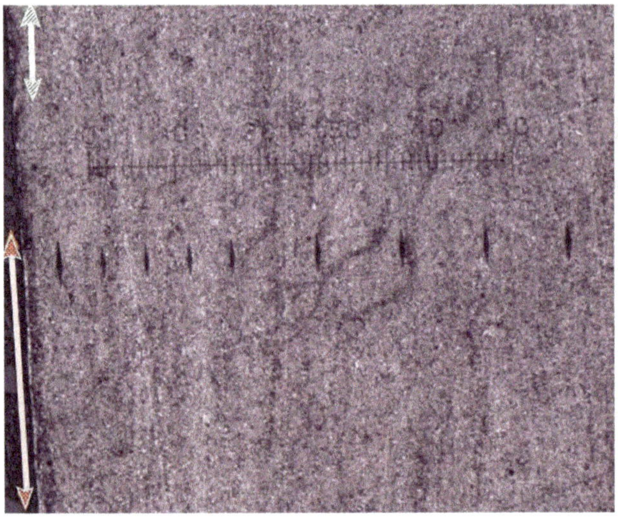

Figure 6.40A: *Wrought material has been hardened by austenitizing, quenching and tempering.*

The very fine band of white at the far left is grinding burn. It is easier to see in 6.39B. The original magnification is 100x. Small divisions of the embedded scale are 10 microns. This is a nital etch.

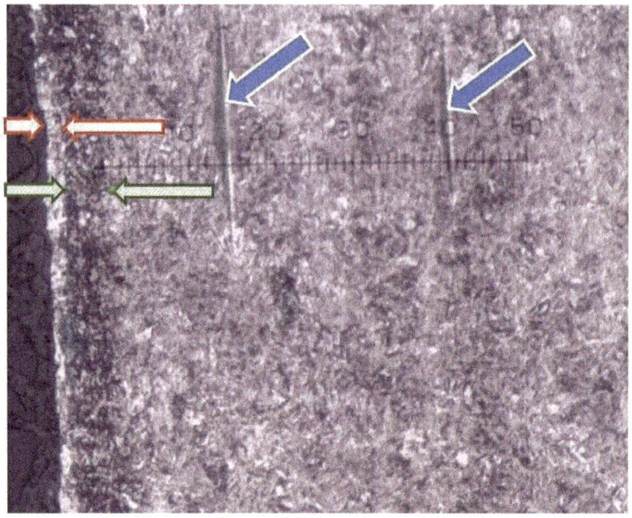

Figure 6.40B: *Detail of Figure 6.40A (nital etch)*

Blue arrows show 500g Knoop indents. Orange arrows show white etching untempered martensite. Green (lower, horizontal) arrows show the over-tempering from the grinding burn. The original magnification is 500x. Small divisions of the embedded scale are 2 microns.

If we have done our jobs during the scope negotiation portion of the project, we have a detailed description of the process steps. This part was ground to size after heat treating. But there might have been a breakdown in the coolant or some other intermittent problem with the grinding set up. Then the surface layers overheated enough to be re-austenitized (heated to cherry red) and then immediately quenched to near room temperature, creating fresh martensite. But because the hardening (quench from cherry red heat) in this case was not followed by a temper (reheat to moderate temperature), we are left with a very hard surface layer of untempered martensite. But immediately underneath this layer, the material that did not quite get to red heat and was strongly tempered.

> **The microstructure provides a way for us to check that the specified processes were properly performed.**
>
> **This is a key reason metallurgists get involved in failure analysis. The part carries this unwritten "historical" record with it wherever it goes!**

Note that it would be difficult or impossible to actually test the hardness of the white layer in particular, but also the dark layer to its immediate right. It's way too narrow compared to the width of the Knoop indentations, and too close to the surface of the part.

Without etching the cross-section, the grinding burn near the surface would be impossible to diagnose. Many medium carbon steel parts are subject to decarburization during hot rolling, which is done in the normal 20% oxygen containing atmosphere. The oxygen finds the carbon atoms at the part surface and converts them to CO or CO_2 gas, which evaporates. This sets up a carbon gradient with lower carbon at the surface. Mother Nature doesn't like having such a gradient, and diffusion is the method by which new carbon atoms from the interior of the part are supplied to the outer layers in a thermodynamic effort to reduce the gradient.

> *Conceptually, the thermodynamic urge to reduce composition gradients is similar to that which drives the net radiation balance from a heated house to result in heat loss to the cooler environment (or vice versa).*

The higher magnification image of Figure 6.40B provides evidence of what happened to this component. This photo allows us to infer the cause of the low hardness value. It is grinding burn, not decarburization. If you did not know that the parts were ground after heat treating, you might not have thought about this possibility. In any case, low hardness at the location where the service stress is highest is generally undesirable.

"Generally undesirable" does not mean it is the main cause of a failure! In some cases, it may be completely benign.

In closing the analysis of this pair of images, it is important to remember that the grinding damage (or grinding burn) was done unintentionally! The microstructure provides a way for us to check that the specified processes were properly performed. It is worth repeating: this is a key reason metallurgists get involved in failure analysis. The part carries this record with it, wherever it goes!

6.25 Microstructure Evaluation for Failure Analysis

Chapter 4 introduced the idea of process condition. This is shorthand for the cumulative effects of all the individual processes used to manufacture a raw material, and by extension, a part. We emphasized how what appear to be small variations in a process or a dimension may have significant consequences for the durability of the product.

Microstructure is the most important, and in many cases, really the only way to determine whether the part was processed according to the design intent. Sometimes, even microstructure analysis leaves us with an incomplete picture of how the part was manufactured.

When manufacturers are in a rush (and who hasn't been?), they may shorten processes. Was a belt furnace loaded more heavily than usual? If that happens, some parts might not get to their required temperatures.

I recently heard a story about a furnace operator at a heat treat shop who loaded three baskets of parts on top of each other so he could extend his smoke break.

Years ago, I was invited to give some training at a small plant that was forging and heat treating their product in-house. The parts were long and skinny. When the plant needed to save money due to a slow economy, management had a great idea to close for a week. They told the furnace department to turn the red heat zone of the furnace down to 1200°F (650°C) from 1500°F (815°C). From Friday evening, when the regular production was completed, until Sunday evening, ten days later, the parts sat at 1200°F (650°C). To make a long story short, the parts no longer had the required straightness.

They thought it was funny when my jaw dropped to my waist upon hearing the explanation of why they thought their parts were not straight. I was younger then. I guess that jaw dropping reaction allowed me to save my words. It was clear to them that this economizing measure had been a bad idea.

When I got back to my lab, where I had been doing ongoing work to try to help them determine why the parts were not hard enough, I noticed the presence of what are known as temper carbides in the microstructure. Usually temper carbides are very small, and when I first started seeing them before my visit to the plant, I had asked them if they had left the parts in the furnace longer than usual. The answer had been "No".

After the visit and the explanation about the ten-day heat-treating program, I brought it up again. It turned out that "No, we didn't temper the parts longer than usual" actually meant "We, as usual, left the parts in the furnace over the weekend when the plant was shut down" But this time, the two-day weekend turned into a ten-day weekend.

The microstructure had changed in what appeared to me as a young, semi-experienced consultant, to be noticeable. But I did not have the experience to realize how important the difference was. To the non-metallurgist plant personnel, the process had not changed. They did not grasp the importance of "time at temperature".

In this case, they had much longer time-at-temperature, the opposite of the rush situations that are much more common.

Steel microstructures reveal a lot about their processing and composition; but aluminum alloys do not give up their secrets as readily. Features that create strength characteristics in the most common wrought aluminum alloys are too small to see, even at 1000x in a well-prepared metallographic cross-section. But it is still worth doing metallographic work because if something really went wrong, it will often be visible due to a change in some other microstructural feature.

If you are working with a material as a regular part of your work routine, you will likely rapidly gain expertise with the recurring microstructures you see. It will become fairly straightforward to see, often in a few minutes or instantly, if the structure is normal or not. But if you are examining mounts from an unfamiliar alloy system, more work and time will be required for the evaluation.

To help others, over the years, many experts have contributed to a vast body of published microstructures of common materials in common process conditions. There are many published atlases of microstructures. If you don't have a known good part, the atlases will be worth seeking out. In fact, even if you do have a known good part, it is worth checking these resources. The published images provide additional components to compare-contrast.

6.26 Micro-Indentation Hardness Testing: A Powerful Tool for Failure Analysis

In the text related to Figures 6.39(A and B) and 6.40 (A and B), we briefly introduced micro-indentation hardness tests. Earlier in the book, we even more briefly introduced the importance of ensuring our proposed protocol included a method of comparing component strength to anticipated design stress (or actual stress). This is especially important when we evaluate mechanical damage (including fracture, deformation and wear).

Tensile, Charpy and various fracture mechanics tests are the best way to do this. However, sometimes there is not enough material to perform those tests, and even if there is, the scope of the project (time and money) may not allow such expensive and time-consuming evaluations. This is exactly when micro-indentation hardness testing comes to our rescue. Very little material is required. That also makes it possible to do the test at the area of greatest interest.

Microhardness testing is now properly called micro-indentation hardness testing. This is because it is the indentations that are small, not the hardness values! Micro-indentation hardness testing reveals the presence of both hard and soft, or high and low hardness value materials. Micro-indentation hardness testing may usually be performed on the same coupons used for the analysis of the microstructure.

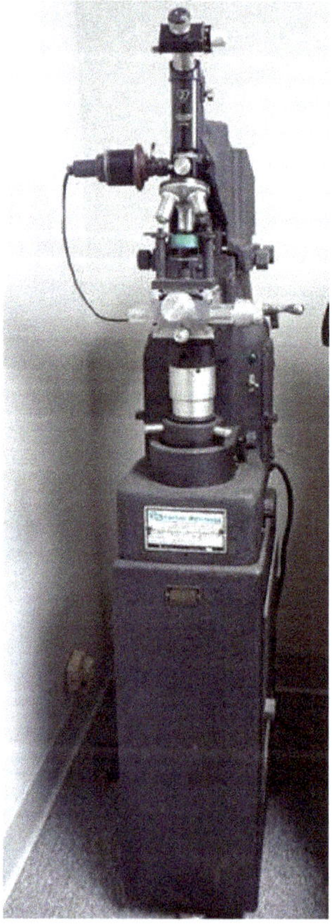

Figure 6.41A: *This is a Tukon hardness tester with the green mount visible in the observation position.*

Even my old Wilson Tukon microhardness tester (circa 1960s model) had a precision moving stage so indents could be placed at features of interest. Figures 6.41A and 6.41B show this once popular older model Tukon tester and a 1981 semi-automated Buehler tester. The newest micro-indentation hardness testers are PC controlled.

The inset view (lower left) on the video monitor in Figure 6.41B shows parallel black vertical lines, which are 800 microns (0.8 mm) apart. The Knoop diamond, which has a length to width ratio of 7:1, allows measurements to be placed very close together. Maybe those shown are a bit too tightly spaced, since the test method says the spacing must be 2.5 times the width of the indentation. Of course, you don't know what the width will be until you run the test. (In this case, I should have known, since this is a calibration test block.) These indents were made 0.05 mm apart.

In order to calculate the hardness from the indentation size, the old Tukon testers required writing down the position of the left edge of the diamond point, and then writing down the position of the right edge, and then subtracting the values to get the actual distance. Converting the length of the indentation into a hardness value required a separate calculation.

The new instruments have two marker lines. And the instrument is smart enough to directly calculate the distance between them. So, all the operator must do is align the movable lines (called "filars") with each end of the indentation, push a button, and the hardness value is calculated.

The newest instruments can generally read the values as well as place the indents.

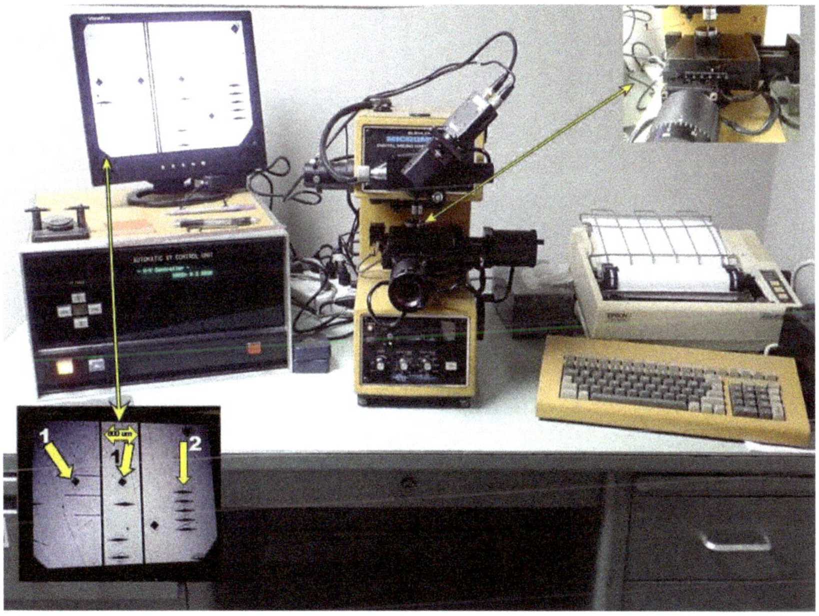

Figure 6.41B: *Shows a 1981 era semi-automated micro-indentation hardness tester*

Measurements are made on the screen (saving eyestrain) and a single distance is directly measured. The video monitor views (upper and lower) show two types of indentations. Arrows depicting diamond pyramid (Vickers #1) indents, which are square and a series of Knoop indents (#2), which are elongated. The distance between the two black vertical lines is 800 microns.

The box at upper right shows a closer view of the stage. A calibration block is being viewed. The Vickers indents were made on a different instrument. The box between the two views of the screen is the control box allowing the instrument to make a series of indentations while the operator is performing another task.

I often use this time to polish the next set of metallographic mounts. I routinely use the micro-indentation hardness tester for most of my metals-related failure analysis work. Having a quick way to get some idea of the strength is something that I have taken advantage of for a long time.

Think about the units of tensile strength: a force divided by an area. A hardness test is also a force being pressed into an area. Of course, the area changes as the hardness changes, whereas with the tensile test, it's the force that changes! Both tensile tests and hardness tests cause some cold work, and a definite change in shape (deformation). So, hardness and tensile strength are both influenced by the yield strength and the strain hardening behavior. Having said that, it is not possible to extract the yield strength from the hardness; it is only possible to estimate, using empirically derived conversion equations, the tensile strength from the hardness value.[140]

Look again at Figure 6.39A (repeated here for your convenience). See how close to the crack profile we may place our hardness indentations.

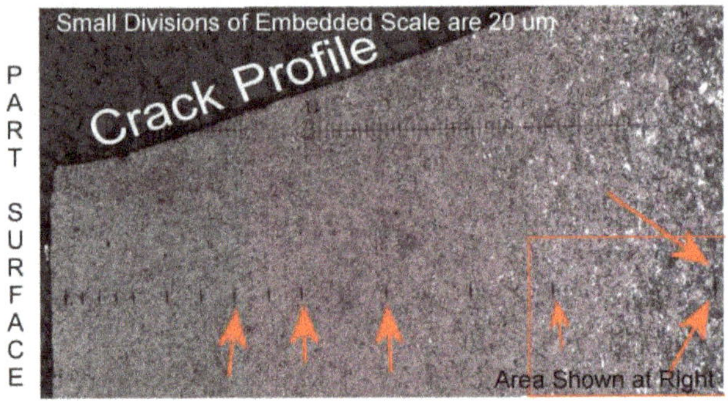

This method may also be used to determine hardness response to welding. Indents may be placed in the heat affected zone near the weld toe, often a location of high stress. Very small wires may be evaluated using this method. If the indents are too large to meet the requirements of being 2.5 widths from another indent or the edge of the cross-section, the test load may be reduced.

For Vickers indents, loads of 300 grams and up are readily convertible to other more commonly used scales. The same is true for Knoop tests made at 500 grams or higher.

Hardness conversion methods (charts or equations) are material-dependent and completely empirical, rather than theoretical. The depth of penetration is dependent on the modulus of elasticity, the yield strength, the work hardening characteristics, and other properties. That said, the correlation of hardness to tensile strength is very good for many materials. ASTM offers hardness conversion charts for different materials in ASTM E140.

Figure 6.42A (below) shows a piece of low carbon alloy steel that has been carburized. This was a test coupon that accompanied a custom-made piece of machinery through its heat-treating cycle. A separate coupon was made since the component itself was too large, critical and expensive to conveniently sacrifice for evaluation. The coupon was a one-inch diameter piece of bar stock.

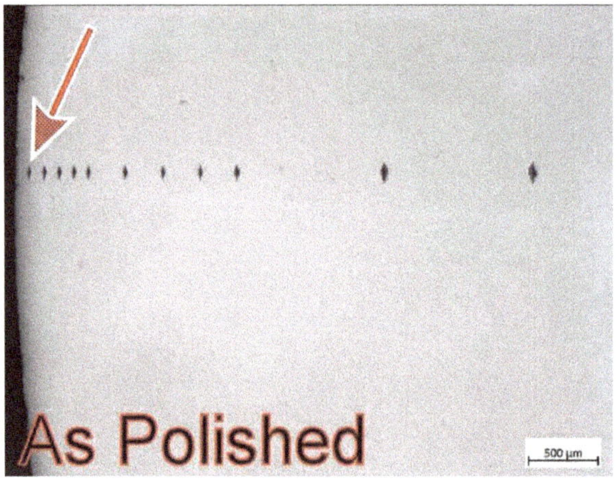

Figure 6.42A: *As-polished carburized low carbon alloy steel with a Knoop 500-gram micro-indentation hardness profile*

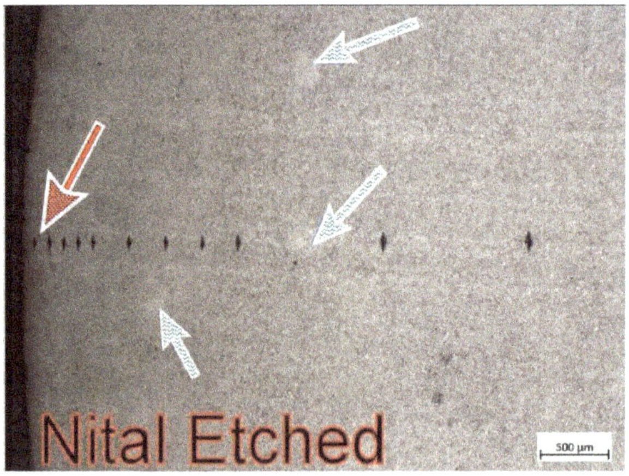

Figure 6.42B: *Nital Etched carburized low carbon alloy steel (1" diameter) with a Knoop 500-gram micro-indentation hardness profile*

CULTIVATE CURIOSITY! *When I performed my usual self-consistency check of the data after running the test shown above, I was shocked to find a big difference between the converted average of the values and the average of the converted values.*

[140] Thank you to Danny Brinkley, a test reader, for calling my attention to the similarity between units of tensile strength and hardness in a private correspondence.

Even as I always tell my clients that hardness conversions are approximate; two different equations that I have used over the years produced this situation.

See the gold highlighted cell in Figure 6.42C, versus the two blue highlighted cells. The blue values differ from the tan by approximately two to four points Rockwell C.

The table shows the depth position of the hardness indent (column 1), raw Knoop 500-gram test value (column 2), and two different converted values. The value at far right (column 4) is based on a logarithmic equation and the second to far right (column 3) is a complex polynomial fit.

True, ASTM says that the inherent test method error is approximately 2 points. But getting this difference seemed surprising to me. This is not a value I routinely look at. The only reason I calculate the average value for a case-hardened part is to make sure that I have not made any typographic errors when entering the data from my printer sheet into Excel. The printer calculates an average, so it's a quick check.

Position, Depth, mm	Knoop Hardness Values	ASTM E140	Metals Handbook 9th Ed Vol 8
mm	ROUND Transverse 1" Diameter	RC Conversion	
0.10	641	55.5	55.3
0.20	657	56.3	56.2
0.30	636	55.3	55.0
0.40	638	55.4	55.1
0.50	614	54.1	53.7
0.75	579	52.2	51.6
1.00	550	50.6	49.7
1.25	529	49.3	48.3
1.50	477	45.9	44.6
2.50	330	32.4	31.2
3.50	297	28.0	27.4
4.50	295	27.7	27.1
5.50	291	27.1	26.6
AVERAGE	503	43.37	44.76
Std/ Dev/	147	0.55	
HRC Conversion of Avg.	47.6		

Figure 6.42C: *Self-consistency check for converted values from Knoop 500-gram test (1" Test Bar)*

Due to this confusion, on another unrelated project (not Carburized, but Through Hardened), I added the comparison of the converted average to the average of the converted values. They were in line with expectations.

The point I am trying to make here is that even after doing hardness tests and conversions for about 45 years, I still have things to learn.

This type of data (hardness profiles) is routinely used to certify conformance to heat treat requirements. By common use, the depth position where the hardness gets down to Rockwell C 50 equivalent is the effective case depth (ECD). Technically for low carbon steels, lower values of hardness are used; but in my industry experience, few people bother with this subtlety. So, if we had a minimum requirement of 1.00 mm for ECD, the ASTM E140 equation would show that this component met the requirement, while the Metals Handbook equation would show a nominal failure.

I remember a few years ago, I changed from routinely using the Metals Handbook equation to the ASTM equation. I started using the new equation for all my clients' work. I thought the ASTM equation was "better". And it appears more consistent with the industry consensus of

542HK500 = HRC50. Maybe that is why someone decided to make up the very ugly (complicated) polynomial equation to replace the simple and elegant logarithmic fit. I don't know the history here.

Getting back to the more general topic of interpreting hardness data, in the case of a narrow test failure, I would remind my client of the test method error and state that a definitive conclusion about the conformance cannot be made. I might suggest that the part not be rejected, unless it is a highly critical component. The fact is that few people engineer their products to the degree needed to even see if the hardness difference will result in a measurable performance difference.

This fairly wide difference between the conversion of the average when compared to the average of the conversions was also found in a larger test coupon (4 inches in diameter) when evaluated using a longitudinal cut. The heat treat shop used a different lot of steel for this coupon; a necessary compromise if we wanted to get the parts into service in a reasonable time frame. See Figure 6.42D.

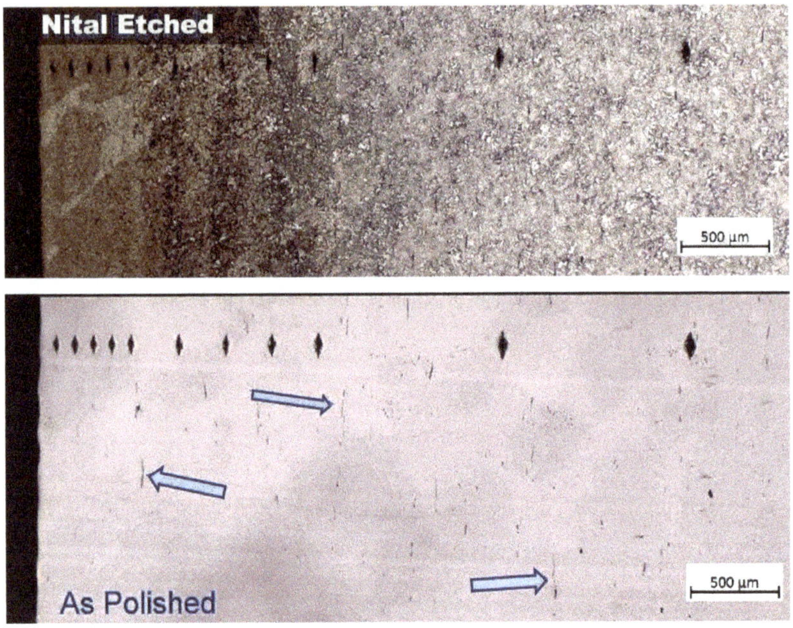

***Figure 6.42D:** Carburized low carbon alloy steel (4 inch diameter test bar) – nital etched is above; as polished is below.*

The original magnification is 25x. Knoop indents get larger from Left to Right as the carbon content returns to the base alloy level and the cooling rate slows.

Note the grey stringer features at the blue arrows (described in the text.) Cause of light etching streaks on the left side is unknown. It is probably caused by bad metallography (a stain), but not necessarily. Note that the high alloy bands making martensite in the deeper layers are more than making up for the slower cooling.

Note Figure 6.42D in the As Polished condition has fine grey stringy features that we did not note in Figure 6.42A. There are two potential reasons for the difference. These features are mainly manganese sulfide inclusions that form when the steel solidifies. As the steel is squeezed and elongated to form a bar from a billet, the sulfides get slimmer and longer. The elongated dimension is not visible when a transverse cross-section is examined.

The part of Figure 6.42A was prepared as a transverse cross-section. Note that we can see the surface curvature at the left edge of the photo, while the part edge is straight in Figure 6.42D.

EXERCISE 6-H: Compare-Contrast – Examine Figure 6.42D (above); you are looking at a plane that is oriented along a diameter or chord of the cylindrical part. Thus, the elongated features are revealed for what they are. Had we made a transverse cross-section, the long stringers would appear round. Try to imagine this if you are not familiar with the difference between longitudinal and transverse cross-sections. Write your Compare-Contrast notes here.

If you are having trouble visualizing this, go get a carrot. Cut the carrot in half. Now cut half of the carrot lengthwise making two long skinny pieces. Cut the other half into circular disks. Compare the pieces. Note how the core and outer layers appear in the two types of cuts.

Now put your carrot pieces in a salad and enjoy as a curious Failure Analyst!

Position, Depth, mm	Knoop Hardness Values	ASTM E140	Metals Handbook 9th Ed Vol 8
mm	ROUND Transverse 1" Diameter	RC Conversion	
0.10	641	55.5	55.3
0.20	657	56.3	56.2
0.30	636	55.3	55.0
0.40	638	55.4	55.1
0.50	614	54.1	53.7
0.75	579	52.2	51.6
1.00	550	50.6	49.7
1.25	529	49.3	48.3
1.50	477	45.9	44.6
2.50	330	32.4	31.2
3.50	297	28.0	27.4
4.50	295	27.7	27.1
5.50	291	27.1	26.6
AVERAGE	**503**	**43.37**	**44.76**
Std/ Dev/	**147**	**0.55**	
HRC Conversion of Avg.	**47.6**		

Figure 6.42E: *Hardness profile data for self-consistency check.*

Hardness profile data is for the longitudinal cross-section made from the larger test coupon (4") heat treated with Subject of Figure 6.42A. Again, note the difference between the conversion of the average and the average of the conversions.

Another feature to observe in Figure 6.42D is the dark and light bands in the nital-etched view. They are at the deep end of the case. The light etching bands are likely martensite that has hardened more fully due to alloy rich layers, while the darker bands probably have more bainite, a softer phase, which forms more readily in lower alloy material. Note that the high alloy bands making martensite in the deeper layers are more than making up for the slower cooling rate.

Why do we care? As Failure Analysts, when investigating mechanical failures (fracture, deformation, wear) we are always interested in where and why the stress exceeded the strength. Micro-segregation has consequences for the uniformity of the resulting heat-treated hardness.

Before leaving this section on micro-indentation hardness testing, look at Figure 6.42F (next page). It is very helpful to plot the data, as has been conveniently done in the spreadsheet software. For most of my billable work, I test more than one area per part, so I can compare-contrast the areas with each other and with theory. If one area is thinner than another, I would expect higher hardness in the thin area (if the parts are made from hardened steel).

This graphical construction in the Figure 6.42F allows us to estimate where the hardness dips below HRC50 equivalent, using the ASTM E140 polynomial conversion equation. If the data were not reasonably smooth at this depth, it would be preferable to get an additional data point close to the expected HRC50 position.

Figure 6.42G combines the data for both test coupons. It shows that the surface of the larger diameter part is a little harder than the smaller diameter part. This adds confidence to the earlier speculation that the low hardness surface value in the 1-inch diameter bar is due to a lower carbon level, perhaps due to decarburization of the original bar stock. Or perhaps it could be a result of retained austenite, which is a common feature in carburized steel. Higher magnification microstructure evaluation would be required to distinguish the two potential causes.

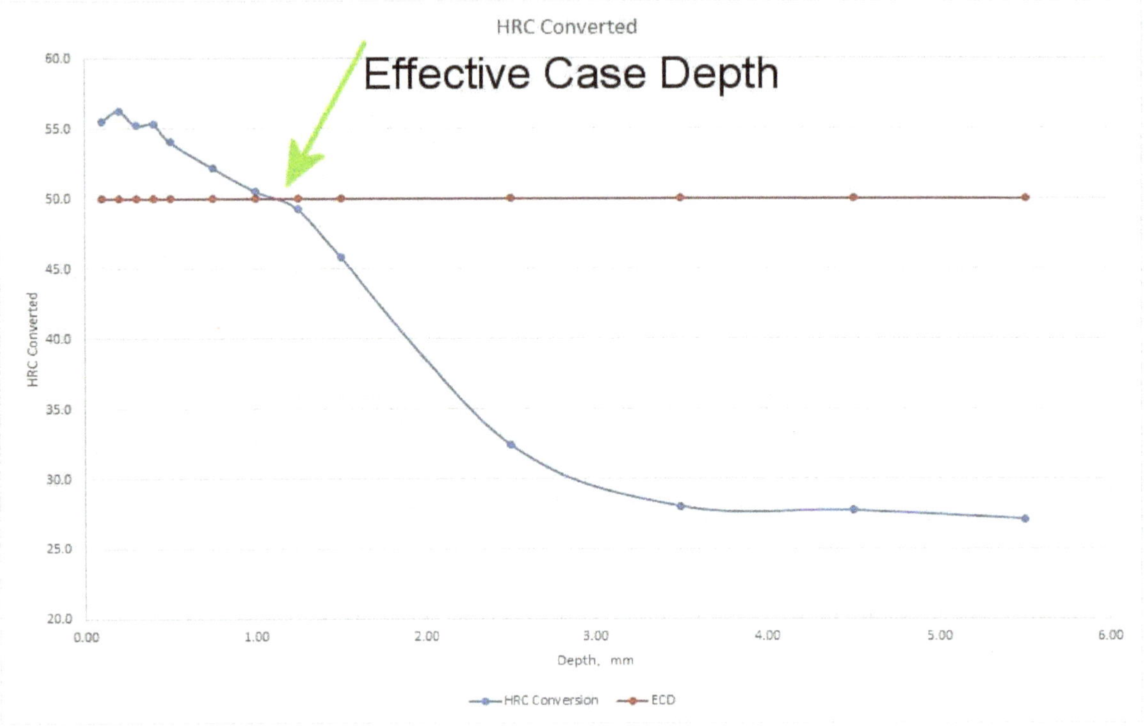

Figure 6.42F: *Graph of data from Figure 6.42C (transverse cross-section of 4 inch diameter bar) – note the slight dip in hardness nearest the surface, which is likely a result of slight decarburization as opposed to grinding damage that we studied in Figure 6.40A and B. The red line is the specified minimum value for the effective case depth.*

In the graph above, the main point to note is the substantially thinner Effective Case Depth (ECD). While the smaller diameter part has an ECD of a little over 1.1 mm, the larger diameter part crosses the HRC50 line at ~0.7 mm.

The fact that the small diameter part has a deeper case than the large diameter part is consistent with expected heat extraction rates for large (heavy) and small (light) steel sections.

Stresses are often depicted from Finite Element Analysis software on the surface of the part. But stresses are transmitted into the core of the part as well. We would need a cross-section view of the stress model to see how the stress changes with depth.

In pure torsional loading, the stress tends to zero at the center of the part.

By studying the FEA cross-section and comparing it to the hardness profile, we can evaluate whether a design is adequate.

- We would have to convert the Knoop value to an approximate tensile strength, useful if the part is subject to fatigue (cyclic loading).
- Or we could attempt to find a way to estimate a yield strength if the part is subject to monotonic (single or sustained) loading.

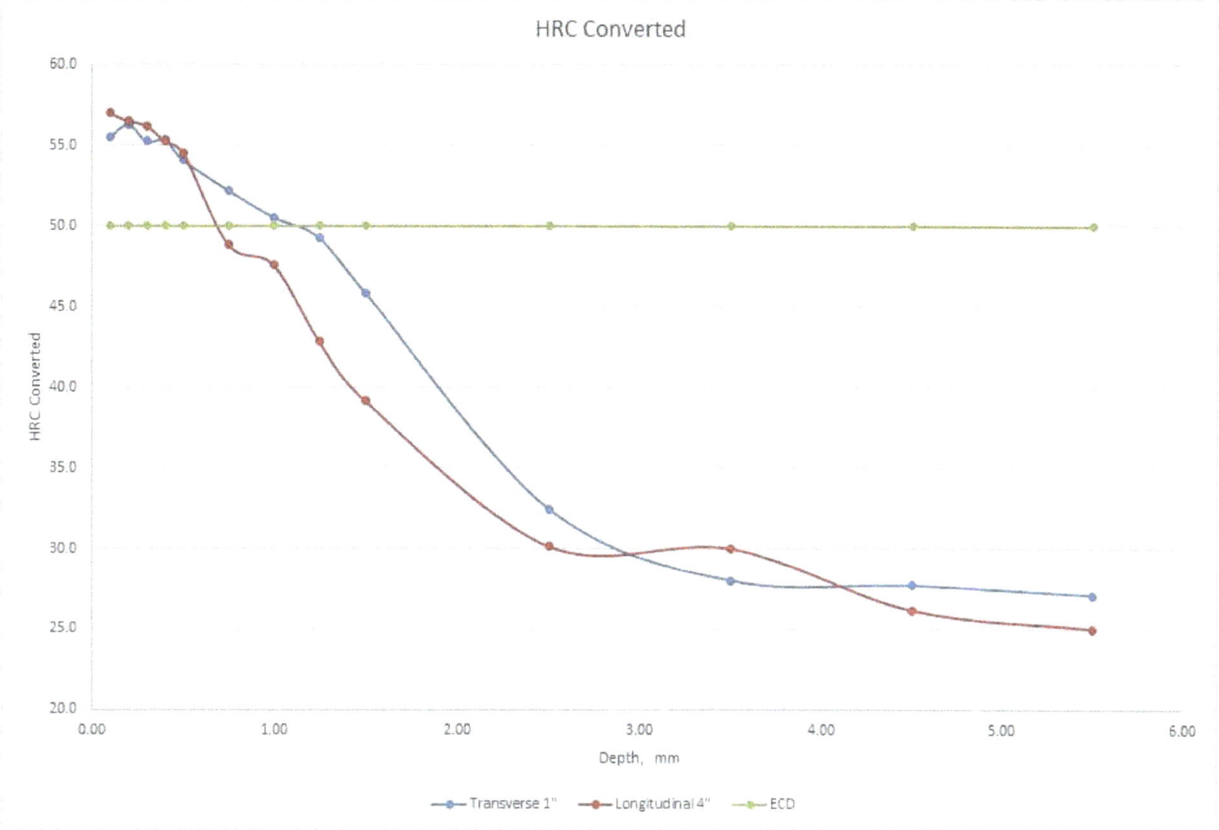

Figure 6.42G: *Graph shows the surface of the larger diameter part is harder than that of the smaller diameter part.*

6.27 Bulk Composition Analysis

Here's a quick overview of bulk composition analysis. Bulk composition analysis is performed in order to determine if the material used to make the Subject of the investigation was "in spec".

We must exercise extreme care when interpreting chemical analysis results during failure analysis work. Minor deviations from specified composition are rarely the sole cause of the damage or failure. All chemical analysis methods have their limitations. The most common problems with bulk composition analysis in failure analysis work are related to sampling. An extended discussion of these issues follows.

Most of the analytical methods commonly used for bulk analysis are instrumental methods whose accuracy depends on calibrations based on alloys of similar composition. Be sure to provide the chemist with a general idea of the amount of any expected element. If you know the specified grade, this will be enough information for the chemist. It the material is truly unknown, given that enough material is provided, the chemist can repeat the test several times. Each time the chemist may more closely approximate the proper calibration setup. However, this is costly in time and the risk of missing a specified element exists. If the analyst does not trust the chemist to actually perform the test and provide the results obtained, then another chemist should be found. Do not expect a competent analyst to work blindly out of a misguided belief that this will produce objective and correct results.

A quick EDS can shed some light on the alloy family, if no other information is available. Obviously, the experienced metals chemist will be able to tell that something is a piece of aluminum or a copper alloy or a piece of steel. More unusual alloys may cause problems. If the investigation team wants to know the levels of trace elements that are not generally specified or reported on the certification documentation, this must be communicated to the chemist.

Many standard and economy[141] methods will not detect relevant amounts of residual elements, even though trace elements can, in certain circumstances, cause major degradation of properties. Very high concentrations of elements can also cause false readings with many of the common test methods. If the chemist has not calibrated with a similar specimen with a higher content than the specimen in question, an analytical error could easily go undetected. I ran across at least two situations where spectrographic results were later found to be incorrect because the tested material was outside of the instrument's calibration range and the technician performing the work was unaware of that fact. In neither of these cases was the technician a degreed chemist; however, both individuals had been accredited by internationally recognized accreditation bodies.

Another issue relates to the material; is it wrought or in the as-cast condition? As-cast products usually have more segregation than wrought products so the sampling needs to be performed from more material layers than the minimalistic approach legitimately taken for wrought product. A way around the segregation issue is drilling through the entire thickness (or at least deep enough to dilute skin layers). Chips produced by drilling must either be tested by wet (more expensive) methods, or the chips need to remelted into a block, which must be done in a protective atmosphere. **NOTE:** Unidentified carburized steel can result in an excessively high carbon value.

Determining the bulk composition is not always a key piece of information in a failure analysis, but it is often a useful piece of information! The potential consequences of exceeding impurity limits for zinc, aluminum and magnesium alloys can be severe; so, know your alloy and be sure you have all potentially relevant alloying elements evaluated.

Remember however, in service, some elements may be removed in greater amounts than others as a result of particular corrosion processes. An example is "dezincification" of brass. The fact that the composition does not meet the specification today does not prove that it did not meet the specification at the time of manufacture!

THE BOTTOM LINE: We Failure Analysts need to communicate with the chemists doing our work, and be very thoughtful before "blaming" the mill, foundry, or material producer for shipping out-of-specification material. Thus, great care is required when interpreting results of bulk-composition tests performed on actual products, especially when they have been in service.

This is even more true for polymers than it is for metals. With polymers, the composition refers to the types of molecular bonds, in addition to the types of atoms present. If you want to analyze polymers for small traces of plasticizers, or even high concentrations of multiple resin blends, you may need some pretty fancy and expensive equipment. While Fourier Transform Infrared spectroscopy (FTIR) instruments are fairly common, Gas Chromatography Mass Spectroscopy units (GCMS) are not. Finding a competent chemist to work with is a must for those who are working to produce useful failure analysis data.

We must also remember that there may be multiple versions of materials specifications. There may be versions for virgin material, for large batches (ladle analysis for steels, for example), and for smaller batches of finished product. Be sure to select the relevant version for your project. There might be one for reviewing the paper trail (ingot, for example) and another for determining conformance of a particular piece of the Subject Component (Product).

For carbon and alloy steels as well as zinc casting alloys in product form, a broader range of allowed concentrations of intentional and residual

[141] ASM Metals Handbook, Vol. 11, 2002 | *Chemical Analysis of Metals in Failure Analysis by Debbie Aliya* | Standard and Economy Methods are explored in this article.

elements is published by the Society of Automotive Engineers. Failure Analysts need to understand that many product-form metal specifications have other requirements than the composition. Conformance of a product-form to mechanical property requirements may be much more significant to the failure investigation than the composition.

I choose economy methods for most of my work, when it is appropriate for the material and geometry of the component to be analyzed. Economy methods can be performed on a small chunk of material. The benefits of economy methods compared to referee methods include: no time is spent milling, drilling, weighing, or dissolving the specimen. As long as a clean flat area is available on the surface of the part, analysis may proceed after a quick grinding operation on a special sanding belt.

When you, the Failure Analyst, are choosing a chemical analysis service, one of the first important questions is, "How much material is available to sacrifice to the composition analysis?"

Referee methods for metals, such as ICP (Inductively Coupled Plasma), usually allow smaller specimens to be analyzed. But the time to perform the analysis is often significantly longer. The cost is usually higher. **Referee methods are considered suitable for dispute resolution because the sample is more representative of the material, not necessarily because the instruments are inherently more accurate or precise.**

6.27.1 Metallic Materials

Composition analysis of some components is not necessarily more problematic for failure analysis than for other endeavors. But if the part has been subject to corrosion or heat, as we have already noted, the composition can change. This complicates assessments of specification conformance at the time of manufacture. After all, it is not the material manufacturer's fault if the material degraded over time during service. But the corrosion process may make it look like the wrong material was used.

Copper alloys, of the common metallic materials, can definitely change composition during use. That is why judgement must be exercised, and the report must contain a discussion that the material possibly did meet the specification at the time of manufacture.

Other materials may have coatings or surface treatments applied. Some of the coatings are difficult to remove, especially if the Subject Components are thin or small. Remnants of the coating could complicate analysis of the base material. This geometric situation creates similar challenges for those who are doing reverse engineering work or comparing two different suppliers' products to each other.

Removal of carburization, nitriding, and other surface treatments for steels introduces another step in the composition specimen preparation process. The (small) size of many product forms makes it difficult to perform an analysis. Economy composition analysis methods include optical emission spectroscopy (OES), glow discharge spectroscopy (GDS), and x-ray fluorescence (XRF). For example, a flat circle of 6-12 mm diameter with a remaining minimum thickness of approximately 0.5 to 1 mm is generally required for these methods. There is also a maximum size for some of these instruments.

There are some small section components such as screws, small coil springs, or other parts made from small diameter wire, where the minimum size requirement leaves no possibility for a direct component analysis with these economy methods.

Some labs will melt multiple small parts together in order to create a larger disk that they can then use on the economy method instruments just mentioned. However, they generally need 30 grams of material. If you have one small broken coil spring, there will not be enough material. In my experience, only the most trustworthy labs will inspire confidence in this type of specimen preparation. If the protective environment breaks down during high temperature exposure, the specimen may be oxidized, with the resulting oxide floating out of the test slug. Since some alloying elements oxidize more readily than others, another source of potential error is introduced.

If you have small diameter cylindrical parts, such as pins, and they have been carburized, you will need to cut off the carburized end and drill down into the center of the cylinder and collect the chips in order to know the original carbon content of the raw steel.

Maybe you want to know, on the other hand, how much carbon there is in the outer layers of the pin in order to determine how effective the carburization was. Usually, as we saw earlier, a micro-indentation hardness profile is the easiest way to gather this information.

But sometimes people want to know the carbon content itself, rather than the resulting hardness. This is usually part of a heat-treating process troubleshooting activity. That's when the cylinder must be put in a lathe and the depth of cut controlled to obtain the layers of interest; and collect the chips. The chips from both the carburized case and the core must be separately tested using the combustion method described later on this page [*]. When interpreting the data, we must consider the accuracy and precision of the feed on the lathe that is used to remove the outer layers. Since nitriding results in thin cases, this technique is most frequently used for carburizing.

In all such removals of thin outer layers, the machinist must take particular care to inspect the tool for freedom from chips both before and after collection of the specimen; because inclusion of even a small chip from the tool in the specimen could cause a major error in the analysis.

Now we continue to look at referee methods, but in this case, a remelted specimen is not to be used. That is when we use "wet chemistry". All coatings and surface contamination must be removed before weighing and then the remaining material is dissolved in acid. The acid solution is then analyzed.

There is no reliable instrumental method to perform this for two very important strength related elements (carbon or sulfur) in steels. Instead, these elements are usually analyzed by combustion.

[*] Combustion Method: ***Small bits (usually totaling ~1 gram) are heated in the presence of oxygen. The quantities of resulting gas molecules that form (carbon dioxide or sulfur dioxide) are evaluated. The entire process of doing wet and combustion composition analysis is obviously more time consuming and thus more expensive than simply supplying some electricity to a chunk of material and using an instrument to "look at" the resulting light wavelengths.***

Powder materials, including additively manufactured components, also present challenges, especially if they are not fully dense. Some powder metals are intentionally impregnated with lubricants. These lubricants can make it extremely difficult to get a reliable analysis.

> *I ruined a long weekend struggling to match some unknown powder metal parts to different specifications. All of the specifications were very broad as the material supplier wanted to keep the actual recipes secret from the end customers. In addition, the final alloy was composed of particles of three or four different compositions.*
>
> *Working with the microchemical EDS method, I tried to analyze a particle to avoid the lubricant. Even averaging many particles, I had a non-representative specimen. It was an exercise in frustration - or it could be considered a failure of a laboratory quotation and materials characterization analysis!*

There are various instruments sold to the maintenance and nondestructive testing communities that use x-ray fluorescence. These instruments purport to positively identify many alloys. The problem is that many of them (as of the year 2020) do not have a way to determine carbon. Extreme caution must be exercised when making grade identifications based on this test method. Instead, all an analyst may say is, "If the major alloying elements matches the specified grade, then the material is consistent with the specified composition, for the alloying elements tested."

When I am teaching classes, I usually tell students, "I hate these instruments." I don't really hate the instruments themselves. However, my inner geek experiences extreme stress when companies that sell these instruments imply that they can provide a positive material (grade or alloy) identification.

The fact is many alloys require restricted ranges of carbon and other elements.

The fact is many XRF units cannot detect or adequately quantify carbon.

The fact is other alloys require specific ranges of strength or hardness or microstructure - in addition to the base composition requirements - prior to being "identified" as being within (or not within) a particular family of alloy material specifications.

The good news about metallic material composition analysis is metallic alloys are often defined based strictly on the relative number of different types of atoms. There is rarely an effect on composition determination because of how the atoms are bonded to each other. Polymer composition does change as a result of molecular bonds.

Economy methods are not inferior in accuracy to the more expensive methods, when used on appropriate specimens.

6.27.2 Polymeric Materials

Composition analysis of polymeric materials is much more complicated than what we just described for metals.

The basic modern test method for polymers is Fourier Transform Infrared spectroscopy (FTIR). Metals methods look at the atomic species. That is not what FTIR looks at. FTIR evaluates ***the bonds between the atoms.*** Polyethylene, polypropylene, and polystyrene, for example, have only carbon-carbon or carbon-hydrogen bonds. However, the ratios and geometries of the carbon-carbon and carbon-hydrogen bonds differ in the three materials. The FTIR method is able to distinguish these variations one from another.

WHAT IS A POLYMER?

Polymers include the materials commonly known as plastics. They can be heated, reshaped, and cooled multiple times.

Polymers also include thermosets (such as traditional rubbers), which undergo a chemical process that essentially turns the entire component into a single molecule of fixed shape. Even if reheated, it will burn or break down into other substances before it deforms into a different shape.

All plastics are polymers. All polymers are not plastics. Plastics are a sub-set of polymers.

This method works by sending infra-red light through the polymer. The molecular bonds absorb certain wavelengths of the light and let others pass. This creates a spectrum of absorption that acts like a fingerprint for the resin.

Other common resins such as acrylonitrile butadiene styrene (ABS) are more difficult to "fingerprint". When analyzing the popular ABS material, the styrene absorbance peak is so large that the acrylonitrile and butadiene peaks are practically indetectable. But if the chemist knows that the material is supposed to be ABS, they can match the unknown to a known spectrum of ABS material.

Nylons (a type of polyamide) are easy to identify with FTIR, but it's much more difficult to distinguish the sub-types of nylons from each other. Nylon 6 (PA6), for example, has a very similar FTIR absorbency spectrum to Nylon 6,6 (PA 6.6) or Nylon 15 (PA15). Yet, these different versions of the molecule have different architectures, or shapes, and different service characteristics.

Occasionally, old school chemists will analyze pyrolyzed material. To produce pyrolyzed material, a resin sample is heated in the absence of oxygen in order to break down the molecules.

The resulting material is more readily attributable to an original specimen of a particular sub-type of nylon that the solid resin was.

More frequently, however, the material is tested by a totally different method called Differential Scanning Calorimetry (DSC). This test is performed to determine a nylon's melting temperature. The melting temperatures fall into distinct ranges for the different versions of nylon; that allows the sub-category to be readily identified. It is sometimes possible to detect small amounts of contaminant mixed in with a nylon material using the DSC method.

Different grades of Polyurethane, a common material, can also be very challenging to distinguish using FTIR.

Other problems with FTIR analysis relate to intentional additives used to customize resins for specific applications. For example, additives to stabilize a resin during heat exposure (molding) are present in very low concentrations. They are undetectable by the FTIR method. If it is really important to know for sure that the correct additive package is present in a plastic component, the extremely expensive GCMS method must be used. "Extremely expensive" is relative to FTIR method comparisons. The GCMS method includes separating the molecules of a sample into groups based on their molecular weight.

But perhaps the biggest challenge of polymeric material identification is determining the molecules' sizes.

The size of a molecule is generally measured by molecular weight. For example, if a polyethylene molecule has 10,000 carbon atoms, it will have 20,002 hydrogen atoms and the theoretical molecular weight may be calculated readily. In a given batch of resin, the producer can influence the average size of resulting molecules. The larger the molecule, in general, the greater the strength will be in the solid state and the greater the viscosity will be in the molten state.

Now, we get back to our original subject, which is identifying grades of commercial polymer resins. The name of the resin indicates the molecule's structure, such as polyethylene or polyamide. But for certain resin types, it's also important to know the sub-category. This is often related to how big the molecules are. Knowing the (average) molecule size will be an important factor in determining if the correct resin was used, or in other words, if it conforms to the specification.

Many resin producers make multiple sister grades of resin, with increasing strength levels. The higher strength is generally associated with greater difficulty in molding. So, determining whether the part was molded using the 1000 MPa tensile strength version or using the 800 MPa version is not going to be possible with the FTIR alone. Resin manufacturers rarely give out their secret data on the distribution of molecule sizes. If in fact, they have even determined it for their own use.

In general, the stand-in for measuring molecular weight is to determine viscosity. You can determine the viscosity of either the molten material (Melt Flow Index) or a solution consisting of a small amount of polymer in a liquid (solvent). This means, in case you had not already guessed, yet another test.

Melt Flow Index[142] (MFI) is often performed during a failure analysis to determine whether the molecules have been broken down to smaller sizes as a result of improper processing. This could include molding wet material or excessively hot material. Even normal processing will break some molecules into smaller sizes. So, interpretation of the data is key.

We should not assume a material is inconsistent with its specification because the Melt Flow Index is 10% higher than the material data sheet indicates. A Higher Melt Flow indicates more

142 Melt Flow Index is a simple measurement indicating the amount of material flowed through a specified hole dimension at a given temperature, load, and time conditions.

material flowed through a hole in a mold at the same temperature and load and time conditions. That means the viscosity is lower and the molecules are degraded. That means the material is likely not as strong as the designer anticipated. And that could be a result of improper processing or aggressive service condition. A 10% higher Melt-flow Index is enough to notice and consider but may not be enough to trigger an alarm.

Melt Flow Index testing can usually be performed with about 10 grams of material. Much more would be required to perform a direct strength measurement. Thus, Melt Flow is often a stand-in for strength measurement during failure analysis work.

The other method used to evaluate molecular weight is solution viscosity. Testing the solution viscosity is done by dissolving a small amount of polymer in an appropriate solvent. This is more expensive and less common than MFI testing, but might be appropriate for some resins. I have used it for glass-filled nylon. I request that the lab personnel filter out the glass. In this manner, I gather data about resin properties without interference from the glass fiber reinforcement.

Solution or resin viscosity measurement is often more practical than direct measurement of tensile strength. Note that mechanical properties of polymer resins are certified based on test bars that may be cut from flat plaques or individually molded. They are not certified based on test bars cut from molded parts. Yet in a failure analysis, we are usually stuck with the parts.

If we have a large number of small parts, they could potentially be melted and molded into a plaque. But remember, every time the material is reheated, it degrades - at least to some degree.

So, if molecular weight is required to determine strength to identify a grade of polymer, Melt Flow testing is usually the method of choice. We need to keep in mind that the measured response on an MFI test performed on a molded part will differ from a measured response obtained from virgin test resin pellets. If the part was subject to a severe environment, we have ever more layers of complication to consider when evaluating the data.

"To Observe is to Disturb," as David Levy says in his book, *Tools for Critical Thinking.*[143]

Generally, in material identification work, chemists perform metals composition tests, while metallurgists perform (or at least supervise) the microstructure and mechanical evaluations. But for polymers, chemists are generally the ones performing all of the tests we have been discussing in this section (except tensile tests).

But wait! There's more.

Sometimes the price of resin is too high to use in a particular application. That's when polymer resin formulators mix in various cheaper materials as filler. Fillers may include various powdered or chunky minerals, for low strength requirements. On the other hand, if greater strength is required than the polymeric material can provide, high strength fibers may be mixed into the polymer matrix. This will provide greater stiffness, too.

Often glass fibers are used for this purpose. So, in order to know if the material had 30% glass fill, you have to find a way to separate the polymer resin from the glass. This is usually accomplished by weighing a small chunk of the material, burning off the plastic, and weighing the remaining glass. In the old days, this was done with a balance and a furnace. Now, most labs want to charge you for using their Thermogravimetric Analyzer (TGA).

If you want to know if your piece of Nylon 6,6 (PA 6,6) 30% Glass is really the specified grade of Nylon 6,6 30% Glass (PA66G30), you need to:

- perform an FTIR analysis,

[143] David A. Levy, *Tools of Critical Thinking: Metathoughts for Psychology* | Chapter XXI

- follow this with a DSC analysis to determine the melting point,
- perform a TGA to make sure that there is really 30% glass,
- then add the Melt Flow to get some idea of whether the material has been degraded.

In 2022 in the USA, you just spent 800^{\underline{00}}$ at a customer service friendly ISO accredited test lab. It probably took at least a week to get your results. Compare that to your 25 mm diameter, 12 mm tall disk of 4140 steel. It was analyzed in two days, at an equally customer service friendly ISO accredited test lab, for a price of 150^{\underline{00}}$. Having said that, if you want to analyze the small carburized cylinder (described earlier) using wet and combustion methods, and separately analyzing core and surface carbon content on tediously collected chips as described earlier, you probably just spent a significant portion of the amount paid to the polymer analysis lab. It pays to know your costs and the cost of outside expertise.

> *HERE ARE A FEW FINAL THOUGHTS ABOUT CHECKING THE GLASS-FILL CONTENT OF POLYMERS. I have had around 25 specimens of glass-fill polymers checked out over as many years. I don't think the lab has ever found a deviation of more than 1 or 1.5% from the specified glass fill level.*
>
> *Usually, I ask for the glass powder back so that I can look at it and make sure it looks like glass fibers. I can measure some fiber lengths with an electron microscope.*
>
> *Again, longer glass fibers are generally associated with higher resin strength.*
>
> *Improper processing can break down the glass fibers.*
>
> *One thing that I began to notice around 2010 was many glass fibers were not burned clean after the ashing process or the TGA. I looked into this and found out polymer resin suppliers started treating glass fibers to improve their adhesion to the matrix resin. I have not been able to determine whether I might ever need to worry about this when interpreting my polymer test results.*

6.28 Odd Composition Analysis Bits and Pieces

We will not discuss contamination or debris analysis in great detail. It's been a fairly significant part of my professional work, and maybe the second edition will include a Chapter 10 on the topic of "How to Identify Unknowns."

6.28.1 Debris Analysis

Mixed small particulates can be very challenging to analyze and very important to identify. The EDS microchemical analyzer that we discussed in the Scanning Electron Microscope section is one powerful way to approach debris or particulate analysis.

Another good resource for your further education on the subject can be found in the *ASM International Metals Handbook Volume 11 2002*; read the article on Microchemical Analysis in Failure Analysis (Page 432).

Some FTIR instruments are equipped with microscopes useful for analyzing small amounts of unknown organic materials. These may include: polymers, oils, bits of insects or plants, etc. Using the FTIR method requires a bigger particle size than the size requirement for the EDS, but it can still be a very helpful contribution to the task.

> *Years ago, I came across plating blemishes caused by debris. It turned out to be bits of insect carcasses. The EDS showed high carbon (it was an older EDS that could not detect nitrogen). The main analysis was performed by shape recognition. Good thing I was paying attention in my biology class.*

6.28.2 Lithium Analysis

In the last few years, I've been asked to provide analysis of several different lithium materials. Since lithium only has three electrons, theoretically it could provide an EDS signal. But for now, it is only practical using newer technology.[144] To fulfill my obligations, I use specialized chemical analysis services that own the equipment to detect and quantify lithium.

6.28.3 Solving the EDS Conundrum

Nickel plated components from two entirely unrelated clients, from different industries, providing different components suddenly started showing up with a significant concentration of Ytterbium. Since I have never seen Ytterbium specified on any part I have worked with, I became suspicious.

My National Institutes of Standards and Technology 316 stainless steel test block analysis remained unchanged. No Ytterbium showed up. Of course, the amount of nickel in 316 stainless is only 10%, while nickel plating may be almost 100% nickel. After a couple of hours of research (not billable!!), I concluded that I did not believe that there was Ytterbium present. It was not obvious to me what the problem was.

Eventually, my EDS service provider confirmed that the calibration had drifted.

144 My 2015 EDS system can detect Beryllium, which has 4 electrons and 2 electron shells. Lithium has 3 electrons and 2 electron shells. Both are detectable in high concentrations using an EDS with newer technology.

CHAPTER 7

Negotiation of Failure Analysis Project Scopes

Business Aspects of Failure Analysis

7.1 Overview

First things first; what exactly do we mean by the "scope" of a project? The scope of any project is essentially the range of defined tasks that both parties agree to perform. This is important.

From a high-level point of view, our customers want their suppliers to commit to deliverables. They want to know exactly what they will receive for their money as well as the timeline of expected results. From the Failure Analyst's high-level point of view, we need to know what we are expected to provide and we must lay out the customer's responsibilities. For example, they need to collect required data, provide test specimens, and pay for the project.

Most failure analysis clients[145] want to know, "Why did the subject of the investigation fail?" But this question is not very helpful to either party when trying to pin down a specific list of tasks. We need more specific questions in order to intelligently allocate a reasonable set of resources to the job. Reasonable, in this case, means appropriate given the severity of the consequences (or potential consequences) of the failure.

Scope negotiation involves ensuring that we will have the time and budget to perform all the work

---

[145] What is the difference between a customer and a client? Many consultants say, "The customer is always right." That may or may not be good business practice. Professionals working on matters that are related to public safety must embrace a different reality. Sometimes clients are wrong. The reason they hired you (the specialized professional) in the first place is because they lack the required knowledge or credibility to make a trustworthy decision. Meditate on the doctor-patient relationship for a Compare-Contrast exercise.

that will produce the reliable, relevant information required to achieve our goals. Common project goals may include:

- Preventing recurrence
- Accurately pointing the finger of blame
- Securing data to allow estimates of remaining life of similar equipment
- Preparing a report

I will cover "Preparing a report" in greater detail in Chapter 9. The rest of the goals listed above are generic. It is generally a bad idea to promise to definitively answer any specific questions until the project is completed. Specific project scopes will result in the best chances of success when the particular details surrounding the history of the design, the construction or manufacturing sequences, and use and service history are reviewed, at least superficially, during the negotiation.

It is generally a bad idea to promise to definitively answer any specific questions until the work is done.

Before moving ahead, it should be noted that approximately 50% of the time spent in a typical failure analysis project is preparing the report. It's critical to make sure you have agreement on how detailed and formal the report needs to be.

One of the aggravations of this work for me is when I find out that the purpose of the report is to put it in a file and forget about it, just so it's on record in case someone asks.

If the work involves physical object failures that caused human injury, the work must ensure that a sound understanding of the sequence of damage is part of the analysis. In this case, the physical-cause layer of the analysis is fundamental to the rest of the investigation.

The easiest way that I have found to explain how to negotiate failure analysis work is to use case studies illustrating a practical philosophical approach to doing the minimum work appropriate to the situation. We will get to some case studies shortly. But first, let's discuss an approach that is particularly useful for larger, catastrophic problems. In fact, this approach is very simple! Set the investigation budget to 10% of the estimated total cost of the failure. That does not include fixing the problems. That is for the investigation itself. If people were killed or injured, that value may be disputed.

- To get negotiations started in disputed situations, determine if the injured or deceased individual was employed. If so, we might consider adding 10% of the total wages they would have earned had they remained alive.
- When it is an automotive safety issue that might require a product recall and management will not agree to that type of budget, I sometimes feel like saying impolite things! It's true that management people who are facing a recall often want to believe that all will be fine and the problem will go away on its own. It's difficult to manage an automotive manufacturing operation without a basic optimism. Most people don't care to pay heed to Cassandra-type-warnings.[146]

In any case, deciding on the budget does not relieve the team of preparing a detailed protocol proposal with a list of the actual steps to be taken.

Of course, sometimes the specific physical reason a failure occurred is obvious as soon as the evidence is viewed for the first time. But I have learned over the years it is best to keep my mouth closed and not share my opinion about the cause. This hint is for the benefit of the client as well as the potential failure analysis service provider. Giving a potential client a little piece of data out of context will not allow them to make the best decisions about the situation moving forward.

[146] Cassandra was a Trojan princess who was cursed with perfect prophetic vision and the pain of nobody ever believing her. For more general information, go to the following url: https://en.wikipedia.org/wiki/Cassandra

And besides, the client might feel stupid for asking for help for something that is obvious. Best to take things one step at a time!

On occasion, clients will start arguing with me immediately when I suggest the tasks required for a failure analysis. They think they only need a certification check. By now, if you have been studying this book, you will already know that certification tests may not destroy the entire part or assembly but are usually destructive of evidence.

7.1.1 Scope Negotiation Bottom Line

Scope negotiation includes educating the client about what their options are, and sometimes, what they are not.

It is helpful to remind potential clients that most failure events are not a result of specification compliance failure. Most failures are due to:

- unanticipated service conditions,
- unclear material specifications,
- manufacturing process variations,
- maintenance issues, or
- a constellation of factors that may be difficult or impossible to completely untangle.

But that doesn't mean it's not worth doing a real failure analysis. Almost always we learn something to help us understand **what happened**, **how it happened**, and **why it happened**. Usually there will be answers to those questions at multiple levels. It is by defining the project scope that we may influence the levels of insight we may gain.

After we review general project planning case studies, we will look at three other ways to make sure we don't overlook something during the scope negotiation. Finally, we'll look at some common questions that might help us focus our work in order to ensure the specific client questions are answered to the best of our ability.

Before we move into the case studies themselves, let's remind ourselves of the activities required to Construct Competence in Failure Analysis.

> **Constructing Competence in Failure Analysis**
>
> 1. **Cultivate Curiosity:** Without it, you will never rise to competence in failure analysis.
> 2. **Clarify Context:** Determine the Frame of Reference (the situational picture painted during Scope Negotiation)
> 3. **Compare – Contrast:** Determine what reference samples or historical data are available.
> 4. **Connect – Create:** Data interpretation; how many iterations of testing and analysis will be included? How do the data sets help shed light on each other?
> 5. **Contemplate Complexity:** Review possible inter-dependencies of factors/causes revealed by the evidence. Give time for reflection to reduce the chance something important might be overlooked.
> 6. **Calibrate Your Confidence:** Is the data adequately self-consistent given the consequences? Are there any unexplained anomalies left? Any lingering doubts?
> 7. **Convey the Findings:** Present Data, Conclusions, and any Recommendations that were requested in the agreed format.

Ok, now let's listen in on a conversation between a client and a Failure Analyst:

Client: *"Why did it break?"*

I have a smart aleck answer to that question. My

answer is always true! When someone asks me why something broke, I can tell them…

Failure Analyst: *"The stress exceeded the strength!"*

Standing by itself, at a literal level, that answer is an example of poor use of thinking skills. It is an example of tautologous reasoning. Basically, since strength is the definition of the maximum sustainable stress, my answer contains no new information. I am acting like I am answering, without answering! So why would I give this answer?

Well, first of all, it's fun to see how people react! Since I am telling them a true fact, I don't have to feel bad about tricking them! In addition, maybe they will like my sense of humor and realize that it is possible for a nerdy engineer to have a sense of humor. This answer gets their attention. Most people respond to this answer with silent attentiveness. They are now ready to participate in the conversation, and they realize that they are going to have to be thoughtful about their end of the conversation.

Another reason I like to tell people that their part broke because the stress exceeded the strength is because, at this point, they usually have not agreed to pay for my services. I do not like giving my technical services away for free. It is unfair to me and unethical to undermine the ability of others to earn a living in the field of engineering.

If you are working within a manufacturing company analyzing your own company's products and you receive a salary, then this is not an issue.

Once you have the other person's attention, you can follow up with a more productive question: "What else would you like to know?"

This is a key moment in the dialogue. Before I go on to the list of case histories, it's important to note that this basic answer (with a slight word change) applies directly to other mechanical damage categories (deformation and wear). It also applies to environmental damage (with a slight change of words).

The equivalent answer to "Why did it corrode?" is "The environment was more severe than the corrosion resistance of the material/component could tolerate." We follow up in all cases with, "What else would you like to know?"

7.2 CASE STUDY: Broken Bolt, Highway Vehicle Application

Client: The bolt was low carbon steel. Why did it break?

Failure Analyst: The stress exceeded the strength. What else would you like to know?

Client: Why does this crack look so unusual? Why isn't it at the smallest cross-section?

DISCUSSION: The language chosen to populate the project scope proposal is intended to clarify the deliverables. This is for the protection of both parties, Failure Analyst and client, so agreement on completion of tasks is properly and readily documented.

Now, at this point, we have a focus to our investigation. We know that the client is aware that this is an unusual crack. The next step is to gather the usual background information, including:

Failure Analyst: Is this a new design?

Client: No, 30 years of successful history.

Failure Analyst: When did the crack occur?

Client: Several on installation, a few in service.

Failure Analyst: Is this the only broken piece?

Client: No. (**Note:** Failure Analyst was not properly listening to the previous answer!)

Failure Analyst: What happens to the vehicle if the bolt breaks when it is going down the road?

Client: It is undesirable, but not catastrophic.

Failure Analyst: Do we need to worry about the parts that were manufactured as part of the same lot, but are not broken?

Client: Yes.

DISCUSSION: The conversation is open at this point. The back-and-forth conversation will allow the analyst to understand what the potential client needs and will educate them to understand what tests and evaluations are available to help reveal the causes of the fracture or the factors that contributed to it. Remember, the analyst must create a test protocol proposal.

Figure 7.1: *The dark bands of discoloration on the broken bolt are a result of a friction weld process. They are visible on both sides of the shiny area where the weld flash was machined away.*

Because this is a crack, we want to understand the loading that caused the crack so we continue to gather the usual information, including:

Failure Analyst: Was it a single load cycle, or a fatigue crack?

Client: The part was subject to fatigue in-service, but some of the cracks happened on impact during installation.

Failure Analyst: Was the load imparted with a high or low strain rate?

Client: Unknown.

Failure Analyst: What was the temperature at the time of the crack?

Client: Not recorded for in-use, but ambient for indoor assembly fractures.

Failure Analyst: We will need to determine where the crack initiated. We can do that with a visual inspection.

Client: Ok. Is there anything we can do to reduce the chances that additional parts will crack (etcetera)?

Failure Analyst: We will have to do the lab-work first, before we can figure that out!

Client: Ok. Good. We'll want to know that.

MORE DISCUSSION: Other concerns that can be addressed in the opening created by the smart aleck always true but on-its-own useless answer include "what type of stress exceeded what type of strength?"

Was it the fatigue stress that exceeded the fatigue endurance limit? Was it the impact stress that exceeded the impact resistance?

Did the stress exceed the maximum sustainable value in an aggressively corrosive environment?

Was the part a high-strength component contaminated with hydrogen and then loaded under a sustained tensile load that promoted hydrogen embrittlement?

A flexible dialogue format will generally reveal what the client wants to know and how important it is to determine. That is the essence of determining a protocol and budget that meets requirements.

WHAT PROTOCOL WAS FOLLOWED?

Negotiated Activities

- Visual Exam
- Documentation Review
- Scanning Electron Microscopy
- Optical Metallography
- Tensile tests on core of threaded section
- Bulk composition analysis (comparison to certification documents)
- Tensile Tests on Full Size Components

Follow-up Activities

- Experimental heat treating followed by Tensile Tests on Full Size Components
- Recommendation for heat treat fix for affected lots

WHAT DID WE FIGURE OUT?

Micro-fractographic features (cleavage with ductile tear ridges) revealed by the scanning electron microscope indicated that the fractures were due to a combination of residual stresses in the cold formed head of the bolt and a sudden impact load.

Visual examination revealed that the crack started in the core of the part, not at the surface. The Subject Welded Assemblies were very brittle. They could be broken by tapping one end on a concrete floor.

After stress relief at 450°C, this was no longer true. Furthermore, when the parts were tested in tension, there was localized necking (deformation) and then separation (fracture) in the small diameter threaded portion. This was the expected crack location.

PUSHBACK

Many years later, a few colleagues told me that they thought my diagnosis was incorrect. "We think," they explained, "the problem was due to strain-age embrittlement, a result of the friction-weld heat acting on the cold worked microstructure."

I countered by saying, "The micro-features of the crack surface show cleavage."

They said, "No, they were intergranular; but looked flat like cleavage because of the severe cold deformation required to shape the head of the bolt."

I pointed to some tear ridges on the flat grains.

"Well," they said, "maybe there's a mixture of cleavage and intergranular."

"I understand," I countered, "but, my fix worked. The stress-relief eliminated the brittle condition."

They answered, "The stress-relief would also disperse the nitrogen which had been collecting along dislocations, thereby preventing ductile behavior."

Or as David Levy says in *Tools for Critical Thinking*, "The cure doesn't prove the cause." These parts had been produced for decades before the problem arose. After the fix, the problem never happened again.

We did find some differences in the mechanical behavior between the problem lot and other lots, but we never completely understood why these parts behaved so differently. Maybe the aluminum used to "kill" the steel and refine the grains was ineffective. Maybe there was more nitrogen than usual. We'll never know for sure!

7.3 CASE STUDY: Cracked Stainless Steel Tank, Desalination Process

Client: Why did it break?

Failure Analyst: The stress exceeded the strength. What else would you like to know?

Client: We have successfully used this design in other locations. We had this unit, and four more, in service in very salty water with low pH. This crack is extremely premature. We may need to redesign the product for this application.

DISCUSSION: Again, now we have something to go on. If the product is to be redesigned correctly, we must understand how the crack propagated and what stresses acted on the material. This influences the scope and thus client satisfaction.

The part is cracked, but the discoloration indicates that corrosion may have been a contributing factor to the early demise of the tank. Also, the client has told us they suspect corrosion is the problem so we want to confirm or rule out that suspicion.

We will need more information about the composition of the salty water in contact with the tank as well as what stress levels were present.

WHAT PROTOCOL WAS FOLLOWED?

Negotiated Activities

- Visual Exam
- Scanning Electron Microscopy with EDS microchemical analysis
- Optical Metallography
- Bulk composition analysis

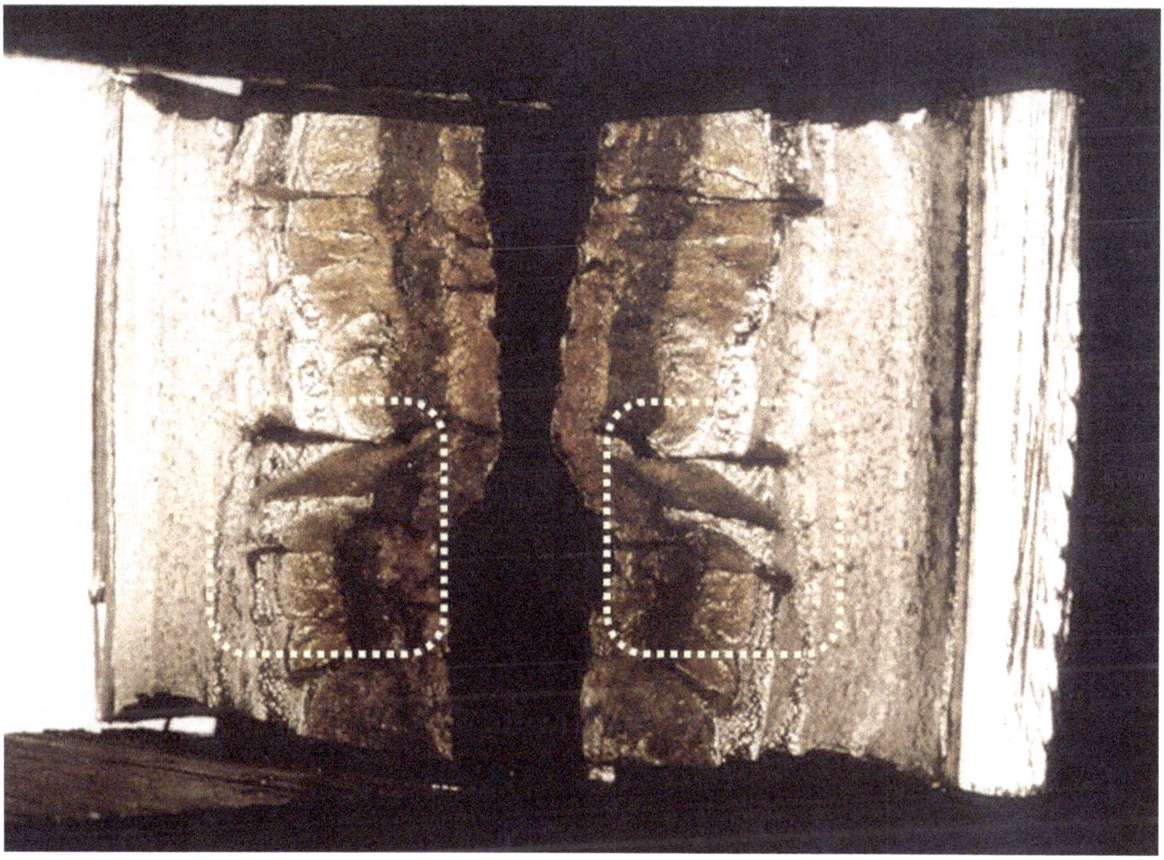

***Figure 7.2:** A portion of the stainless steel tank body is shown. The inner edges of the opened crack are to the left and right of the black empty space in the center. The far left and right edges depict the outer surface of the tank. The crack arrested before reaching the outer surface, and the remaining thickness was cut with a band saw. The encircled features are highly anomalous for fatigue, but make sense for a corrosion enhanced cracking process.*

PUSHBACK

I first thought this was a corrosion-enhanced fatigue crack due to the stripes on the crack surface. But some colleagues challenged that final conclusion. So, I performed a more careful examination. During proofreading of this manuscript, 15 years later, I realized my colleagues were correct. It was not fatigue. Note the encircled features.

WHAT DID WE FIGURE OUT?

Even though there were cyclic stresses on this assembly, with more careful examination, the stripes do not really look like fatigue beach-marks. The color within each band is too uniform. It's more likely the different colors are the result of changes in the water composition at the crack-tip over time; and the crack steadily progressed as a result of the stress created by the pressurization of the tank.

Metallography revealed corrosion pits covered with corrosion product at many of the crack origins along the plate inside surface.

FINAL FEEDBACK

With the client's suspicions confirmed, they decided to use a more corrosion-resistant alloy going forward. Stopping the formation of the pits due to the corrosion reduces the effective local-stress level. It is expected that this will prevent cracking.

For the replacement equipment, the client selected a material with significantly better pitting resistance in the specified environment.

7.4 CASE STUDY: Welded Industrial Fan Wheel, Low Carbon Steel

Client: Why did it break?

Failure Analyst: The stress exceeded the strength! What else would you like to know?

Client: When this ventilation fan broke, a chunk of steel tore itself off of the fan and exited the ducts. The chunk narrowly missed hitting one of our employees. We have four more of these in service. Aside from the safety concern, we now have reduced our production because we can't operate our wood-cutting operation without a fan to collect the sawdust.

Failure Analyst: Do you have any other background information you can provide?

Client: Well before the catastrophic crack with the liberation of a dangerous piece of flying steel, the fan developed an unusual vibration. But the vibration analyst could not find any problems, so they kept operating. About one week before the final separation, the vibrations became more severe and they decided to cut back on operating hours. But the final crack happened without additional noticeable warning (the vibrations did not become yet more severe).

DISCUSSION: Note that this design was not so easy to subject to an intuitive stress analysis. For a propellor, the stress is highest at the location where the blade joins the hub.

The linear velocity of the material is higher at the outside edge of this fan than the inside. Would that make the hoop stresses higher at the outer edge? The vibration analyst told the client that the crack started at the outer diameter. But he was incorrect. So why did the crack start at the inside edge?

The two important key factors affecting scope were: a crack initiation at the inside diameter and the desire to calculate the crack propagation rate. Analysis of the crack propagation rate was needed to epistemologically confirm the discoveries in previous activities and recommend an inspection frequency. Manufacturing was halted during the investigation.

WHAT PROTOCOL WAS FOLLOWED?

Negotiated Activities

- Visual Exam
- Visit to the fabrication shop to review documents
- Optical Metallography including weld cross-sections, as the crack was found to

have initiated at the inside diameter collar to plate weld

- Analysis of crack propagation rate was based on operating history and visual examination of beach marks

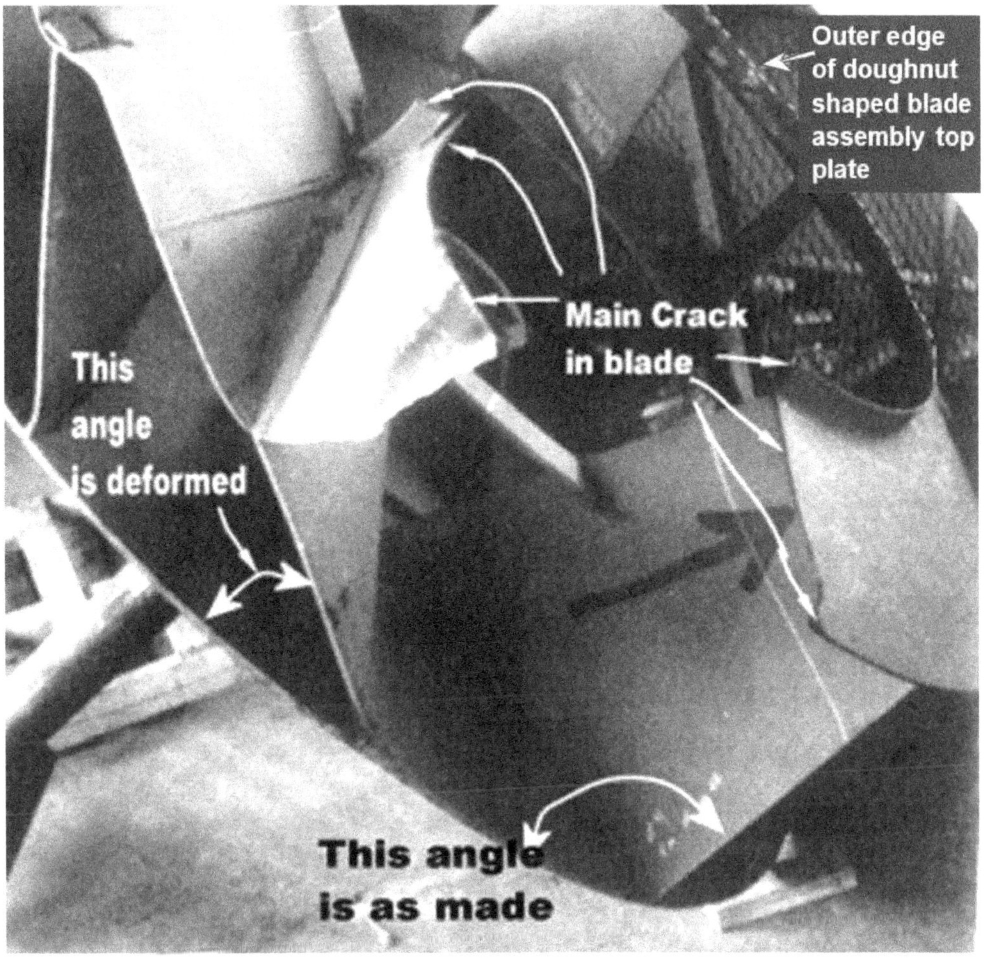

Figure 7.3: *Fan Wheel Weldment*

Note the missing piece of material at the outer diameter of the fan (left side of the crack).

Additional details about this particular broken fan are provided in Chapter 9, Figures 9.8A-C.

WHAT DID WE FIGURE OUT?

The crack initiated because the weld was very poor and concentrated the stress into approximately 1 mm of weld length. Normally the stress would have been distributed over several centimeters of weld length.

The weld had no penetration and also suffered from lack of fusion. The welder, who was not required to maintain certification for this type of work, probably had a bad day. Maybe the welder was upset about having had a fight with their partner the night before!

Especially at the end-of-service life when the fan was being turned-on and used for only one shift per day, it was easy to correlate the beach-mark growth to the hours of operation. The analysis allowed us to conclude that THIS fan likely started cracking the day it was placed in service.

While the application in question was not a pressure vessel, or a highway bridge, or some other structure that is usually regulated by government, the failure had safety-related consequences. For this type of fabrication shop, there might be intermediate level actions that they could take in the future. One such action included: magnetic particle inspection of some of the welds.

At least for the replacement fan, this would perhaps give the end user of the fan a little more confidence in the assembly. The client decided to continue monitoring the remaining fans for excess vibrations.

7.5 CASE STUDY: Part Failed During Production Quality Control Test

Client: Why did this break?

Failure Analyst: The stress exceeded the strength! What else do you want to know?

Client: The part is allowed to deform during our quality control check, but it is not supposed to break. Also, the load attained before the crack happened was below requirements. This is a critical safety part for an automotive application.

DISCUSSION: Multiple parts had failed this production inspection test, but the test caught the production problem before any parts had been shipped to end users. The quality system was vindicated and the manufacturing process was adjusted. This was a big project that went on for months because of the criticality and complexity. To protect client confidentiality, we are sharing only limited details.

The reason that I like to include the image (below) is there is evidence of a long, slightly bent but square edge (thus macro-brittle) crack going from left to right as well as a shorter crack that is at a 45-degree angle (thus macro-ductile) to the plane of the material.

Also, there are shear lips at the end of the short crack that are also at 45-degree angles to both the plane of the material and the first 45-degree angle features. These features allow us to understand the order in which these three cracks formed! We have shear lips on shear lips!

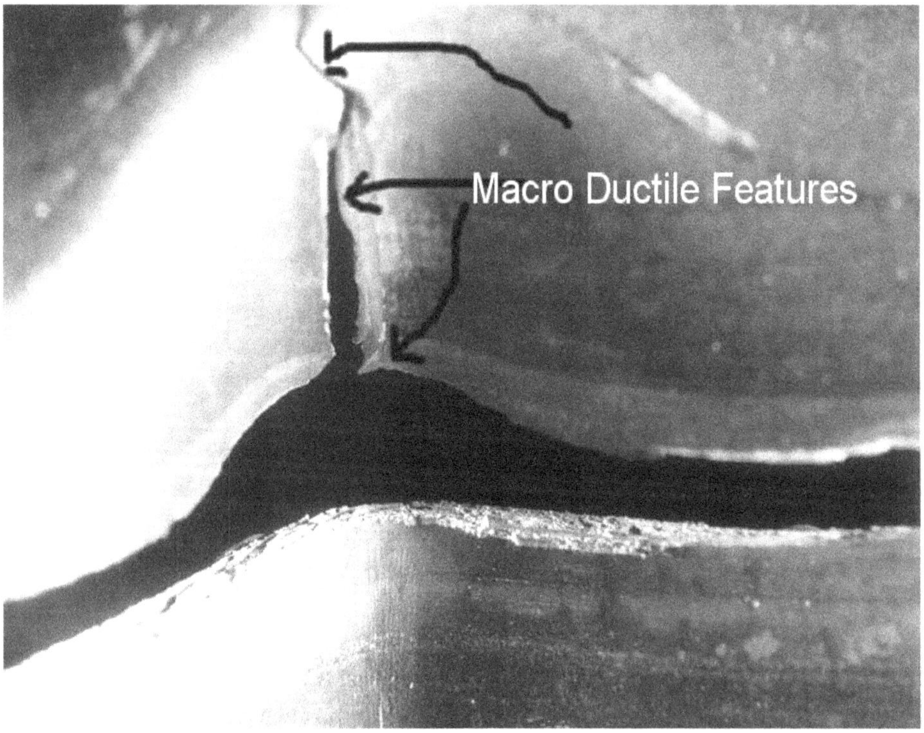

Figure 7.4: *Crumpled sheet metal component with multiple cracks*

WHAT PROTOCOL WAS FOLLOWED?

Negotiated Activities

- Visual Exam
- Scanning Electron Microscopy
- A visit to the steel mill in which the material was made in order to view the production process
- Optical Metallography
- Micro-indentation hardness

DISCUSSION: Due to the sensitive nature of this investigation, that is all that may be said! Sometimes that happens in failure analysis work!

I once worked on an investigation related to a building that was under construction. After the workers went home, the building collapsed. It was a very interesting scene, but I was not allowed to use any of the photos to illustrate the points in an article that I was writing at the time.

7.6 CASE STUDY: Shaft Fracture, Manufacturing Machinery

Client: Why did this break?

Failure Analyst: The stress exceeded the strength! What else do you want to know?

Client: This machine is only three months old! We bought it from Europe and the company wants to charge us \$20,000^{00} for a new shaft. They also say it will take three months to get the replacement to us! We want to make our own! We need to understand if there is something wrong with the machine! See Figure 7.5 (below).

Failure Analyst: This crack looks suspicious. It appears to be due to reversed bending stresses rather than typical rotating bending stresses. Are you sure the shaft was aligned properly?

Client: We'll check the alignment.

Failure Analyst: We can determine the composition of the steel that they used and we can perform hardness profiles and strength tests so you can specify the replacement material.

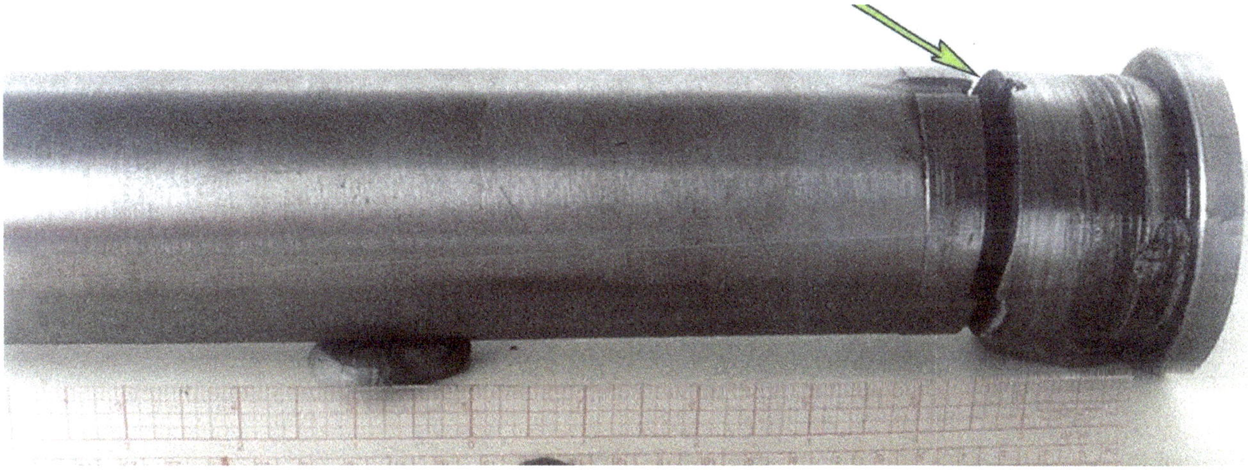

Figure 7.5: *Cracks initiated at both ends of the through hole, which was positioned at a diameter change. The people who machined the shaft did not keep very good control of the surface finish of their products. This is clearly poor workmanship. So much for the supposedly great quality of European craftspeople.*

WHAT PROTOCOL WAS FOLLOWED?

Negotiated Activities

- Visual Exam
- Rockwell Hardness Tests
- Bulk Composition Analysis

DISCUSSION: The damage to the shaft and a \$20,000^{00} tab for a replacement part with a three-month lead time drove the decision to hire me. The project considerations were to specify the right material for the service condition.

This was a minimalistic project. And the vendor's price for a replacement part was outrageous. This is a beautiful example of the power of the visual examination and an understanding of macro-fractography. I could clearly see the shaft was revealing inconsistent loading geometry compared to the known operating conditions: it had reverse bending fracture features. They were in a bind. They were hurting with production being down and I believed I could get them up and running. They were willing to take the limited risk they saw.

WHAT DID WE FIGURE OUT?

The cracks started at opposite ends of a cross drilled-hole. The center of the hole was aligned with a small step (diameter change) which was roughly machined. Thus, without having information on the entire shaft assembly, bearing positions, etc., we can still note that the initiation of the cracks at a diameter change and through-hole is in line with the crack being at least nominally at the expected position. See Figure 7.6.

The proposed scope for the job was acceptable for producing client results. Thank you, macro-fractography and loading geometry knowledge! We discussed the machine operation: the shaft was subject to rotating bending and not reversed bending. Understanding the machine operation and thus confirming the loading geometry was important; and gathering data from the strength, hardness and material composition tests for specifying the new steel helped the customer regain control.

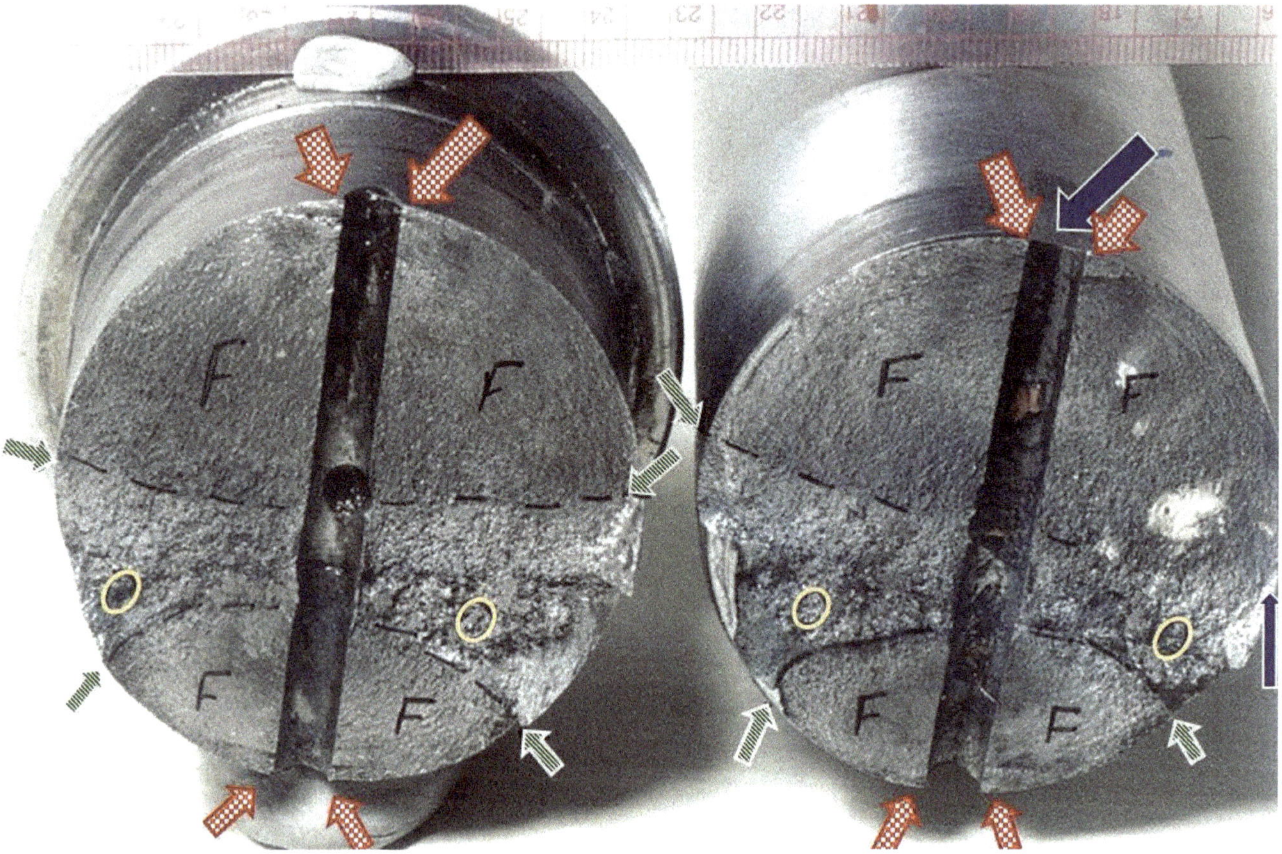

Figure 7.6: *Rotating bending fractures rarely have such symmetrical fracture surface features. "F" indicates the fatigue area and "O" indicates the monotonic overload (final separation).* ***Note*** *that there are few to no visible beach marks. There are some visible ratchet marks on the inside diameter surface of the hole (above the red arrows at the bottom of the image).*

FINAL FEEDBACK

The client did find the source of misalignment that was bending the part back and forth, and the odd shape of the beach marks was not a fluke.

They made sure to correct the alignment issue before installing the new shaft. The machine supplier should have paid them for all of their trouble!

7.7 CASE STUDY: Stainless Steel Pipe Fittings from Chemical Process Plant

Client: Why are these pipes leaking?

Failure Analyst: We'll have to determine if this is corrosion damage or cracking damage. Here are some questions I likely asked, but it has been too long and I don't remember the answers.

Failure Analyst: What was inside the pipes?

Failure Analyst: How long have they been in service?

Failure Analyst: How many of these leaking fittings do you have?

Failure Analyst: What are your options for repair or replacement?

Figure 7.7: *Several stainless steel pipe fittings were found leaking in service. I cut strips out of the full circumference specimens. It was possible to break the cracks open by hand revealing the crack surfaces.*

DISCUSSION: The scope was negotiated based on the initial assessment that leaking was due to corrosion or cracking damage.

WHAT PROTOCOL WAS FOLLOWED?

Negotiated Activities

- Cutting of strips from fittings
- Visual Exam
- Scanning Electron Microscopy with EDS microchemical analysis
- Optical Metallography
- Bulk composition analysis

WHAT DID WE FIGURE OUT?

After small strips were cut out of the full-circumference specimens of several fittings, I bent the pipes outwards and away from the flanges, by hand, to reveal the crack surfaces.

The damage was found to be stress corrosion cracking in the weld heat affected zones. The fairly thin walls would make NDT somewhat challenging on the pipes that remained in service.

7.8 CASE STUDY: Premature Tie-Rod Fractures

Client: Why did this break?

Failure Analyst: The stress exceeded the strength! What else do you want to know?

Client: These hydraulic press tie-rods failed prematurely in service. They were on an injection molding press. The failures continued to occur after replacement by the Original Equipment Manufacturer. It's interfering with our production. Can you explain why and make a recommendation?

DISCUSSION: The cost of the production interruptions meant that this company was open to a comprehensive investigation to understand the issues. They wanted to solve the problem.

Knowledge of the loading geometry, crack initiation, crack path[147], and the manufacturer provided Goodman diagram, helped me to negotiate critical elements of the project scope.

WHAT PROTOCOL WAS FOLLOWED?

Negotiated Activities

- Detailed Visual Exam
- Mechanical testing
- Metallography
- Bulk composition analysis
- Documentation Review

WHAT DID WE FIGURE OUT?

The cracks were found to be initiating at the known high stress location (at the diameter change), in the presence of abusive machining that created a thin layer of untempered martensite. This was revealed by metallography. This crack has some unusual features, which I cannot explain. See Figure 7.8.

7.9 CASE STUDY: MIG Welded Low Carbon Steel Auto Chassis Frame Structure

Client: This welded assembly failed a durability test that it was expected to pass. It's only four weeks before we are supposed to go into production. We think it is a weld problem, but we must be sure. Focus on the weld quality. We need this fast and correct!

DISCUSSION: The client was adamant, knowledgeable and under time pressure. So, I either had to work within these constraints or turn the job down. On occasion, I charge a client for their need to rush forward quickly. For most labs, the standard turn-time is 3 weeks. The rush fee in standard labs can vary from 1.5x to 3x the normal flat rate fee or the hourly rate.

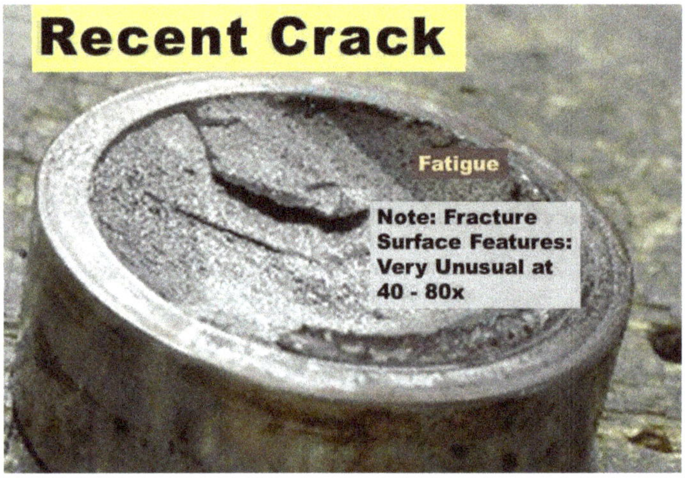

Figure 7.8: *This fatigue crack compromised over 90% of the cross-section prior to final separation.*

My lab performs only custom work, and usually more rapidly than others. But sometimes we must charge a rush fee. In those cases, I usually make our lab exclusively available to the rush client, and then request a minimum purchase order amount of about $5K (2016) and bill in 6-minute increments. Even though I dislike watching the clock, this is what I feel is justified, especially when delaying other work. I feel that rush fees must serve everyone's interest.

WHAT PROTOCOL WAS FOLLOWED?

Negotiated Activities

- Gathered background information and context (fracture surfaces were not visible)
- Weld Evaluation by cross-section at multiple locations

---

[147] This all happened at a glance during the initial contact with the client. Because I visited their plant, I was able to epistemologically confirm the loading geometry by viewing the replacement part in service. It is very valuable to understand loading geometry for negotiating project scope.

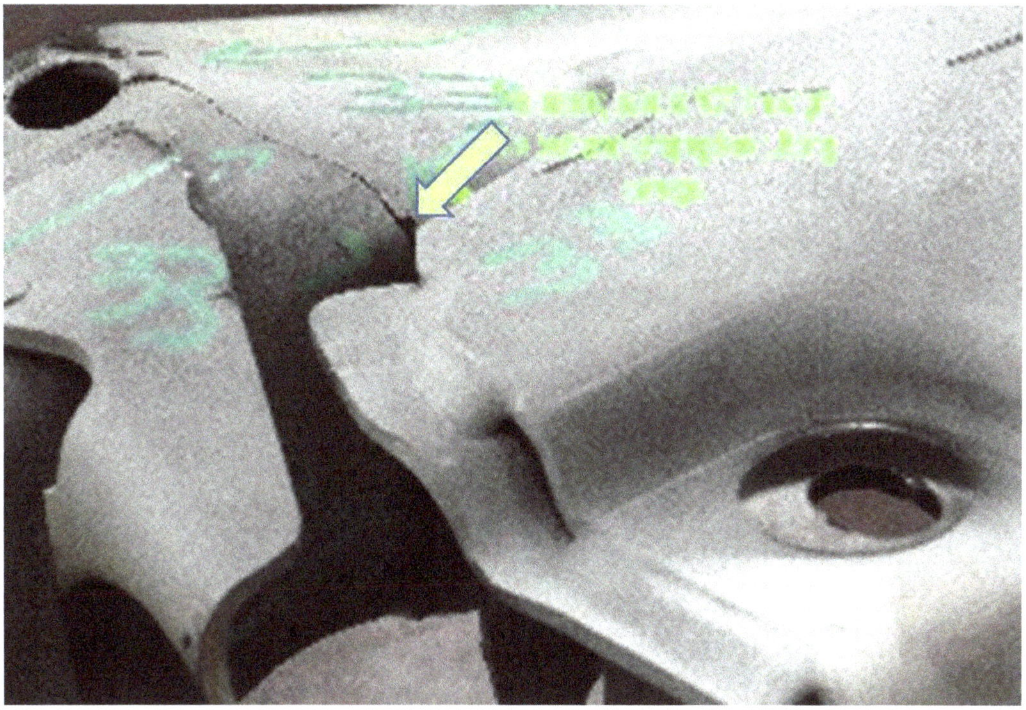

Figure 7.9: *Sheet metal weldment with multiple cracks revealed after durability test – green text is intentionally blurred for confidentiality.*

WHAT DID WE FIGURE OUT?

The investigation was confined to weld evaluations by cross-section at multiple locations. There were multiple cracks in the sheet metal weldment.

Had something other than weld quality been found to be a major contributing factor to the early demise of the structure, the pre-production team would have become very stressed-out! **My investigation confirmed their suspicions.**

7.10 Key Points on Negotiating the Scope

POINT: There are different types of strength in materials. The investigation, when possible, must attempt to produce strength data that reflect the conditions in which the fracture happened. Some choices of strength classifications are: tensile strength, yield strength, impact strength, fatigue strength, and strength in the presence of aggressive substances. Sometimes we need to consider something a little more complicated. This is damage tolerance. Damage tolerance allows us to determine the effect of a pre-existing discontinuity in the structure. One of the tools we use to evaluate these types of situations is called fracture mechanics. The key parameter is fracture toughness – not a simple strength value. Sometimes the visual inspection will reveal the presence of a pre-existing discontinuity and a knowledgeable fracture specialist will alert the stakeholders to the potential benefits of fracture mechanics data. All of these considerations will affect your negotiation of scope.

POINT: The protocol for a failure analysis needs to be well thought out so that an appropriate amount of testing and evaluation will be performed to achieve the client's goals and needs.

POINT: The appropriate amount of testing varies with the complexity of the subject of the investigation and the consequences of both the failure that is being investigated and any subsequent failures that might follow.

POINT: For production parts that have a fairly long history of successful use, getting reference parts from older lots is extremely helpful. This allows the Failure Analyst to evaluate any changes in composition or other characteristics of the parts.

POINT: The analyst's needs are as important as the client's. Here's an enlightening case study that we touched on in Chapter 4.

> *An incorrectly written request-for-quote (RFQ) ruined the budget for my project. I created a project scope based on information that turned out to be wrong. The client requested surface composition analysis by EDS on two sets of five parts. It was during an economic downturn so I quoted it low wanting to get some work rather than no work.*
>
> *When the client arrived, she had nine unique specimens. This was a totally different project which took much longer. I was stuck with the quoted price.*
>
> *Testing was required on each of four sets of parts, and two potential contaminants, and three additional test spots on assemblies that had to be cut to fit them in the electron microscope.*
>
> *The work actually went okay. But the report did not. It's a lot more time consuming to explain seven different types of specimens than two. The data differed widely from group-to-group and it took me a lot longer to explain than I had anticipated when the plan was to cover two groups in detail. The ultimate happiness of either the client or the analyst was in jeopardy, as the two groups to analyze turned out to be nine.*
>
> *Had I done my usual interrogation of the client before accepting the work, this would not have happened. But business was very slow and I wanted to be doing something rather than earning no income. This project is a prime example of scope creep, or change of definition, communication breakdown and poorly applied logic.*

POINT: In order for the Failure Analyst to recommend an appropriate test plan, the situation surrounding the event must be explained by those who are requesting the work.

There is an old wisdom story about five blind men who go to a nearby village to experience the excitement associated with a visiting elephant. None of them know what an elephant is.

The first blind man (it's always a man in these old stories!) touches the side of the elephant and proclaims "I know what an elephant is. It's like a wall!"

Then the second blind man steps up and feels the elephant's trunk. "No," he cries. "An elephant is very similar to a snake!"

The patient elephant is now approached by the third blind man, who grasps the tail and exclaims, "This elephant is surely just a rope."

The fourth blind man moves forward and touches the very patient elephant's ear between his hands. "A fan! It's a fan," he informs the bystanders.

Finally, the fifth blind man grasps the leg of the extremely patient elephant and states: "You are all wrong. An elephant is just like a tree trunk!"

Figure 7.10: *Photograph taken by author on a trip to Assam, India*

But we are not done with the elephant story. A great Failure Analyst knows, of course, an elephant is more than its body parts. An elephant is a member of a social group. Any specific elephant may be a mother, a father, a friend, or a nursing baby. All elephants exist in an environment and have needs to roam, drink water and bathe, and eat plants. In other words, we need to inspect an elephant in its context. See Figure

7.10, from my February 2020 trip to Kaziranga National Park in Assam, India.

We can learn more about elephants by studying them in their environment. It's a very parallel situation when we get the background information for the failure events we study.

If we know that we are describing an elephant, we need to carefully look at the mental picture of the elephant that we've created and determine that all of the details we have uncovered are consistent with what would be associated with an elephant; not a wall, snake, rope, fan, or tree trunk.

If we can tell a realistic story that ties all of the details together, then we can be satisfied that we have done our work well. If the failure was something related to human health or safety, environmental integrity, or even a costly mechanical problem that could affect the viability of a business and the livelihoods of those who work there, it's important to do enough work that the typical person of sound mental health won't keep having bad dreams about additional failures or additional unforeseen consequences of the original failure. If we tell such a story, we justifiably can welcome the warm, fuzzy feeling we get as we drift-off to sleep after a good day of work well-done.

Doing enough work to generate this type of Data-Based Warm Fuzzy Feeling is my criterion for calling my work failure analysis, rather than damage characterization, mechanical testing or some other more limited type of analytical activity.

POINT: Sometimes even a detailed and prolonged failure investigation fails to answer all of our questions. Here is a funny quote, attributed to Anonymous that has rung true for me over the years:

We have not answered all of your questions. We have not addressed all of your concerns. In some ways, we feel that we are still as confused as ever. However, we are confused at a higher level, about more important things!

Getting the project scope right will enhance your odds of getting the failure investigation correct – of being confused at a higher level and about more important things. The object is to ensure that the client's big picture priorities are aligned with the negotiated activities in your proposal. Think about this as you read below. This is very important.

Why did the screw break? …Because it wasn't heat treated properly. Why was an improperly heat treated screw used? Because the people who specified the screw did not understand that they needed to also specify the testing that would ensure that an improperly heat treated screw would not be used. Therefore, there was no testing performed. The heat treater was allowed to certify that they made the part the same way they always make it.

So, if the company that bought the untested screw now understands that there are different levels of quality control systems at different heat treaters, and the next time they procure a heat-treated part, they negotiate the specific testing that they think will be needed to ensure the desired level of product durability, then they have become confused at a higher level, about more useful things!

POINT: Sometimes, during the original negotiation, clients specify that a recurrence prevention suggestion must be provided. After the investigation the Failure Analyst discovers there is no pathway to an economical fix. Sometimes the only way forward is to make peace with living with the problem. If there are no major safety concerns, this is a legitimate business decision.

FOOD FOR THOUGHT: Lesson from the point of view of Quantum Physics: If you don't *look-for* it you won't see it! If you are *looking-for* waves, you won't see particles...

If we don't specify the test or evaluation when we define the scope, we may lose the opportunity to do that work.

7.11 Detailed Investigation Planning

This is a good time to review three different methods for defining investigation purposes to determine the source of a failure/damage. All three methods might be necessary to perform in complex situations.

It's important to know all three methods because reviewing all of the ways failure sources are evaluated often helps Failure Analysts notice important insights revealing critical information.

7.11.1 Categorization Scheme 1 for Sources of Failures

What I call "Scheme 1" is laid out in Figure 7.11. Where do communication failures fit into this scheme? The depicted overlaps acknowledge some situations are difficult to classify.

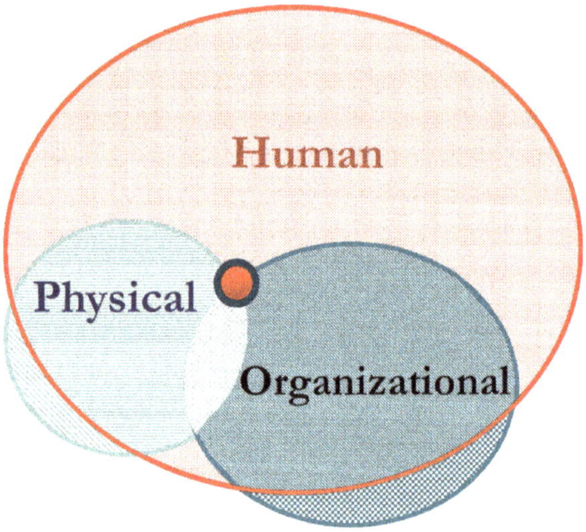

Figure 7.11: *Scheme 1 for understanding failures and planning investigations – human influence dominates. Read or study Bernardo Kastrup (The Idea of the World), Edward Deming, Kanada, and Ramana Maharshi to understand human influence on physical causes.*

Overlapping ovals represent different ways to define sources of failures. First and most obvious, we have physical causes. I'm going to refer back to a previous example (the drinking water example in Chapter 4 – Complex Systems). As I am writing this, the lead contamination crisis of the drinking water distribution system for the City of Flint, Michigan USA has come to international attention. So, let's look at the overall situation.

> PHYSICAL CAUSES EXAMPLE: *The City of Flint was experiencing financial problems for quite some time. They could not pay their bills. The State of Michigan stepped in, took away the powers of the elected mayor, and put in place an Emergency Manager. The Emergency Manager studied the finances of the city and decided that they could save $5,000,000^{00} annually by switching to water from the Flint River instead of buying treated drinking water from the nearby City of Detroit.*
>
> *It turned out that the water from the Flint River was chemically aggressive. As soon as the new water source was tapped, it started to attack the oxide deposits that lined the pipes.*
>
> *An unknown number of pipes in the Flint area were made from lead. Additionally, many service lines leading off the mains and going to homes, schools and business had leaded brass fittings.*
>
> *Lead oxide began flaking off of the pipe interiors. Because the water flowed through the lead and leaded brass plumbing after it left the treatment plant, there was no way to filter-out the flakes. Rather, lead came out of faucets in Flint homes, kitchens, and bathrooms. It is unknown at this time how many children have lead poisoning. Citizen groups hired researchers from Virginia Tech University; the report stated they had never seen such high lead levels in drinking water anywhere in the world.*
>
> *So here we have arrived at the basic physical source of a failure in Flint's drinking water system. At first glance, it is a corrosion damage process. That is what the news reports stated. In fact, if you think back on the damage categories that we learned about in Chapter 5, this is happening because the lead oxide is being worn away (eroded) from the inner surfaces of the pipes. Yet, the pipes were not manufactured with oxide coatings. The coating formed as a result of corrosion (in this case, a beneficial surface oxidation) over many years. As long as the oxide remains in place – no one gets sick. Once the oxide is damaged, wear starts. We*

can see that this is a complex damage process, which may be called corrosion-erosion.

Properly maintaining modern water systems that are using lead pipes includes treating the water with chemicals that keep the lead oxide on the pipe surfaces in an adherent layer. Properly treated water does not carry the lead oxide into homes.

In many older water systems, solder joints may also be made with lead-based solder alloy. If this was the case in Flint, this may be another physical contribution to the problem.

But every failure has both physical and human sources. Let's continue with the Flint, Michigan water example to examine the interface between physical and human sources.

The previously excellent Flint economy was destroyed because of automotive globalization, which resulted in widespread poverty within the city. This destroyed the tax base, and eventually, Flint was bankrupt.

It is still not clear to me why the appointed emergency manager decided to adopt a bean-counter attitude[148] *and completely ignore the advice of the water scientists. Why did this emergency manager make such poor technical decisions?*

When I first heard about his actions, I assumed he was a racist white person because many of the affected children were poor and African American. I had originally fallen into the trap of thinking that the Republican governor appointed a crony; but I was wrong. I thought this, because most of the poor people in Flint were African American and probably did not vote for that governor, and let's face it; I was not a fan of this governor either!

But no, it turned out that the Emergency Manager was not a racist white person. He appeared black when I finally saw his photo. His biography shows he was a former president of a national organization of African-American City Managers. Clearly, he had excellent credentials and this was not a political hanger-on appointed to office without regard to qualifications. Therefore, for myself, I assign his actions to fall within the category of Human Factors. Bean-counters will be bean counters!

In closing this section on Scheme 1 Failure Sources, we've identified Physical and Human. But, wait! There is more to this story.

The water that was taken from the Flint River would not have attacked the piping had the water treatment been performed according to the general good practices and applicable laws and regulations of the State of Michigan and United States Federal Government agencies. The reason it was so aggressive is because the emergency manager refused to approve the $37^{\underline{00}}$ per day expense for a corrosion inhibitor required by law.

Is this due to the general ignorance or hostility, in some American demographic groups, towards science? Or of "bean-counters" toward anything other than "the bottom line"? This is a costly example of foolishness. It was a business decision which we may categorize as an organizational factor.

The destruction of the tax base and city's bankruptcy also fall within the third category: organizational factors.

We've now identified Physical, Human, and Organizational as potential sources-of-failure categories. I would like to mention that most people use the term "cause" rather than "source" or "factor". But we'll soon clarify why the word "cause" must be used with a high degree of caution, if at all.

148 The term, "bean-counter" may have been first used in print by Hans Jakob Christoffel von Grimmelshausen in *Simplicissimus'* in 1668 according to Gary Martin, the author of the Phrasefinder website (founded in 1995). His URL is https://www.phrases.org.uk/meanings/bean-counter.html. Basically, the term stands for a precise, narrowly-focused, miserly accountant who is more interested in saving pennies than considering anything that would shift their focus (such as the big picture). I believe "the bean-counter methodology" is almost universally hated by engineers and scientists.

The vast majority of the investigations I have participated in focused on physical sources. A few investigations enabled new approaches to design, which would be a way to address organizational sources. Those few investigations were further limited to the Subject Product. Companies did not undertake larger changes in how they conduct their affairs. Many companies leave behind opportunities for improvement.

> *Most industrial component failures don't seem to lead to introspection, even when the failures lead to bankruptcy. Years ago, when I was a member of the lively Yahoo Group on Root Cause Analysis, we had a discussion about how the personality types most frequently found in company leaders rarely engage in introspection. They find it very hard to even formulate a question about their personal contribution to the problem. It's much easier to blame outside influences.*[149]

7.11.2 Categorization Scheme 2 for Sources of Failures

What I call "Scheme 2" is laid out in Figure 7.12. Scheme 2 offers another method to specify a failure investigation and ask questions that focus information in useful ways.

Raw material issues and problems with the manufacturing processes are usually at the top of a client's list when they ask investigators to *Look-For.* Of course, they are also open to learning about service condition problems.

It's this traditional, dualistic industry mindset that prevailed for decades: Was it defective or was it abused? However, framing the question this way leaves out many chances for product improvement.

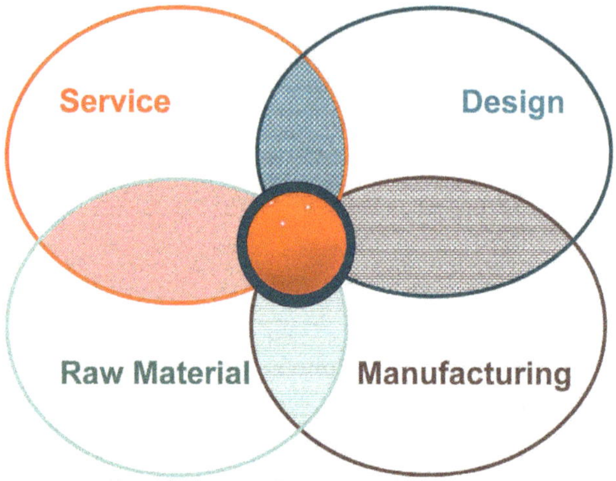

Figure 7.12: *A classical way to think about sources of failure*

One reason the dualistic mindset has hung on for so long is because experts responsible for the design and those requesting the investigative work often are at odds in motivation. Designers may not think to ask for help critiquing the design. They likely think that the design process they used and the result of that process are both fine!

The way I frame it for companies so they are more open to my help for design is to deal with design authorities. I suggest tightening material or process specifications. Or, I suggest bringing material or process specifications in line with standard industry practice. This seems to make experts more interested in soliciting help and refrains from pointing fingers.

Alternatively, I offer to help them clarify material specifications. Often my clients are design and quality engineers and they are generally open to receiving advice on materials specifications from a materials engineer.

Of course, there are times no one can get around the fact that geometric aspects of the design have a serious problem. You may find it hard to believe there are still engineering prints being released right now that have a sharp, square fillet radius because a detailer overlooked the need to specify a value for the radius.

> *In 2020, a client brought a part that had a fillet radius that was actually specified as "sharp".*

---

149 This is perhaps a manifestation of The Fundamental Attribution Error: a feature of human thought that allows us to reason that when we do undesirable things, it's because we were in a tough spot, but when others do something bad, it's because they are bad. | *Tools of Critical Thinking: Metathoughts for Psychology* by David A. Levy, Chapter XV

When a fillet radius is sharper than needed, the design concentrates stress to exceed the inherent value naturally needed for a specific application.

Some day in the future, we might see all education on component design efforts emphasize the benefits of using the largest fillet radius possible. In that way, stress is reduced; which makes it less likely that the stress will exceed the strength!

7.11.3 Categorization Scheme 3 for Sources of Failures

What I call "Scheme 3" is laid out in Figure 7.13 (below). Scheme 3 is perhaps the most useful method to specify a failure investigation and ask questions that focus information in pertinent ways. Even if you, the analyst, do not create a diagram for your specific project, I believe it would be helpful to think about how such a diagram might look during scope negotiation and when writing the final report.

Figure 7.13 shows two types of factors: those mostly inside the outer-red cause-ring and those mostly outside of it. This is a way of showing how influential we think specific factors might be.

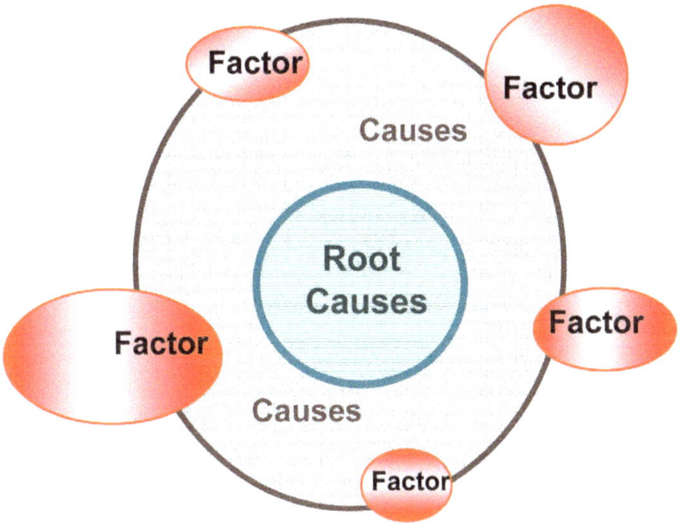

Figure 7.13: *A useful aid to structuring the scope negotiation, the data interpretation, and communication processes*

This is a useful as a tool for reflection on key factors that influenced the failure and that are important to discuss in the final report. I have never put this graph in the client's report.

Often when I contemplate this diagram as preparation to writing the report, I put a factor for which I've found no evidence on the perimeter. See the ring in Figure 7.13 at one o'clock. I put this factor entirely outside of the cause-ring because it was considered and subsequently ruled out. Figure 13 is a way of asking "What are we going to look for?" It's a way of asking "What did we find?" as we are evaluating the data writing the report.

In planning our investigation, we often start with beliefs and uncertainty. For example, if a threaded fastener failure was thought by the client to have been influenced by improper installation, resulting in insufficient clamp load, this is where we may start. In figure 7.13 this is indicated by a factor that is large and overlapping into the cause-ring. But if, during the investigation, this was ruled-out, we could indicate that "Clamp Load" had been considered, but then ruled out. In this manner, those reviewing the final report don't conclude that the investigation was incomplete. Be sure you are negotiating for the resources to be able to conduct the investigation and give the client what they need with an appropriate level of confidence. I am not necessarily a very disciplined person, but I am disciplined about making sure I have a record of what I agreed to do, and disciplined about making sure I report my findings, positive or negative, for the factors we agreed to examine.

Damage analysis is foundational to the failure analysis work we are learning to perform in this book. The goal of any damage analysis includes revealing a sequence of unusual events that ended with the demise of the Subject. This is how we define the causes of the damage.

For example, a rolling-mill roll may have spalled, which led to bumps being formed on a strip of product. Then, when the strip was squeezed through the next set of rolls, the bumps were flattened and created a second layer over the small affected areas. As the strip gets longer and thinner through every subsequent set of rolls, the double layer areas get longer and longer.

Now this flat material is used to make deep drawn cups; the double-layers may split apart and thin spots are revealed at those positions. If the minor double-layer ended up on the inside of the deep drawn cup, it might not be seen by any inspection process. Alternatively, the double-layers may remain close to each other and be difficult to see. In either case, if the deep drawn part is pressurized, the thin spot could be a critical weakness and result in a catastrophic damage event. This may be determined to be a failure. The links in the chain of causes have thus been identified. See images of features of two unrelated parts, Figures 7.14 and 7.15.

Figure 7.14: *This image shows a deep drawn cup with u-shaped feature. This is the edge of a small area that has two layers. The outer layer has not visibly separated from the underlying material. This is a raw material discontinuity called a lamination or lamo.*

In the thinner area of Figure 7.15 (next page), the stress level will be higher than anticipated based on the nominal wall thickness. If the safety factor is insufficient, the part could deform or break. This was a stainless steel part that was annealed after cold forming. The grains are equiaxed in this plane.

If we define Root Cause or a cause as an actionable recommendation that reduces the chances of repeat failures, then not all failure analysis work will reveal a true root cause. That's because in many cases, management will not allot resources for such a scope.

A cause that might be proposed in the above lamination case might be *inadequate inspection.* But if the company involved has a culture that doesn't require inspection for the subject types of assemblies, then this is not actionable, so it may not be very helpful to designate inadequate inspection as a cause.

This is why it is so important to collect information to help you scope a project. You need to know your customer. What were the consequences of the failure? If there were no injuries, the costs of a small number of events may not appear to justify the added expenses of additional inspection. These are business decisions, not strictly technical ones.

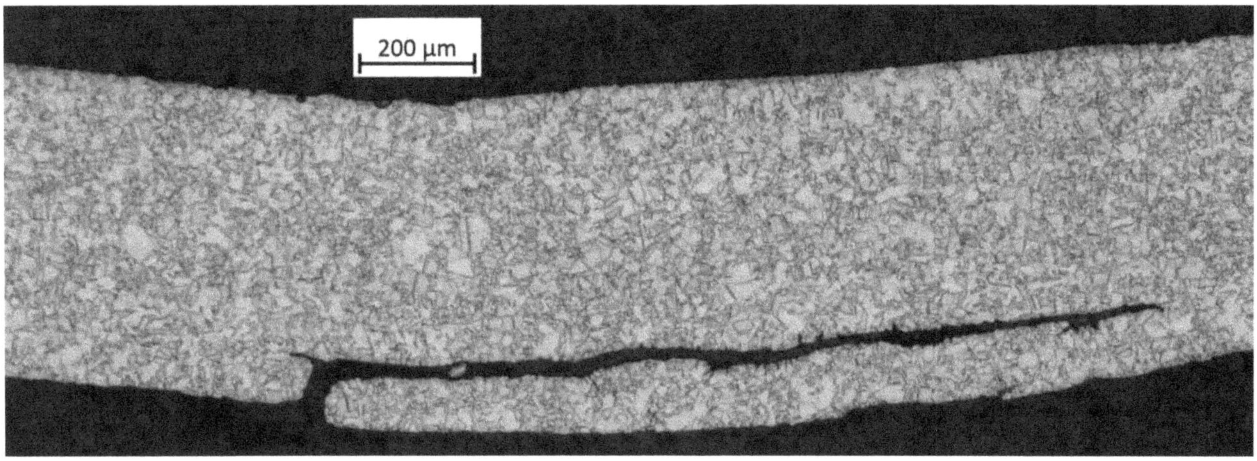

Figure 7.15: *A transverse cross-section through a lamination in a deep drawn part shows its nature. It is clear that if this material is stressed, there is less continuous material available to transmit a circumferential hoop stress.*

On the other hand, if the management team decides there's no need to change inspection procedures and a subsequent lot of material arrives with more severe but still undetected lamination discontinuities, and human injuries result - or perhaps catastrophic property damage occurs - then the company could come under a barrage of criticism and other consequences.

Now we can look at the factors that might have influenced the damage sequence. This could include a decision to make the part more lightweight, by using thinner material. Perhaps there was a change in lubricant in the deep drawing operation that made it harder to notice surface irregularities. Perhaps the steel mill had been persuaded to perform a new type of automated inspection, which did not perform as well as anticipated. Perhaps the usual inspector on the stamping press line had just retired and an inexperienced person was performing the inspections. We can go on-and-on with various speculations.

In a real investigation, the scope empowers or disempowers the analyst. The scope defines what the analyst can *Look-At* or, alternatively, what the analyst cannot *Look-At*. Access to background data and dialogue with the client in an open ended way is critical. Certainly, it is possible that additional factors might also be revealed during the course of the investigation. Also, these may be presented in project documentation.

7.12 Common Information Desired from a Fracture Analysis

In closing this chapter, I'd like to share some questions commonly asked by people specifically seeking fracture analysis services. Even if they don't ask these questions, it can be helpful for us to make sure they are not the real questions behind the spoken or emailed ones.

> **These questions, why did "THIS ONE" break or why did "IT BREAK NOW" have analogues for the other damage categories.**
>
> **Why did the deformation happen now?**
>
> **Why did this part corrode/wear out so soon?**

7.12.1 Better and Clearer than "Why did it break?" is "Why did this one break?"

Usually, the implication is that there are scores, hundreds, thousands, or millions of these components out in the world and all are successfully doing their thing. This is a better question because it focuses on a specific line of inquiry and provides instant suggestions for control or reference specimens. Also, it can provide guidance on what to look for in the historical record relating to the production of the Subject Component or those from the same lot.

Another implication is that this is a new design, but nominally similar designs are out in the field and doing well. Again, understanding the specific concern allows the investigation to proceed in a more direct manner.

Ultimately, the answer is still, "Stress exceeded the strength." However, it's a far easier task to find a deviation from a known good design and manufacturing process than it is to definitively determine the stresses and strengths in a quantitative way in order to numerically demonstrate that the stress exceeded the strength.

> **HOW IS YIELD STRENGTH DEFINED?**
>
> **We need to remember that the 0.2% offset yield strength is an arbitrary number that people agreed to use for standard material certification purposes. It is not a universal truth about the objective nature of the material behavior itself.**

7.12.2 Better and Clearer than "Why did it break?" is "Why did it break now?"

Usually, the implication is that the failure was premature or happened during what was considered light duty service. Because this is also a good question, it might lead us, relatively quickly, to look in greater detail at fracture processes that are cumulative. Cumulative fracture processes might include: fatigue or creep.

This type of investigation is often full of opportunities to increase understanding and awareness of subtle interacting factors in the life of the component. Here is an opportunity to contemplate complexity!

At one level, a true answer is, "Stress exceeded the strength." But in this case, we will most likely focus on the service history in detail. Often the service history leaves evidence of deviations from normal operations. These may be subtle, and scrutiny of the service history may allow the investigation team to obtain or rule out epistemological confirmation of significant suspected unusual events.

It's important to remember or investigate fatigue cracks (especially with low-cycle fatigue in steel parts) in view of the "S-N" curve. This refers to the stress level ("S") and the number of cycles ("N"). If the stress is near or at the tensile strength, then "N" is 1. In other words, only one cycle is required to break the component.

If stress is lower, then more cycles are required to break the part. Once the stress gets below 80% or 70% of the tensile strength, the number of cycles might be in the hundreds.

But at these high stress levels, when we are testing parts at given repeated force or deformation levels, all of the N values will not be identical. Even up to a hundred thousand or more cycles, the range of individual "N" values for a given load may be wide.

We say that fracture is a stochastic process. It has randomness in it. A single grain is often where the crack initiates. Is that grain in a particularly unfavorable orientation? That part may be more short-lived than other, nominally identical parts. Each part has variation in thickness, width, hardness, and position and orientation of grain boundaries, etcetera. All of this can contribute to its overall durability.

> *It is common in the transportation industry to have life cycle requirements in the tens or hundreds of thousands of cycles. Suddenly, a part that has been routinely passing ongoing quality durability tests has a run of failures. What has changed? Sometimes, nothing. Or at least nothing that makes the part fall outside of its range of specified characteristics. What has changed? The manufacturer / designer ran out of luck!*
>
> *I consider these situations opportunities to help my clients clarify their specifications. For example, a hardness range may need to be tightened.*
>
> *Most people don't like being told that the root cause of their failure was that they ran out of luck.*

But the stochastic nature of fracture is an old story. Even Leonardo DaVinci experimented with the tensile strength of wires and found that the longer the wire, the lower the strength. It is more likely that a significant discontinuity will be present in a longer piece of wire.

We can think of this in terms of lot size. I have worked on assemblies that required sets of 10 parts to undergo durability testing of several hundred thousand cycles. The fifth time they ran the test, one of the ten parts failed. In this case the part was considered to be safety critical. I had to shut down my lab to all other clients for 3 days while we investigated. The conclusion was: they simply ran out of luck! We could not find anything "wrong" with the part. And we tried!

7.13 A Good Way to Lose Business – Being Too Specific

A final word of caution: do not be too forthcoming; rather be generic, in your initial communication and proposal with the client. Too much information could be a disservice to the client in setting them along a path they cannot properly or safely execute. Also, it can be a disservice to your own business, in that you could reveal critical knowledge for which you deserve compensation. This is a million dollar lesson. Take heed!

On Scope

Lesson From Quantum Physics

If you don't look for it you won't see it!

If we don't specify the test or evaluation when we define the scope, we may lose the opportunity to do that work.

CHAPTER 8

Fractography, Loading Geometries, & Mechanical Behavior: The 3 Act Method

Learning to read fractured parts

As an extra resource on the new fractography method, please write to Debbie Aliya at DaAliya@itothen.com or visit the following link (QR code) for a 20 minute video on this valuable topic. I encourage you to do this especially if you are new to these ideas. The video will help you get oriented.

8.1 Fractography – The Three Act Method

Often, people say Fractography is the study and documentation of *fracture surfaces*. In this book and in my work, I use a more comprehensive and three-dimensional definition.

A NEW DEFINITION: Fractography is the study and documentation of *the appearance of fractured parts*. In this chapter, I'll introduce and promote the use of a three-act method of fractography (macro, meso, and micro) which includes a separate (integrated) three-step method of macro-fractography. This new method is more powerful than the classical way the subject has been taught.

MACRO-FRACTOGRAPHY relates to the shape of the fragments. It tells us the loading geometry and allows us to assign the crack to the ductile or brittle categories.

MESO-FRACTOGRAPHY relates to the coloration and texture of the fracture surface. It tells us the crack direction, the environmental and loading conditions in which the crack grew.

MICRO-FRACTOGRAPHY relates to how the crack interacts with the microstructure. Pattern recognition plays a strong role in helping differentiate types of crack paths. Some people call this morphology or topology of the fracture surface.

Fractography exists to answer the question, "How did it break?" But what people usually want to know is "Why did it break?" And we already know the answer. It is always, "The (local) stress exceeded the (local) strength." Therefore, to more deeply understand, it behooves us to start with an analysis of the stresses. But stresses are invisible. So, how do we proceed?

We can imagine the loading geometry, which influences the stresses. What exactly does the term loading geometry" mean?

I'm sure you can imagine a rod being stretched or twisted, bent or squeezed; these are different types of forces or loads. I'd also bet you are imagining that twisting, bending and squeezing from the point of view of a person looking at the rod – watching it being stretched, etcetera. But loading geometry is more than the way the load is transmitted into the volume of the material.

Loading geometry also refers to the experience of the thing being stretched, bent, twisted, or squeezed. Failure investigators must learn to imagine *existing as the object being acted upon* to gain insight into the part's behavior.

The object itself doesn't have a nervous system. Even if it could feel our hands or the tools being used to transfer the loads or forces into the material, the points of contact are not necessarily or usually the most highly stressed locations. This will hopefully become clear soon, if it is not already.

NOTE: *These paragraphs are short. That is because each one contains a very important idea; each is worth thinking over for a few minutes.*

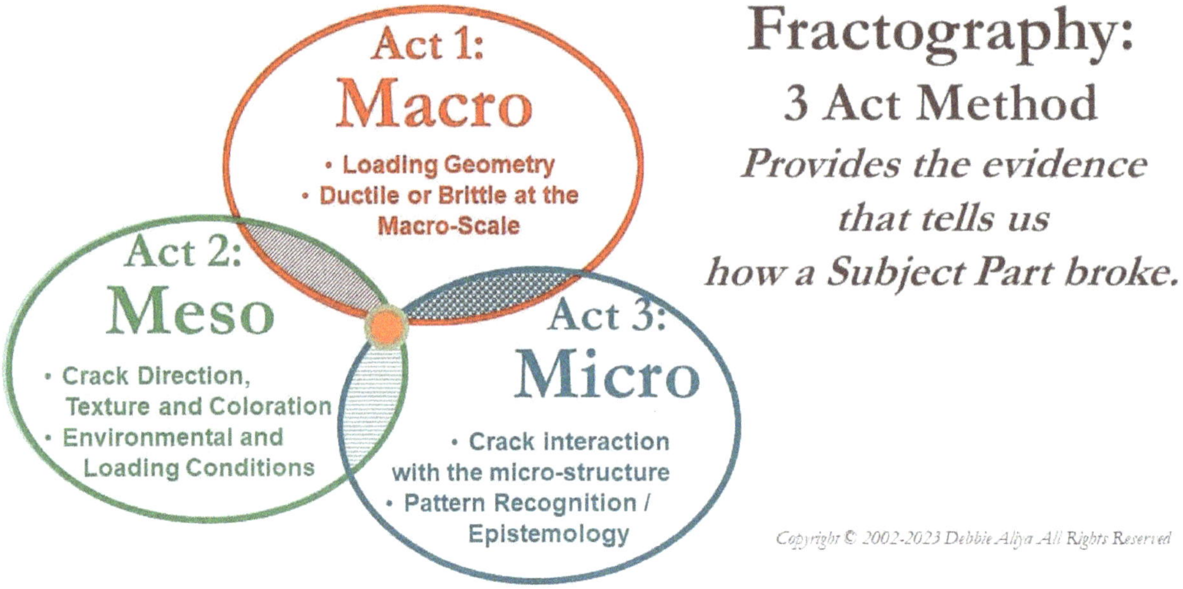

Figure 8.1: *The 3 Act Fractography Method provides the evidence to understand how a Subject Part broke. We start with "The Big Picture" (macro) and then move to subtle surface features, and finally drill down to the details (micro).*

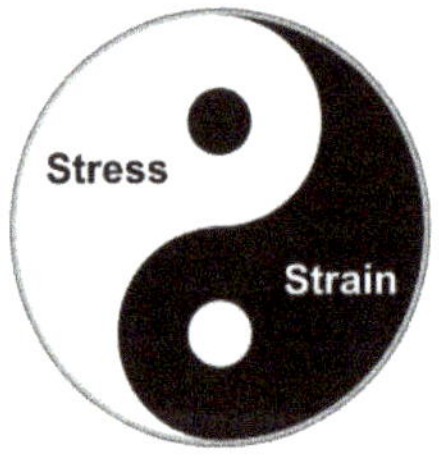

Figure 8.2: *The Yin Yang symbolizes the dynamic, interactive process experienced by Subject Parts.*

This image[150] contains dots of contrasting color, which remind us that the experience of the part results from an interactive process. Stress and strain are two ways of looking at the same component and serve different purposes for the engineer.

[150] Image by Klem; vector image was created with Inkscape by Klem and then manually edited by Mnmazur, Public Domain, https://commons.wikimedia.org/w/index.php?curid=3213322. The text, Stress and Strain, was added by Aliya.

The experience of stress is created as forces act within the volume of the material causing atoms to be displaced from their rest states. This may result from mechanical or thermal factors.

EXAMPLE: If a volume of material has an at-rest atomic spacing of 2 Angstroms, and as a result of carrying a load, some of the atoms are 2.0001 Angstroms apart while others are only 1.9999 Angstroms apart, this distortion is known as strain.

Stress is numerically related to the forces; strain is numerically related to the shape change; and the two are themselves inextricably entwined and entangled. Stress & strain and forces & shape change are two ways of looking at a single phenomenon. More detail on forces, stresses and their visualization will be discussed later in this chapter.

In a simply-shaped component, the loading geometry that resulted in fracture is often immediately discernible in the fragment shape *(at least to the owner of a knowledgeable mind!)*. Determining the orientation of the fracture surface with respect to the original part surface features is the first step in macro-fractography. *Take the time to contemplate this for a few moments.*

In some cases, when questioning which loading geometry created a specific crack orientation, you might not find a unique answer. In other words, there might be two (or occasionally even three) choices. Thus, we may need to go a step further along the decision tree to get our answer. We will need to classify the material behavior during the fracture event itself as either ductile or brittle. The final part of this first "act" of fractography is asking, "Is the crack at its expected location?"

One benefit of this method is that it separates ductile and brittle material classifications based on their behaviors during standard mechanical tests from ductile and brittle ***component*** behavior during crack processes in real service conditions.

This method also separates the behavior leading up to the crack event from that during the crack event, and that after the crack event.

The behavior leading up to and following the crack event both allow the fractographer to understand the material's capability for ductile behavior in potentially more favorable conditions than those actually driving the crack. The behavior during the crack event itself is more likely to be brittle. The classical way that ductile "material" was separated from brittle was based on the presence or absence of readily visible deformation, seen in a broken tensile bar. This is the method that almost all engineering students are familiar with. Post-test necked (pinched) coupons are considered ductile. Undeformed coupons, which usually have a crack perpendicular to the length of the coupon, are considered brittle. This is fine. However, when we extrapolate this categorization to components that have been loaded in an infinite number of conditions we run into trouble. For this reason, in this methodology, we are separating material capability in test from component behavior. It is the component behavior that we must identify in order to properly determine the loading geometry and then understand how it broke. Furthermore, we need to separate out three time frames:

1) The behavior leading up to the crack (elastic deformation only, or including permanent deformation)
2) The behavior during the crack event
3) The behavior after the crack is completely through the thickness and one other dimension (width or length)

Thus, potentially ductile material may act brittle or ductile as influenced by loading rate, temperature, and service environment. Brittle material always acts in a brittle manner.

8.2. Macro-Fractography - ACT 1 of 3

Macro-fractography is broken down into three sub-steps leading to four outcomes. For a visual representation see Figure 8.3 (next page).

1) Define crack orientation with respect to the part shape.
2) Determine if crack is ductile or brittle at the macro-scale.

3) Determine the loading geometry resulting from iterations of 1 and 2
4) Determine if the crack is at the expected location.

Framing Fractography Act 1, or macro-fractography, specifically in this way opens the door to productive discussions with mechanical design engineers. It forces materials engineers to speak the language of mechanical engineering. The basic language of loading geometries is much easier to explain to the general population than the arcane terminology of phase diagrams, micro-constituents, Beraha's or Villella's Etchants, Greek letter alphabet soup, grain texture, X-ray diffraction, and isothermal cooling curves.

- As a practitioner of failure analysis, help yourself by taking my advice here!
- As a "consumer" of failure analysis, demand that your fracture analysis specialist explain the loading geometry first. If they don't know how, tell them to read this book!!!

Now, to show how simple this method is, let's look at the short list of actual loading geometries.

- Was the part being stretched (tension), bent (bending), twisted (torsion), or squeezed (compressed) when the crack started? Or was it a combination?
- Direct shear and Hertzian stresses will be discussed later.

Next, you pose a question to yourself. "Based on the answer to question 3 was the actual loading geometry what the designer intended?" Get help from someone familiar with the assembly if you need it. See Figure 8.4 (next page).

That's it! That is macro-fractography. That is Act 1 of the three-act process of fractography. By looking at the part and the fragment shapes, it is possible to determine whether the loading was as the designer intended. Sometimes, we need loading geometry to be epistemologically confirmed later in the fractography process. But usually, macro-scale features suffice. Then we move to Act 2, which is meso-fractography.

The Essence of Act 1: Macro-Fractography

Determination of the loading geometry that resulted in the crack is arguably the most central factor in determining whether the design use was normal.

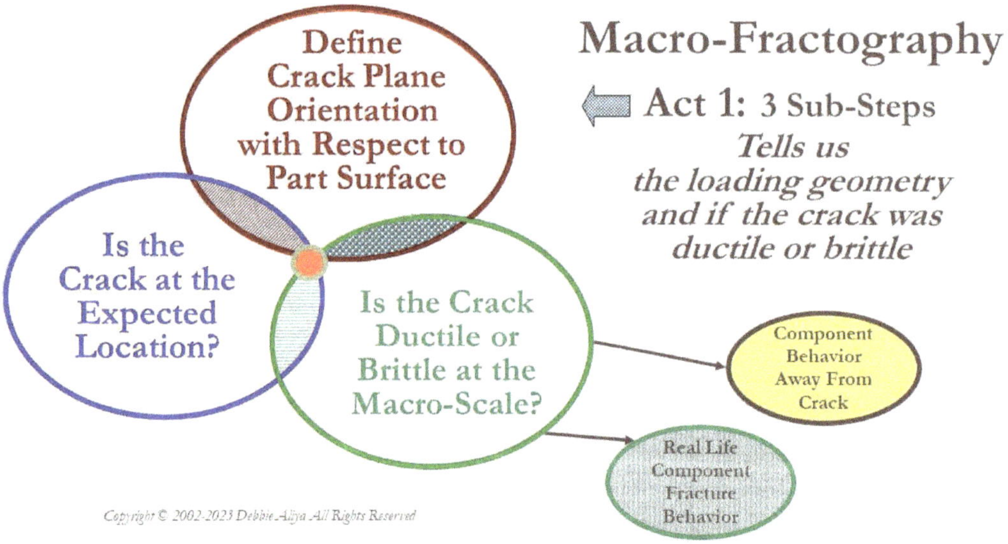

Figure 8.3: *Act 1 of the Fractographic Process has three sub-steps. Whether the crack is ductile or brittle at the macro-scale is sometimes ambiguous. The fracture and component behavior away from the crack can reveal that ductile materials may break in a brittle manner. Act 1's goal is loading geometry knowledge.*

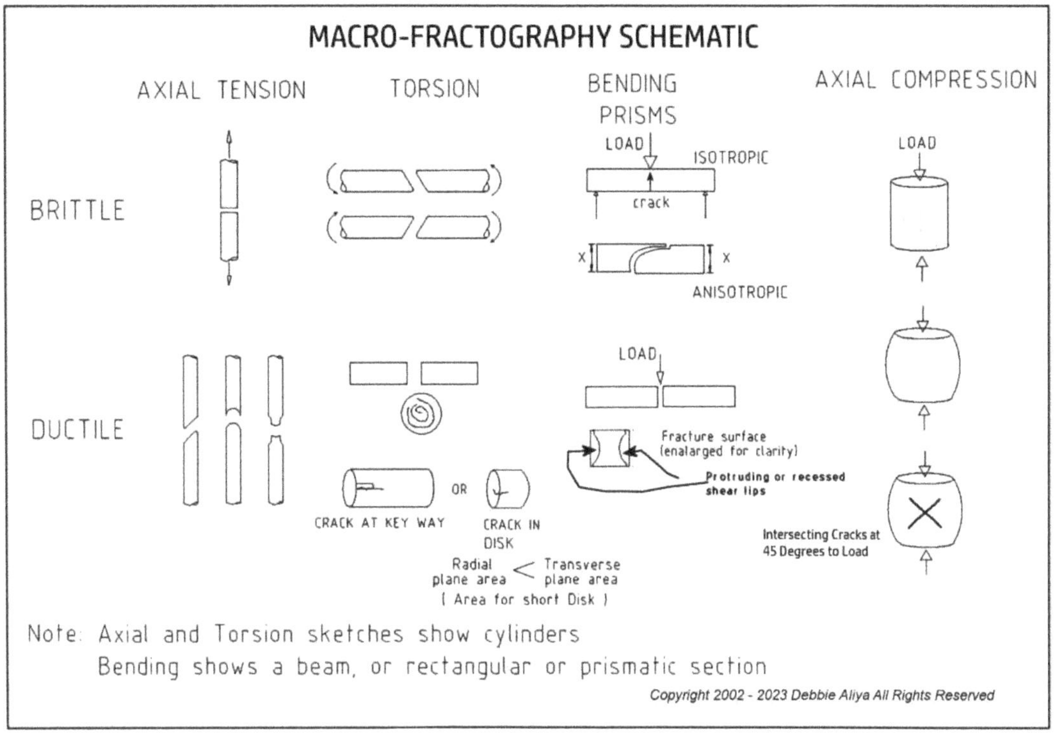

Figure 8.4: *Loading geometry relationship to fragment shapes and mechanical behavior*

All of the cracks in the top row (Macro-Brittle) are created by normal stresses (with the possible exception of the more complex anisotropic bending). This is the definition of macro-brittle. Some of the cracks in the bottom row (Ductile) are a result of shear stresses (left example of Axial Tension and all torsional examples). The other sketches require a little more explanation. Author acknowledges Abijeet Vaze for help preparing this drawing.

8.3 Meso-Fractography - ACT 2 of 3

As we move into Act 2, which is meso-fractography, we will interpret ridges, coloration, steps, texture, fan patterns, beach marks, ratchet marks, and other features commonly described in the classical fractography literature. Don't worry if you don't know all of these terms yet. You will learn many of them here, and other failure analysis books explain them as well.

The most important goal of meso-fractography is determining the initiation and final separation positions, which automatically reveals the crack direction. Finding the crack initiation allows us to plan the rest of our destructive testing.

Identification of the crack initiation position(s) allows us to properly evaluate the microstructure at a relevant location on the part, and thus, determine whether the part was manufactured properly. This is a key aspect of what lets us determine if there was a damaging discontinuity at the initiation.

Details of the fracture surface can be damaged, rubbed, scraped, or even corroded. This is when meso-fractography comes to our rescue with the "Smooth-to-Rough" rule. The Smooth-to-Rough rule tells us that cracks start out smooth and get rougher as they grow. Of course, there are some exceptions, which we will encounter later.

Meso-fractography also allows us to understand whether the loads were repeated (cyclic) or monotonic (single load). Meso-fractography finally gives us clues about the strain rate or temperature that influenced the crack propagation. We must realize that at some level low temperature and high strain rate are experienced by the part in the same way. High strain rate and low temperature may be indistinguishable from the point of view of the part. This may complicate the diagnosis in the absence of background data.

NOTE: You can print out Figures 8.4 and 8.5A as reminders for your fractography practice. As usual, please include the copyright.

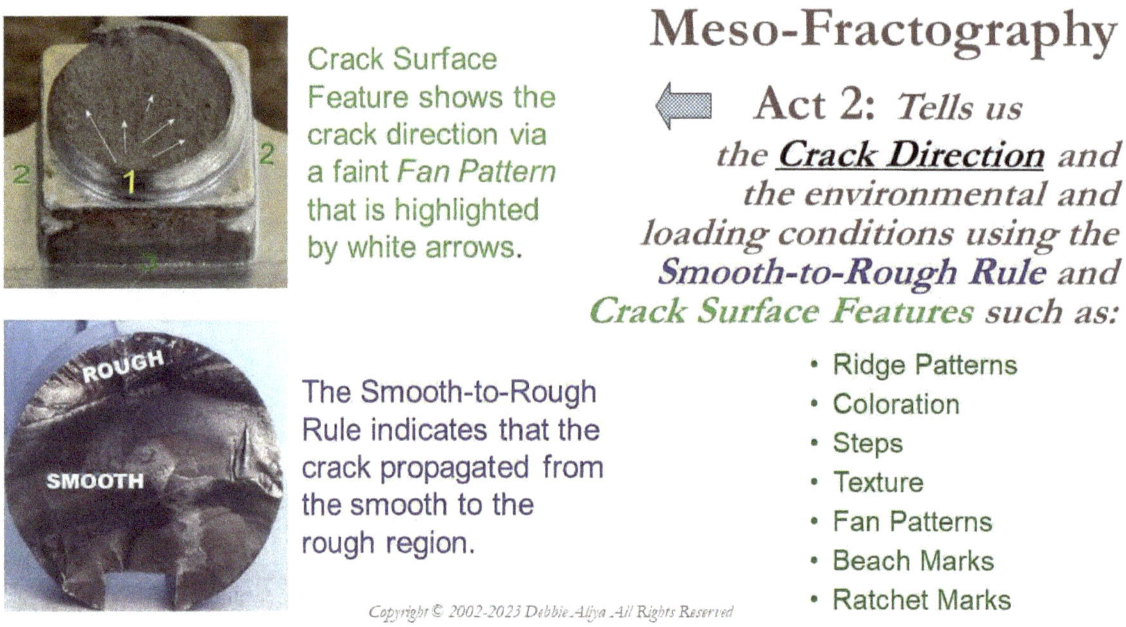

Figure 8.5A: *Meso-fractography Act 2*

The Smooth-to-Rough rule comes to the rescue for damaged fracture faces. The figure on the bottom illustrates beach marks, as well as smooth-to-rough. The figure on top has a shear lip opposite the initiation.

Figure 8.5B: *Close-up of images – Fan Pattern and Shear Lip (left) and Smooth-to-Rough (right)*

8.4 Micro-Fractography - ACT 3 of 3

Finally, Step 3 is Micro-fractography, which is often performed with the aid of a scanning electron microscope and usually focuses on the suspected initiation and final separation areas determined in the meso stage, but intermediate zones also contribute useful confirming data. In particular, evaluation of fatigue cracks in the intermediate zones of the crack may allow estimation of the time it took for the crack to grow. Micro-fractography helps us understand how the stresses and environment interacted with the microstructure of the material. Epistemological confirmation of our meso-fractographic conclusions is often revealed by micro-fractography.

> *All three stages of fractography provide the opportunity to notice discontinuities that could reveal manufacturing process errors.*
>
> *These may include porosity or oxidized cracks (castings and weldments), unincorporated powder particles (additive manufacturing or 3D printed), and discolored cracks (heat treated steels). Some evidence is visible to the un-aided eye, but some requires magnification.*

Using this three-step method of fractographic inspection and analysis is inherently epistemologically confidence inspiring. When correctly performed, it gives the practitioner a rightful claim to having practiced one of the major foundations of skillful thinking.

Using this three-step method of fractography (including Macro, Meso, and Micro) minimizes the chance of making major mistakes. This includes misidentifying the crack initiation. Using the three-step method of fractography (including Macro, Meso, and Micro) also maximizes your chances of getting the most useful information from available data, even when the fracture surfaces are severely damaged.

8.5 Key Fractography Concepts Acts 1-3

Fractography is an integral part of visual inspection for fractures.

KEY CONCEPT 1: Macro-Fractography Relates Part Features to the Fracture Event

Let's now refine our definition of macro-fractography and link it back to earlier discussions on visual inspection. Macro-fractography is the subset of visual inspection that relates the features of a broken part to loading geometry and the mechanical response of the component.

KEY CONCEPT 2: The 3-ACT Fractography Approach is a Breakthrough

As we have mentioned previously, in this book (as early as Chapter 1), we are exploring an approach to macro-fractography that differs from how it is often taught in universities and through professional societies. In universities, the primary people who teach mechanical behavior of materials are (of course!) Materials Scientists and Engineers. Generally, universities ask these professors to focus on basic materials research rather than component design or failure analysis of parts.

This means they often dig-in to material behavior in a test coupon rather than in a part. They are materials experts especially knowledgeable about microstructure and the associated physical metallurgy; and perhaps, they are less expert in macro-scale behavior of a beam, for example. At universities, beams are primarily the business of mechanical and civil engineers, not materials engineers. Unless professors of materials possess strong interdisciplinary mechanical or structural experience, they may never have considered the best way to teach and perform fractography.

The 3-ACT method unifies materials science with mechanical engineering and is integral to the way I teach and do business. I invite visionary experts and interdisciplinary professors to embrace this powerful macro-fractography method and begin teaching the next generation!

KEY CONCEPT 3: Know the Major Capabilities of Fractography

- Macro-fractography sheds light on loading geometry, which is crucial for understanding whether the object might have been subject to improper use.

- Meso-fractography sheds light on the crack growth direction.
- Micro-fractography sheds light on how the stress fields associated with the crack interacted with microstructural features. This is crucial for identifying and understanding manufacturing problems and variations from anticipated environmental conditions.
- Macro- and micro-scale features must be interpreted in light of mechanical behavior. In its most basic form, this means determining whether the behavior was ductile or brittle, independently for each of these two scales.
- Macro, meso and micro features may be self-consistent with respect to each other (ductile/brittle behavior) - or not. In either situation, documenting these facts is an important aspect of leading the stakeholders in the investigation to a clear understanding of what happened, and how it happened.

KEY CONCEPT 4: Best Practices Improve Clarity, Reputation, and Communication

Since categorizing macro-fractographic features into ductile or brittle is based on their macro-scale appearance, anyone can see what the fractographer is describing. As previously noted, this invites consensus, and reduces the chances that the investigator will be perceived as an arrogant bamboozling know-it-all who is trying to coerce agreement. Soon, we will explore how specific crack orientations reveal whether the crack was being pried open (by normal stresses) or was created by concentrated shear (sliding). This is important because it allows us to create a simple mental structure that, with a bit of practice and a bit of patience, clarifies loading within both very complex and very simple geometries. In this book, I'm using simple geometries and this 3-ACT system for elementary teaching purposes.

For example, take a plain cylindrical or rectangular bar made of materials forced to display (in two separate loading events) both ductile and brittle characteristics. If the material breaks via a known loading geometry, such as tension, bending, torsion or compression, we can easily summarize our findings on a one-page document. Figure 8.4, which shows several examples of loading geometries and example crack orientations, confirms this simplicity.

Learn a few more facts about how contact (Hertzian) stresses and direct shear loading work and you will have a very solid foundation for your career as a Fracture Analyst.

KEY CONCEPT 5: Non-Technical People CAN Understand Your Role

Here is a pithier way to define fractography, which is useful when you must describe your work to the interested lay person:

Fractography is the art and science of interpreting the appearance of broken objects.

Broken objects, by definition, result from mechanical damage, although environmental factors may contribute to the ease with which the crack propagates. Most historical fractography literature focuses on a simplified dichotomous view of the purposes of macro- and micro-fractography. However, while meso-scale features (which are called macro-scale in most of the existing literature) are often sufficient to determine the crack initiation and final separation locations, sometimes micro-scale features are needed to confirm the particular initiation and/or final separation positions. This is particularly the case when the meso-scale (or classical) features are damaged.

The simple definition of Fractography presented above combined with supporting evidence from the macro-, meso- and micro-scale will help a non-technical, interested lay person understand the failure analysis investigation results.

KEY CONCEPT 6: The Effect of Strain Rate and Temperature on Mechanical Behavior

From the viewpoint of the Subject Component, ambient temperature high strain rate fractures are often indistinguishable from low temperature moderate strain rate fractures. That is because both high strain rate and low temperature reduce the chances for diffusion, which is necessary to facilitate plastic deformation and ductile fracture.

Diffusion requires time. At low temperatures diffusion requires more time. Low temperature and high strain rate both restrict diffusion, and increase the chances of brittle fracture.

Macro-fractographic features fall into two groups based on their macro-scale mechanical behavior appearance, as determined by the angle between the crack plane and the normal (crack opening) stress.

Once you master this material you will have a simple tool kit (mental structure) to directly extract information about the loading of both complex and simple geometries.

If you are really an enthusiastic go-getter: try this at work! Machine plain cylindrical or rectangular bars and test them in varying loading geometries, strain rates and environmental conditions. This will hopefully result in the display of both ductile and brittle characteristics.

Then if you got lucky with your material choice, and selected something inherently ductile at room temperature and slowly applied loading, you will see and measure characteristics that demonstrate a transition to brittle behavior with rapid loading, with low temperatures, or in the presence of certain chemicals. If you do this, please contact Aliya Analytical or the publisher if you are willing to share your results!

KEY CONCEPT 7: Our End Goal

The basic goal of fractography is determining and visualizing the sequence of forces and environmental conditions that led to the fracture. Fractography is a subset of visual inspection, performed with the naked eye and with microscopes. Fractography focuses on *Looking-For* features during visual inspection.

KEY CONCEPT 8: Fractography is a Family of Related Visual Interpretation Skills.

Macro-, Meso-, and Micro-fractography work together to strengthen or weaken the tentative conclusions we made based on our first impressions. This practice of macro/meso fractography is inherently epistemological in nature. The family of related visual interpretation skills include *Looking-At* and *Looking-For*:

- the shape of the fragments created by the fracture event in order to get an idea of what loading geometries possibly did, or definitely did not, create the crack. This is macro-fractography.
- the patterns of ridges and steps on the fracture surfaces or along the edges of the fracture surfaces in order to determine the crack direction. This is meso-fractography.
- the color, texture, and reflectivity of the fracture surface in order to envision crack history and timing as well as the environmental conditions in which the crack happened. This is also meso-fractography.
- the micro-scale topological features reveal whether the material was properly processed.

NOTE: We use *Looking-At,* an open-ended process, to see what is there; but when there are ambiguous situations, we need to force a decision. We need to be able to see what is present and know what types of things COULD be present. That is when we *Look-For* features from a "pick list". Alternating between *Looking-For* and *Looking-At* can help us resolve ambiguity.

KEY CONCEPT 9: The Traditional Definition of Macro-Fractography is Currently Inadequate

In the "old days", many people defined *macro*-fractography as evaluation of visible features on a fracture surface or those revealed with a low power optical microscope. They defined *micro*-fractography as evaluation of features that were too fine to see without use of a scanning electron microscope. This is outdated.

These definitions may still be found in some references, but a new definition was required (especially with the advent of micro-devices).

Imaging micro-scale devices requires a scanning electron microscope to determine even the basic macro-fractographic features.

These new and more useful definitions of macro-, meso-, and micro-fractography are related to *what we are trying to understand* rather than what instruments we may be required to use to aid our senses.

The modern definitions of macro-, meso- and micro-fractography are valuable and powerful ways to understand the art and science of fractography. We are trying to independently determine loading geometry, nominal mechanical behavior, crack growth direction, and microstructural influences on crack behavior. Those are our primary goals. That is why we now categorize fractography into macro, meso and micro to encourage and support clearer thinking.

We'll return to micro-fractography later in this chapter.

8.6 Macro/Meso-Fractography Practice

Macro-fractography allows us to compare and contrast the actual loading geometry with the design intent. One of the main questions we are trying to answer is whether the component broke during normal use. Being able to determine the loading geometry that resulted in the crack is arguably the most central factor in determining whether the use was normal. It is the essence of Act 1.

> *EXAMPLE: Imagine a hand tool involved in a part fracture. "Did the craftsperson apply too much force?" This question is often asked but difficult to determine.*
>
> *Inadequate development of strength in the alloy could look the same as "Hercules was too forceful". Only detailed evaluation might shed light on this issue.*
>
> *But "Did the craftsperson apply force in an unintended direction?"; answering this question is often straightforward.*

Fragment shape reveals information about the loading geometry, but may not be definitive. That is because for example, in simply shaped parts, ductile torsion, brittle bending and brittle tension resemble each other. Orientation of the crack plane with respect to the part surfaces is identical. Fragment shape gives a lot of information, but like any other fact, its interpretation must be supported by other knowledge.

Fractography Practice

- **Fragment shape reveals loading geometry and ductile or brittle behavior. (Macro)**
- **Patterns of ridges and steps reveal the crack direction and specific initiation position. (Meso)**
- **Color, texture, and reflectivity of the fracture surface reveal the crack history. (Meso)**

Another important assessment is sorting-out ductile and brittle at the macro level.

Each major loading geometry has two possible macro-scale expressions on the way to fragmentation: ductile and brittle. *Because this is an in-depth and exciting field of study, of course, there are examples that have both ductile and brittle features. We call them "mixed mode".*

If the subject is large enough to see with the naked eye, it is often possible to categorize both the loading geometry and the mechanical behavior during the preliminary visual inspection. We must also be sure to identify the mechanical behavior during (versus before or after) the crack event or process. Refer to Figure 5.11 chain link fracture. We must separate the fracture behavior from the material behavior away from the crack. See the yellow oval in Figure 8.3. The behavior immediately before or after the crack may be ductile even when the crack is brittle. Should the aspiring Failure Analyst choose to prove that the material is ductile despite the brittle fracture, a tensile test coupon could be prepared and tested. However, there are many reasons that this coupon may fail to reveal much ductility. Thus, the visual appearance of the deformed component takes precedence in identifying the inherent ductile capability of the component, despite the brittle crack.

Despite such caution, definitive assignment of loading geometry and mechanical behavior must sometimes await further progress in the investigation. During certain service conditions, inherently ductile materials may behave in either a ductile or a brittle manner. Some materials are inherently brittle and will always behave in a brittle manner, at least at the macro-scale. Many ceramics almost always behave in a brittle manner at both macro- and micro-scales. Using visual interpretation skills, the Failure Analyst can discover and reveal information about crack direction, initiation, and crack history. This can be facilitated with the example of Figure 8.6 (below).

8.6.1 Example of Visual Interpretation Skills

To the trained eye, the perpendicular orientation of the fracture surface relative to the cylindrical part surface may indicate macro-brittle tension or bending. The small steps at the (medium) pink arrows indicate multiple initiations. The step at the large green arrows at 10 o'clock shows the final separation. This provides information on the direction of crack propagation. Also take note of the darker and lighter bands at the small black-outlined beige arrows. At this point, we take fatigue as our working theory of loading history. Fatigue cracks result from repeated loading over time.

Based on the red rust and adherent dark oxide shown in Figure 8.6, we can determine the environment was likely at least intermittently moist. However, the red rust is not present as bands so it is most likely a result of moisture in the environment following the final separation. If we had the mating piece, we could look to determine if the pattern mirrors this one or not. If it does, we could conclude the rust happened prior to final separation; if it did not mirror this, we would determine rust happened after final separation.

Figure 8.6: *Perpendicular orientation of the fracture surface relative to cylindrical part surface indicates macro-brittle tension or bending. There is also information on crack propagation. Dark bands running across the crack face, near the diameter, indicate progressive cracking. Our working theory is fatigue for the loading history.*

Note that this could also be ductile torsion, but we would have circular smears, not a series of parallel lines as the "meso" characteristics, were that the case.

One additional important feature in Figure 8.6 that is often revealed on fatigue crack surfaces allows us to estimate the effective safety factor baked into the design. The uppermost of the small black-oulined beige arrows points to a fine black beach mark (crack arrest mark, or witness to the position of the crack at some point in time). That indicates that the crack was still progressing in fatigue when there was barely any load bearing cross-section left. This indicates that the design load capability in the intact part was much greater than the load that this particular fastener experienced. In this case, the tiny final separation / shear lip feature may indicate a nominal safety factor of up to 15.

8.7 Definitions for Macro-Features Based on Loading Geometry

NOTE TO READER: This section's content is repeated from short similar sections in Chapters 1, 3 and 5 because

- it's so vital to your understanding, and
- it is an interdisciplinary breakthrough, and
- most people need to hear it multiple times to grasp it & accept it, and
- some people might open to this specific chapter without reading the book in order.

DEFINITION: Macro-Brittle cracks are formed by normal (crack opening) stresses and Macro-Ductile cracks are formed by shear (sliding) stresses.

Many people get up in arms when an expert using this system points out the flat, central portions of a Classical Ductile Cup-and-Cone Axial Tensile Test as an example of "macro-brittle".

"WHY," they cry out "are you calling this 'macro-brittle' when the micro-features are dimples?"

"Hello, Wake-up!" I have pointed-out on multiple occasions, "It's because we're not talking about ***micro*** *features. We're talking about* ***macro*** *features".*

We are defining macro features by the type of stresses that created them. There are only two types of relevant stresses in this case.

- Normal (crack opening) stresses, which create brittle-macro features, and
- Shear (sliding) stresses, which create ductile-macro features.

Okay, what about mixed-mode? In this system, the classical cup-and-cone fracture with a waisted (necked-down) geometry surrounding a crack is considered mixed-mode. It is macro-ductile everywhere except the flat central area, which is macro-brittle. Another way to say this is the sidewall of the cup is a shear lip.

Another point is that fatigue (cracks from repeated, or cyclic loading) is called macro-brittle because by definition, a fatigue crack propagates below the nominal yield strength.[151] Yes, yes, at the micro-scale, fatigue is often ductile. But again, that's a separate classification procedure.

Here is one more confusing situation. Study Figure 8.4 (repeated below). Notice the cracks are on the radial shear planes in the Ductile Torsion lower sub-row. In practice, these types of cracks are usually found in fatigue at excessively sharp corners of keyways or where there is a material discontinuity - for example, at or along a heavy stringer inclusion. Monotonic cracks from torsion on a radial plane are vanishingly rare to non-existent. So, these cracks are on shear planes, but are macro-brittle by virtue of typically occurring below the yield strength; so, no macro-scale-deformation will happen. This, to my mind, is the only true internal contradiction within this system. We just have to live with it, because...

DAH dah DAH (drum roll, please)...

The benefits of linking the crack appearance directly to the macro-stress state (planes of maximum normal or shear stresses) are so abundant for every other aspect of

[151] A colleague or two have pointed out that some alloys will crack in fatigue above the nominal yield point. If you must work with that type of material, you need to keep this fact in mind. However, in historical context, fatigue was originally defined as cracks propagating below the yield strength. We'll stick with this for the purposes of building our mental structure.

communication in the failure analysis process. First and foremost, it helps us understand the LOADING GEOMETRY.

Figure 8.4 (repeated for your convenience) is a guide to the use of macro-fractography principles, particularly those related to loading geometry. It can also help to reveal the sequences of deformation damage even if fracture is not a result.

Now consider Figure 8.4 from a technical viewpoint; the two ductile axial tension behaviors depicted at the center and right are mixed mode at the macro-scale. The flat central portions of the crack are macro-brittle.

While it is technically possible for a hoop stress crack to open at the bulge line of the axial compression shown here in the ductile row, the crack itself would then be macro-brittle. That's because it's on a hoop stress plane (normal stress). That is because the primary normal stresses parallel* to the loading direction are squeezing any nascent cracks shut rather than pulling them apart. Only the secondary normal stresses at the hoop plane can act to create a crack.

It is possible, usually during forging operations, for ductile cracks to form along shear planes at 45-degree angles to the load. Sometimes this will form a series of intersecting cracks.

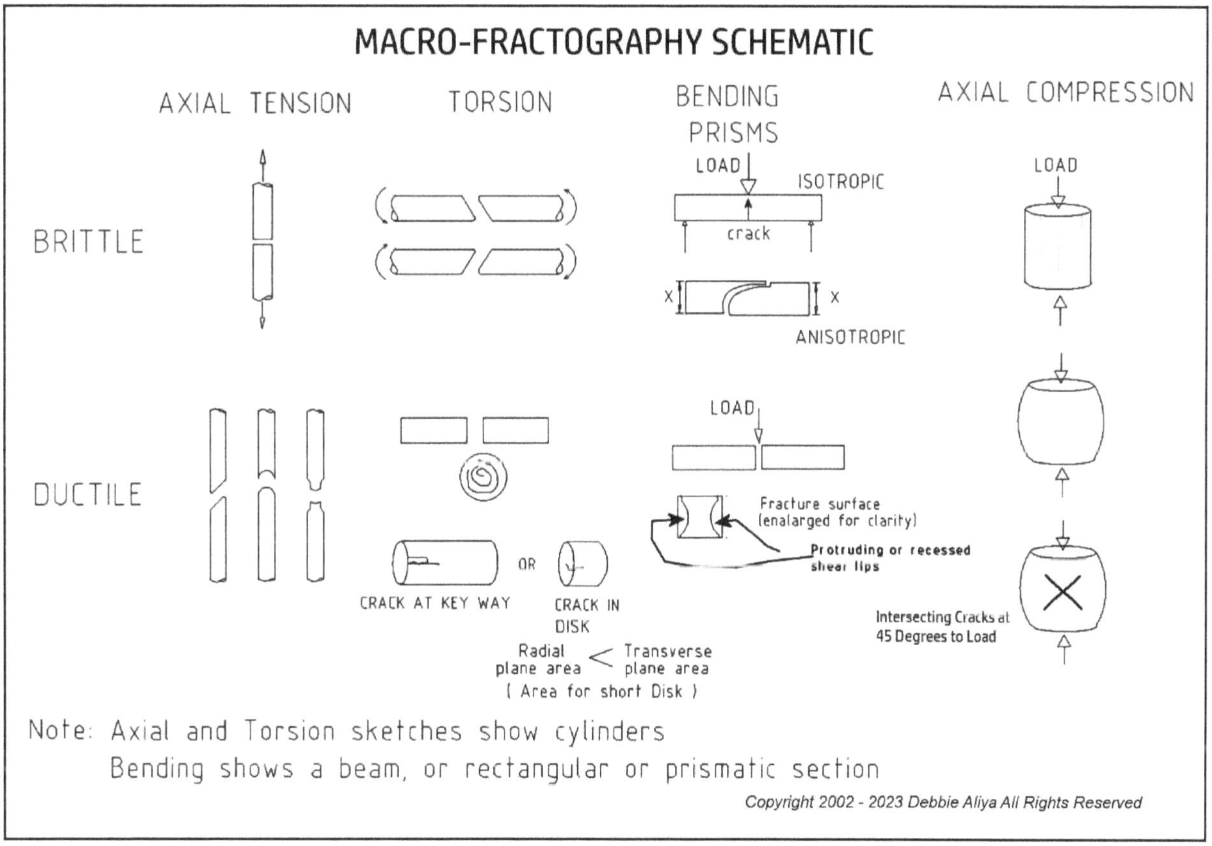

Figure 8.4: *(Repeated) Loading geometry relationship to fragment appearance and mechanical behavior for simple shapes*

EXERCISE 8-A: Hose experiment.

Do this experiment for yourself! Get two pieces of hose about 15 centimeters long each. Almost everyone has an old hose laying around their house just waiting to be thrown away. Ask your friends or, if you are a university student, ask your parent's friends.

Make 4-6 cuts in one piece of hose: Make slanted cuts about half-way through the diameter. Each cut should be spaced about 1-2 centimeters apart.

On the second piece of hose, cut your hose once, perpendicular to the length of the hose, about three-quarters of the way through.

- Now twist each section of hose.
- For the slanted slits, one direction will open the slits. That's showing tensile normal stresses.
- Twisting in the other direction will close the slits. That's compressive normal stresses.
- For the other piece of hose, the one with a transverse (perpendicular) slit, you will demonstrate shear on twisting.

Now we have demonstrated for ourselves the effects of torsional loading! Figure 8.7 (below) depicts my interpretation of this exercise.

Figure 8.7: *These hose examples show compressive normal stresses (top), tensile normal stresses (middle), and shear (bottom) in torsional loading.*

8.8 ACT 2: Meso or Classical Fractography Revisited

NOTE TO THE READER: What we are calling meso-fractography is usually considered in the older literature to be a part of macro-fractography. I call this Classical Fractography.

We may determine the direction of a crack's growth by using ridge and step patterns, beach marks, and variations in surface roughness.

This is critical: whenever possible we want to know exactly where the crack initiated. When we get to the micro-fractography, we need to examine the crack initiation for "suspicious" characteristics, such as a crescent of intergranular features, from which a fatigue crack originates. When we get to the metallographic work, we need to study the volume immediately surrounding that same first separation. This allows us to determine if the material has the characteristics we expect based on the known or specified processing and service history.

> **It's often very important to study the volume of material at the precise position where "the trouble started".**

We need to examine the initiation area for the presence of any potentially damaging pre-existing discontinuities that may have facilitated the crack event. This may be done by eye, by microscope, by metallography, and other methods such as those we have already covered in earlier chapters. Proper use of meso-fractography *is key to planning* the later stages of the investigation.

For many old school Failure Analysts, this step is the most important. But let's not get ahead of ourselves. I'm offering a caution here. Skipping directly to this step wastes data; and once you skip it, you usually can't go back and get the data that it would have provided. Preserve evidence!

Evidence preservation, evidence preservation, evidence preservation!

- Crack initiation may occur in one or more locations and are often microscopic in size.
- If the stress level is high enough, even if the high stress levels are intermittent, the crack will propagate.
- Although some parts fragment completely as a result of a single load event, other parts require hundreds or even millions of cycles for a crack to result in complete fragmentation.
- The forces acting to initiate and propagate the crack(s) may be transmitted to the part by an adjacent component in an assembly or by an object that it was supporting.
- In other cases, the forces may be transmitted to and concentrated in a portion of the component as a result of a thermal gradient within the part itself or from pre-existing (residual) stresses from the manufacturing process.

Are the stresses distributed in the manner we expect them to be? This is a good question and stress expectations are related to design. Whether the forces acting on the part are internally generated or transmitted from an external source, engineered components will have expected stress patterns based on the specific service conditions for which they were designed. These patterns may vary as a result of the usual (planned for) service conditions, such as loading and unloading, temperature fluctuations, variations in the chemical environment, etcetera. As long as the service conditions remain within the design parameters, then Part cracks will usually be located where expected; in other words, in a place the design engineers knew was susceptible to fracture. Their knowledge of the expected crack initiation position will have come from experience, testing, computer models, or some combination of those elements.

If the crack is in an unexpected location, we now have a hint that there might have been either

1. a significant discontinuity, or

2. unanticipated loading conditions or damage scenarios, such as torsion rather than bending, or fretting due to dimensional irregularities.

As a crack propagates, the crack itself concentrates the loads into a smaller and smaller load-bearing cross-section. If you are familiar with fracture mechanics, you might furthermore note that the stress intensity is increasing, due to the presence of the sharp crack tip. When this happens, the material tries to get rid of the uncomfortable strain energy any way it can. One of the ways to do this is by making more surface area per corresponding area of cross-section. Thus, cracks often get rougher as they propagate. See Figure 8.8A (below) for an example of this phenomenon. Thinking back, you may remember this trend was depicted in Figure 8.5B highlighting the Smooth-to-Rough Rule.

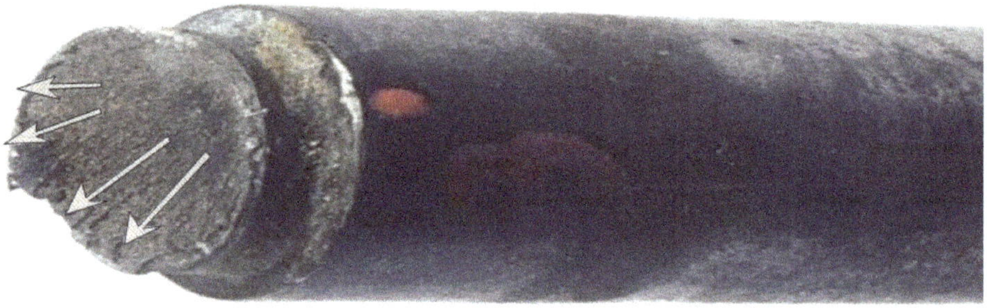

Figure 8.8A: *The fan pattern demonstrates smooth-to-rough crack surface features. Crack initiations are usually smooth. This crack resulted from axial tension in a hardened threaded fastener. The crack initiated at the position shown by the red dot on the circumference of the bolt shank.*

In fatigue where the crack repeatedly opens and closes; the original crack initiation is rubbed to an even smoother condition than when the crack first initiated. This is often the case in shafts subject to rotating bending fatigue. Knowing this can help us determine the crack initiation position in fatigue cracks.

As always in failure analysis, there are aspects of a rule that can fool you. One case where cracks may appear to get smoother as they propagate is in brittle castings which are subject to confined three-point bending. Okay, so let's follow the logic in this case.

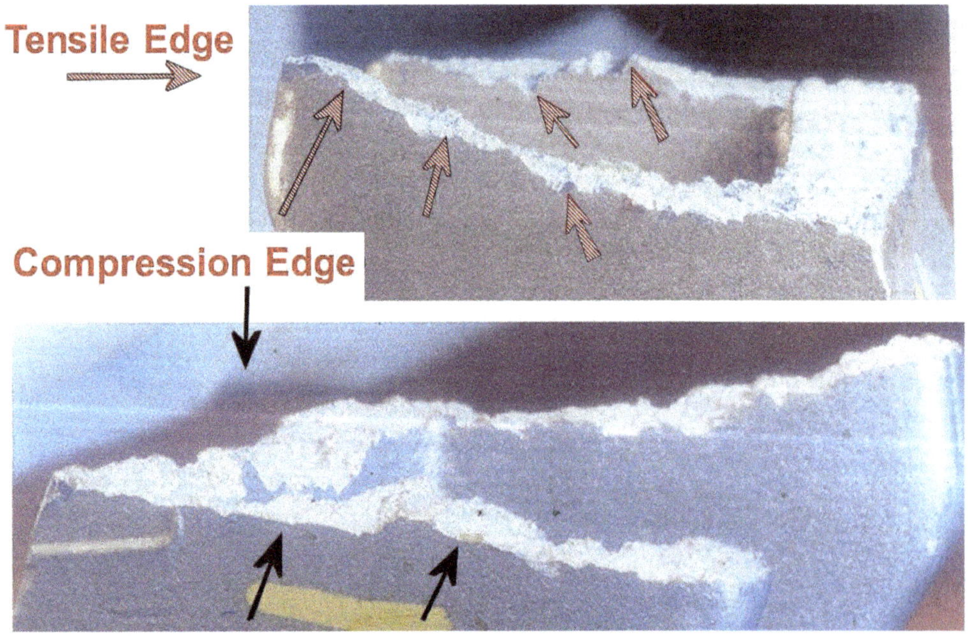

Figure 8.8B: *Cast aluminum beam broke in bending – this is viewed as an exception of the smooth-to-rough rule. Bending cracks generally initiate in tension.*

In general, a crack in the tensile side of such a brittle cast beam loaded-in-bending will seek lines of weakness. It may find shrinkage porosity or lower melting point inter-dendritic phases. If the fragments are in a situation where the far side of the beam is in compression as the crack grows, a relatively straight fracture edge may result. See both parts of Figure 8.8B (previous page). This would not be the case in combined bending and tension. Try to imagine it!

8.8.1 Extracting More Information from Meso-Fractography

Changes in humidity or changes in the levels of aggressively corrosive substances often produce changes in the reflectivity of portions of the crack surface. These environmental fluctuations may result in varying color of the oxides that form on the freshly exposed crack surfaces. Fatigue cracks often provide the classical example of colorful bands that run perpendicular to the direction of crack growth. Review Figure 8.6 (a repeated image is provided below) as an example.

Gleaning information from the colors and textures of a fracture surface is what I like to call "classical macro-fractography". It's what I learned in engineering school. Many mechanical engineers are at least familiar with the basic premises of this aspect of classical macro-fractography, which we are now calling meso-fractography. More details follow later in this chapter.

8.9 Imagining Forces and Stresses

A few preliminary concepts will help us set the stage for further exploration of the fascinating world of fractography.

In physics, we learn that a given force acting on an object will create acceleration proportional to the mass of the object. This is Newton's Second Law of Motion.

- Qualitatively, a force is a push or pull, and
- Quantitatively, a force is measured in Newtons or Pounds and has a direction of action.

A Newton is the approximate force that an apple would exert on your head if it fell from the tree while you were sitting on the ground with your back leaning against the trunk!

A stress is a quantitative way of normalizing a force over the smallest cross-sectional area that has to sustain the force. With practice, in simple shapes, you will be able to visualize the stress pattern throughout a volume of material due to a constant or changing force. The particular local values and orientations of the stresses will be influenced by a number of factors including cross-sectional area, shape, and force direction.

EXERCISE 8-B: IMAGINATION – The simplest way to understand stress is as a way to normalize force over a cross-sectional area. Take a 1-inch by 1-inch square rod that is one foot long. (Feel free to imagine a 1-cm by 1-cm square rod that is one meter long if that is more natural to you. Use whatever units you are most familiar with!)

Now glue the small end to the infinitely strong ceiling with infinitely strong glue. At the other end, glue the harnesses of a group of very calm elephants, who don't mind being suspended this way from the ceiling.

The stress in the rod is equal to the weight of the elephants divided by the cross-section of the rod.

(In this case it is either one square inch or one square centimeter.)

If we have two identical rods hanging side-by-side to share the elephant load, then the stress would be reduced to half of the original value. Stress is force divided by the area that supports the load.

NOTE: The section on mechanical behavior has more information on stress.

For now, we are not going to worry about quantifying the stresses with numbers. We are going to try to acquire or reinforce our knowledge of some concepts about stress.

EXERCISE 8-C: IMAGINATION – Exerting Forces (Magnitudes and Directions) in a Sequence - Opening a Sardine Can. Example of Direct Shear (One of the two loading geometries that has not yet been introduced in detail.)

Imagine opening an easy-open food can such as a sardine or fruit can. It is designed to break around the edge of the recessed flat top, at a prepared score mark; and it has a metal pull tab. What are the forces that must act on this assembly in order to eat the contents for lunch? What are the results of the forces? Walk through the exercise in your mind and record your ideas in a digital document or a notebook. Include tensions, shears, bending, locations, timing, yield strengths, forces, etcetera.

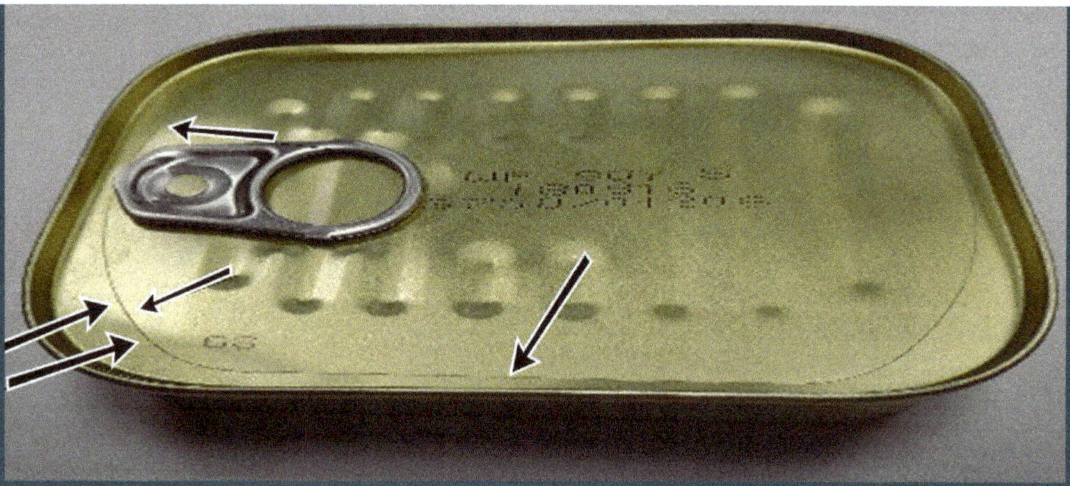

Figure 8.9A: Easy-open can – note the score mark.

Take a look at my description and compare & contrast your record with mine:

EXERCISE 8-C: IMAGINATION – MY DESCRIPTION

Look at Figure 8.9A (above). While my finger exerts an upward force on the ring end of the tab of an easy-open can, the far end of the tab exerts a downward force near the inside edge of the score line in the lid sheet metal. Note that, meanwhile, the tab is rotating around the outer edge of the small inner rivet that holds the tab to the lid.

Now let's look at this from the point of view of the can. See Figure 8.9B (below). The sheet metal responds in a sequence of events. First, the lid material just inside the score line is pushed downward, effectively bending it, while the material outside of the score line comes with it. This is elastic deflection.

The motion is similar to that of a membrane covering a drum as is it being struck. As the force increases, the inner portion keeps moving down, while the outer portion, reinforced by the lip of the can body, remains stiff and resists movement. This creates shear (or relative sliding motion) between the material inside and outside of the score mark. When the sliding motion has proceeded along a distance equal to the thickness of the can lid material, then we have initiated a crack at the desired location.

In Figure 8.9B (below), we see that the tab has partially sprung back toward its initial flat position compared to the 90° required to initiate the crack, as described below. I did not get that photo, so you may experiment or choose to believe me!

Figure 8.9B: While opening an easy-open can, the sheet metal responds to forces experienced and the observer sees the effects of stored elastic strain energy on initial crack growth.

Let's look at Figure 8.9C (next page) as we remember opening a similar can ourselves. (Now is a good time to try this, after you have read this section.) You may have noted how quickly the fracture runs to both sides once the first bit of material is separated (two black arrows at left). When I was taking these photos, I tried to stop the crack as soon as it was initiated. Elastic energy stored in the tab and lid propagated the crack almost immediately to a width about equal to the tab width and much greater than the lid material thickness.

In order to initiate the crack, the tab had to be bent to a position almost perpendicular to the plane of the lid. The difficulty in stopping the crack immediately after it initiated was not due to my clumsiness! It was due to the fact that there was elastic strain energy stored in the system because of the forced displacement of the tab. The amount of energy required to initiate the crack is much greater than that required for the crack to grow - even when it's growing in two directions. This is primarily a result of the geometry and the inherent mechanical behavior tendencies of the material.

Now, if we want to finish opening the can, to eat our lunch, (this long-winded description has allowed me to get really hungry!), after releasing the tab to capture the photo, we will repeat our lifting and rotating motion so that the entire edge of the material next to the tab and inside the score mark is permanently bent downward. We have exceeded the yield strength (resistance to permanent shape change) of the material along an arc that extends beyond the width of the tab. Look at Figure 8.9C and note the two white arrows.

Then we need to change the direction of the force, and still lifting upward, we need to pull the tab back toward the right, to peel the lid away.

The force is upward, and we are creating more direct shear (sliding) stresses at the moving tips of the crack, as well as bending stresses along a line (parallel to the short side of the can) which continuously moves from the left to the right end of the can. See Figure 8.9C (below).

The can lid is permanently curled at the end, a result of the bending stresses having exceeded the yield strength (deformation onset). Both the yield strength and the breaking strength have been exceeded in the thin layers where the shear stresses acted (on both sides of the separation). Careful metallography of a cross-section of the can body and lid sides of the separation would be expected to reveal a thin layer of heavily deformed (locally cold worked) material.

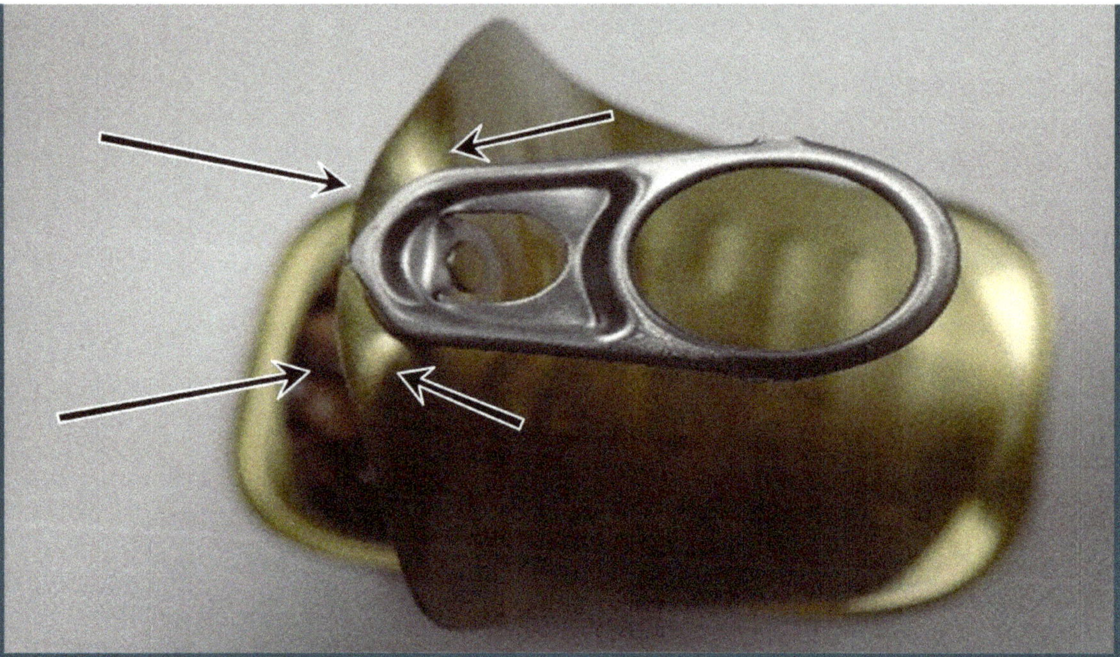

Figure 8.9C: The arc of deformation at the edge of the lid is highlighted by the left arrows.

8.10 Clarifying Your Thinking About Fractography Concepts

Take a moment to remove yourself from the linear flow of the previous sections. Take a deep breath and relax before starting this section.

Consider the following thinking skill exercises as an opportunity to construct your competence in a non-linear way. One exercise overlaps and flows into the next. Observe your thinking as you progress.

EXERCISE 8-D: THINKING SKILL – Ductile or Brittle?

One of the key tools that facilitates clarity of thought is understanding when to classify something as "either/or" and when to use a spectrum of choices. If you are trying to determine ductile-versus-brittle then the choice is inherently phrased as an either/or choice.

A repeat image of Figure 8.4 is included on the next page for your convenience. Use it to determine the mechanical behavior that resulted in a crack. Your choices are EITHER ductile OR brittle.

We've already mentioned the fact that it's possible to have both ductile and brittle features on the same part. Sometimes Mother Nature forgets to tell the parts, "Hey parts, you were supposed to choose EITHER ductile OR brittle."

Another way to think about this decision is whether you will be selecting from a Black-or-White (Either/Or) ***or*** a Shades-of-Grey (A spectrum of possibilities) "palette." Your early selection of a dichotomous or a continuous "palette" will limit what you see. For example, if we don't notice that there are patches of both brittle- and ductile-macro features on some (or most) of the broken parts that we examine, we may overlook other key characteristics that could help us understand the situation.

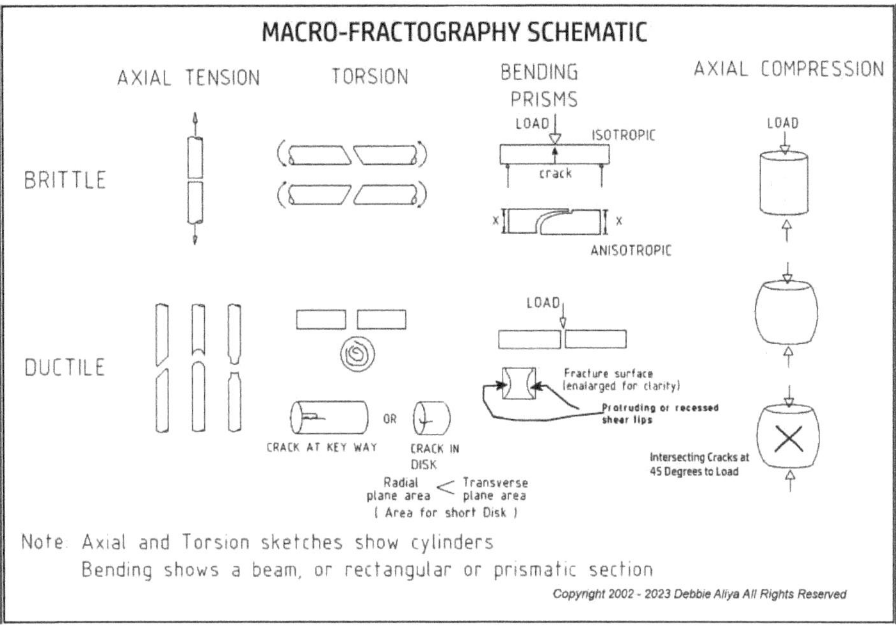

David Levy addresses this issue in Chapter 5 of his book, *Tools for Critical Thinking - Metathoughts for Psychology*. Some of the examples he gives of Black-or-White include:

- a light switch that is either on or off,
- a letter is either mailed or not,
- a person is married or single,
- a gun has been fired or not.

Some characteristics may be commonly phrased as an opposite pair (black-or-white) but they are best assigned to a spot on a continuous spectrum (shades-of-grey); examples include:

- sane or insane,
- democracy versus dictatorship, and
- optimist or pessimist.

Enough of the lead-up to the exercise; can you make an argument for both cases (Black-or-White, Shades-of-Grey) about the characteristics "dead versus alive"?

EXERCISE 8-E: THINKING SKILL – What about dead or alive? Can you make arguments for both cases? Hint: How would you define the words to argue for one or the other?

___

___

___

___

David Levy starts Chapter 5, like all of his chapters, with some pithy quotes. The first quote for this chapter is from *The King James translation of The New Testament*, <u>book of Matthew 12:30</u>, "He that is not with me is against me."

EXERCISE 8-F: THINKING SKILL – How has the idea of "He that is not with me is against me," been used by every nation at war? What about during peace time? How does this phrase minimize the potential for mutual understanding? Is the ideology behind this sentence a "root cause" of war? Define the words to argue for a spectrum of choices or a dichotomous choice.

How does apathy fit in this for and against dichotomy? Could someone be both for and against at the same time? What would that look like?

The consequences in failure analysis may not be as dramatic as the consequences of war, but in the case of an investigation involving one or more high consequence events (such as fatalities) it could have far-ranging consequences.

I have found the closer I look at many dualistic divisions, the blurrier the boundary between them becomes. That said, sometimes a dualistic choice adds clarity, such as macro-scale ductile or brittle cracking.

The more careful we are to cultivate clarity as a part of every investigation we work on, the easier it may become to do it well on the bigger, more consequential jobs, while people are breathing down your neck.

That said, we must always guard against arrogance. We're all human. We can all walk into traps. And the easiest traps to walk into are those set by our own habits.

EXERCISE 8-G: THINKING SKILL – Now, it's time to retreat from the complex world of politics, and get back to engineering. Let's consider a range of loading experiences that can derive from a single part geometry. We will consider the 4 primary loading geometries for a tall skinny column. Think about a straight, skinny, cylindrical pole that is glued at each end with infinitely strong glue to a flat plate (see below).

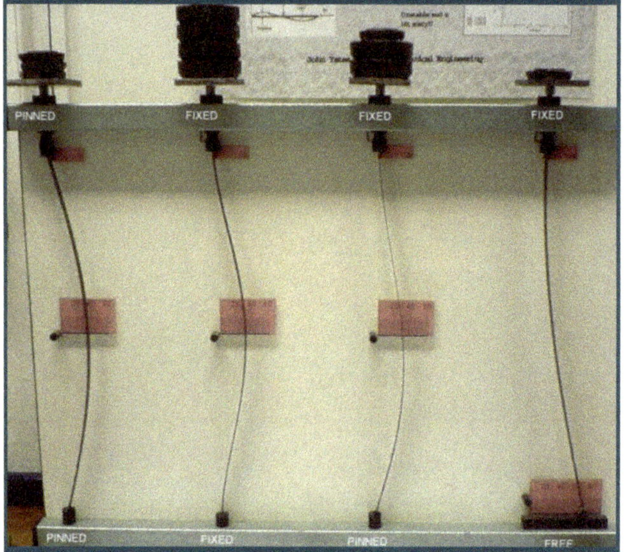

Imagine the plates are moved apart from each other so the pole is stretched. Assuming that the stress eventually exceeds the fracture strength, where will the pole break?

THINK! THINK! THINK! THINK! THINK!

OK! Write your answer below; where will the pole break? Write support for your conclusion.

Are you sure of your answer? THINK! THINK!

My thoughts are written below; compare & contrast your record with mine. The answer is (see footnote)[152]

The stresses are everywhere identical because the pole has a uniform, circular cross-section. So, assuming that the strength is essentially uniform within the pole, the crack could initiate and propagate from any location within the core or at the surface of the pole.

If the material is able to behave in a ductile manner, it may start to deform by pinching down at a particular location. Then the crack will be constrained to that location.

EXERCISE 8-H: THINKING SKILL – Same pole, reverse action

Now, let's reverse the action of the plates and have them move closer to each other. In this case, the ends are pinned to the flat plates (refer to far left pole in the image in EXERCISE 8-G). Remember, the pole is straight, skinny, cylindrical and the pin provides a rotation point. Where will the pole break?

THINK! THINK! THINK! THINK! THINK!

OK! Write your answer below; where will the pole break? Write support for your conclusion.

____

____

____

____

____

____

____

____

Are you sure of your answer?

[152] The answer is: **Anywhere!**

My thoughts are written below; compare & contrast your record with mine. The answer is (see footnote) [153]

The part will bend. In a skinny column, uniaxial compression, taken to an extreme, will induce bending. Here we see another example where loading geometry and stress state must be understood, not simply memorized. Think of the shape of a vibrating guitar string plucked in the center of its length. Depending on how the attachment plates are held in place, the outer edge of the vibration envelope may be similar to the shape the pole takes. Now the crack is most likely to initiate at the center of the length, which is the maximum lateral displacement. The outside of the bend will be stretched, promoting a crack, while the inside of the bend will be squeezed and cracking will be suppressed.

So, we now can imagine not only the general location of the anticipated crack, but also its precise initiation position. And simply by moving from tension to compression, we've gone from all locations equally likely to crack, to a specific location (at least once the unstable bending takes over from the initial elastic change in length).

EXERCISE 8-I: THINKING SKILL – Same pole, new action

What if, instead of moving the plates toward or away from each other, we rotated one of the plates? This could be a rotation within the plane of the plate, which would twist the pole. Alternatively, this could be a rotation to align one of the plates closer to parallel to the original orientation of the pole, which would bend the pole. Or, if we did not move closer to the fixed plate, it would create a bending plus a stretching geometry.

[153] The answer is: **A tall, skinny pole does not have enough rigidity to keep itself straight. It is subject to Euler's Buckling.**

THINK! THINK! THINK! THINK! THINK!

OK! Write your answers below; what will happen to the pole? Write support for your conclusion.

Are you sure of your answer? THINK! THINK!

My thoughts are written below; compare & contrast your record with mine. The answer is (see footnote)[154]

8.11 The Power and Limits of Macro-Fractography

As powerful as macro-fractography is, it has limitations! We would need additional information to distinguish bending from bending plus stretching!

But what about the in-plane twisting? Perhaps except at the very ends, which are constrained by the glue joint to the plates, the torsional strain distribution would be equal at every length position. If the plates were rotated enough to break the pole, it could break anywhere along the length. However, the stresses in torsion are greatest at the surface, so the crack would probably initiate at the surface. In the case of straight axial tension, it could possibly initiate in the core, especially for a nominally ductile material.

[154] The answer is: Torsion would create uniform angular strain. It could break anywhere along the length, but would initiate at the surface. Rotation of the plate from perpendicular to parallel to the original length of the pole will be the outside of the bend near the center of the length.

We now have the conceptual foundation for a system of knowledge that will let us look at a broken part and get an idea of what may or may not have happened. We'll provide examples to help readers get more familiar with this system of knowledge and more familiar with the differences between ductile and brittle behavior.

At the simplest level, ductile behavior results in a permanent shape change, which, in the case of ductile fracture, precedes the change associated specifically with the creation of the individual fragments/fracture surfaces. At some location, the part has to end up with a change in the ratio of the dimensions of two or more features. **NOTE:** Ductile torsion of a smooth cylindrical bar may not have any visible dimensional changes.

Brittle behavior almost always results in cracking with little or no visibly discernable shape change.

For example, see Figure 8.10 (below). It shows a necked area right around the crack. This is usually a reliable sign of ductile pre-fracture behavior. The neck in Figure 8.10 is a result of plastic deformation and the ratio of the diameter to the length of the part has decreased at the neck.

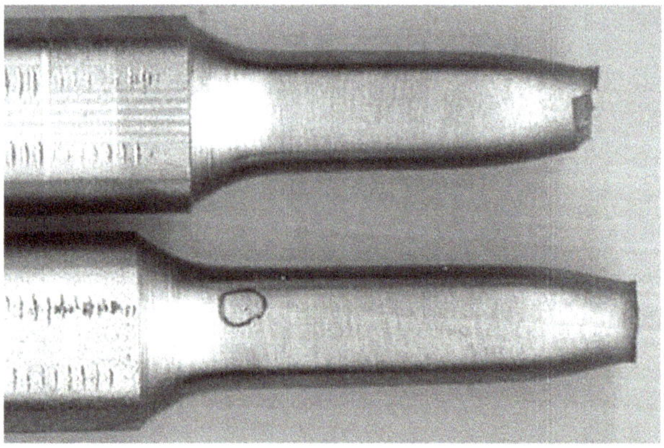

Figure 8.10: *(repeated image) Macro-ductile behavior (formation of a neck) followed by crack*

Many people think that necking or visible deformation of the part means it is a ductile crack, and lack of such readily visible deformation means it is a brittle crack. But this is not necessarily the case. First of all, remember that we must specify macro- or micro-scales as well as the material's general behavior during the specific, actual crack event, versus what happened to the component before and after final separation. Let's look at some tricky examples.

The chain link in Figure 8.11 (below) used to have a symmetrical shape and be a continuous flattened loop. But now there is a crack and the shape has obviously changed from the as-manufactured symmetry. So, this is a ductile crack, right?

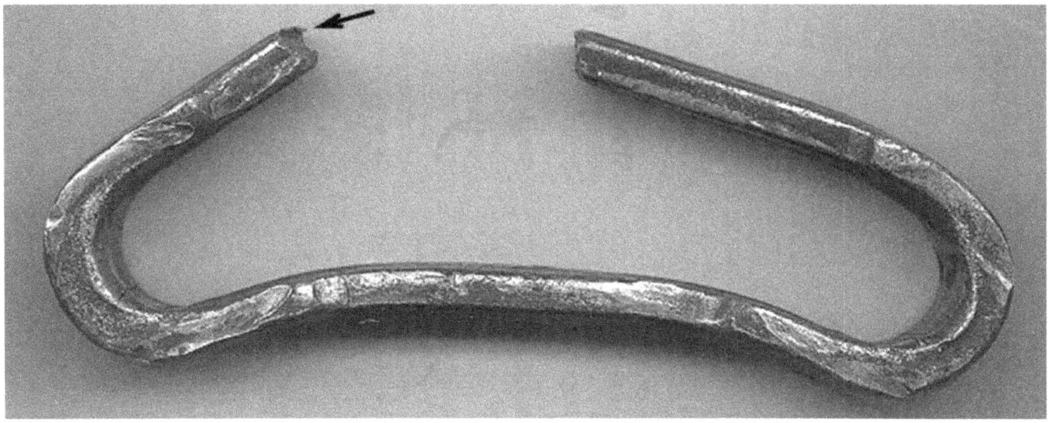

Figure 8.11: *(repeated image) This is mostly a macro-brittle crack followed by ductile behavior in a different portion of the link. The arrow shows a tiny shear lip, which indicates the final separation location.*

KEY POINT: A ductile part may behave in a brittle manner during cracking and subsequently may behave in either a ductile or brittle manner.

The question here relates to the crack process itself rather than what happens after the crack completely separated the previously connected material.

A chain link is loaded in tension, so there would have been tensile stresses acting along its length to pull the material apart. In this case, the majority of the crack surface is more or less perpendicular to the length of the material on either side. There is only a small shear lip along the outer surface of the part, which is a ductile feature (see black arrow in the image above).

The deformation that resulted in the two crack surfaces being separated in space is not associated with the crack process; rather this deformation happened after the crack was completely through the cross-section. This now allowed bending stresses to be created in what is shown as the lower portion of the link. The portions of the link near the crack are actually pretty straight; just as they were at the time the link was made.

It is not possible to see in this single view, but the material adjacent to the crack is not necked down at all. This was a macro-brittle crack, in a ductile material. The fact that the material was ductile is shown by the post-fracture deformation elsewhere in the link.

Of course, it would require a complete analysis to determine if there was some local embrittlement at the crack area. It is very possible, since this chain link was part of a conveyor in a meat processing plant and harsh chemicals were used to clean the conveyor line. A corrosion pit might have helped to initiate the crack and the corrosive media might have generated embrittling hydrogen. Corrosion or embrittlement could have been contributing factors to the ease or speed of crack initiation and propagation.

In any case, the crack was found through additional examination to have beach marks typical of a fatigue crack. Conceptually, a fatigue crack is by definition one that propagates below the nominal yield strength, even though locally the stress must be above the tensile strength or no crack could grow. Macro-scale (visible) deformation is impossible when the stress is localized to such a degree. A fatigue crack has only very limited (micro-scale) ductility in the layers immediately adjacent to the crack.

MORE ON NORMAL STRESSES AND SHEAR STRESS

DEFINITION OF A MACRO-SCALE BRITTLE CRACK: A macro-scale brittle crack is one that grows due to normal stress, which is not the opposite of abnormal stresses! Ha ha ha! Instead, it is the stress that pries the crack open.

DEFINITION OF THE OPPOSITE OF NORMAL STRESS: The alternative to normal stress is shear stress. High magnitude shear stress causes layers of atoms within the crystal structure to slide past each other. New atomic bonds are made as soon as the old ones are severed. In some geometries, eventually the sliding in opposite directions of the material on each side of the shear plane means that there will be no more atoms available to reform the bonds *(again, as in the easy open can, described earlier in this chapter).*

KEY POINT: Shear stresses are what allow deformation to happen, as well as facilitating ductile cracks.

In addition to the normal stresses, the chain link contained shear stresses, both resulting from tension. The presence of shear stresses is why there's a shear lip. The reduced constraint acting on the thin remaining ligament allowed ductile behavior to take over from the macro-brittle fatigue. Many loading geometries create simultaneous normal and shear stresses. At an oversimplified level, if the shear stress exceeds the shear strength, or the normal stress is greater than the tensile stress, a crack may propagate.

In tension, the maximum shear stress is only half of the level of the maximum normal (crack opening) stress. This means that normal stresses will drive the main portion of the crack.

THIS IS IMPORTANT: If the crack is perpendicular to the length of the component, it could be a crack driven by normal stresses due to tensile loads.

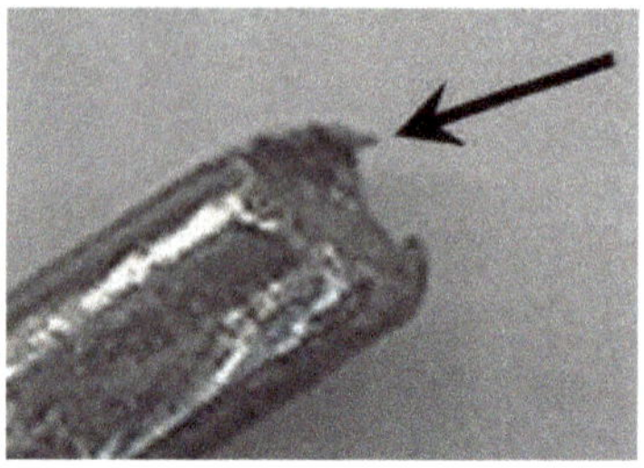

The plane of maximum shear stresses in tensile loading is at 45 degrees to that of the maximum normal stresses. In fact, we can see a tiny shear lip at 45 degrees to the main part length (see repeated image close-up, which also shows evidence of post-fracture impact damage). This is the final portion to separate along the upper edge of the fracture surface; it was a macro-scale ductile crack. In many cases, macro-scale brittle cracks (driven by normal stresses) change to macro-scale ductile cracks (driven by shear stresses) as the final separation occurs. Some people might call this shear lip a "slant fracture".

Until there was very little connecting ligament left, the shear stresses were not able to cause any large-scale deformation. Remember, in tension, they can only rise to half the value of the normal stresses.

Figure 8.12 (next page) shows another tricky part. Learn this knowledge and practice, and it will certainly separate you from the merely competent.

We are looking at the fracture surface of a shaft that was loaded in torsion. We see the smear marks that happened as the crack was in the final stages of fragmenting the shaft. There is no necking. The diameter is unchanged at the crack area. So, this is a brittle crack, right?

Think again! This shaft was loaded in torsion, so the crack opening stresses (the normal stresses) act on helical planes that are at 45 degrees to the axis of the shaft. Go back to look at Figure 8.5B (the hose twisting experiment). Brittle cracks are helical when they are created in torsion.

How are you going to remember this very important point? You will remember it by proving it to yourself with a physical example. Prove brittle cracks are helical when they are created in torsion by twisting a carrot or a piece of chalk. Carrots and chalk are inherently brittle materials. They will not usually break in any manner except a helical shape when subject to torsion. See Figure 8.13.

***Figure 8.12:** Crack surface of macro-ductile shaft fracture that was broken in torsion. This is a classical smear pattern.*

Since carrots are brittle, if you apply enough torsion to crack a carrot, the crack will be helical (similar to a screw thread). In Figure 8.13 (below) you will notice a small vertical shelf on the right; it is a radial plane. In torsion, radial planes are shear planes. Thus, technically, that part of the crack would be considered macro-ductile.

When we look more closely, we see that the radial plane crack is likely mesostructure driven. The core of a carrot has axially-oriented fibers with radially-oriented planes of weakness associated with them.

I'm sure you are a lifelong observer of carrots just like me. If so, you may have noted that sometimes there is a radial crack in the carrot even before it is taken out of the dirt or home from the grocery store! What does this indicate?

Radial cracks in the outer layers of a carrot are usually from hoop stresses; these are due to (or assisted by) internal pressure created by the core diameter growing faster than the inside diameter surface of the carrot's outer flesh. This can happen when there is a lot of rain. These radial cracks are planes of maximum shear stress in torsion; but the radial cracks are normal stress (hoop) if generated by internal pressure.

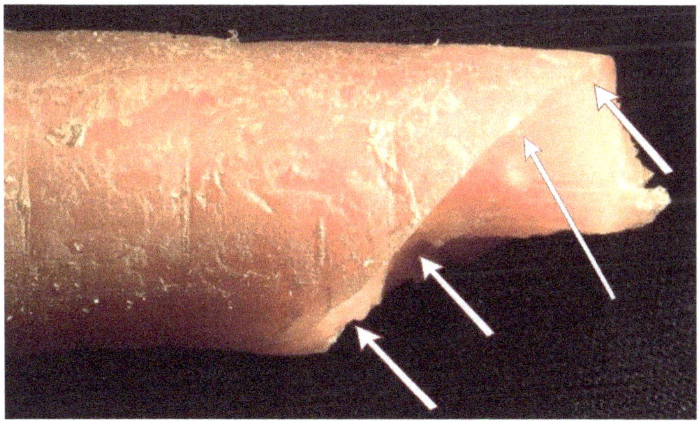

***Figure 8.13:** Brittle torsional crack in a carrot with macro-ductile shelf*

Gardeners can experiment by loading a carrot in torsion in different "manufacturing processes". *Compare and Contrast!* One carrot has grown in relatively dry weather; the other carrot grew during a 3-7-day wet spell toward the end of the growing season. Compare the behavior when you load each carrot in torsion. The one that grew faster (in the wet spell) sustained residual stress (the core added internal pressure) to the torsional forces applied by way of the gardener's hands. The outer layers of the carrot act like a pressure vessel in this case.

Stresses due to torsion are much lower in the core of the carrot than at its surface. Carrots are complicated structures with an interface between the outer flesh and the core. The core is subject to compression while it creates tension in the outer layers.

The crack often changes behavior as it interacts with the core. The core of the carrot has a very fibrous structure and that is weaker along radial/longitudinal planes.

As previously mentioned, we will likely see macro-ductile radial planes in carrot cores subject to torsion, even though the core isn't ductile in the common sense. It is a lesson to remember. The cracks are macro-ductile because they are due to shear stresses – otherwise known as sliding along the radial planes of weakness.

KEY CONSIDERATIONS

Always remember to clarify the context! Are you talking about crack behavior or the material capability in favorable conditions?

Ductile cracks happen unexpectedly much more rarely than brittle cracks.

EXERCISE 8-J: Break some carrots

It's time to break some carrots in torsion and bending. Observe the crack planes and describe them.

____

____

____

____

____

If you have some fresh carrots from your garden, compare them to older ones your neighbor purchased in a market. Experience the force required to propagate cracks in different populations of carrots. Also, observe and describe the fragment shapes. Unless you grow your own, people outside of of North America are perhaps more likely to have or be able to find truly fresh produce in a market.

____

____

____

The planes where shear stresses can act in torsional loading are the circular transverse planes and the rectangular longitudinal (radial) planes (as described above).

Now let's look at Figure 8.14 (next page). It shows a steel shaft. It was also highly stressed. It twisted off in a ductile manner. In fact, if you look at cylindrical surfaces which have longitudinal grinding marks you can often see evidence of twisting even though there is no change in shape obvious to the casual observer. In this figure, we are looking at the edge of the crack, or the crack profile.

Figure 8.15 (below) shows two socket wrench inserts that were the same hardness (HRC41). They were used in similar applications. Due to design problems, there might have been some bending in addition to the torsion. But one showed a nearly classical ductile response, while the other showed a nearly classical brittle response.

Did the part on the left experience a higher strain rate? Was it colder than the one on the right when it was used?

Figure 8.14: *Note spiral (more properly called helical) marks from ductile response to torsional loading.*

KEY POINT
In torsional loading of a uniform diameter cylinder, at any given depth position (distance from the surface of the part), the shear and normal stresses are equal to each other.
Thus, torsional cracks may occur in so many different orientations even within a single part.

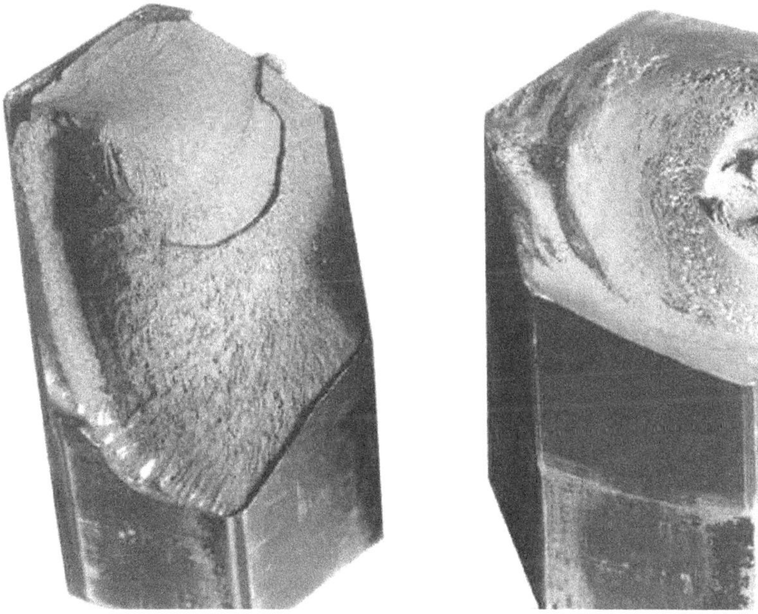

Figure 8.15: *Predominantly macro-brittle (left) and macro-ductile (right) crack behavior in nominally identical hexagonal bars (the one shown at right had a slightly larger "diameter")*

MACRO-FRACTOGRAPHY PRACTICE - If you ever have the chance to walk in the woods after a major storm, you may amuse yourself trying to understand why a particular tree was blown down while a nearby tree remains upright.

You may see trees rotten and broken at the base. The rotten parts of a tree offer their own opportunities for contemplation of wear (insect or bird damage) and corrosion (fungal action).

While falling over, some of my rotten trees took apparently healthy trees with them. You may see apparently healthy trees pushed to the ground almost intact as major portions of their root systems were lifted out of water saturated ground and the remaining roots acted as a hinge.

You may see tree branches in an orchard that were broken off as a result of heavy fruits adding to stresses from the wind or ice. Surely such a weather-related event presents a wonderful (if sad) opportunity to meditate on the relationship between strength and stress in structures.

8.12 Fractographic Revelations for Failure Analysis

Understanding *how* a component fails is an important step in understanding *why* a component fails. In order to understand how a component cracks, it is important to understand what loading geometry or geometries could have been responsible for the fracture. It is equally important to understand qualitatively how high the load was, how fast the component was loaded, and to know what type of material was specified, and present.

The Failure Analyst must strive to learn to read the fragment shapes to determine what loading geometry was actually present. Remember, this is key to being able to properly determine whether the component was manufactured, installed, and used according to the design intent.

8.12.1 Loading Geometries

The simplest loading geometries to understand in failure analysis are axial tension (stretching) and compression (squeezing). Bending creates more complex stress patterns than the axial loading conditions. In common with axial loading, the normal stresses (crack opening) are higher than the shear (sliding) stresses. In torsional loading, the fact that shear and normal stresses are equal creates opportunities for a wider array of possible crack orientations.

These loading geometries: axial tension, axial compression, bending and torsion, are the major concerns of people working to understand fracture of consumer products, most machinery components, and civil structures.

Hertzian stresses, or contact stresses, are another important loading geometry, but are mostly relevant to bearing assemblies and chains.

Direct shear is the final major loading geometry. We met direct shear when we discussed the easy-open can earlier in this chapter. Here we'll focus on the Big Four loading geometries: axial tension, axial compression, bending, and torsion. We will briefly describe Hertzian stress characteristics in a separate sub-section.

8.12.2 Insights from Axial Tension for Failure Analysis

In tensile loading, crack-opening (normal) stresses create cracks that are perpendicular to the loading direction. In Figure 8.16 (next page), we see a crack running perpendicular to the length (or the axis) of the cylinder. This is one side of a broken grey cast iron tensile test coupon.

Aside from threaded fasteners and hydraulic cylinder rods, there are very few actual simple tensile fractures in the parts that have crossed my desk over the years. This is a more common loading geometry in civil structures such as bridges and the spokes in bicycles wheels.

KEY POINT

In axial tension, the brittle crack is oriented perpendicular to the direction of the load that creates it!

Figure 8.16: *Macro-brittle fracture in axial tensile loading – grey cast iron always behaves in a macro-brittle manner at room temperature.*

In Figure 8.17 (below), we see a piece of flat strip steel that has been tested for tensile strength. The angle of the crack is not exactly 45 degrees, which is the theoretical plane of maximum shear stress for tensile loading, but it is clearly not perpendicular. The ratio of the width to the thickness affects the exact orientation of the plane of maximum shear stress. For our purposes, we can use "roughly 45 degrees" as an indicator of ductile cracking in axial tensile loading.

Figure 8.17: *Flat tensile bar – macro-ductile fracture resulting from axial tension*

See Figure 8.18 (below) for a look at the classical "ductile" cup and cone fracture. Tensile testing routinely produces this in properly heat-treated moderate hardness, round steel test coupons. This is, technically speaking (as noted previously), a mixed-mode fracture. The slanted walls of such a fracture are indeed macro-scale ductile. However, the flat bottom center area would be considered macro-scale brittle.

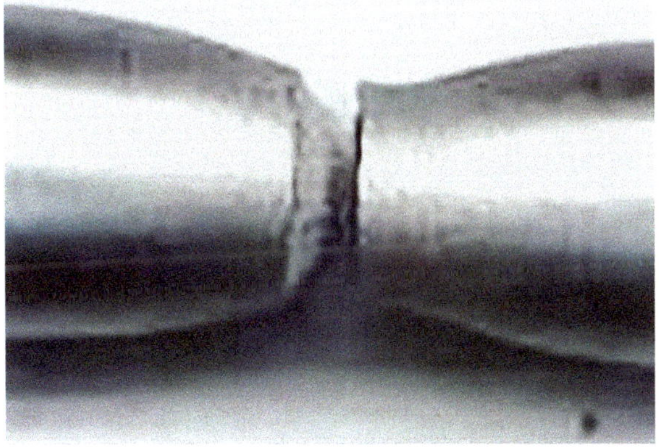

Figure 8.18: *Cup and cone, or classical ductile fracture, is macro-ductile (45-degree angle) around rim, and macro-brittle (perpendicular to axis) in the center.*

What? This does not make sense! It goes against everything most mechanical, civil, and even many materials engineers learn about ductile and brittle cracks. This is not the first time we're highlighting this point. In particular, I encourage heat treaters to learn to think this way in order to learn to distinguish ductile and brittle cracks in a manner that allows them to defend themselves from accusations that they embrittled someone's parts during heat treatment.

As Lewis Carroll wrote:

"When I use a word," Humpty Dumpty said in rather a scornful tone, "It means just what I choose it to mean — neither more nor less."

Well, we don't need scorn here. We are simply cultivating clarity *(neither more nor less)* in the face of the arbitrary nature of the human classification systems that we use to describe aspects of physical reality.

Remember that a major point of macro-fractography is to identify the loading geometry so you can tell if the part was loaded the way it was supposed to have been loaded. If you look at the micro-scale (scanning electron microscope) features on the flat bottom of a macro-scale brittle cup-and-cone fracture, you will most likely see micro-voids (or ductile dimples). Just remember, the *macro*-scale information relates to a totally different set of factors than the *micro*-scale.

- The micro-scale data (micro-voids) tells you how the crack interacted with the microstructure.
- The *meso*-scale tells you the crack direction.
- Macro features derive from the loading geometry and the mechanical behavior.

A small repeated image of Figure 8.8A is provided to the left for your convenience. It shows a macro-brittle fracture of a threaded fastener loaded in an application that did not have any known fatigue loading. However, the Subject Batch of fasteners was not properly tempered and the hardness was approximately HRC45, approximately 6 points harder than the SAE J429 Grade 8 allows. The environment was moist and there was little corrosion protection. Corrosion generates hydrogen, which may embrittle hardened steel. Threaded fasteners do their work by sustaining a clamping load. The fastener is stretched as the things it is fastening are squeezed together.

The crack initiated at a thread root, which is an expected high stress location. This crack shows us an example of a fan pattern. A fan pattern is a meso-scale fractographic feature that *points back to the crack initiation.* The crack deviates from perpendicularity as it follows the thread root helix. The final separation of a crack in a thread root always has a step. The crack has to exit at both positions that are across from the initiation. This creates further deviation from true perpendicularity. However, the deviations are readily understood to be a result of the slightly more complex part geometry.

Note that the student of damage analysis must be on guard to discern when someone "explains something" and when someone "explains something away!"

I hope that you will do the research, if you are a new practitioner, to realize that I gave a legitimate explanation and am not "explaining away" an anomalous feature that should be further investigated.

Note that a fan pattern is an element of classical macro-fractography. We're calling it meso-scale. This is the part of fractography that focuses on surface color and texture, rather than the crack plane orientation (which tells us about the loading geometry). Meso-fractography often sheds light on the crack initiation position, but doesn't necessarily lead us to draw a conclusion about the loading geometry. For that you must consider the fragment shape.

8.12.3 Insights from Axial Compression for Failure Analysis

We were introduced to the idea of axial compression back in the section on deformation damage in Chapter 5. Now, let's consider some stouter (or bulkier) compression members.

In Don Wulpi's book, *Understanding How Components Fail,* he performed an elegant experiment using a chunk of material with hard, brittle upper and lower surfaces, but soft and ductile everywhere else. See Figures 8.19 A-C (below) showing images[155] from the book.

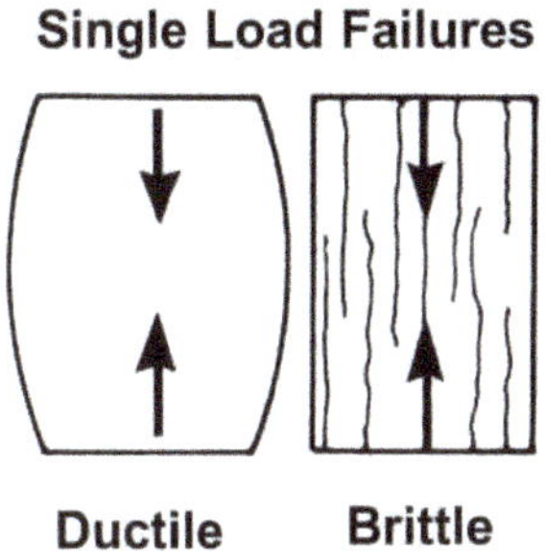

Figure 8.19A: *Illustrations of single-load failures as ductile and as brittle*

Figure 8.19B: *Ductile material with hard surfaces*

Figure 8.19C: *Close-up of Figure 8.19B*

In Mr. Wulpi's Test Specimens, compressive loading resulted in a bulged mid-section with cracks in the hardened surface layers. The cracks are parallel to the direction of the force. Cracks parallel to the force happen because the force is squeezing the material along the direction of motion. A crack can't open perpendicular to the force (which is what happens in tension). However, the elastic response of the material causes it to expand in the two directions that are not being squeezed. Tensile stresses are created in the directions perpendicular to the compression load! Do you appreciate the elegance of this experiment? Are these results what you would have imagined?

Figure 8.19B shows a cube of low carbon steel that was carburized, followed by machining of four adjacent sides to remove the carburized layer. Under compressive axial loading, the soft and ductile steel bulged out. The harder, brittle top and bottom layers cracked parallel to the direction of the force application, as shown in the close-up view (Figure 8.19C).

While this image (Figure 8.19A) does not show it, a ductile part loaded in axial compression could crack as a result of these same hoop stresses. We'd have a macro-brittle crack in a ductile material. The crack would start at the center of the height of a cylinder of uniform cross-section, where the bulge was greatest.

Figure 8.20 (next page) shows a punch that was used in a metal stamping operation. The punch was subject to axial compression. Its shape is of intermediate slenderness. What do I mean by that?

Its slenderness is between a long thin pole (similar to a radio antenna) and the thick block shape of Figures 8.19A-B. So, it is a punch of intermediate slenderness used in a metal stamping operation. The crack was a result of likely at least tens or perhaps hundreds of thousands of compressive loads in a stamping press. The crack has the beach marks classical for a fatigue crack. Beach marks are another meso-fractographic feature. The crack has multiple initiation points at the sharp points used to punch holes in flat steel as well as at a lubrication/vent hole (drilled from the side).

The crack, similar to those of Figure 8.19B-C, is parallel to the direction of force. The material used to make punches is usually quite hard (Rockwell C 60s). Tool steel usually has very limited ductility, so it's no surprise that the punch has behaved in a macro-brittle manner.

[155] These images were modified from Don Wulpi, Page 30 Figure 1 and Page 37 Also see Figure 89 on page 622 of Volume 11 2002 of the ASM Metals Handbook. Copyright ASM International. Used with Permission.

Figure 8.20: *Punch used in axial compression – fatigue with beach marks radiate from lower left point. The crack plane is near the diameter of the punch, not at a random chord. The diametral plane has a stress concentration due to the upward angling vent bore. The diametral plane is also the plane of maximum shear stress in bending. Orientation in use is vertical.*

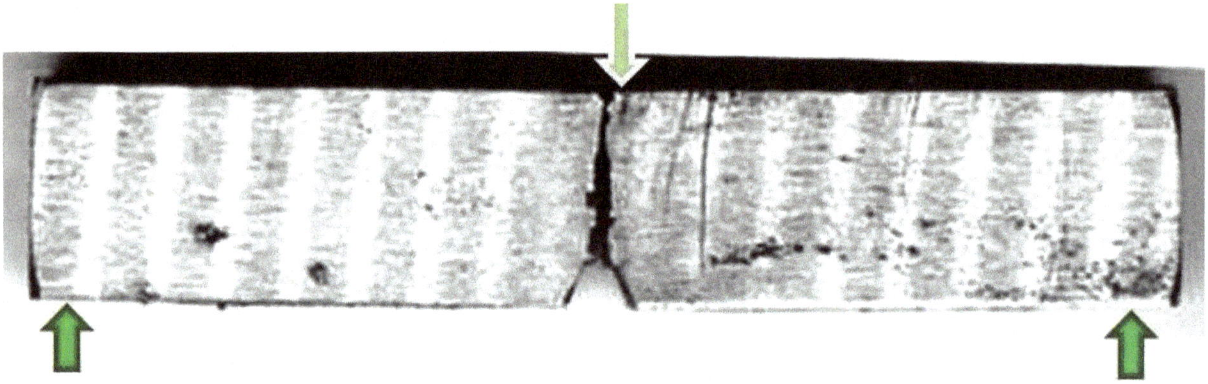

Figure 8.21: *Three-point-load bending of a notched bar (beam) – the crack initiated at the notch root and propagated through. Green arrows show load points. The notch in this test coupon increases the stress beyond what it would be without the notch.*

8.12.4 Insights from Bending for Failure Analysis

At the beginning of this chapter, we learned that bending is one of the basic loading geometries.

In Figure 8.21 (above), we illustrate macro-scale brittle fracture due to three-point bending in a simple beam. The beam is pushed up on the two ends and down in the middle.

Figure 8.22 (next page) shows that the tensile stress is maximum along the lower surface (opposite the load application point in natural gravity loading) and compression stress is maximum at the upper surface. Stress is proportional to the distance from center of height. This image is only showing us the values for the positions directly under the load. The positions directly under the load are the most highly stressed; and this is the expected crack initiation location.

I recommend looking at the Machinery's Handbook[156] for equations for other load conditions and positions.

The shear stresses peak along the neutral axis and they always have a positive value. But the maximum shear stress is approximately half of the maximum normal stress, as shown in Figure 8.22. Thus, unless there is a plane of weakness at mid-plane, at least when the crack initiates, it's likely to do so in a macro-brittle manner, on the tensile surface.

Compare the stresses shown in Figure 8.22 to the crack appearance of Figure 8.21.

Note the crack in Figure 8.21 is in a Charpy V notch bar, which was actually broken by a hammer impact going parallel to the ground at the

---

[156] Machinery's Handbook, Digital ,Edition, 31st Edition | © 2023 Industrial Press. ISBN: 9780831138318 | https://books.industrialpress.com/machinery-handbook/

point of impact! The vertical or horizontal position of the part is not really important when we talk about loading geometries. The stresses that are created in the part as a result of the interaction between the loads and the part shape are what we need to worry about.

Continuing to Figure 8.23A-B (below), we note that any time stress exceeds the strength, a crack can open. The presence of the crack tip makes the loading even more uneven. The stresses no longer vary gradually from tensile to compressive as you go from the bottom to the top surface layer.

The maximum tensile stress is now experienced at the crack tip, which may be anywhere in the thickness at any given moment.

The original lower surface layers now have a near zero stress. But the crack tip location does not experience the same level of stress as the intact material in its original condition. The crack tip location has a much higher stress value due to concentration of the stresses. Most of the shape change (strain) associated with the stress is forced into the small area at the crack tip instead of being spread out. The atoms along the crack tip are *really stressed out*! When the material is very hard and has a high yield strength, deformation is very limited and the crack tip remains very sharp. This creates even higher local stress levels.

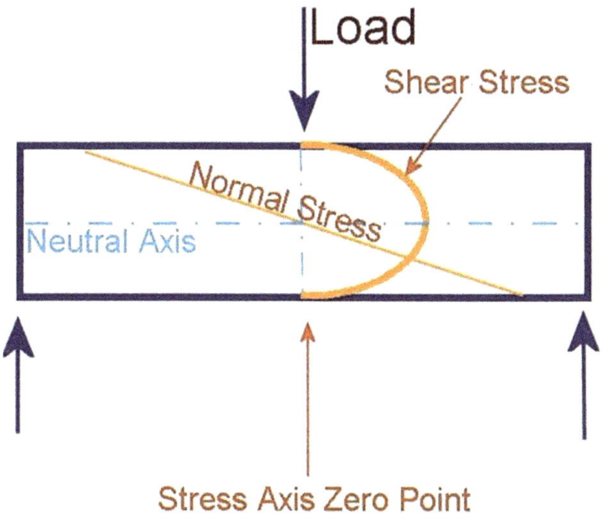

Figure 8.22: *Relative values of normal stresses to the shear stresses (by height distance from the Stress Axis Zero Point) as a function of the height position in the beam*

In Figure 8.23A-B, we can see that the crack started out perpendicular to the length of the beam, parallel to the direction of the load application and perpendicular to the tensile stress; but the crack does not go straight through.

Figure 8.23A: *Wrought steel, hardened (HRC55) with macro-brittle bending*

Note that the white vertical stripes are image artifacts from the optical scanner used to create the image.

Figure 8.23B: *Close-up of wrought steel, hardened (HRC55) with macro-brittle bending*

The swoop is not a shear lip or slant fracture. The crack that creates the swoop is propagating due to normal tensile stresses. Try to imagine the center of the bar going down while the ends push back upward. But the main reason for the swoop is anisotropy. See text.

Something else is complicating the situation, which we can see because the direction of the crack changes. This is a swoop-shaped profile. One reason this happens is because this material does not have the same strength in different directions. This is often the case in wrought materials. Grain orientation and the presence of weakening stringer-type inclusions are two things that can make it easier for the crack to travel along a direction parallel to the beam than it is to travel perpendicular.

Think of a piece of wood. If you split a log, it is much easier to split it along the grain than across the grain.

Figure 8.24 shows a real part rather than a test coupon. It broke due to a bending load and it shows where the crack changed direction - at least in part due to its oriented grain structure.

As already alluded, shear stress are maximum along planes that are oriented differently from the planes of maximum normal stress.

EXAMPLE: *Imagine a deck of cards in your hands. If you look closely while bending or flexing the pack in your hands, the edges of the cards at the ends perpendicular to the length of the bend are now forming tiny steps. This is because the cards have slid a little bit with respect to each other due to the bending.*

The layers of a solid part can't slide as easily as the cards, but the same stresses are acting to make this happen in the solid parts as well. Slip at the micro or nano-scale will be happening along close-packed planes; a topic which is coming up later in this chapter. Shear stress is the technical name for this. We have mentioned this multiple times now.

Figure 8.24: *Real part showing macro-brittle bending fracture with swoop in wrought material*

When an intact beam is loaded in bending, the shear stresses are parallel to the length of the beam - just like in the pack of cards. The shear stresses are highest near the center of the beam thickness.

Figure 8.25: *This leaf spring was broken mainly due to tensile normal stresses, but a few features (arrows) are a result of shear stresses near the mid-thickness. Bright white narrow band at upper edge of crack (not visible at ~25% of width at the right end) is a shear lip (or the final separation feature).*

Occasionally, I have seen a crack make a "hard turn", usually in the center half of the height (or thickness) of the beam. This is due to shear stresses taking over the crack propagation. In the absence of a crack tip, the normal stresses are near zero at the center of the height of the beam. It's possible that the shear cracks happened before the normal stress crack arrived at the neutral (midheight) plane.

Cracks due to shear stresses in bending are at 90-degrees to the cracks created by the tensile stresses. See Figure 8.25.

It is possible that the strength of the mid-thickness plane is lower than other locations due to segregation of low melting point alloys or impurities in the steel. These tend to solidify in the last portion of the slab or ingot to solidify. Obviously, this is in the core, near the centerline. This would modify the ratio of the strength to the stress at this orientation and position.

NOTE: this same anisotropy related swoop phenomenon is frequently found in coil springs because they are usually very heavily cold worked in one direction. Thus, they are very anisotropic (having different properties in different directions and orientations).

Let's explore another more common situation. As the crack lengthens, the stress changes dramatically. The entire remaining ligament may be in tension or compression. The tensile and compressive loading-related shear stresses are at a 45-degree angle *(you may remember this fact)*. These tensile and compressive loading-related shear stresses at 45-degree angles can allow a macro-scale ductile crack to form. This finishes off the part that started cracking in a macro-scale brittle manner. The shear lip along the upper edge of the leaf spring shown in Figure 8.25 represents this type of transition from bending to tension, or potentially to torsion.

NOTE: Do you see the change of direction (protruding to receding) at the far right of the shear lip? What do you think might have caused that?

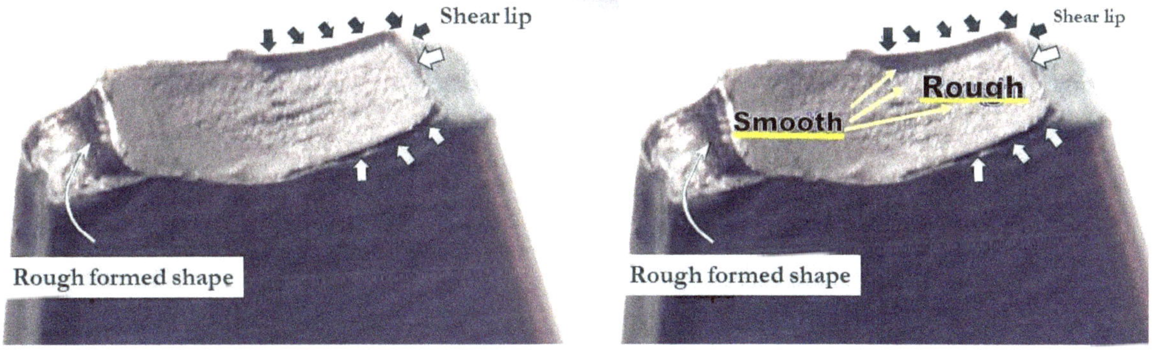

Figure 8.26: *This component was also loaded in fatigue, but it was a durability test in an air-conditioned lab, performed with a constantly alternating load, so there are no easily visible beach marks. The orientation during the loading was such that the plane was vertical. That plane is now the fracture surface. The edge of the crack shown at the left was the lower tensile surface.*

Note in Figure 8.26 how smooth the crack is. There are two major reasons for this.

1) It is the crack initiation area, so as the load was applied and released, the crack's surfaces rubbed on each other and smoothed the surface.
2) When the crack is very short, the stress concentration is lower than when the crack is longer and there is less remaining ligament to distribute the force.

Also, pay attention to the faint fan pattern indicated by yellow arrows in Figure 8.26, which is just beyond the smooth fatigue features. The ridges that create this fan pattern are a result of the more intense concentration of stress. There is more strain energy available to be transformed into surface energy.

The "picture frame" or lip feature around the right side (top in test loading) indicates the final separation. Compare and Contrast this to the feature in Figure 8.25.

Now, please examine Figure 8.27 (below). This was a beam that was used as a mast for a forklift. The plane of the crack would have been horizontal in service. The crack initiated at the lower right shown by the tan arrow at the stress concentration created by the tooth root. The bending load was due to an impact, not gravity.

Note the small dark pie slice shape, with the dark arrow showing the direction of crack (up to the left). This does not appear to be a fatigue crack.

It's rough everywhere.

The flash rust was already present when I first received the artifact. This is typical of rapid monotonic crack surfaces, which have no oxide to protect them. There is no obvious shear lip present.

Figure 8.27: *This beam was used as a mast for a forklift. The forks ran up and down the teeth shown on the right. The bending here was not due to gravity. It was due to an impact.*

8.12.5 Insights from Torsion for Failure Analysis

Torsion is the loading geometry that is experienced by a screwdriver blade shank. See Figure 8.28 (below). As in the case of bending, the stresses are highest at the outer surface of the part. In contrast to bending, the shear stresses and normal stresses are nominally equal to each other at any given location within the object.

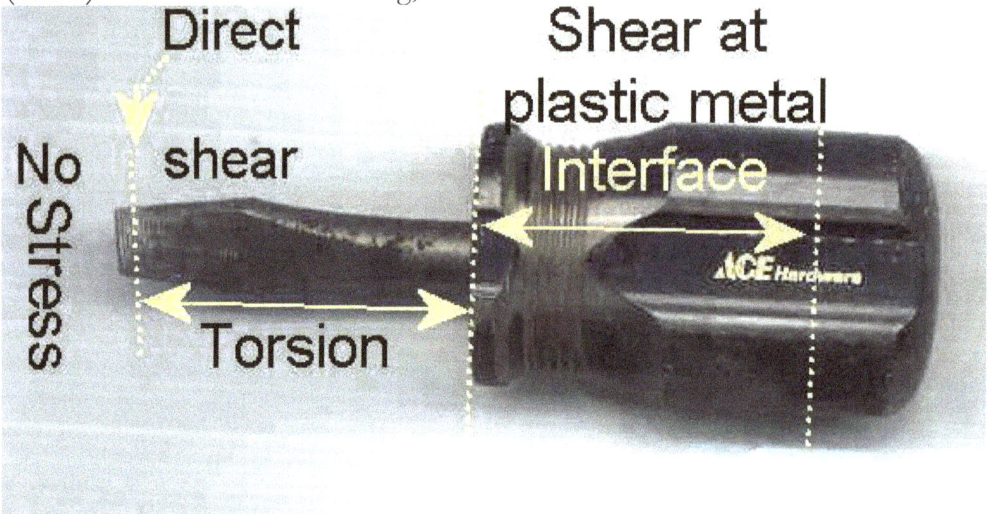

Figure 8.28: *The blade material outside of the slot in the screw is subject to torsion. Within the protruding portion of the cylindrical length of the shank, the stress would be equal at a given depth position. There is a shear stress acting at the interface between the blade shank and the over-molded plastic handle.*

Because of this, as we saw in Figure 8.15, nominally identical loading may result in ductile fracture or brittle fracture at the macro-scale. Ductile fracture results from shear stress. This occurs when the yield strength is locally exceeded. Brittle fracture results from normal stresses, or when stresses pry the crack open.

See Figure 8.29 for the stress distribution in the protruding cylindrical part of the blade shank.

My fractography mentor liked to start all of his macro-fractography lectures with a thought experiment about broken screwdrivers. He had worked as a forensic expert at times. At least once, someone gave him a broken screw driver claiming it had caused an injury. So how could we tell if the injury was caused during normal use?

Normal use is essentially torsion with a steady axial compression force (to hold the screwdriver in the screw slot). So, we would expect either a helical crack like the carrot of repeated Figure 8B or a flat smeared transverse crack such as repeated Figures 8A, 8C, or the right side of repeated Figure 8D. Or, the tip could break off due to direct shear at the top of the slot.

There is one other possibility for a ductile crack. It would be along a radial plane. However, with long slender cylinders, the cross-sectional area of the transverse plane is much smaller than that of a radial plane (the transverse plane is a circle, while the radial plane is a rectangle with width equal to the diameter of the cylinder/shaft and length equal to its length). So, the stress would be much higher on the much smaller transverse plane; it would be much more likely to experience the fracture. If there is a radial crack, it would usually indicate a highly directional texture or microstructure or a major weakness in the longitudinal direction such as an oxide seam (compared to the mid-plane shear fracture in bending).

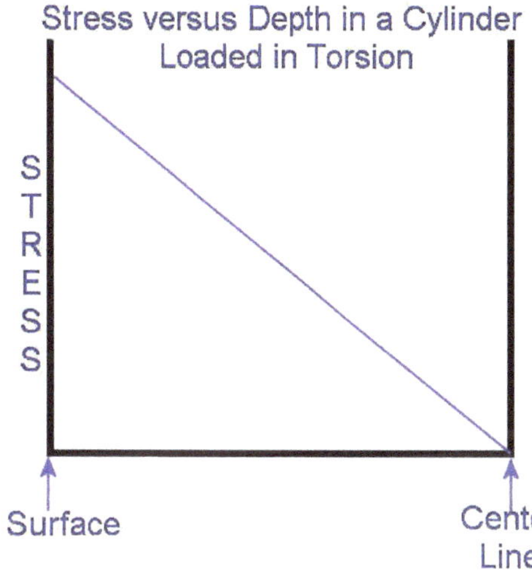

Figure 8.29: *The stress is greatest at the surface and drops to zero at the center line. The values of the normal stress and the shear stress are equal at any given position.*

Figure 8A

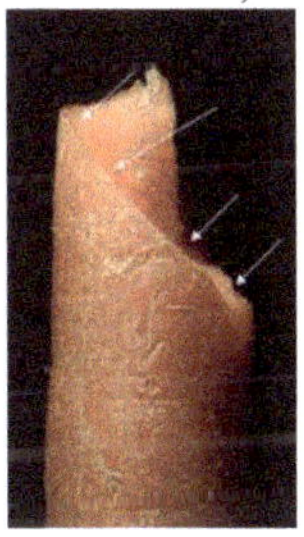

Figure 8B

Figure 8C

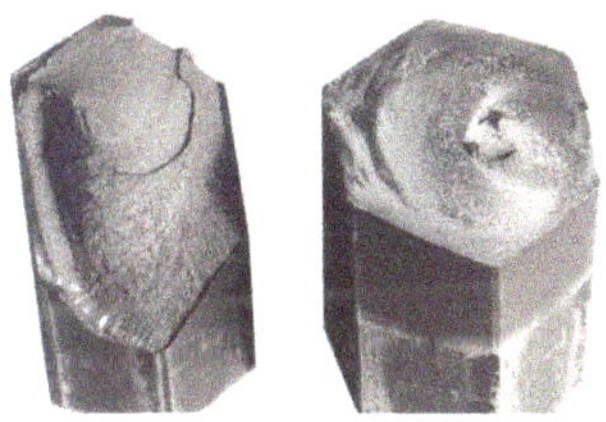

Figure 8D

All figures 8A through 8J are repeated images for your convenience

What if the person who had the accident or injury due to the screwdriver damage event presented a broken screwdriver with a swoop profile such as those shown in Figures 8E and 8F; that would indicate, whatever the individual said, that the screwdriver had been used as a pry bar. It can't have a swoop fracture from torsional loading.

Figure 8E

Figure 8F

If someone used a screwdriver to open a can of paint, it would probably not result in a fracture. If someone used the screwdriver shown in repeated Figure 8G to open a can of paint, it would be quite unlikely to result in fracture. But if King Kong, Hercules, Ma Durga, or Boudicca (not to leave out the strong women!) used a longer screwdriver to break open a steel door from a steel frame, it could result in a fracture. Due to the processing required to create long, small diameter cylindrical material (wire rod), the crack would likely have a noticeable swoop in bending.

Finally, if someone used the screwdriver as a chisel, then the long slender cylinder would be loaded in axial compression, which could result in bending (we saw a long time ago, at the beginning of this chapter).

But it could also result in chipping at the blade tip. Maybe the screw slot was filled with paint, rust, or dirt and user tapped on the handle. Maybe the user struck the handle. Then the handle could chip.

One point to note is that many parts end up being subject to torsion at the final separation into fragments. Even if the part is intentionally subject to bending, as it loses rigidity, torsion may come to predominate. This is sometimes true of rotating bending fractures in shafts. See repeated Figure 8H.

Figure 8G

Figure 8H

See repeated Figures 8I and 8J, which demonstrate orientation of cracks due to stress states resulting from torsion, or twisting.

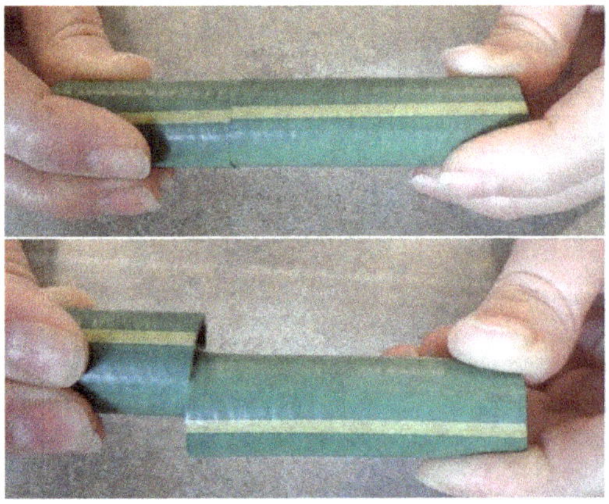

Figure 8I.

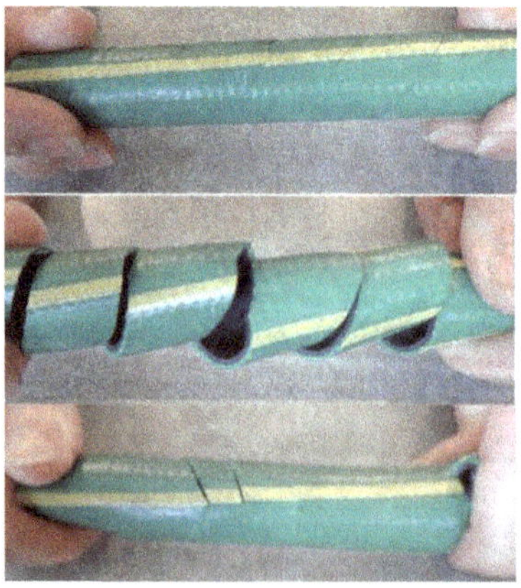

Figure 8J

In repeated Figures 8I (previous page) we see that a shaft with a circumferential groove could end up being stressed along such a plane or any shaft that has a length to diameter ratio that creates the greatest stress concentration in the component along a transverse plane.

In repeated Figures 8J (previous page) we see that the slits are forced closed and the hose must be twisted in the opposite direction to open them. Also note that the two slits at right in the middle inset have not opened as much as the ones to its left, which are closer to the 45-degree plane of maximum normal stress.

Figure 8.30 *Rotating bending is the loading geometry experienced by rotating shafts in machinery including propellers and fans. In this case, the keyway points were the crack initiation areas. This generally indicates an improperly fitted key. The high stress would otherwise be expected at the roots of the keyway or at surface damage elsewhere on the cylindrical surface.*

Now we examine Figure 8.30 (above) for rotating bending as the loading geometry. Notice that the first 40% of the crack is quite flat and perpendicular to the shaft surface. Then the crack starts to deviate. That is when the stress state gets more complex. Some torsional forces might be generated during start-up and stopping.

The last portion to separate is centered around the upper left (from 11 o'clock to 2 o'clock). Note the ridges that are along radial directions. This is the final separation; it's due to (1) unidirectional bending or (2) bending and torsion.

Remember Meso-Fractography Rule of Smooth-to-Rough. The final separation is much rougher than the fatigue area. The crack opens and is squeezed back shut at every location around the circumference in rotating bending. The initiation areas have been rubbed smooth.

8.12.6 Insights from Hertzian or Contact Stresses for Failure Analysis

Hertzian contact comes into play in specific applications. In failure analysis we see this type of loading geometry and resulting stresses when two surfaces, at least one of which is curved, contact each other. Bearing races, chain links, rollers and rotating cams are examples of applications where Hertzian stresses are generated. See Figure 8.31A and B (below).

Unlike the four primary loading geometries, Hertzian stress is unique in that the maximum stress occurs not at the surface but at a slightly sub-surface location. That is the reason that the stress fields in Figure 8.31B are widest at a position within the envelope of the component. Discontinuities such as inclusions can drastically shorten the life of bearings if they are in the zone of the highest subsurface stress. If it is determined that the cracks initiated at the surface, then that generally indicates high friction due to a lubrication problem. It is usually only possible to make this determination through the process of metallography. Visual inspection is not usually a helpful method to pinpoint Hertzian stress crack initiations.

8.13 More on Meso or Classical Macro-Fractography

We are now going to Contemplate Complexity.

We have just spent quite a bit of time looking at simple-shaped components that broke due to readily identifiable, simple loading geometries. But broken parts come in all shapes and sizes, and machinery components are used in many different types of environments. They are made from many different types of materials with widely varying microstructures. This complicates interpretation of fracture features. It can also be very challenging to evaluate a fracture in complex-shaped brackets, tooling with hidden, internal features, or any parts with heavy corrosion deposits.

Yet, some fracture surface contaminants give helpful clues about how a crack started and grew and how long it took to grow. These contaminants may have been deposited during service, manufacturing, or inspection. We must assume that fracture surface contaminants contain valuable information. So, it's important to leave the dirt (clues) on the crack surface until the appropriate time to clean has arrived.

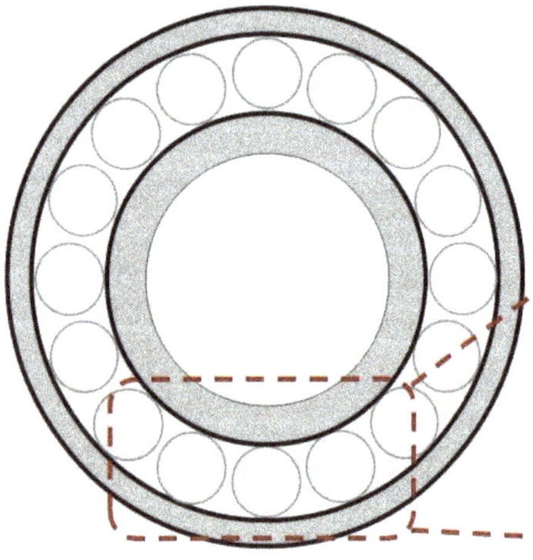

Figure 8.31A: *Ball bearing assembly schematic shows a cross-section at the center of the race width. Races and balls are subject to Hertzian stresses.*

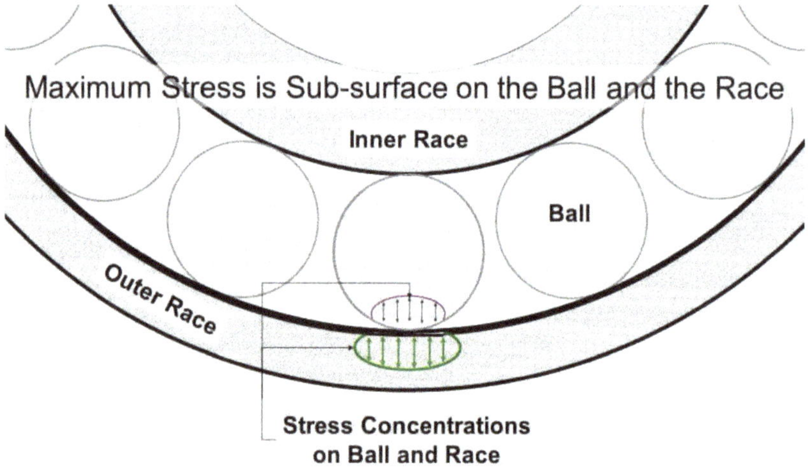

Figure 8.31B: *Maximum stress is sub-surface on the balls and races. In service, the stress fields shown will be developed when the ball contacts both the inner and outer races. The footprints will be elliptical in this scenario. The long axes on the inner and outer races will be perpendicular to each other.*

Parts having relatively simple shapes, such as straight shafts or beams, have fracture behavior that has been well-documented in specialized technical literature. Parts with complex shapes, such as intricate castings or brackets, require the investigator to think a little more about how the part would be stressed in normal use, and how those stress conditions might change if something was abnormal. For example, if a component was attached by a bearing and the bearing wore out (or was misaligned), what would the stress state look like? Often, I ask my client for help in understanding any complex loading situations.

As we have been learning, the features of a broken part itself do provide clues as to what types of stresses were present. It is not always possible to confidently know exactly what types of stresses were present. But even in these cases, it is often possible to know what stresses did not cause a particular crack to happen. For example, the usual stresses experienced by a rotating shaft will create one of a group of particular appearances on the fracture surface. If the fracture surface does not resemble any of these classical appearances, then it is safe to assume that something else besides routine operational stresses was going on. This "something else" may be unusual operating conditions or material problems. Fractography can shed light on how to tell these two apart.

In addition to the loading geometry, mechanical behavior characteristics, and crack propagation direction – that we've already covered – **visual examination** and **fractography** give you the tools to determine whether or not the design appears to be basically sound. In other words, fractography helps to determine whether the component is more or less adequately sized. Specifically, we can calculate a rough safety factor by comparing the proportion of the load bearing cross-section that breaks in the final loading event to that broken in repetitive cyclic-loading or that compromised due to corrosion.

Other features can give you an idea about whether a manufacturing problem has created a brittle material where a ductile material would have been expected. Brittle materials tend to be much less forgiving of unusual operating conditions than ductile ones. Still other features such as multiple crack initiations can shed light on whether an excessively sharp fillet at a section change is creating a higher stress than necessary.

8.14 Visual Examination of Broken Components: More Features

When performing a thorough visual examination, it is important to look at the whole component as well as any other components with which the broken part was assembled. Information about the orientation of the stresses can often be obtained by looking in three dimensions at the overall shape of the fragments and trying to imagine how the loads might have been imparted to the Subject Part from its fellows in the assembly. In other words, over the course of the visual inspection, back out and get the larger view before you zoom back in to examine a detail.

We have discussed this at length already in this book. However, very complex parts may require FEA to help visualize the high stress areas. And visualizing the high stress areas is not enough. It's important to recall that stresses have directions as well as magnitudes. Stresses are represented by vectors. Matching the crack orientation to the FEA data requires knowledge of the vector direction.

Way back in Chapter 3, we reviewed the importance of taking photos of broken parts from multiple directions. It's imperative to determine and document the angle of the crack as it relates to the loading geometry. Heroes of failure analysis resist the temptation to allow repair to precede thorough visual inspection and photo-documentation.

If you are ever in the situation where a costly unexpected fracture has occurred, it would be very helpful to subsequent investigators to have a detailed series of multi-angle overview and closeup photographs of the scene before any parts are removed, disentangled, subjected to a burning torch, etcetera. Even better would be to have someone experienced in performing engineering investigations of catastrophic events take the photos or guide the photography.

It is of the utmost importance to understand how the part was supposed to be loaded or stressed when analyzing a fracture. Even if a quantitative stress analysis is not available, with some basic knowledge, it is usually not that hard to figure out if the part was rotating, reciprocating, being bent repeatedly; or if it was subject to temperature fluctuations or gradients that might have induced stresses due to differential thermal expansion, for example. This is the main reason for the loading geometry content. You may need to ask many questions, especially when you are starting out, or encountering a new situation. In most situations, people will try to help. Find someone who can help you understand the operation of the component. As we have highlighted in this chapter so far, the key to understanding why something broke is first and foremost to determine if it broke in a manner that could have been expected, or whether, by contrast, it broke in a surprising or unusual manner.

8.15 Review Keys to the Puzzle

Let's review some potential classical fracture analysis findings, or "keys" to the puzzle, one by one. These are things that the experienced practitioner knows to look for.

KEY 1: Preservation of the Physical Evidence is Part of the Visual Exam

Preservation of evidence is often best undertaken, at least in situations of major consequence, with the help of a specialist.

- For normal machinery components, keeping the components dry is usually desirable.
- It is always desirable to keep the fragments protected from mechanical damage to the fracture surfaces.
- Do not try to remate broken fragments.
- Important information, that is not available to the eyeballs, can often be obtained by microscopic analysis of the fracture surfaces at a later time.

Touching the part with fingers can be extremely damaging to features that might have otherwise helped figure out what went wrong, since the salts contained in perspiration are often transferred with the oils of the fingerprint area, and again, the features can actually be dissolved by the corrosion processes.

KEY 2: Determine the Expected Fracture Location

Figure 8.32 (below) shows a broken externally threaded fastener. The crack location follows the thread root(s). This is not a surprise. This is as expected. The load bearing area is smaller at the thread root, so the amount of material available to support the load is lower, and thus each atomic bond has to carry a greater load at this position than at the thread crown. The thread root creates a stress concentration.

Note there are multiple, closely spaced initiations along the arc going from 4 o'clock to 7 o'clock. Multiple closely spaced initiations often indicate the presence of a significant stress concentration.

Figure 8.32: *This threaded fastener cracked through the thread root, the expected high stress location. There are multiple, closely spaced initiations along the arc going from 4 to 7 o'clock. Multiple closely spaced initiations often indicate the presence of a significant stress concentration.*

We will introduce the idea of ratchet marks later in this chapter (in section KEY 7: Ratchet Marks Give Clues to the Initiation Locations). We'll use this same fastener so if the initiations are not visible to you here, don't worry.

But now, look at Figure 8.33, the mating part, still embedded in the nut. Surely this is not the "expected" fracture location, since the nut adds rigidity to the assembly at that location. Our conclusion based on our initial inspection would probably be that something unusual was going on in this situation. Please note how easy it would be to completely miss this important point, if we had only the end shown in Figure 8.32 to evaluate.

PLEASE NOTE: according to a discussion with Dr. W.T. Becker, quantitative stress analysis of properly installed threaded fasteners has shown that the stress peaks somewhere in the roots of the first two or three engaged threads. Thus, this fracture location may not be as unexpected as I first thought. Threaded fasteners that are installed with insufficient clamping load will usually crack outside of the nut.

Figure 8.33: *This is the mating fracture surface to that shown in Figure 8.32. Note the importance of preserving the relative position of all fragments and components. The nut has retained the end fragment of the bolt.*

Please also note that I decided to "accept the explanation from the expert", rather than "figure this out for myself". Fifteen years later, I have still not investigated the situation for myself. This remains the only threaded fastener I have ever seen that broke in the nut. It was on a fancy machine, so maybe they made sure to install it properly.

Figure 8.34 shows a shaft. The crack clearly goes through a drilled hole. This is the expected fracture location, based on what we can see in this photo.

Note also the hole is precariously located through a stepped-down diameter of the shaft.

Figure 8.34: *The fracture goes through the natural weakness created by the hole in the shaft.*

This part was shown in Figure 7.5 from another angle.

Figure 8.35 (next page) shows another threaded fastener. This time, the crack runs in a transverse direction, through the head. Clearly, this is not the expected fracture location. Something unusual is going on here. For some reason, which is not at all obvious to the casual observer, the local ratio of stress–to-strength in the head of the screw must have been higher than the local ratio of stress-to-strength in the threads during whatever loading resulted in this crack. Remember that cracks initiate where the local stress exceeds the local strength.

No matter what loading scenario is considered, it is impossible to imagine a service or installation load which can create a higher stress in the head than in the threads. Something must have weakened the material in the head.

Figure 8.35: *The unexpected crack location in this threaded fastener head is a signal that something unusual is going on.*

This could be the case if the threads were induction hardened, for example, and the head was not. There is another possibility. The internal stresses (frequently called "residual" stresses) that were left as a result of the manufacturing operation are adding themselves to the applied or service-related stresses. Since threaded fasteners are often made by a cold heading process, this third possibility would be a useful clue to follow up. Alternatively, the material in the head might have been weakened through an embrittlement process. Or, there might have been a pre-existing crack from the cold heading (forming) operation used to make this part.

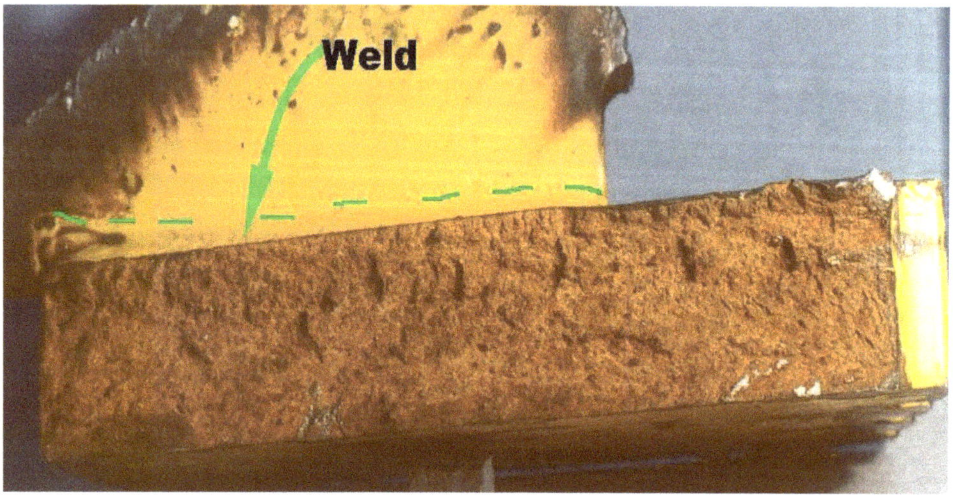

Figure 8.36: *The crack appears to follow the weld in this view. This is the same fracture as shown Figure 8.37. This view is misleading. It even confused me when I found it a few years after I had performed the investigation. Then I found what is shown as Figure 8.37.*

Figure 8.36 shows a welded fabrication with a crack in it. This is the same fracture as shown in Figure 8.37 (next page). This color version of the image shows that the crack itself is very rusty.

This did not greatly interfere with my determination of what the owner of this weldment wanted to know.

However, had the client wanted a more thorough investigation, some information might have been limited by the presence of the rust. This is because the rust was created by oxidation of the layers of steel that made up the original fracture surface. Rusting not only covers up the fracture features, it consumes the fracture features! Thus, de-rusting a fracture surface does not reveal the original features.

Look again at Figure 8.36. There appears to be a fillet weld running along the top of the cracked member. Anyone who works with welds knows that weld problems are rather common, so maybe it is not a big surprise that the crack is at the weld. I expected the crack to be at the weld. That is why I was surprised that the initiation was away from the weld. But in the next section we are about to see the danger of looking at something from only one viewpoint, especially in a photo. Figure 8.37 shows the same crack from a different angle. Here we see that the crack is not following the weld at all. So, what caused this crack to happen?

We see that the right edge of the horizontal plate has a row of teeth. Here we can see that the crack does go through the "root" of one of these notches. If we take the next step in fractography and ask not only what the crack location is, but more precisely, where the crack started, or initiated, then we are moving forward in our understanding of the subject.

A useful "trick" to find the crack initiation site in some cases is to trace the radiating fan pattern back to its origin point. When we were introduced to Meso-Fractography, we learned about fan patterns.

Figure 8.37: *The radiating "fan" pattern indicates the crack initiation location. This is a very good illustration of the "smooth-to-rough" rule: cracks often start out smooth, and get rougher as they grow.*

KEY 3: Fracture Features Often Show the Precise Initiation Site

This is important because if we think that there might have been a manufacturing imperfection of some sort that facilitated the fracture, we need to show that the location of the discontinuity is the same as the location of the crack initiation.

Note that the radiating lines point back to the root of the edge notch. In this case we have determined two things:

1) Although there is a weld present, it did not ease the crack initiation, and
2) A machined notch concentrated the stresses enough to facilitate the crack.

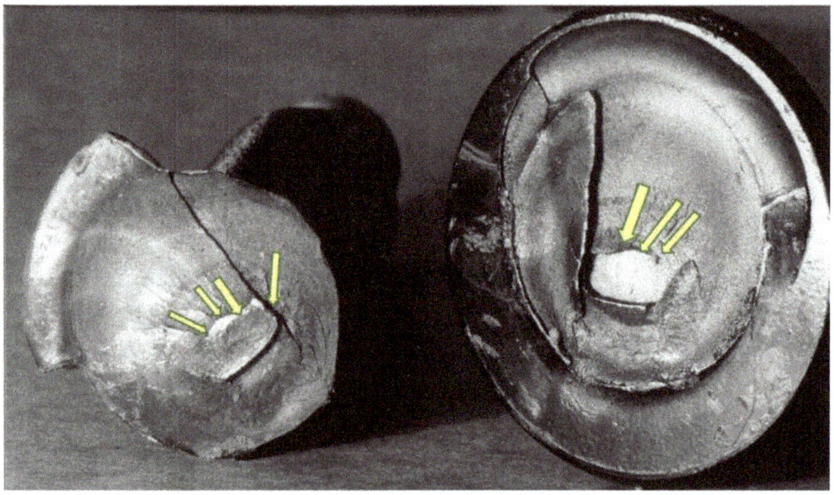

Figure 8.38: *It is important to distinguish the crack initiation site along with the more obvious crack location.*

Figure 8.38 (above) shows another component that has some interesting fracture features. Part of the crack hugs (follows) the fillet between the small and large diameters. Diameter changes and fillets are frequent locations for cracks, because many loading geometries allow significant stress concentrations at these features. But careful examination of the features on the crack surfaces will reveal radiating fan patterns that point back to a position near the CENTER of mass of the large diameter flange feature. What does this indicate?

This indicates that the main portion of the crack initiated close to the center of the cross-section, as shown by a fan pattern. Since the stress due to most loading geometries is generally highest at the surface, it is likely there was a discontinuity in the core.

KEY 4: The Crack Starts Where the Local Stress Exceeds the Local Strength

The total stress experienced by the component at any location is equal to that created by the applied loading, plus any stress left in the part by any manufacturing operations - particularly casting, molding, forging, stamping, welding, and heat treating. Please note that some manufacturing operations do not necessarily impart stresses, but unbalance them so that they are more noticeable. For example, if a large cavity is machined out of a piece of heavy plate, stresses that were balanced in the initial configuration may become unbalanced, allowing the component to become distorted if it is allowed to "move". In machine shop parlance, "move" doesn't mean it sprouts legs to walk! It means it gets deformed or distorted. Residual stresses are hard to see and are thus frequently overlooked as factors that might have contributed to a fracture. Use unusual fracture features as a trigger and remember that residual stresses need to be considered.

KEY 5: Fatigue Cracks are Sometimes Easy to Identify by the Presence of Beach Marks

Another common type of feature on fracture surfaces is what we call "beach marks". This is not the first time we are mentioning beach marks. Figure 8.39 (next page) shows a fracture surface of a leaf spring. The concentric dark and light bands radiate out from the initiation site. This is a classic sign of a fatigue crack, or a crack that grew due to dynamically varying (cyclic) loads in varying environmental conditions.

> *Some corrosion processes can also create concentric rings of dark and light areas. It is always important to have several features confirming any diagnosis!*

Figure 8.39 has four arrows that show the four locations where a small crack initiated. As the initiating cracks at locations 2 and 3 grew, they merged into a single, larger crack. Cracks 1 and 1A likely initiated much later than 2 and 3.

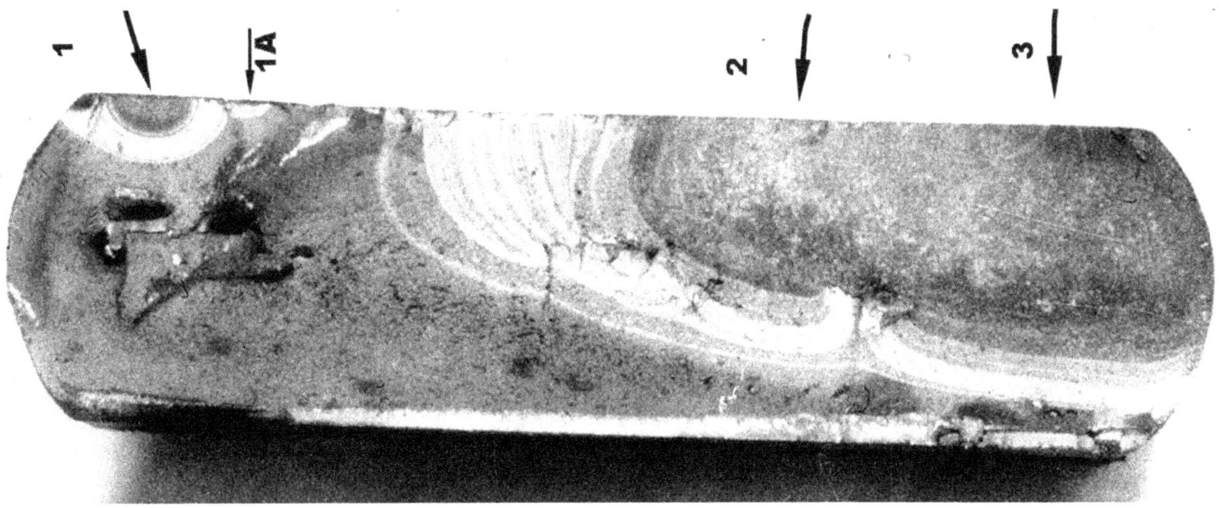

Figure 8.39: *The arrows show the areas where the fatigue cracks initiated. We've seen this crack before!*

Figure 8.26 showed a fatigue crack with no beach marks. Beach marks require varying loads and varying environmental conditions to form. Just like the fastener in Figure 8.32, it was a lab test fracture. The air conditioning and uniform loading prevented the formation of beach marks.

Not all broken components with the alternating concentric rings known as beach marks break due to excessive loads. If it is believed that the loads were not excessive, microscopic analysis can be used to confirm that corrosion processes from aggressive chemicals were or were not contributing to the ease of crack initiation or propagation. Corrosion processes can lower the actual strength by as much as 90% of that expected in susceptible materials.

Figure 8.40 (below) shows another fatigue crack, this time, in a piece of plate. Often the concentric ridges, or beach marks, are more difficult to see than here. Please note these features were actually very difficult to see on the part itself. The reason they appear as prominent as they do here is that I "played around" with the angle of the light to show them off as clearly as possible and then enhanced the contrast.

The beach marks shown in most fracture analysis books are chosen because they are easy to see. They are classical. The real world is not always so helpful. Showing that beach marks are often difficult to discern will hopefully encourage the student or practitioner to keep looking!

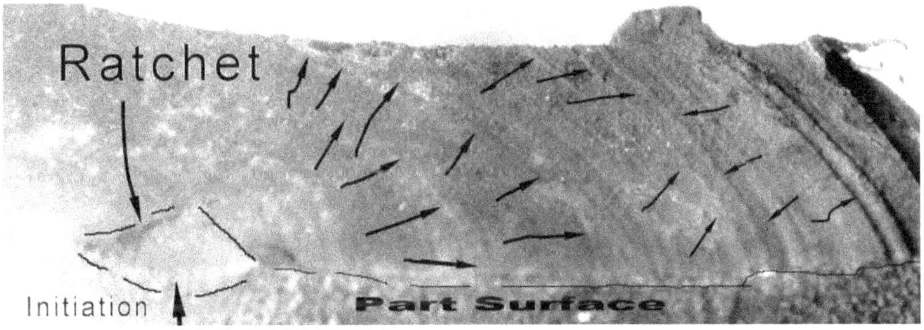

Figure 8.40: *The concentric rings in this piece of plate show that this was a fatigue crack. It initiated in the triangular area at the left.*

KEY 6: Progressive Cracks Can Reveal a Rough Safety Factor

We have mentioned "the rough safety factor" calculation earlier in this chapter. If you have a fatigue crack, and can see the entire fracture surface, and the crack has fully propagated to complete fragmentation, then you may have another very significant key to understanding why the part cracked in addition to where the crack started.

Estimate the relative amount of the beach mark area compared to the load bearing ligament that broke all-at-once during the final load cycle. For example, the alternating concentric bands on the leaf spring shown in Figure 8.40 (previous page) take up a little more than half of the total fracture surface. This tells us that the crack had eliminated over half of the load bearing area before the stresses became so high that they actually caused complete separation of the spring.

We can thus calculate a very approximate safety factor of two, in the presence of a fatigue crack. Note that this is different from the safety factor of the intact part.

To understand this, we need to consider that fatigue crack tips are very sharp. They create a very significant stress concentration. Yet the part still survived until half of the cross-section was unable to transmit any loads. In many applications, the engineer would acknowledge, based on this visual examination, that this leaf spring had an adequate cross-section for the service conditions in which it was working. Please note that this does not mean that other aspects of the design were all implemented in an adequate manner.

For example, the presence of a high-quality surface condition is generally critical in leaf springs. If the crack started in multiple locations, as is the case depicted in Figure 8.40, this may be a hint that surface condition was of insufficient quality, or it was damaged in service.

KEY 7: Ratchet Marks Give Clues to the Initiation Locations

Fatigue cracks do not always have obvious beach marks. Another telltale sign of a fatigue crack is the presence of step like features along the edge of the fracture surface, which are more or less perpendicular to the component surface.

As promised in Figure 8.32, we are revisiting this fastener in greater detail. See Figure 8.41 (next page). The broken threaded fastener, shown in Figures 8.32 and 8.33, is a fatigue crack. Let us look at this same part but with a totally different focus. Go to Figure 8.41 (next page). Note the fine step-like features at 7:00 to 8:00 o'clock along the edge of the crack surface. These are called ratchet marks and they indicate multiple independently originating crack initiations. These fine steps (or ratchets), form due to tearing of a thin membrane between two nominally parallel cracks. Thus, they occur between pairs of initiations, and are not at the exact initiation position.

There are actually some very fine beach marks running parallel to the chord (not circumference) from eleven to one o'clock. The beach marks were probably made more visible by the fact that the local stresses were so high by the time the crack had grown to a size of approximately 90% of the original load bearing area.

Why is the final overload area so small? The "flaps" centered around 3 o'clock and 9 o'clock provide a hint; this is made more obvious when combined with the view of Figure 8.33 of the mating fragment trapped in the nut. These flaps likely were created as the head end fragment pulled out of the nut. There were likely additional fatigue cracks under the "ears", but the thin rim of material between the primary and lower cracks helped the little arc of remaining metal hold on.

We need to realize the presence of the crack usually acts to increase the stress the part experiences. Without being able to recognize ratchet marks as clues to crack initiation sites, it might have been tempting to incorrectly think that the visible beach marks were the initiation site. Fracture analysis is not for those who don't care for details!

We may wonder why a crack ever started in the first place if there was enough cross-section to sustain the working loads with only perhaps 5-10% of the original cross-section remaining. This is an excellent question, and it might lead one to speculate that something unusual was going on, such as spike loading, for example, or a low-strength surface layer. Decarburization of medium carbon steel during heat treatment is not an uncommon cause of fatigue initiation in high strength steel fasteners.

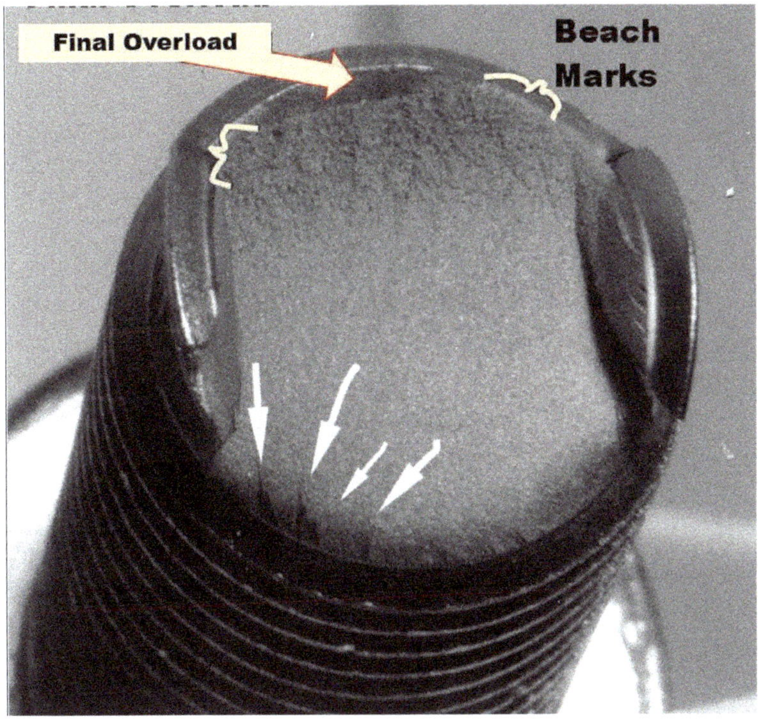

Figure 8.41: *Ratchet marks are a reliable sign of the initiation area of the crack.*

This is Figure 8.32 repeated with annotations on the photo. You may wish to zoom in on an electronic version of this image, to see the fine steps in the white arrow location more clearly. They are perfectly obvious to the author!

Before leaving the subject of fatigue cracks, this is a good time to remind ourselves that they are brittle at the macro-scale. In traditional fractography, this means that the fragments look like they can be reassembled into their original shape. **But please do not do this! They may need further analysis.**

KEY 8: Take Note of the Traditional Way of Distinguishing Ductile and Brittle Cracks

Cracks that happen all-at-once due to a catastrophic load excursion can be either brittle or ductile.

- Ductile cracks may (or may not) create fragments whose shape is noticeably different after the fracture than before. See Figure 8.42 A-B (below).
- Fractures that propagate in a brittle manner will look like they could be reassembled into the original shape. Again, please do not do so!!!

In Figure 8.42A and B (next page) there are threaded rod and standard round tensile bars that were forced to break due to axial loading at the tensile strength of the material. Both specimens show the classical plastic deformation associated with ductile fracture.

While some materials are inherently brittle, most structural materials are processed so that they have significant amounts of ductility.

As we have mentioned on several occasions, Fracture Analysts need to determine early on whether the crack was macro-ductile or macro-brittle. If something was expected to be macro-ductile and it is not, it is important to note that for follow-up.

A confusion that can happen here is that those trained only in this traditional fractographic method may think that macro-ductile torsion, which often has no visible deformation, is macro-brittle. Avoiding this error is another reason to use the epistemologically informed method of classifying fragment shapes that is promoted in this book.

That would mean determining fragment shape in light of knowledge of the loading geometry, not simply on appearance.

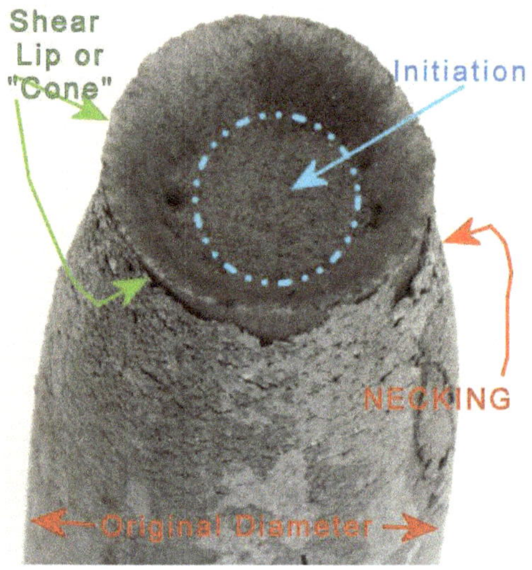

Figure 8.42A: *Close-up of the upper tensile test coupon fragment shown in Figure 8.42B*

The central area of the fracture surface is expected to have ductile-micro-voids. At the macro-scale this portion of the crack is brittle, because it is perpendicular to the axis. The shear lip cone and the necking indicate ductile behavior at the macro-level. Thus, at the macro-scale the fracture is mixed-mode.

KEY 9: The Surface is Often the Most Highly Stressed Location

This brings us to another key point, which is that most service fractures initiate at the surface of the component. There are two major reasons for this:

1) In bending and torsion, the most common types of loading geometries, stresses are highest at the surface of the component.
2) Because the stress is highest at the surface in the most common loading geometries, the surface of the part is most frequently subject to crack initiation. By virtue of being the surface, it is also most subject to mechanical or environmental damage. The damage can create a stress concentration, raising the stress even higher than anticipated.

This reminds us of the importance of good machining practice, which brings us to the second reason most service fractures initiate at the surface of the component. A smooth surface will often go a long way toward preventing fatigue cracks. It has been well known for years that surface finish is critical to the performance of structural components.

Figure 8.42B: *Halves of two tensile test coupons*

Both specimens show the classical plastic deformation associated with ductile fracture.

There are many types of imperfections that happen only on the surface of the part, which increases the frequency of crack initiations on the part surface.

Other than torsion and bending, the remaining types of primary loading are axial compression, tension, and shear. In a section of uniform cross-section loaded in axial tension, until a significant amount of permanent deformation has taken place, the stress will be the same everywhere. Compression loading is more complicated, as we saw earlier in this chapter.

Thus, for components of uniform cross-section which are loaded in tension, theoretically, the crack could initiate in the core, rather than at the surface. See Figure 8.42B to view two broken cylindrical tensile test coupons. Both materials were ductile and the resulting necking prior to fracture is apparent at the left side of the image. Figure 8.42A shows the smooth broken tensile bar, which is close-up view of the upper image in Figure 8.42B showing a different angle. Note the detail of the separation features and fracture surface. The flat "cup bottom" in the core is where the crack initiated. As the part was stretched to fracture, it necked down and eventually broke.

The classical cup-and-cone fracture surfaces on straight tensile bars are among the times that internally initiated cracks are frequently found. Surface hardened (carburized or induction hardened) steel components may also be prone to internal crack initiation in some cases.

Most real machine parts are subject to a combination of bending, torsion, and axial loads. Bending and torsion have the highest stresses at the surface, making surface-cracks more common than internal cracks.

See Machinery's Handbook[157] or a mechanical engineer's handbook for the equations which allow you to calculate the stress at any location in simply shaped beams or shafts of various configurations. For more complicated shapes, performing a Finite Element Analysis will be a more realistic option.

KEY 10: Crack Initiation Patterns and the Relative Amount of Fatigue or Overload

RECOMMENDATION: For more on this subject, it is well worth looking up Page 111, Figure 18 in the 1986 Edition of the ASM Metals Handbook Volume 11. There are eleven rows of schematics, containing six versions each of either cylindrical or prismatic sections which have been subjected to axial, bending, or torsional fatigue, at high or low stress concentrations. The relative amounts of fatigue versus monotonic (final separation) areas are depicted, along with the number and relative positions of anticipated initiation areas.

Years ago, at an ASM International Fall Conference, retired professor Dr. William (Bill) Warke, who had worked for years as a rotating machinery expert, showed me a set of pencil drawings on onion skins, which looked very similar to those in the 1975 Edition of Volume 10, on page 102, and in a slightly different format to those in Don Wulpi's Understanding How Components Fail. The full committee list of authors in the original 1975 Volume 10 does not show Dr. Warke as a contributor. In completing the text of this book, I had a phone call with Bill Warke, who clarified that he added detail to sketches he had seen in an older edition of Wulpi. Chasing the references, we got to a 1961 edition of Republic Alloy Steels. Well, I have an original copy of the beautiful 1949 edition, but it does not have these particular images.

I obtained a newly printed copy of the 1961 version from a New Delhi bookdealer and this version refers to a 1950 article in Machine Design, by C. Lipson (September 1950, pp 147-148). Here, C. Lipson credits (an otherwise unspecified) "Bacon", in his Figure 10-35, for establishing "the symbols and nomenclature" used to describe fracture surfaces.

Because the ASM policy for attributions is problematic for the handbooks (any change in wording for subsequent editions means that the original author's name is removed), it is not surprising that this trail is longer than I anticipated. Maybe I will write a future article to address the origins and history of these very helpful drawings.

The bottom line is that ideas get recycled. Of course, we all want to contribute to the profession, but it's nice to get credit for our work. Dr. William Warke requested that he be credited with adding some of the details of the graph that appears on page 632 of the 2002 Edition of Volume 11 of the Metals Handbook. I'm doing that here and now.

KEY 11: Concerns with FEA Modeling

FEA modeling is a very powerful tool. It is especially useful for complex shapes, and where stress fields are generated as a result of dynamic thermal gradients. It's going to be nearly impossible for anyone to perform the "intuitive stress analysis" that I have been promoting in this chapter for those situations.

[157] E. Berg, F.D. Jones, H.L. Horton, and H.H. Ryffel, Editor R.E. Green, *Machinery's Handbook, 24th Edition*, Industrial Press Inc., New York, 1992, Pages 92 - 351

FEA modeling software is generally capable of providing output showing the direction of the local stress. However, I have only worked with one mechanical engineer who bothered to provide this data. It seems that most FEA modelers are "looking for hot spots". Identification of hot spots alone can not help you predict the orientation of the crack. Hopefully by now you are thoroughly convinced that understanding the orientation of the crack is exceedingly important.

KEY 12: Multiple Branched Cracks Can Be a Sign of Contamination by an Incompatible Environmental Substance

Stress corrosion cracking (SCC) is a designation that applies only to metals.

Similar types of problems also happen to plastic parts, where it is known as "Stress Cracking". Do NOT use this a term when referring to metals, since all true cracks are due to stress and some metal cracking experts think that the use of the term "stress cracking" shows ignorance.

Stress corrosion cracking is a phenomenon that can cause the strength of a susceptible metal part in an incompatible environment to be reduced to values as low as 10% of that expected. SCC often has a particular appearance, defined by the presence of multiple, branched cracks. The crack surfaces are often covered with some type of corrosion product.

Figure 8.43 (below) shows a fragment of an austenitic (300 series) stainless steel part that was in an environment with a much higher level of chlorides than advisable for this type of material. Solving SCC issues usually comes down to figuring out what in the environment is causing the problem, and substituting a more resistant material, or changing the environment, if possible.

> **VISUALIZING STRESSES & MATCHING TO RESULTS**
>
> **Knowledge of how a component experiences usual operating stresses is important in fracture analysis because it can be used to determine whether the features of the crack are consistent with expectations.**

Figure 8.43: *This crack geometry, with its multiple branches, is a classical indication of Stress Corrosion Cracking. I broke this section of stainless steel off of the rest of the component with my bare fingers!*

KEY 13: Sub-Surface Crack Initiations are Important to Identify

As we have previously noted, the stress in most common loading geometries in real components is higher on the part surface. There is also a higher chance of having many of the common manufacturing discontinuities on the surface of the part. This includes surface roughness from improper machining, metallurgical damage from improper grinding, unintended carburization or decarburization due to heat treating conditions, and unincorporated "globs" of material in additively manufactured (AM) components, particularly in downward facing surfaces in Laser Powder Bed Fusion processed parts. Such internal surfaces are known to be associated with premature crack initiations in some industries. In practice, most widely experienced Fracture Analysts will find the vast majority of cracks initiating on a part surface. If they don't, there is usually a specific reason.

For example, shot peening is often done on steel springs, in order to create a compressive stress that counteracts the natural development of superficial tensile stresses. However, if there is a large non-metallic inclusion in the surface layers, it may create a stress concentration which nullifies the benefits of the shot peening. Most springs, whether of coil, leaf, or torsion design, are loaded to levels near their yield strength. So even fairly small sub-surface inclusions can move the location where the local stress exceeds the local strength.

The other situation where crack initiations may be found at a sub-surface location is when a surface hardened steel part has an inadequate case depth. See Figures 8.44A-C for an example.

Contemplation of Figure Series 8.44 is a key to understanding how to quantitatively select a case depth specification for surface hardened materials. Normally this is done by "experience" or the "Guess and Test" method. But here, combined with a Finite Element Analysis, an educated guess could be made regarding the case depth requirement.

The sub-surface crack initiations are shown in Figure 8.44A at arrows A and B. Note the fan patterns and the Smooth-to-Rough features. The crack propagated around the case from the sub-surface initiations. Note the chevrons or fan patterns. Initiation B is near, but not at, the boundary between the case and the core. There may be an inclusion at this position that made the stress higher here than at the case core boundary.

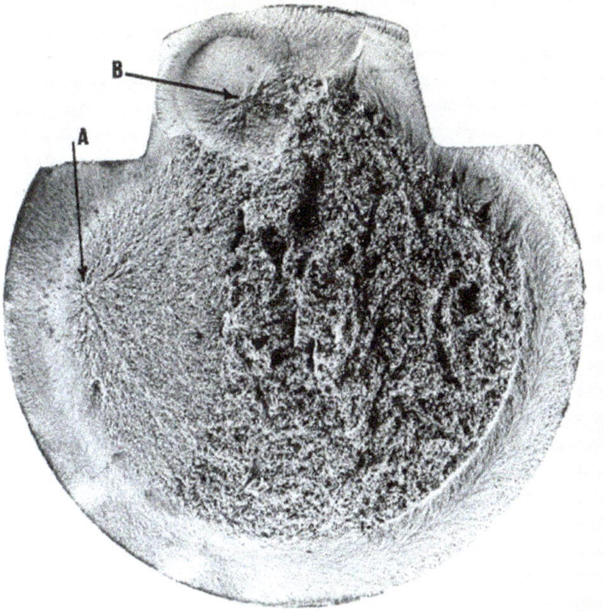

Figure 8.44A: *A direct view* [158] *of a fracture surface of an induction hardened steel shaft*

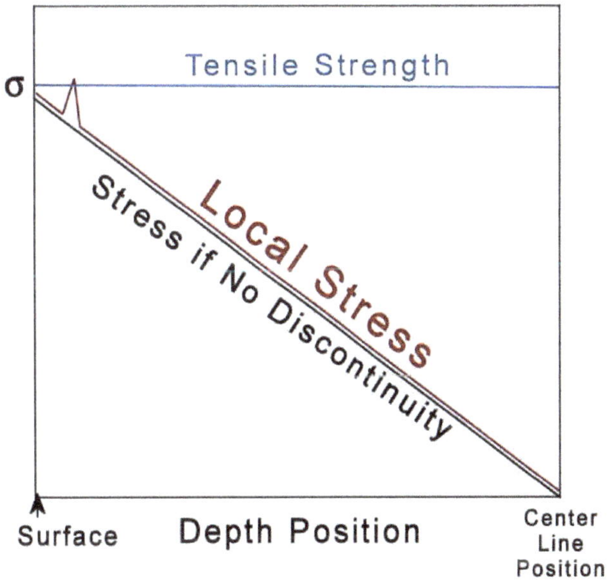

Figure 8.44B: *Conceptual explanation of crack initiation at Location A in Figure 8.44A – stress exceeding the strength at a discontinuity may have initiated the fracture.*

[158] Image courtesy of ASM International | *Understanding How Components Fail*, Donald J. Wulpi, 1985, American Society for Metals, p94 | Used with permission.

The harder and more brittle case material does not allow for a rough crack surface to develop, while the core material is a bit rougher, even in fatigue.

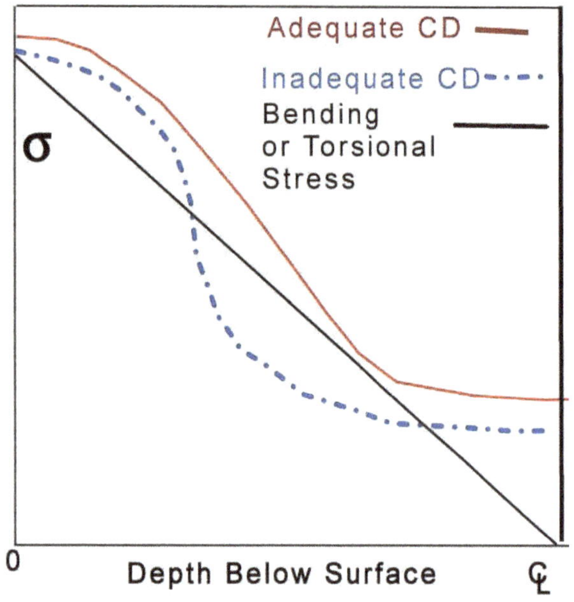

Figure 8.44C: *This figure may conceptually explain crack initiation B (in Figure 8.44A). It may have resulted from an inadequate case depth. The crack initiated at the case core boundary, as shown by the change in texture of the crack surface.*

8.16 Mechanical Behavior and Micro-Fractography

At the beginning of this chapter, we learned to recognize the basic fragment shapes produced by different loading geometries. We saw that in materials that were behaving in a macro-scale ductile manner, there was one set of expected fragment shapes, while for macro-scale brittle behavior, a different set of shapes was expected from the same loading type.

We have been introduced to these ideas several times now (starting with the Visual Examination material in Chapter 3). Take another look at Figure 8.4, which shows a summary of all the major loading geometries as well as the expected fragment shapes for both ductile and brittle behavior at the macro-scale.

Properly performed fractography can often allow the analyst to create a mental time-lapse-movie, which shows the progress of the crack.

In Chapter 6, we briefly described the manner in which crystalline metallic materials deform in a plastic (permanent) manner. We showed ears of corn with irregular lines of kernels to help us understand the concept of dislocations. Dislocations make it easier for planes of atoms to slide past each other. This is inherently an atomic-scale behavior, not visible even with an optical or scanning electron microscope. Although we can sometimes see metallographic evidence of the presence of what researchers have shown to be high concentrations of dislocations.

When teaching heat treating of steels to non-metallurgists, I have long used marshmallows and toothpicks as the structural units to build open models of the two basic crystal structures found in steels. See Figure 8.45A (next page).

I suggest that you attempt to make one of these. However, I also see in a brief Google search that despite the origin of sweet marshmallow confections in ancient Egypt, Americans are the primary consumers of marshmallows[159] *today. I like using marshmallows because they are inexpensive and generally readily available. Making a marshmallow model is a tactile, kinesthetic learning experience. It's easiest to make the model with one toothpick equaling the center-to-center atomic distance between the atoms along the edge of the cube.*

Making these models also reinforces a practical understanding of the Pythagorean Theorem. People who were not paying attention in their geometry class

[159] According to https://candyusa.com/marshmallows/ Americans buy more than 45,000 tons of marshmallows per year. It is thought that the Girl Scouts may have given the candy a boost with the invention of the camp fire toasted marshmallow featuring in the "S'more". Inquiring minds want to know! Go Girl Scouts! Suggestions for other options are welcome!

realize that they need more than one toothpick to hold a marshmallow (atom) in a stable manner at the center of the cube, or the center of one of the faces of the cube.

Those who make a larger model can readily see that some orientations have more "atoms" packed along a given plane than others. See Figure 8.45B. The idea of a close packed plane becomes clear with this "prove it to yourself" [164] *activity. The sliding between atomic planes that facilitates plastic (permanent) deformation in metallic crystals happens parallel to close packed planes. Doing the marshmallow exercise allows us to see the process of plastic deformation in our minds eye. Of course, if we are using small marshmallows, the perimeter of the individual marshmallow represents a distance somewhere between the outer "boundary" of the nucleus and that of the "electron cloud". Read further to gain understanding of the actual approximate relative sizes of the individual atoms compared to the size of an individual "unit" crystal.*

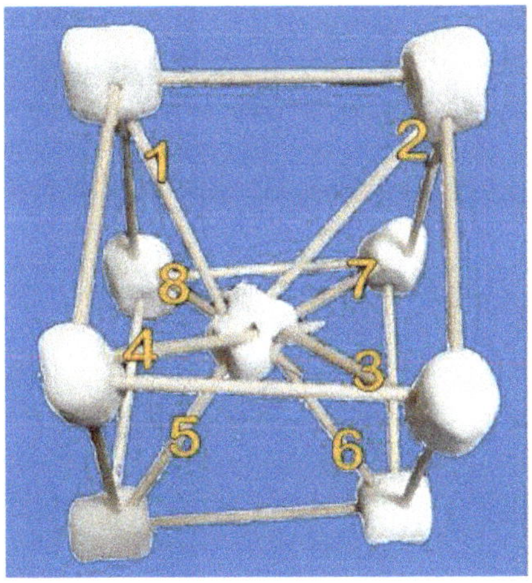

Figure 8.45A: *This is a representation of a Body Centered Cubic crystal structure. Each marshmallow represents an iron atom and each toothpick represents an atomic bond.*

In Figure 8.45B (below), the red "x's" show that if you traverse the cube diagonal, you are on average closer to two atoms than if you travel along one of the 12 cube-edge toothpicks. The black circles

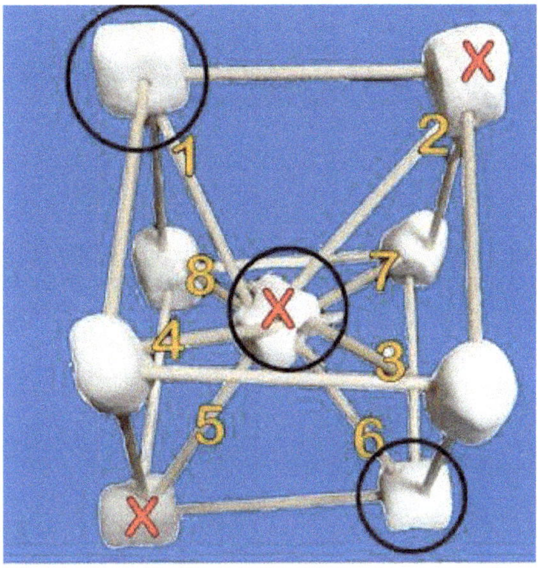

Figure 8.45B: *The representation of a Body Centered Cubic crystal structure: each marshmallow is an iron atom, toothpicks are atomic bonds, and numbered atomic bonds represent the eight "nearest neighbors" to the center atom.*

show an additional close-packed cube-diagonal. In reality, the ratio of the sizes of the atoms (based on their electron clouds) is much greater than the size of these small marshmallows compared to the length of the toothpicks.

Imagine, in your mind's eye, the marshmallows expand beyond the size of the black circles shown in Figure 8.45B. If you keep expanding them, the atoms will touch each other as you travel along any cube diagonal. At that point, you are imagining a more direct representation of the atomic packing.

A careful observer might notice that when traveling along planes that contain two mutually perpendicular "close packed directions" (the red x's of Figure 8.45B plus atoms 4 and 7), then the marshmallows are more closely spaced. **These are the closed packed planes. The sliding that allows deformation takes place between**

[164] This is a key point in Michael Gelb's *How to Think Like Leonardo da Vinci- Seven Steps to Genius Every Day.*

parallel close packed planes. This is facilitated by the presence of linear discontinuities known as dislocations. We saw a conceptual example of this in the corn (maize) cob images earlier in this book (Chapter 6).

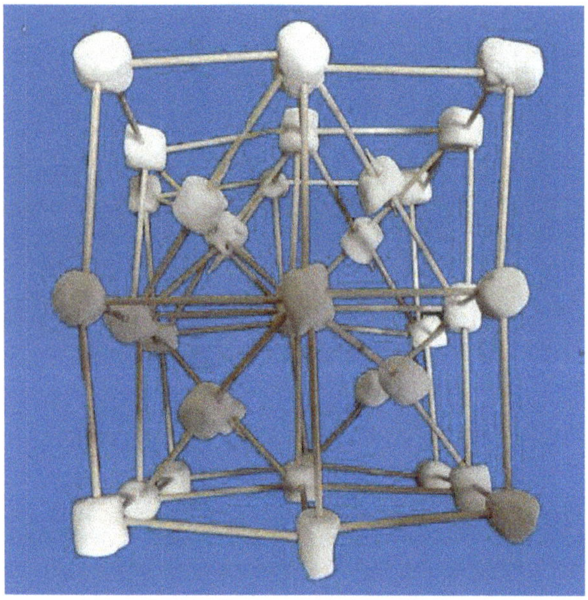

Figure 8.45C: *This structure is made of 8 of the individual cubes shown in Figure 8.45A. I did not include all of the toothpicks. Just enough to keep the structure more or less "square". We are looking down on this cube. Slightly stale marshmallows do a better job of keeping the structure square.*

In Figure 8.45D, the planes and directions shown by the dotted red lines along the cube edges are not as tightly packed with the marshmallow atoms. The planes of these cube-faces do not include the atom at the center of the cube volume.

The point to take away is that we must have a basic understanding of what happens within the individual crystal, before we can gain an understanding of how most real materials deform.

Most real materials are not single crystals, but are made up of many crystals. These many crystals are often oriented in multiple directions and are sometimes randomly-oriented throughout. But if crystals can only deform in certain directions (see Figure 8.45C) when we have randomly-oriented crystals, then some can deform readily and other cannot.

Figure 8.45D: *Planes shown by the dotted red lines along the cube edges are not as tightly packed with the marshmallow atoms.*

Below is an extremely oversimplified image showing how single, differently-oriented crystals will respond to an axial force. In the example on the next page (Figure 8.46), the forces are vertical.

NOTE: if the planes are perpendicular to the vertical force (or parallel, which is not shown), then plastic deformation will be suppressed. Those oriented at angles near 45-degrees will be relatively easy to deform.

In reality, the grains within polycrystalline materials rotate as the slip proceeds.

We may have some circumstances where virtually all of the potential deformation possibilities are suppressed. This could be the case in cold environments, or when the forces are applied so rapidly that the atoms do not have time to slide.

Sliding is essentially a diffusion-controlled process. It may not seem obvious that this sliding motion is directly related to the random motion of air molecules bouncing around inside a balloon, or water molecules moving around in a teapot. But the equations for both processes are very similar. Materials scientists like to say, "Deformation is diffusion activated."

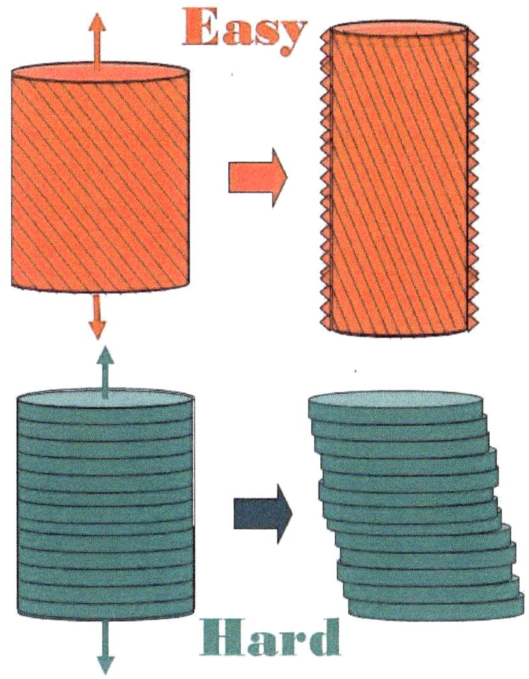

Figure 8.46: *These drawings show how differently oriented single crystals respond to a vertical, axial force, via sliding along close packed (easy shear) planes. Note the reduced diameter and rougher surface of the deformed upper cylinder, and the tiny shear displacements of the lower cylinder trying to respond to the vertical force, in the absence of diameter change.*

The next point to take away then, is that even though a material is theoretically able to deform, in certain circumstances it may be unable to do so. Reflect on Figures 8.46 and 8.47.

KEY POINT: This is why we need to always keep in mind brittle fracture behavior does not necessarily mean that the material itself was inherently brittle. The chain link example in Figure 8.11 is a reminder of this. The real life component behavior at the crack may be very different from the material behavior away from the crack, in position or time.

All of this detailed information about the crystal structure we have been discussing was figured out by scientists who used x-ray diffraction and math. For our purposes here, dealing in a pragmatic way with routine failure analysis work and getting a basic understanding of how and why things break, this very basic introduction to crystal structure features will help us interpret what we see at the coarser scales visible with microscopes.

To repeat, the individual atoms will not be visible with the usual tools of the Failure Analyst, which are scanning electron or optical microscopes. But sometimes, with an understanding of physical metallurgy, we can see evidence of the influence of these submicroscopic processes and features on the appearance of the Subject Part.

Using a metallograph on a prepared metallographic coupon (see Chapter 6) we can often reveal the deformation via the material's response to an etchant. Chapter 6 had a series of metallographic images with elongated grains. The elongation is what usually happens during a manufacturing-related deformation process.

Many polymers also can reveal evidence of solid state shape change. Sometimes a visible color change tells us the polymer part has experienced fairly extensive plastic deformation. In some resins, whatever the color was when the part was molded, the plastically deformed material tends to whiten. If it was translucent, it may become more opaque.

8.17 Predicting Ductile or Brittle

Inquiring minds want to know, "Is there a way to predict whether a component will break in a ductile or brittle manner?"

If a material is inherently brittle, such as grey iron or many ceramic materials, then the simple answer is yes. We know it will be brittle at the macro-scale. Also note that temperature effects must be considered when making this judgement of "always brittle".

Many such macro-brittle materials will surprise us with some micro-ductile features when we examine them closely. If the crack is propagating in fatigue, as previously explained, we inherently have a macro-scale brittle crack. But we may see some ductility at the micro-scale such as stretch marks (kind of like the edges of micro-voids, but straight).

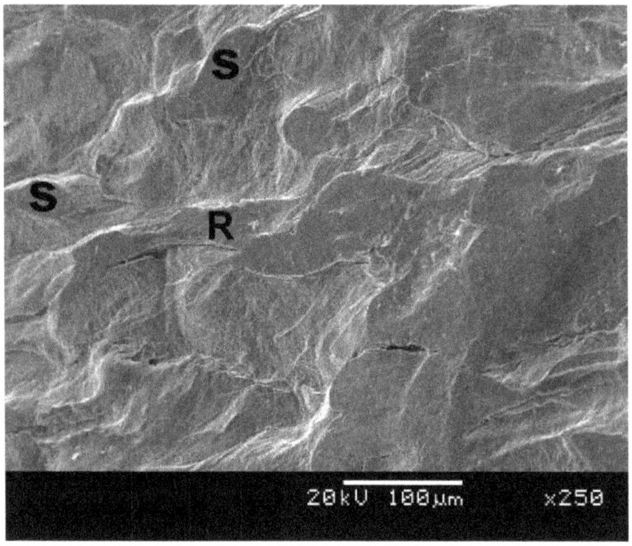

Figure 8.47: *This image shows the inside of a deep drawn 400 series stainless steel cup. The surface has become roughened as the grains rotate.*

Imagine what would happen to the lower crystal of the sketch in Figure 8.46 if there were no empty space for the planes to slide into.

*You would see what we see here! The grain marked **R** is especially elongated, and it, like **the two S** grains, had their peaks flattened by the tooling used to make the cup.*

For materials that are not inherently always brittle, the short answer is we can't predict the behavior so easily. But we can understand trends.

- Rapid loading and low temperature suppress deformation and promote brittle cracks.
- High temperature and slow loading facilitate deformation and promote ductile cracks.

Hydrogen embrittlement in steels is an exception to this last "rule" about slow loading. Hydrogen embrittlement is facilitated by slow or even static (sustained) loading and it promotes brittle (often intergranular) cracks. Intergranular cracks may also be created by shear along grain boundaries. In that case they would be considered micro-ductile. Intergranular cracks created by normal stresses pulling the grains apart would be micro-brittle. People have been arguing about the semantics and associated concepts of ductile intergranular cracking for decades.

High temperature creep damage can be intergranular, and may be produced by shear or decohesion. Decohesion is basically normal stresses acting perpendicular to grain boundaries. Again, these are areas of advanced research. The Failure Analyst needs to be aware of the controversies to avoid sounding ignorant.

- Low temperature suppresses deformation and promotes brittle cracks.
- High temperatures in the range for creep (which involves sustained loading at temperatures over 40% of the absolute melting temperature) for the material in question may allow macro-scale deformation, but show intergranular cracking at the micro-scale.

Micro-fractography is the realm that allows us to understand these behaviors in a more detailed way. In order to understand how micro-scale crack features develop as a crack propagates, we need a little bit more materials science than we have been exposed to so far in this book. Don't worry. It will be fun!

8.18 How Micro-Scale Crack Features Develop as a Crack Propagates

In this section we will discuss micro-scale crack processes: micro-void coalescence, cleavage, and intergranular cracking,

Let's start the discussion of micro-scale crack propagation with the classical cup-and-cone ductile fracture of a cylinder loaded in axial tension. See the lowest image of Figure 8.48 (next page). As is often the case, the top edge of the cup is rather ragged with the protruding side of the cup split between the two sides.

Since the cup walls are at an approximately 45-degree angle, if we consult Figure 8.48 (next page) and the schematic (Figure 8.4), we will see that the cup walls, both positive and negative portions, are macro-ductile. Furthermore, since the cup bottom is flat, and perpendicular to the cylinder direction, and loading direction, then the central area is macro-brittle.

Figure 8.48: *Prior to cracking the material behaved in a macro-ductile manner, as we can see the obvious pinching at the break (upper and middle views). The "positive" (+) side of the cup walls is split between the side of the fracture seen at lower left and its mate.*

But what micro-scale features will we see? Rather than being lazy and digging up some old photo showing micro-voids, parabolic for the walls and equiaxed for the flat bottom, I turned on my SEM to get actual photos of THIS broken tensile bar. See Figures 8.49 A-D (next page).

Image A (top left) shows an overview of the tensile bar's flat bottom and cup wall.

Image B (lower left) is a detail of Image A and shows the equiaxed micro-voids at the center of the crack surface. Some of the micro-voids have hard (undeformed) particles stuck at the void bottoms as shown by the arrows.

Image C (right top) shows the outer rim of the slanted shear lip (cup wall).

Image D (right bottom) outlines the edges of the occasionally parabolic micro-voids in light yellow.

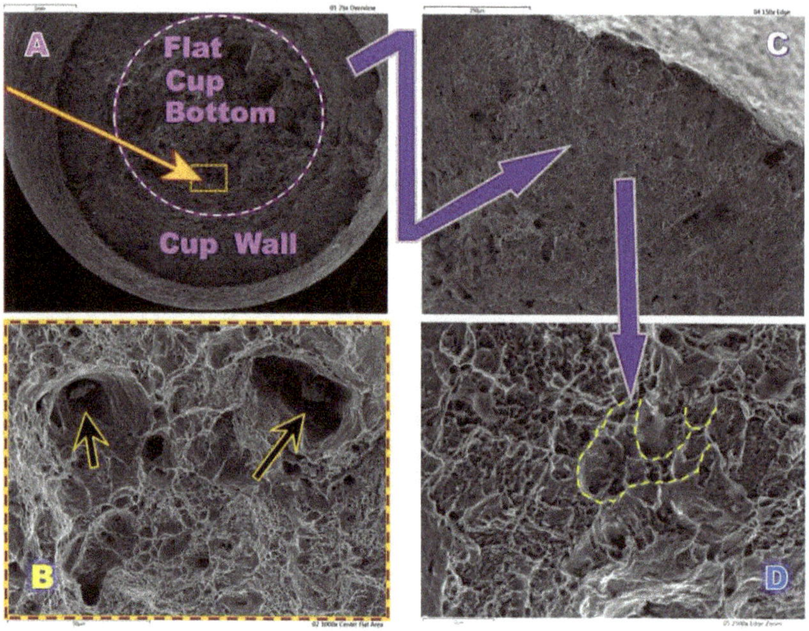

Figure 8.49A-D: *I used my SEM to get actual photos of the Figure 8.48 broken tensile bar.*

Figure 8.50 shows parabolic micro-voids from another specimen. The voids were created during the deformation process that preceded the final separation of the two fragments.

It is likely that there were microscopic second phase particles present in the raw material. The envelope of the matrix atoms surrounding the hard particles expanded during the deformation process. The interfaces of the particle and the matrix got separated as the material deformed. The matrix simply pulls away from the second phase particle.

Many of my clients do not understand this the first time I say it; so, I will repeat.

KEY POINT: Seeing micro-voids such as those shown in the Figures 8.49 and 8.50 does not mean the material had voids when it was manufactured. These voids are formed during the plastic deformation process.

> *Take a look at a nut or raisin filled chocolate bar. If you break it, you can sometimes see the nuts or raisins protruding from the crack surface. Chocolate behaves in a macro-brittle manner at room temperature. Thus, the void opposite the protrusion will be the opposite (mating) shape of the nut or raisin. This would also be the case in cast iron, where we can sometimes see the impression of a graphite flake or nodule on the crack surface.*

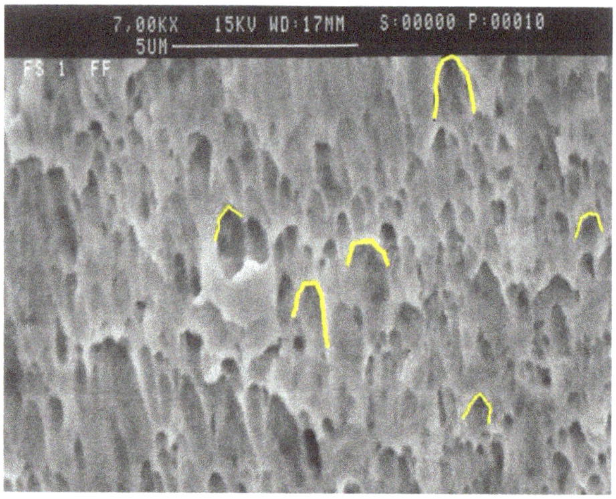

Figure 8.50: *Strongly parabolic micro-voids – this fracture detail is from a different part from that shown in Figures 8.48 and 8.49.*

Sometimes, in ductile engineering metals, we are able to see the particle that initiated a void. See the white arrows at the upper left side of Figure

8.51 (MVC Ductile). In these cases, the void surrounding the particle is much larger than the particle. The void grows and grows during the deformation process. Sometimes, the particle has fallen out of the now open "half-void" before we can take a photo. There are three rows in Figure 8.51. The bottom two rows of the figure show the two types of classical brittle micro-scale features: cleavage, which is always micro-scale brittle; and intergranular (which we have discussed) and is usually considered classical micro-scale brittle, but could be ductile if formed by sliding (shear) rather than micro-scale crack opening stresses.

The bottom two rows (last one is on the following page) of the figure show the two types of classical brittle micro-scale features: cleavage, which is always micro-scale brittle; and intergranular (which we have discussed) and is usually considered classical micro-scale brittle, but could be ductile if formed by sliding (shear) rather than micro-scale crack opening stresses.

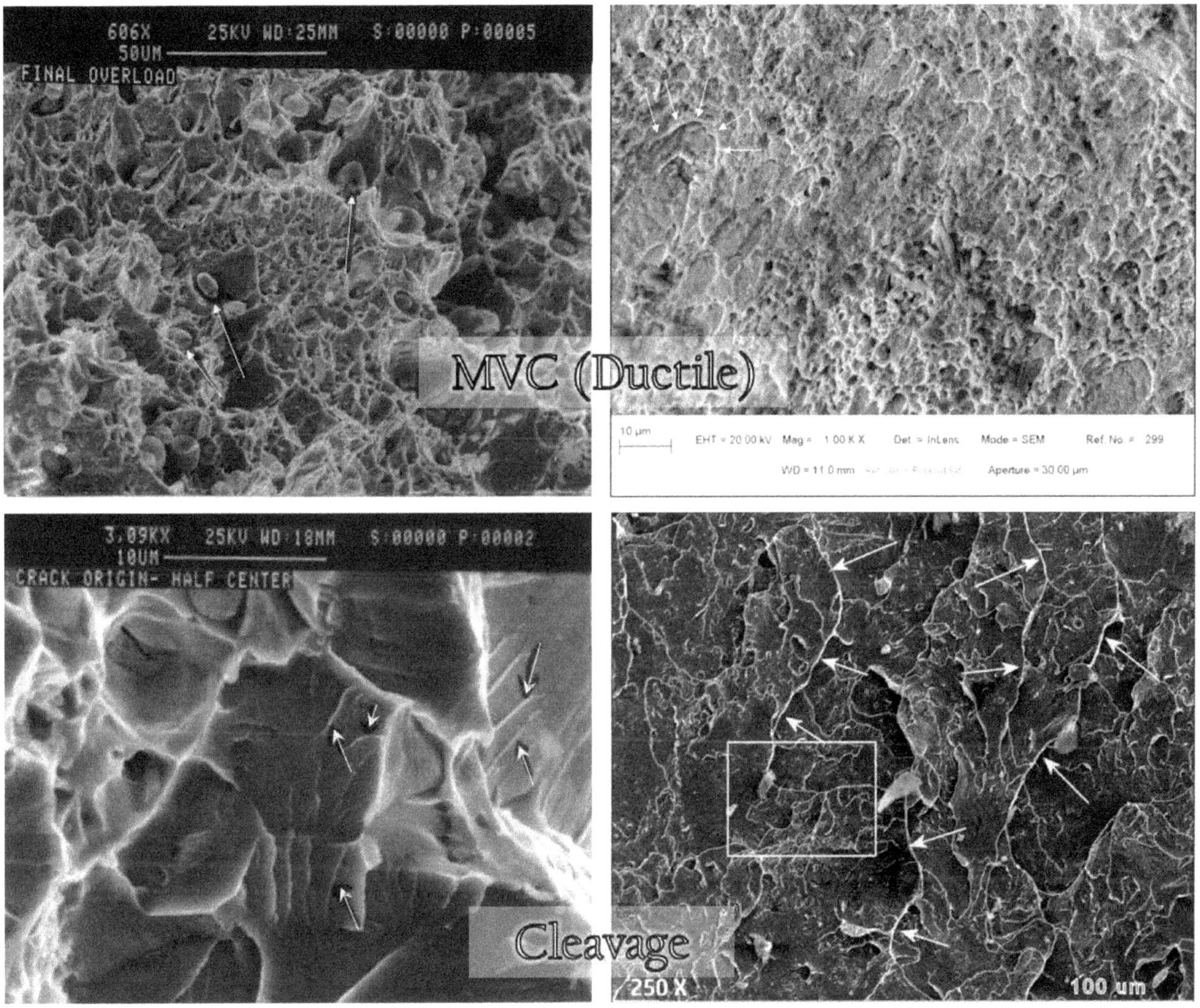

Figure 8.51: *SEM images show classical micro-scale fracture paths: MVC, Cleavage, and Intergranular cracking.*

MVC *is micro-ductile. On the left-hand side is predominantly equiaxed. On the right-hand side is parabolic.*

Cleavage *is the classical transgranular micro-brittle crack path. The left view shows the most classical form where the red/white arrows show small steps that provide clues about the local crack direction. The all white arrows at right show micro-scale ductile tear ridges. This type of cleavage depicted at right may be referred to in older literature as "quasi-cleavage". However, it has been determined to be true crystallographic cleavage.*

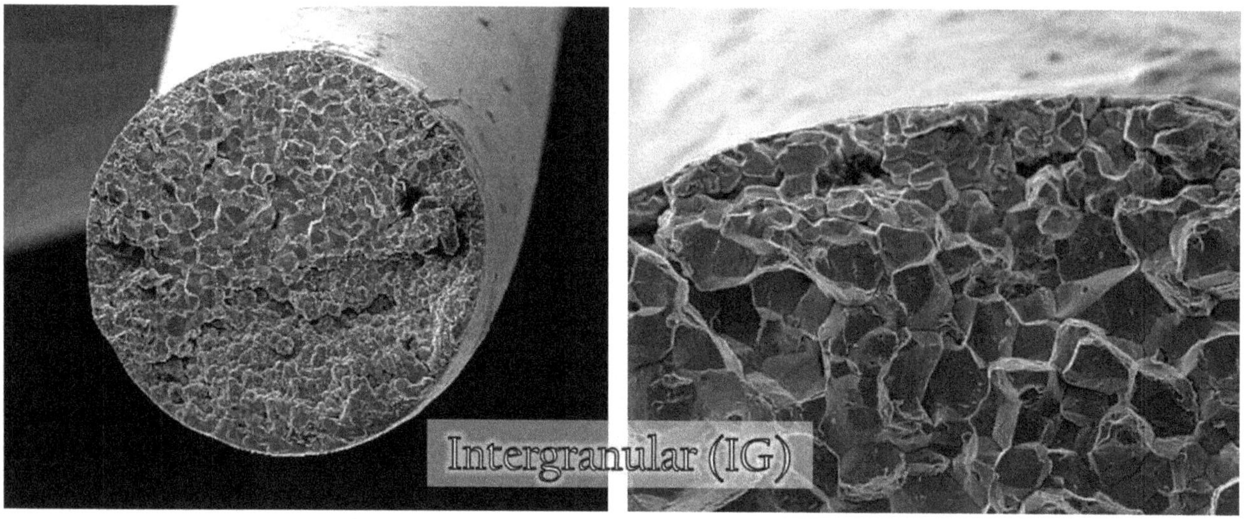

Figure 8.51: *SEM images show classical micro-scale fracture paths (Continued)*

Intergranular cracking *is usually considered the "other" micro-brittle crack type. The grains may come apart due to micro-scale tensile stresses. However, during high temperature creep, separation may occur by sliding, or shear, and may be accompanied by macro-ductility as well. If you want to call intergranular cracks ductile, you need to have some explanatory verbiage to avoid having some experts think you don't know what you are talking about, when really you actually are being more careful than most!*

8.18.1 A Little More Detail about Micro-Voids

What can we learn from the appearance of the micro-voids? Noticing that the walls of the voids are tapered to a knife edge, we can imagine them stretching out as the entire volume of the part is stretched. Don't forget about the planes of atoms sliding past each other. Plastic deformation usually happens at constant volume. That means when we stretch one dimension (length) the other (diameter, and all that is in the plane of the diameter) must be reduced. Since the hard particles that initiated the micro-void in the first place can't deform, the total deformation required is confined to the material surrounding them.

The series of photos in Figure 8.51 shows the features of multiple specimens, but they are all steel. The top row is ductile micro-voids, but mineral filled polymers can sometimes behave in a similar manner.

At some point, the individual walls of what has progressed to resemble a foam can't support the load anymore, and they give way. In uniform and homogeneous materials, the central area generally finishes tearing first. Once the central area gives way, the reduced thickness of the remaining "tube" supporting the load allows macro-ductile behavior. For those with knowledge of fracture mechanics, the material is going from plane strain to plane stress. This tube or ring is loaded in shear. The "positive" and "negative" shear lips slide past each other, creating open (parabolic) voids, rather than equiaxed. Then we can see the details of the fracture with our physical eyes (assisted with an SEM), rather than relying on our mental eyes.

Note that the cup walls are essentially a shear lip feature.

Seeing the entire fracture surface covered with micro-voids helps us understand the term "Micro-Void Coalescence". All individual voids eventually coalesce into a single large nominally (bumpy) two-dimensional surface. Hence, MVC!

Next, we will be going deeper into cleavage and intergranular cracking. But, keep reading for a report of my taffy experiment at the end of this chapter.

8.18.2 Going Deeper: Cleavage and Intergranular Cracking

If micro-scale shear is suppressed due to high strain rate, low temperature, or geometric considerations[165], then we won't see any micro-voids.

Let's leave fatigue aside for now. Let's assume we are discussing a load that is sufficient to cause immediate separation. In the areas that have sound metal, the other two choices are cleavage and intergranular, as shown in Figure 8.51 (previous page).

If something has weakened the grain boundaries, then the material can come apart at the grain boundaries. Since it is generally a shorter distance for a crack to travel straight-through a part rather than for it to wander around following the grain boundaries (it requires less creation of new surface area), intergranular cracking is not the default choice of either the crack or the material being cracked! Intergranular cracking in many materials indicates that the grain boundaries have been weakened. New surface area requires energy. But, if for some reason the grain boundaries are weak, then the crack follows the grain boundaries. For example, hydrogen embrittlement in quench and tempered or strain hardened steel tends to weaken the grain boundaries. Copper-brazed stainless steel joints can result in copper infiltration into the grain boundaries. It does not take much copper to weaken the grain boundaries of ferrous metals. See Figure 8.52.

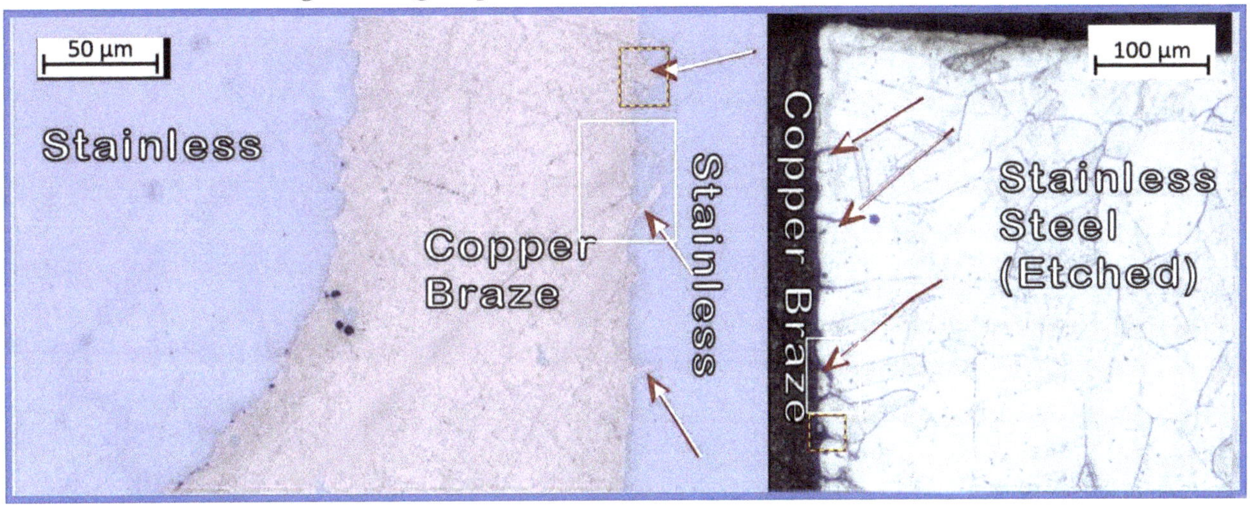

Figure 8.52: *These images are optical micrographs of cross-sections, not fractographs. On the left is the as-polished view and on the right is an etched view.*

On the left, the copper-braze looks copper-colored while the stainless looks blue (as-polished). Note the copper has penetrated into the stainless.

In the etched view (right) the white and red-dashed rectangles are shown in a lower magnification. In the etched view where the grain boundaries are visible, the copper, which now looks very dark, can be seen to follow the grain boundaries. The stresses were too low at this location to cause a crack.

---

165 If an externally threaded fastener is stretched along its axis (outside of the threads captured by the internally threaded mating fastener), then the thread crests have zero stress. The stress is concentrated into the root areas.

Figure 8.53: *An intergranular crack surface from a piece of anodized aluminum is shown. It never made it into service.*

It had been cracked prior to anodizing (as shown in another area on a cross-section). This was determined as a result of the presence of the anodized layer on a portion of the crack surface.

Even though the distance that the copper has visibly penetrated into the stainless is on the order of a single small-grain diameter, it's possible that an EDS line-scan across deeper areas along the grain boundaries would show elevated copper concentration. Figure 8.53 is another scanning electron micrograph showing intergranular fracture. Examine this to compare and contrast with the lower row of photos of Figure 8.51.

Figure 8.54 offers two more views of cleavage fracture to compare and contrast with the middle row of Figure 8.51. The red arrows (with dashed outlines) show classical cleavage steps on the main facets.

The pure cleavage features are most classical in Figure 8.51 Cleavage Row, on the left side, and in Figure 8.54, on the right, between the white dashed lines. The red arrows (with dashed outlines) show classical cleavage steps on the main facets. The preferred cleavage planes are not the same as the close packed planes. The green arrows (with dark outlines) show tear ridges, which likely are following close-packed planes. The blue arrows (textured with white outlines) highlight parabolic tear ridges.

These tear ridges are ductile features. We are starting to see that many fracture surfaces do not "know" that they are "supposed" to be completely ductile or brittle at the micro-scale.

FOOD FOR THOUGHT

How does cleavage of a wood log relate to cleavage of metals as described here?

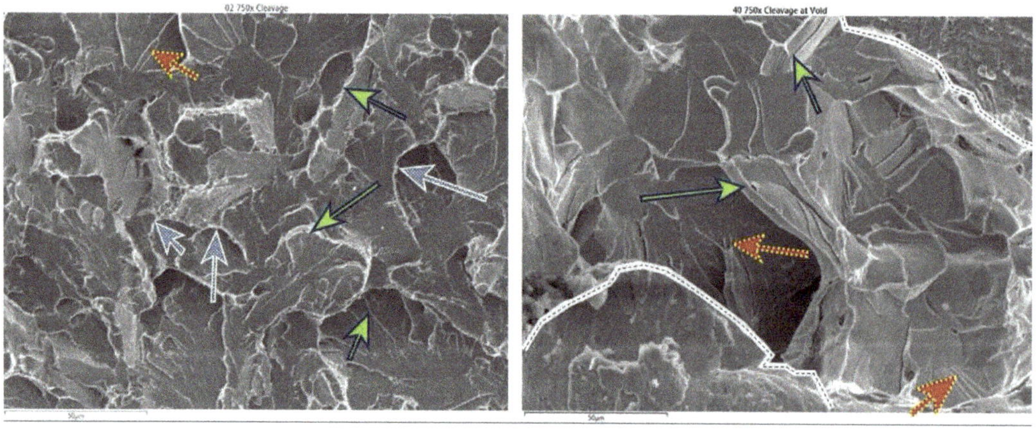

Figure 8.54: *Although these fracture features are considered cleavage, and thus very brittle, there are some features that are likely created by very localized ductility. That includes both the green arrows (with dark outlines) pointing at straight ridges and the blue arrows (outlined in white) pointed at parabolic ridges. The original magnification was 750x. Two totally different components. The image at right is from a material that had coarser grains than the one shown at left.*

8.19 Fatigue Micro-Scale Features

Now for the last of the major classical micro-scale fracture features, we will explore some SEM photos of fatigue cracks. Fatigue cracks come in two basic flavors. The first is classical, which have identifiable crack arrest marks associated with individual load cycles; and the second is microstructure influenced. This second version can range from fairly straightforward to recognize for a semi-experienced Failure Analyst to exceedingly difficult for anyone to recognize as fatigue. The determined analyst will find known reference images of the particular alloy in a similar process condition, size, and geometry.

If you are employed at a manufacturing company that has a durability test lab, you are in a good position to access or make a set of reference samples broken in fatigue. Then you can repeat the loading process using various strain rates to see how the material responds to various monotonic loading situations. If you are not equipped with such a lab, you can review available literature or pay an outside service to do the work.

Brittle castings and multi-phase (or multi-constituent) microstructures tend to obscure the classical striation features. Unfortunately, some of the most common materials, such as heat-treated medium-carbon steel (plain carbon or alloy), are among these problem materials. Cast aluminum is another problematic material, especially when un-modified for toughness.

Then, there are a lot of other features that can be confused for striations.

Figures 8.55A-C depict classical fatigue striations. See the figures below.

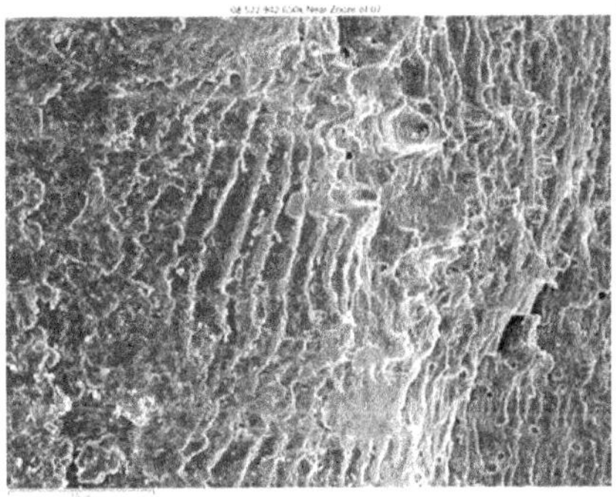

Figure 8.55A: *Fatigue in austenitic stainless steel that was annealed for an unusually long time – the thinned down edges of the striations indicate that this is fatigue very near the yield strength. Original magnification is 650x.*

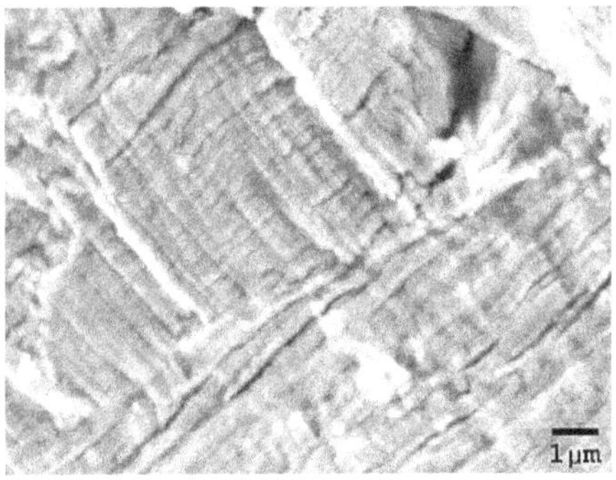

Figure 8.55B: *Fatigue in stainless steel – crack growth is parallel to a line going from lower right to upper left. Striations are sometimes faint and sometimes prominent lines going from lower left to upper right. Original magnification is 10,000x.*

Provided courtesy of Larry Hanke.

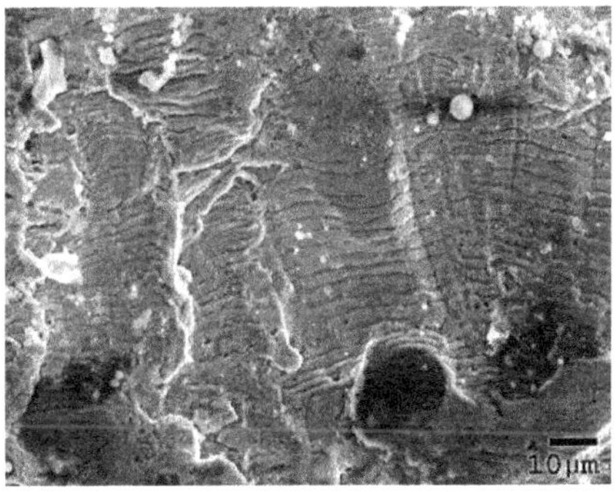

Figure 8.55C: *Fatigue in copper tubing – original magnification is 1000x. Note that the striations, more or less horizontal, are a little wavy.*

Courtesy of Larry Hanke, Materials Evaluation and Engineering. Used with permission.

Figures 8.55D and 8.55E (next page) have fatigue features that are more difficult to recognize because there are other features that are more prominent than the striations. The light purple box below shows an area of very faint striations. Large arrows show patches of more visible striations some of which are potentially associated with secondary cracks (going into the depth of the material).

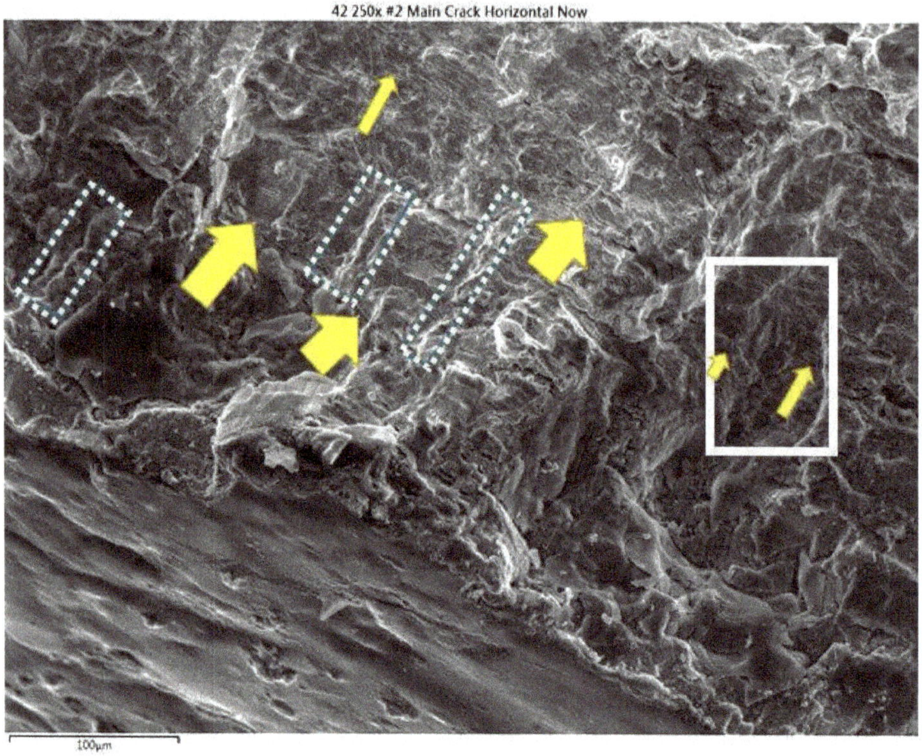

Figure 8.55D: *Fatigue in low carbon steel*

Large yellow arrows show what pass for "prominent" striations in this specimen, small yellow arrows show fainter striations. The green dashed boxes show tear ridges that are parallel to the crack growth direction. Striations are perpendicular to the crack growth direction.

Original magnification is 250x.

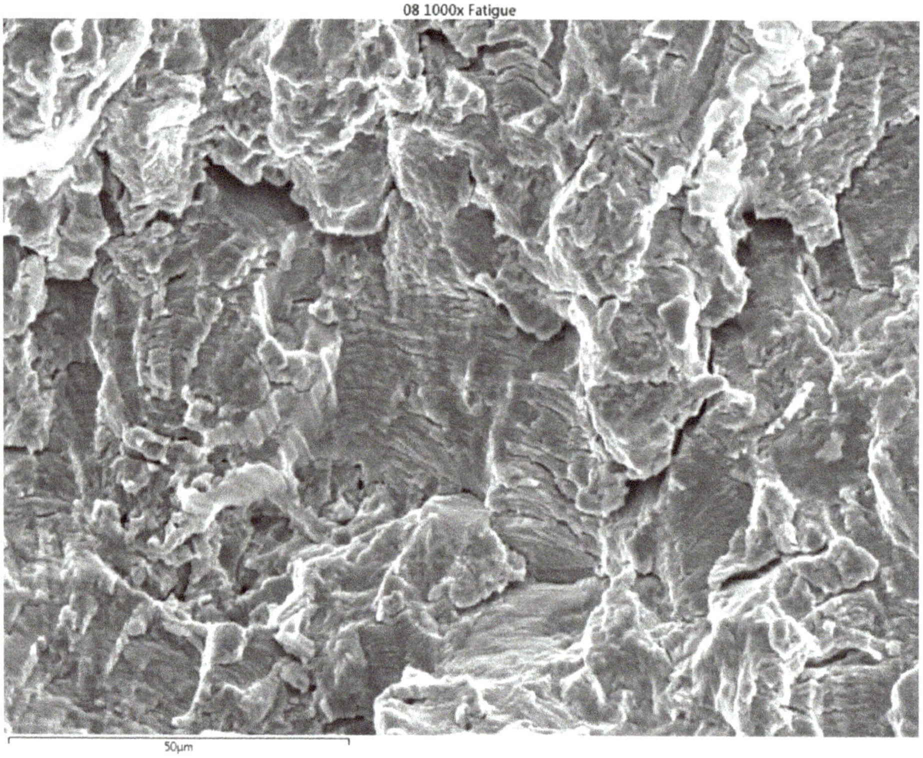

Figure 8.55E: *Fatigue in low carbon steel*

The striations are faintly visible in patches running at various angles near the horizontal in this image. There are various other microstructurally associated tear ridges and secondary cracks that confuse the eye. The large rectangular area in the left-center of the image shows parallel horizontal striations, and there is a nice patch to its lower right, as well as at the lower right corner. Original magnification 1000x.

Let's take a breather from the technical content for a moment. I am going to share a valuable life experience and then explain what that may mean for you.

Years ago, one of my adult education students (Introduction to Metallurgy at a community college) told me that most people need to review something seven times before they learn it. I'm sure that depends on whether it is something entirely new or something already familiar to some degree.

When I have taught fractography in a live course, it usually takes two full days for most working engineers to be able to use the ideas of fragment shape and loading geometry and to understand how they relate to ductile and brittle behavior.

So, I am going to reinsert the key points that started us off at the beginning of this chapter. Take some time to think it over. See if you can find some broken things to examine, in order to try your new skills. It's only by doing that you will master this content. The best of luck in your fracture analysis endeavors. It's an endlessly fascinating field.

WHAT IS THE PURPOSE OF FRACTOGRAPHY?

1) Macro-fractography sheds light on loading geometry, which is crucial for understanding if the object was subjected to improper use.

2) Meso-fractography shows crack direction, crack initiation and final separation positions, evidence for monotonic or cyclic loading, and provides clues about the environment of the part at the time of the crack propagation.

3) Micro-fractography sheds light on how the stress fields associated with the crack interacted with microstructural features. This is crucial for identifying and understanding manufacturing problems and variations from anticipated environmental conditions.

4) Both macro- and micro-scale fracture features must be interpreted in light of mechanical behavior. In its most basic form, this means determining whether the behavior was ductile or brittle and it must be determined independently for each scale.

5) Macro and micro features may be self-consistent with respect to ductile/brittle behavior, or not. In both cases, documenting these facts is important for leading stakeholders to a clear understanding of what happened and how it happened.

8.20 The Taffy Experiment: Final Thoughts on Micro-voids

I had originally written the book suggesting that the readers who want to understand the evolution of micro-voids make some taffy and mix some nuts or little hard candies into it. I thought that round sprinkles (used for cake decorations) would be an excellent choice. This is a way to "see it for yourself" rather than "believing the authority". It's a good thing I decided not to be lazy. I'm glad I tried my own advice, because basically, it didn't work well at all. Here's the story:

So, I made some taffy. The recipe was called "Old Fashioned Molasses Taffy". I had to purchase a candy thermometer to do this, because, being my first try at making taffy, I had no confidence in my ability to distinguish "soft ball" from "hard ball" stage of cooking. I had to aim for a specific temperature range.

The taffy came out very tasty - nice and sweet, and good and chewy. However, I was only able to reveal one micro-void associated with a sprinkle; and I am not sure if that formed while I was stretching the taffy "tensile bar" or if it formed earlier in the process when I was folding and stretching.

If you have never seen taffy being made, a regular part of the taffy manufacturing process requires folding and stretching the taffy. The big commercial taffy manufacturers use machines. This folding and stretching process left a lot of fine air bubbles in the taffy. I have since realized that these air pockets are part of the process that creates the chewy consistency of the taffy, which is desirable for the person eating the taffy.

On the next page is a photo (Figure 8.56) showing all I learned from this experiment, which mostly has to do with the strain rate and temperature sensitivity of Old-Fashioned Molasses Taffy.

NOTE: my hand rolling and stretching of the "test bars" did not result in a very uniform diameter.

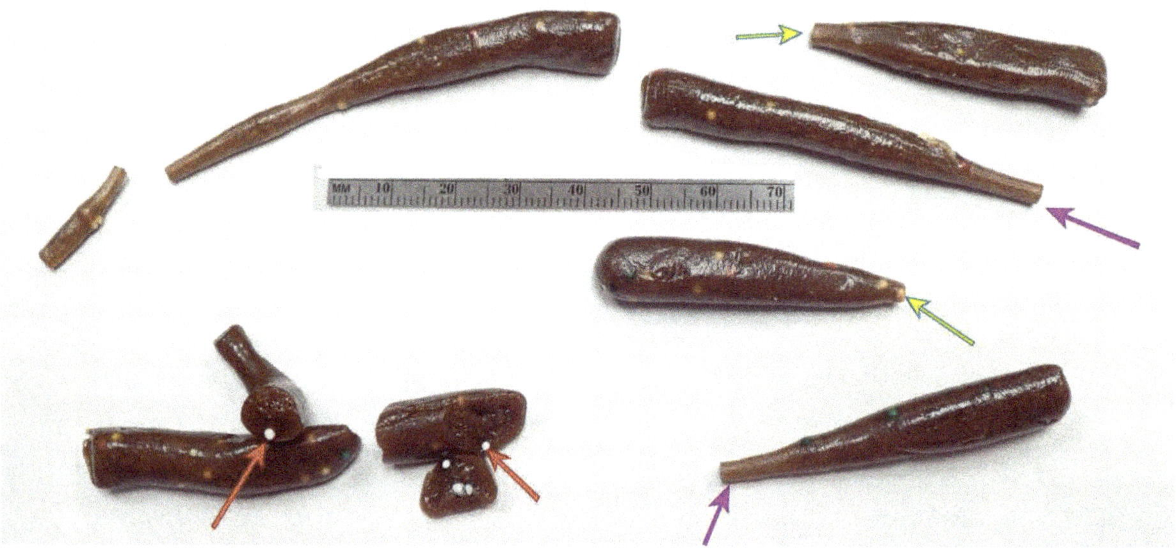

Figure 8.56: *Taffy-making – matched arrow colors show mating crack faces.*

*The lower left (**red arrows**) broke right through one of my "sprinkle inclusions". So much for making a micro-void.*

*The lower right and its mate (**purple arrows**) have no sprinkle inclusions anywhere near the crack.*

*The upper right and its mate in the middle of the image (**yellow arrows**) do have a small yellow sprinkle at the crack. However, I forced the material to break at that location by rolling it to a smaller dimension before trying to break it. There was no deformation (it was a brittle crack) and no void at the yellow sprinkle inclusion. Basically, my problem was that if I pulled slowly, the taffy stretched to a point, and if I pulled a little bit faster, it broke in a brittle manner. It was very hard to get a sprinkle inclusion at the smallest diameter.*

Figure 8.57: *Note the specimen broke at the sprinkle inclusion. There is a void on the mating side. However, the rest of the taffy fracture looks brittle. Likely this void happened as the test bar was being made.*

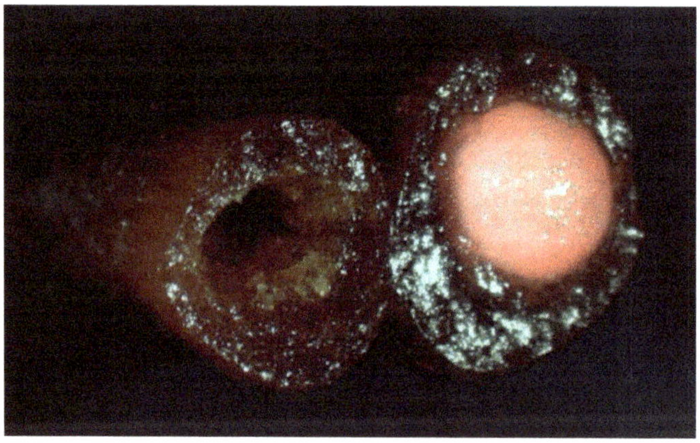

Figure 8.58: *This is the same "tensile bar" as shown in Figure 8.57. This image was obtained with an optical microscope. The shiny fracture surfaces are due to the brittle crack facets and the butter, which was separating out of the matrix as a result of the heat from my fiber optic lights!*

Figures 8.57 and 8.58 show results from additional tests from those shown in 8.56. Both Figures 8.59 (below) and 8.60 (on the next page) are additional documented details.

Now, I know that one of the most annoying phrases appearing in my college engineering text books was, "It is immediately obvious to the most casual observer." I won't claim that here. I had to think about it. When I saw the void in Figure 8.57, I wasn't sure that it had nothing to do with the inclusion. Maybe it did. But once I saw the tunnel in Figure 8.58 (above), it became clear that the best explanation for both was air entrapment from the folding.

And why did the sprinkle aspect of this experiment fail?

I speculate that: the sprinkles are much larger in proportion to the diameter of the "tensile bars" than a contaminant inclusion would be in any commercially produced metallic material.

The ratio of the hardness of the sprinkles to the taffy is not amenable to producing the range of complex behaviors found in commercial materials.

Fortunately, the experiment was a culinary success, even if it was a didactic failure for engineering learners.

On the other hand, philosophically, it's ok for a Failure Analyst to admit to failure. And, scientific research is having a moment where some scientists have realized that failure to publish their failed research experiments wastes the time of others who get the same bad idea, and holds back the progress of humanity. Publishing failures is brave, advances knowledge, and lets others know what did not work. Cheers to failure!

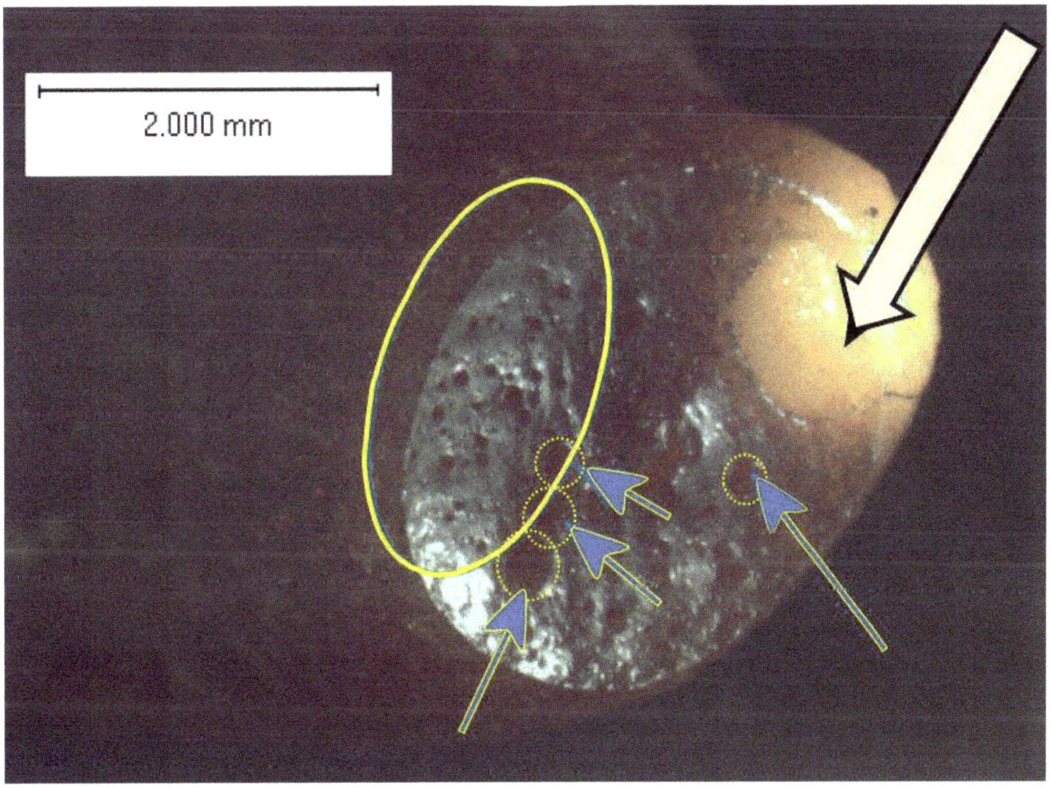

Figure 8.59: *Detail of the large fragment immediately below the scale in Figure 8.56 – this image, like Figure 8.58, was obtained with an optical microscope. I got the angle of the light adjusted to show the voids. The blue arrows show larger voids, and the large yellow ellipse encircles a lot of smaller voids. The crack was brittle so these voids were a result of the fold-pull-fold manufacturing process, and not a result of local deformation around non-existent sprinkle inclusions. Large arrow shows a sprinkle. The crack is not at a diameter.*

The lighting I used in Figure 8.60 illuminates the voids more clearly than the lighting used in Figure 8.59. This demonstrates how important it is to know what you are trying to communicate to the client and then to use the correct tools and methods for that messaging.

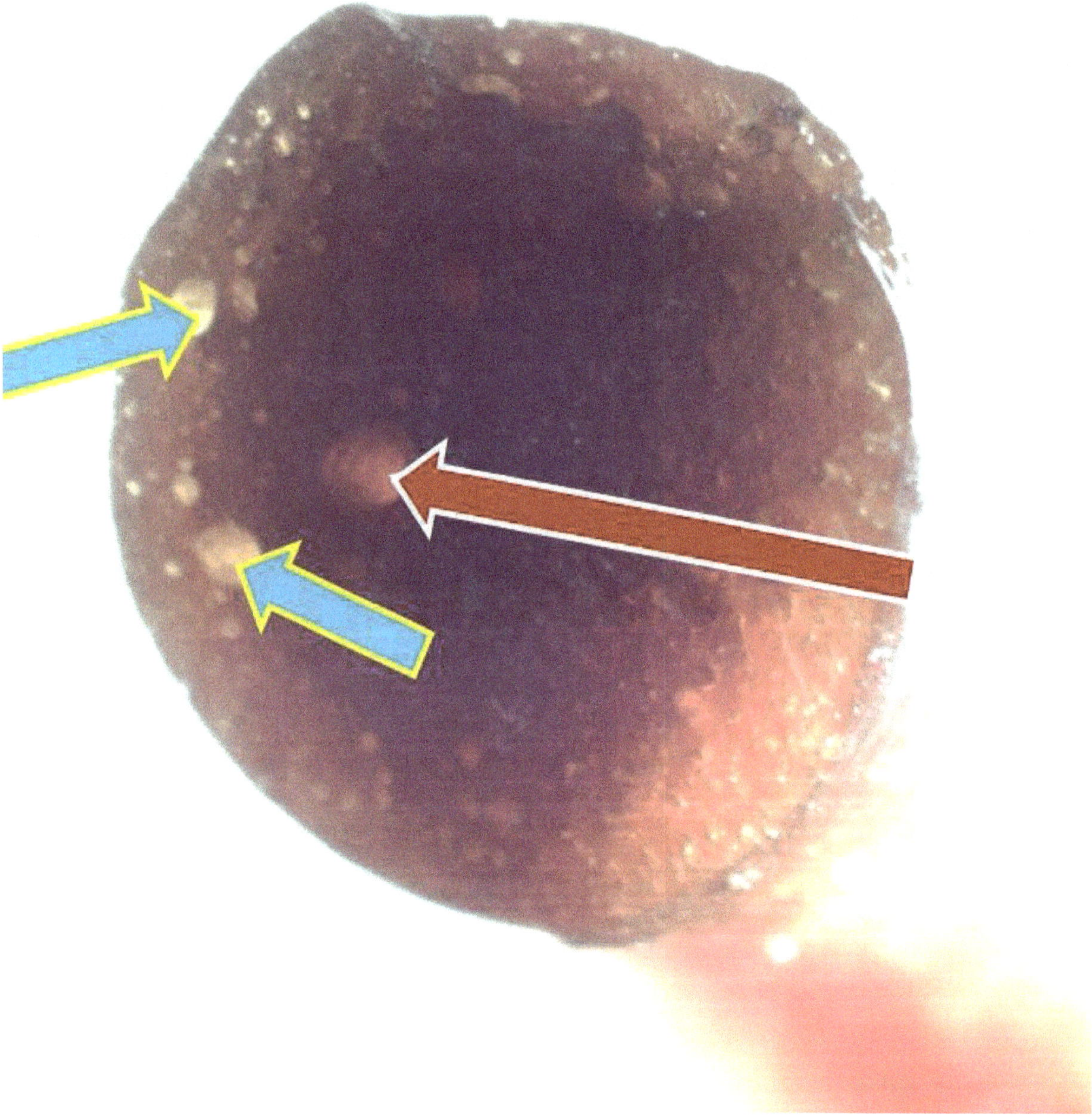

Figure 8.60: *Again, we see a wide distribution of void sizes, but all too small to have been made by a sprinkle inclusion. The red-and-white arrow pointing to an area nearest the center of the crack face reveals that this void is tubular, similar to the one we saw in Figures 8.57 and 8.58. Again, this is clearly due to the folding and pulling, not the presence of a hard particle.*

CHAPTER 9

Data Evaluation & Communication

Step 3 of the "Step Zero Method"

Figuring It Out - Telling the Story

9.1 Overview and Review of the Step Zero Failure Analysis Method

Failure analysis is a type of work that allows the practitioner to use the full range of mental skills that humanity has acquired.

By cultivating curiosity in the natural and human constructed world around us, we simultaneously set foot on the path to compounding competence in our chosen profession. A basic foundation of knowledge, about the nature of both the material world and of the psychological environment through which we experience it, is required in order to perform this work.

When we start a project, we must ensure that we understand what the stakeholders want and need to learn. Defining the frame of reference for the project in a manner that will promote that outcome is one way to ensure that we have accurately confirmed the context of our work. This also allows us to define a non-arbitrary end point for the investigation, which was the topic of Chapter 7.

Practicing rational thought by comparing and contrasting appropriate pairs of features allows us to focus more clearly on the actual characteristics of objects that come into our fields of perception. It also improves our ability to push distorting filters out of the way.

New thoughts and ideas naturally arise once we have a robust foundation of reality-based facts and theories at our disposal. Exactly how we gain those facts is one of the main concerns of this book. Discerning useful from irrelevant or even incorrect thoughts will occupy us soon. Connecting those thoughts and ideas to create potential scenarios of how damage happened is the central part of our failure analysis work. This is the purpose of Chapters 5, 6, and 8.

Stopping to contemplate the complexity of situations that contributed to a scenario we are evaluating is a necessary step toward ensuring that nothing was overlooked. Commit to practice waiting for a few new ideas to pop into your consciousness before rushing to write a report. Chapter 3 perhaps presents the most specific tactics to help us discipline ourselves to stop and think, mull, and look from yet another angle. But we must develop the habit of stopping to contemplate complexity during every step of the way along the path of the investigation.

All of those individual steps lead us to a place where we have a chance of accurately – even if qualitatively only – calibrating our confidence. We will have data. Turning the data into information through the use of inductive, deductive, and abductive logic and epistemology was discussed way back in Chapters 1 and 2. It was covered in various ways, in detail again and again, throughout the book. Unless you have a tendency to allow compulsive personality traits to rule your life, it rarely hurts to look at the data one more time in order to see what deviations from self-consistency they reveal. Often, by the time the end of the investigation rolls around, we will have fairly high confidence in our conclusions. That does not necessarily mean that we have eliminated all significant doubt about the most effective way to minimize future similar failures. Knowing that we are uncertain is the first step toward welcoming any new knowledge that might become available in the future.

Finally, the time has come to learn some techniques to evaluate failure analysis data and other techniques to communicate the findings, conclusions, and, if relevant, recommendations. Before moving into those tasks, we'll take a look at the mysterious processes of thinking. Communicating clearly will be the reward to the person who has cultivated practice of the methodology we have been learning.

EXERCISE 9-A: Contemplating Complexity

Way back on Labor Day of 2007, I was preparing an interfaith prayer service. I ran across a poem that celebrated the work of a carpenter. There was a line in the poem about the nails that the carpenter was about to use to build a house. I liked the poem because it was an exercise in the interconnectivity of all life and the planet Earth. But it didn't go far enough, in my opinion, to expose the myriad of networks at different levels that are necessary to support this life we live. So, I wrote my own poem about this topic (September 1, 2007).

Read, contemplate, and select your own simple object, physical or conceptual, (perhaps it's a song, or a story, or poem, or a person...) And as you read my poem. think about writing your own.

HERE'S MY POEM called, **When I see nails...**

When I see nails, I see the ore brought from the earth, and miners who extracted it. I see the molten iron at the center of the earth, and the cracking of the crust of the earth that let it rise to the surface.

I see the nuclear reactions in the stars that created the iron.

I think of iron meteorites that fell to earth and were sacred stones for early people.

I think of the beginning of the physical world where energy started being transformed into matter.

I see that in the nail.

I see the transportation workers bringing the ore to the blast furnace.

I see the blast furnace crew tapping the molten red stream of metal from the furnace to make pigs and ingots.

I see the solid iron being remelted and oxygen blown into it to turn it into steel. I see the aluminum jacketed workers pouring the steel into the caster.

I see the engineers who designed the caster.

I see that in the nail.

I see the sales people who sold the caster to the mill.

I see the cast billet being made into rod, and the rod being drawn into wire, and the wire being cut and headed and pointed so that it is now a nail. I see all the workers making the billets, the rod, the wire, and the nails.

I see the people loading the trucks with nails to take them to the retail store.

I am going to shut up soon so that we can remember the carpenter who used the nails to make the house. But don't forget the carpenter's father, and his mother...

I see all that in the nail.

Okay, now it is your turn. The assignment is to contemplate the poem and then select your own simple object (something other than a nail; something physical or conceptual; perhaps you will choose a song, or a story, or poem, or a person). After your selection, write your own poem. Continue to share with friends and family until you get at least two people who acknowledge that your poem brought to mind connections that they had not previously noted.

9.2 Data Evaluation and Communications Overview

In most cases, our data must be evaluated before we can communicate it to others.

Data evaluation starts with all the steps we have taken thus far to ensure that we are performing the tests, step-by-step, in a manner conducive to inspiring confidence. A big part of calibrating our confidence is checking for consistency. We have noted the importance of self-consistency among the data sets multiple times.

One of the benefits of using the methodology promoted in this book is that it inherently builds legitimate, knowledge-based, confidence-inspiring insights.

Over the course of the investigation, the Failure Analyst makes a series of decisions as thoughtfully or carelessly as the personality and training of the individual allows. One of the earliest decisions is often what overview photos to take. What will the series of images focus on? You might have seen a feature that is of unknown importance. Taking that photo will allow you to ask someone with more experience what they think it shows. Of course, it would be good to tell them what you thought was noteworthy. Sometimes you might say, "This *xyz* feature looks suspicious/different to me. What do you see? Etcetera."

As noted multiple times throughout this book, a very small fraction of people naturally recognize the importance of the negative finding. Said another way, few people might naturally take a photo to show there was no damage in a particular area. But that might be the very view that allows someone to later rule a particular scenario in or out.

Even if you are not sharing the data with the entire group of stakeholders, sharing it with a colleague, boss, or collaborator will often be extremely helpful. If you start to have a theory of what happened, and you share it with someone who is more familiar with the details of the event, they might steer you away from an impossible scenario sooner rather than after you have gotten attached to it and spent the project limit of time and money trying unsuccessfully to prove it. Nobody likes to hear the phrase, "I could have told you so."

If you can save or collect your data in a manner geared to collaboration, it will be ready to present at a moment's notice. "Data" include(s) images and numerical values (or their depiction in graphics, tables and charts). Even the latest collaboration software will not be used to its greatest effectiveness if you do not think ahead to the endpoint.

9.3 Introduction to Data Interpretation

There is no hard boundary between data collection and data interpretation. Everyone involved with data collection must have the basic knowledge required to judge the quality of the data that they are producing.

For example, soon after an image is captured, it needs to be examined to ensure that the features of interest are visible. Was the end of a feature cut off in the image? Was the brightness inadequate to document the feature of interest or its surroundings?

Some situations do not allow for the images to be evaluated until later. That is a consequence of the amount of time allotted to the project. Even with the use of digital cameras that have fairly large screens, some details will be difficult to see without taking the time to zoom in and make sure they are of adequate quality. Usually, microscope images will be "quality assured" immediately at the time of capture.

As each type of photograph has been separately discussed in the relevant earlier chapter, we will consider our photographic data interpretation here complete.

9.4 Holistic Data Interpretation

Data interpretation, like most of the rest of the failure analysis investigation, is an intellectual activity. That means we use our thinking skills to accomplish our goals. Thinking is a difficult word to define. We all probably believe we know what we mean when we say, "I think so," or "I don't think so."

Is "I believe so" different from "I think so"? Presumably, if you think so, you have more reason (more cause) to hold the subject thought; more confidence that others will also think so, at least after explanation. Holding a belief may be a result of lots of research, but it may not. Beliefs may fade into the idea of faith. I'm not necessarily talking about religious faith here. It's legitimate for anyone to have faith in their professors, for example; or have faith in their mentors. It is sometimes impossible to explain to a two-year-old in a short enough time why they should not run into the street. And the two-year-old should have faith that when we yell, "Stop!" they should freeze in position. Sometimes our professional mentors may give us guidance that they don't have time (or maybe even the skills) to explain how they know that what they are saying is correct. They may be acting out of experience.

So, what are the criteria we use to judge ourselves so we can determine that we have thought adequately about a particular situation? The use of the word "adequately" refers back to building enough confidence for the given situation. That is what we addressed in Chapter 7 on scope negotiations.

We have gone over the basic compare-contrast and connect-create mental processes, and how individuals can nurture their abilities to develop these mental activities. We have reviewed the use of logic and epistemology in order to sort the data so that we can build a sequence of events that realistically represents what most likely happened, and that finally resulted in the damage/ failure.

But we have not covered the topic of reason – or reasoning. We have also not come right out and addressed how to tell if something is a fact. First, reason; and then facts!

During a presentation on critical thinking that I gave in May 2022, someone asked me, "But what does reasoning really mean?" Wow. I was speechless. I had never thought about REASON. I answered that I would investigate the subject. Reason and critical thinking are two increasingly used buzz words that some people employ primarily to complain about their absence in the perceived mental actions of others. Many people say that reason and critical thinking are superior ways to make decisions. Why? Because they avoid emotion.

I cannot state this simply and simultaneously politely; but correctly stated, "Emotionless reason is poppycock."[166] Modern neuro-scientists have shown without a shadow of a doubt[167] that without emotion, no decisions will be made at all.

The great Scottish philosopher David Hume (1711–1776) is shown in Figure 9.1[168]. He worked and wrote well before the advent of modern brain science. Famously he said, "Reason is... the slave of the passions." Even back then, Hume was reacting to the incorrect idea, held by many at the time, that reason is purely rational and logical.[169]

Because what does reason mean? Simply, and politely here, reason means: (as in "The Reason for something") ***the cause***. In human terms, it is the motivation for action.

Figure 9.1: *David Hume had a long and abiding interest in epistemology (how people know what they think they know).*

> *Here's more or less what I sent to the group of people who participated in the on-line discussion about reason:*
>
> *One definition of Reason is Cause (e.g., the reason why, the cause).*
>
> *In human affairs, cause means motivation (e.g., I was hungry, so I ate some food; I was annoyed with the fly, so I killed it).*

166 If you look up the etymology of poppycock, it's exactly the impolite meaning I was trying to avoid. However, it has lost those indelicate associations in recent times. In usage here, Poppycock means non-sense.

167 Ok. The whole point of science is that we always keep enough doubt to allow a more comprehensive or correct explanation to replace the currently accepted "truth". But some things do still seem self-evident, and this is one for me.

168 Image of David Hume from Wiki Commons: Attribution by Allan Ramsay - https://www.nationalgalleries.org/art-and-artists/60610/david-hume-1711-1776-historian-and-philosopher-1754, Public Domain

169 https://1000wordphilosophy.com/2021/11/17/reason-is-the-slave-to-the-passions/ is a wonderful 998-word explanation of Hume's ideas by Daniel Weltman.

Motivation is a type of desire (or is associated with desire) at a basic level; the results of signals generated in/by the body (Review Chapter 1, Section 1.14 to augment your understanding).

Desire is a type of emotion.

So, emotion cannot be separated from reason. Many Buddhists teach that it is very difficult to escape from the twin sufferings of anxiety and longing. Clearly, these are both emotions. And thus, reason is inherently dependent on emotion.

This is what Niels Bohr said about the origin of the Yin Yang symbol[170]; he put the symbol and the following verbiage on his coat of arms, "Opposites are Complementary." He also said, "The opposite of a great truth is another great truth."

Reason is the black dot in the white fish of Emotion in the yin yang. Emotion is the white dot in the black fish of Reason in the Yin Yang.

Emotion and Reason are two sides of the same coin. Reason is about justifying (arguing for) our emotional desires through logic.

Logic is no way to verify truths. It starts from known truths. Sometimes we may conclude that the "truths" we had accepted are not true after all; this happens when we are confident that we have used unflawed logic but, subsequently, arrive at something that we cannot accept.

Furthermore, reason and logic can never lead us to truth about what we should do, because someone can always find another "reason" to add to the losing side. At some point, we make a value judgment, which emerges from our emotions, feelings, or intuition. Hopefully, the value judgment we make is weighted toward the humanitarian end of the greed-versus-humanitarian spectrum.

Getting back to Hume, the only reason we even need reason is to achieve our desires. Our desires might be high and laudable. Our desires might be low and mean. But desires and emotions are the motivations (the causes) that drive our skilled reasoning, logic and epistemology; they drive us toward a conclusion about what to do.

Speaking in terms of the short pithy sentences that people who teach logic favor, we can condense the argument about the complementarity of reason and emotion as follows:

Reason is Cause.
In human affairs, cause is motivation.
Motivation is desire.
Desire is emotion.
So, reason is emotion.

Only emotions, modulated through compassion, can form a foundation for ethics and morals.

This discussion about reason and emotion is very important, because it highlights, as Hume argued, that reason can never tell us what we ought to or should do. And as we noted in Chapter 1, everything that we know most surely is something that we sense or feel as a result of bodily signals. Most of these listed experiences that humans have are the source of our most basic motivations. They are familiar to us by the time we start our formal education.

Only emotions modulated through compassion can form a foundation for ethics and morals. Why should you care to know that emotion is the basis for reason?

If you are one of the intellectuals who has been seduced by teachings that say reason precludes acting emotionally, please consider the following.

170 The Yin Yang symbol is from a traditional Chinese system of thought known as Taoism.

- Of course, it is possible to act thoughtlessly due to untamed emotional currents.
- Of course, this is generally undesirable.
- But emotions are not all dark and evil.
- Being able to experience the range of emotions available to humanity is certainly one of the major essences of the human experience.
- Only people who accept the relationship between reason and the heart will rise to the highest levels of competent communicators. This is the case for both written and verbal communication.

Every communication educator emphasizes the need to know the audience. While it might not be obvious to the casual observer, knowing the audience includes awareness of the basic operation of the human mental landscape, both emotional and logical (or rational).

Here's one of the first poems I ever wrote, at the beginning of a burst of creativity in 1995.

DARKNESS

Without darkness to dispel, what value has light?

Without suffering to assuage, compassion has no task.

Without the female, what defines the male?

Without evil, who can recognize good?

NOTE: *A feminist version that jumped into my head while finishing this book substitutes the following for the third stanza. Times have changed!*

Without the male, what defines the female?

Science has shown that up to 10% of the population may have non-binary gender identification. In honor of those people, perhaps another line is needed.

Who can even say what gender means, when only the ends of the spectrum are known?

This is an example of understanding that we can compare-contrast binary descriptions of phenomena allowing for a full spectrum analysis and understanding.

Data interpretation, like all mental work, is a process that has been shrouded in mystery for as long as humans have been thinking about thinking. Progress is being made. We can learn to think more clearly, and doing this type of failure analysis work allows us to keep improving our mental skills, as others respond to our efforts and point out our errors.

Reason, logic, epistemology, knowing-how-to-*look-at* and knowing-what-to-*look-for*, along with the knowledge of how to use the instrumentation that facilitates our observations, these are the basic skills that allow us to competently obtain, evaluate and communicate our data in a way that allows it to become useful information.

Before moving on from this mysterious process we call thinking, we need to acknowledge that much of it is subconscious. The great chemist Auguste Kekule (1829-1896) had the structure of the benzene ring come to him in a daydream, in the form of a snake swallowing its tail. This is an ancient symbol with many hidden meanings. Figure 9.2 (next page) shows a portrait of Kekule and Figure 9.3 (next page) shows a sketch based on his story of his vision.

Getting the answer to a problem in a dream, at night or during the day, is not a fluke. When we relax our minds, and especially when asleep, new neural connections form in our brains. Most people who use their brain / minds to earn their daily bread have had the experience of awakening with the answer to a question. How does it happen? Who knows? But once we have the answer, we can often use logic to reason back to why the answer makes sense.

Figure 9.2: *Auguste Kekule[171] had a dream that gave him the key to understanding the molecular structure problem he was studying.*

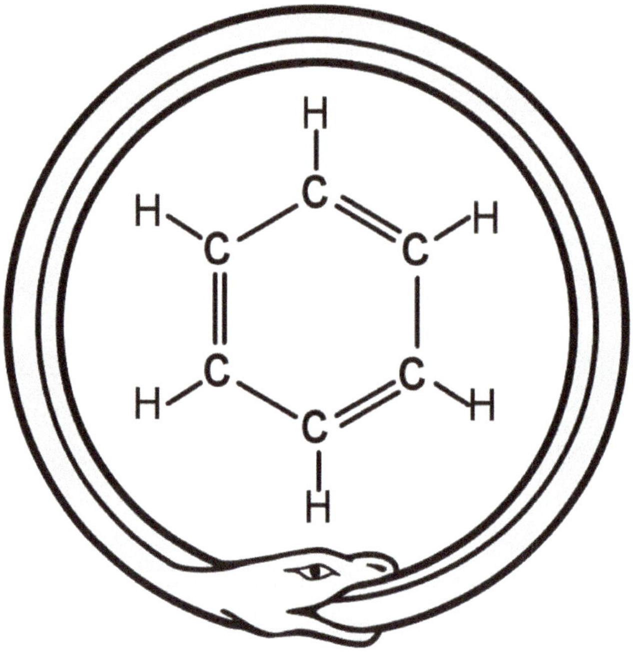

Figure 9.3: *The snake eating its tail is an ancient symbol. It gave Auguste Kekule the clue he needed to understand the structure of the benzene molecule. Attribution: By Haltopub[172]*

9.5 Explaining Intuition as Well as Logic

Why should you care to learn how to list the steps used in data evaluation? One good reason is so that when you grow old, and hopefully wise, and some young pup comes to you for mentoring on data evaluation, communications, and reporting, you will be able to clearly and concisely explain what you do. You will not be one in the majority of people who acquires a high-skill level just prior to retiring, but still has no vocabulary or mental structure to be able to explain how you do what you do. If you are similar to many experts, your knowledge of how you reason will be mostly intuitive. Explaining intuition is exceedingly difficult for most people. How do you explain how to recognize a "sensation in your gut" or how to distinguish it from "another sensation in your gut"? Good luck!

Learning, practicing, and sharing the 7-Step Method of Constructing Competence will keep you moving toward higher levels of competence in all of the steps. Logic involves making use of comparisons and contrasts, metaphors, and classifications, which are all part of rational thinking. Practicing rational thinking will increase your frequency of experiencing creative insights. Creative thinking happens when we cultivate the other steps. Experiencing how these seven steps work together over a period of years will give you the ability to teach and mentor others to a greatly enhanced skill level. It's not a recipe, because it's not possible to write down all the details that you will learn. Each person has to hone their own skills and gather their own lessons from

171 By Unknown author – image is now in the public domain | https://commons.wikimedia.org/wiki/index.%20php?curid=167877 | Image has been modified

172 Attribution by Haltopub - The ouroboros, Kekule's inspiration for the structure of benzene, CC BY-SA 3.0, https://commons.wikimedia.org/w/index.php?curid=29661966; Sources: Benzene Structural diagram.svg; Ouroboros-simple.svg

experience. Then you will truly be able to pass on your knowledge in an effective manner.

And you will be able to communicate to clients more effectively as well.

9.6 How to Recognize a Fact

Now, before we move on to the nitty gritty of writing the report, let us take a short detour to explore the definition of a fact. This has become a more urgent question in a modern, or perhaps post-modern, world where different groups adhere to different fact sets. Just as I had never thought about what the word "reason" really meant, I discovered during the run-up to the United States presidential election of 2020 that I did not know how to tell someone clearly and simply how to recognize a fact. This was despite having taught people to distinguish facts from opinions for years as part of failure analysis seminar.

This aspect of distinguishing various levels of facts from opinions based on layers of specialized knowledge was discussed back in Chapter 3.

> *I was (to be polite) fairly aggravated with my neighbor when he told me I did not know what a fact was. Later, I had to tell him that he had been correct. Eating crow never tastes very good.*

But as I have now researched the topic and presented multiple times on this topic. I feel much better prepared to help others lay a foundation for their own exploration of how to recognize things that are closer to the objective end of the spectrum of data than they are to the purely subjective end. Perhaps that is a good working definition of a fact.

9.6.1 Fact One

A fact is something that is close to the relevant objective end of the spectrum of data.

However, facts about our own personal subjective experience are truly facts. You cannot prove that I am thirsty if I say I am not.

9.6.2 Fact Two

A fact is any true statement about a personal sensation or emotion.

As noted previously, it's generally impossible to know whether the second type of fact is an actual fact or a lie. So, here's another chance to reinforce the teaching of Niels Bohr: The opposite of a great truth is another great truth. If we want to eliminate the second type of facts, we would have to modify the definition of a fact. Perhaps…

9.6.3 Fact Three

Something that is objectively demonstrably close to the relevant objective end of the spectrum of data

However, it is clear to most people who have really thought about this subject from a philosophical view, pure objectivity cannot exist in humans. There are many people who still use the term "objective", and believe that they *can be* completely objective. While it is true that we can learn to clear our mental filters to some degree, anything we see or hear or experience or describe is always done so from our place in the world. Our place in the world is physical, familial, cultural, social, and emotional. The great philosopher Immanuel Kant famously pointed out: there is a disconnection between "the thing-in-itself" and how we experience it. We can only know what our senses can tell us about it.[173] The title of a new book by Bernardo Kastrup, who bases his creation of Analytic Idealism on the work of Arthur Schopenhauer, is called *Why Materialism is Baloney*. Kastrup goes even further, demonstrating objectively, philosophically, and scientifically that there is no physical world to measure or characterize.

Think about five people standing around a small table with a bowl of mixed fruit. Each person is looking from a particular angle. Maybe only three can see that there's a banana. Maybe one is a four-

---

173 Immanuel Kant | https://en.wikipedia.org/wiki/Thing-in-itself

year-old and can only see the legs of the table. Maybe one is color blind, and cannot tell if the apples are red, yellow, green or gold. Had all five people standing around the fruit table been trained as Fair Witnesses[174] and they each reported only what they can directly sense, we could get a partial view of the fruit bowl on the table. But whatever we had would be a reliable description.

Another example is the story about the blind men and the elephant we discussed in Chapter 7. The blind men are all reporting their conclusions, while speaking as if they are reporting facts. Each one has a partial view. The points of view in these examples refer to actual line of sight, or tactile impressions converted into spatial models in the case of the elephant.[175] When we carefully consider how much our particular, personal experience determines what we are able to recognize, we see how fruitless reliance on assumed objectivity really is.

> *I have a large garden. I prefer to grow my own vegetables and fruits. I try to avoid using any synthetic chemical pesticides.*
>
> *I have arthritis. It's not as bad as it was a few years ago, but I cannot do all the work to maintain the garden and orchard myself.*
>
> *I live out in the country. There are lots of practical people. But how many can tell a lettuce seedling from a weed? How about a corn seedling from grass growing in the corn row? How many understand that if they don't take off some of the baby peaches (when they are ~1 cm diameter), the branch may break when the peaches are bigger? Or the tree may be exhausted and make no peaches the following year? The answer to all of these questions is "Not many".*
>
> *How many are capable of following my list of tasks? Very few. Why? They want to do the tasks in the order they want to do them. They don't understand even when I tell them do this or that task first. Why? If they don't weed the grass out of the strawberries NOW, they will disturb the roots of the strawberries when they pull up the crabgrass, as the roots will be longer, and thus the strawberries will be weakened.*
>
> *You want someone to fix your car? Great. There are plenty of people who can do that. But gardening, like doing materials engineering or failure analysis or fixing a car, is a learned skill. We tend to forget all that we had to master to become as competent as we are. It's difficult to pass knowledge on to others.*

That each of us sees the world in a unique way is part of what Francis Bacon was getting at with his Novum Organum. He was trying to help us see how we see things differently from each other because we live different lives. Yes, there are mistakes that almost all humans tend to make in the same way. But that doesn't make all the wrong conclusions objectively true just because it's what most people carelessly think.

> *I often discover too late that my garden helpers don't hear an instruction because they don't have the knowledge that lets them realize it's important!*
>
> *My new peach trees are going to have to be dug up and replanted, because my helper buried the graft. If you care, look up how to plant a grafted fruit tree.*

With objectivity so questionable, is there ground left for us to correctly assume that almost all of us will agree?

Yes, there are a few things that almost everyone almost always agrees on. When on Earth, gravity encourages things to fall toward the center of the

[174] Robert Heinlein, the great science fiction writer who wrote the book, *Stranger in a Strange Land*, introduces the concept of people who are trained to avoid leaping to the usual conclusions that most people make. You can see that two sides of a house are painted white. Do you assume that the rest of the sides are painted white? Would the Fair Witness remain agnostic about even the existence of the other sides of the house?

[175] The combination of simultaneous seeing and touching (a ball, a box), seeing and hearing (lightning and thunder), seeing and experiencing (a fall from a high tree branch) are the original foundation of natural epistemology.

earth. We open doors before trying to walk through them. Beyond that, it's hard to find commonalities. Lee Beaumont [176] writes:

> *"Building on these assumptions, it follows that reality is our common ground, even though any number of worldviews is possible. It is likely that each of us hold a worldview that is somewhat different from others. Of all the possible worldviews, one worldview is especially important. That is the worldview that corresponds to reality as closely as our best current understanding of reality allows. Because reality exists, we can examine reality, and we can align our worldview with reality. Because we all live in one world, reality is our common ground."*

The referenced assumptions are listed at the link shown. I used to agree with the idea that there was one reality. But the more I think about this, the clearer it becomes that reality is a construct rather than an object. I agree with Bernardo Kastrup, Immanuel Kant, and Arthur Schopenhauer. That does not mean I totally construct a complete fantasy. "Materialism is Baloney" and "Because we all live in one world, reality is our common ground" are great truths that are both simultaneously true and false, but useful in different circumstances.

We each construct our own reality. Since constructs (as we have discussed previously at the beginning of Chapter 4 and in Chapter 5 in the discussion of damage categories) are best judged by their utility, I think it is a lost cause to expect my neighbors to ever embrace the same description of reality as I do. Isn't it better to simply acknowledge that we individually construct our reality and acknowledge that the neighbor's reality is their reality? I'm absolutely not saying that I approve of the reality of a person who is aligned with a group whose leaders choose to deceive. I believe, and have evidence, that my reality is more closely based on what humans have been able to glean about the world we inhabit.

But as we learned at the beginning of this chapter from David Hume, knowledge of reality can never give us the answer about how we should use our resources or treat our neighbors. Those are value judgments that can only come from emotional responses. Since there are so many different groups that take advantage of culture and emotions for accumulating power, they will need to be considered to gain agreement on any subject. Objectivity cannot exist in human affairs. Even if it did exist, it would not be an effective way to live.

In closing this muse on objectivity, let us remember that at the very least we can all experience some common aspects of the world we inhabit. It's much easier to get agreement on those shared aspects of our lived reality than on others. We can build from that group of agreed facts; however small it might be. Starting out by assuming that everyone knows that $E=mc^2$ or that $C = D$ can only lead to misunderstandings.

Discussions that go along smoothly and come to an abrupt roadblock are often a signal that the discussion has revealed an end to the set of overlapping reality constructs. It's a signal to backtrack and figure it out.

We must now clarify the difference between logical and critical thinking.

- These are not, as many imagine, the same thing.
- As we have previously noted, logic uses assumptions.
- We have also previously noted that epistemology is what is required in order to evaluate the correspondence of the assumptions with some aspect of shared reality.

[176] This was discussed with Lee Beaumont (Wisdom Researcher and retired engineer) in a private communication. Lee wrote a Wikiversity course on Epistemology. It includes this lesson: https://en.wikiversity.org/wiki/Finding_Common_Ground

Stop to think for a minute here and recall the definition of critical thinking as confirming the appropriate context for the thinking exercise about to be undertaken.

- Critical thinking and logic are essentially at the opposite ends of the scale of objectivity.

Because what is logic really? It's a form of associative thinking. Associative thinking is the original thinking process of humans. It's how we learn most of what we know. We see things together and jump to the conclusion that they are causally linked, when often they are only casually linked.

- George Bernard Shaw[177] is credited with saying, "Statistics show that of those who contract the habit of eating, very few survive."
- Many people have compiled statistics showing that rain and the rate of traffic accidents is correlated.

How can we know whether they are causally linked, and if so, in which direction the causation runs? My electric power company says that "Trees cause power outages."

We must learn to discern CAUSAL from CASUAL links!

- Really? I prefer "Trees being stressed by the wind, sometimes in combination with the weight of ice on the branches, as influenced by gravitational forces, falling on power lines, are factors that may increase the local frequency of power outages."

Here are three ideas about the correspondence that has been found by many social science researchers between watching television shows with violent episodes and acting violent: Watching violence on TV may cause aggression. Or it may simply appeal to aggressive people. Or perhaps family members who are angry with each other act out aggressive behaviors and prefer violent TV shows?[178]

Ralph Waldo Emerson pointed out that "The ancestor of every action is a thought," while Benjamin Disraeli mused that "Thought is the child of Action." To my mind, these are two quotes that are clearly great truths and thus both may be true. In fact, they may be simultaneously true. Both imply causation. Neither can prove it!

To close our warning on relying too simplistically on logic and objectivity in the absence of critical thinking and epistemology, and to get back to some things perhaps more directly related to something a Failure Analyst might actually get paid to consider: Does paranoia "cause" secrecy? OR… Does secrecy "cause" paranoia? AND Does curiosity "cause" knowledge? OR… Does knowledge "cause" curiosity? Determining the "true" direction of causality is often difficult, sometimes impossible.

Does paranoia "cause" secrecy? OR Does secrecy "cause" paranoia?

In closing the section on associative thinking, and its relationship to logic and causality, it's important to understand that associative thinking is also the foundation of superstition. For example, athletes (amateur, professional, and even spectators) often have a lucky charm. How did they acquire their lucky charm?

Usually, a lucky charm is something they had with them when they won a game or couple of games. But it could be gift from someone close to them for the purpose of bringing luck. Can wearing a bullet that hangs on a chain around your neck

[177] George Bernard Shaw was an Irish playwright who enjoyed controversy and political activism.

[178] Ibid | David Levy discusses this in Chapter 10 of Tools for Critical Thinking, Correlation Does Not Prove Causation.

really make you faster than a speeding bullet?[179] But competitive sports are, we are told, as much about mental fitness as physical fitness. Olympic gymnast Simone Biles withdrew from one of her scheduled competitions due to a case of "the Twisties", which is a mental issue at root.[180]

Figure 9.4: *Many Western-educated people think superstitions are bad. But are they always evil in intent?*[181]

Superstition is given a bad name by scientists and philosophers, and even some groups of theologians. Conservative Christian theologians, for example, generally frown on using astrological horoscopes, encouraging the faithful to rely on their deity for strength to meet whatever challenges lie ahead, rather than trying to escape destiny and lose the opportunity for enhanced faith. But as we have seen, acting on our superstitions are sometimes a relatively harmless and useful mental approach. Were that not the case, superstitions would have disappeared from humanity's toolbox long ago.

9.6.4 The Bottom Line On Associative Thinking

Both logic and superstition are fruits of it. The human condition is difficult. Some crutches are helpful.

Finally, as a last tidbit of information for your contemplative consideration about causation and correlation and the dangers of using logic without epistemology in your data interpretation, see Figure 9.5 (below).

Based on the evidence of Tyler Vigen's data-driven graph, one could predict whether it would be a bad year for venomous spider attacks based on the length of the winning spelling bee words at the Scripps National Spelling Bee. These two trends are closely correlated for a decade. Can you see how important it is to be rigorous with our thinking?

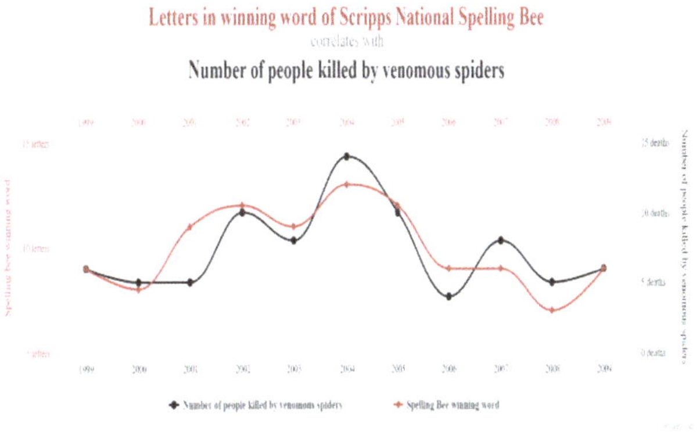

Figure 9.5: *Tyler Vigen used a bot to dredge data on the internet to find unrelated data that that seem to have a cause-and-effect relationship.*[182]

Tyler Vigen used a bot to dredge data on the internet to find unrelated data that that seem to have a cause-and-effect relationship. It's obviously a coincidence; but with so many things happening in the world, it is easy to find unrelated data that shows correlated trends.

179 Attributions: https://parade.com/488766/samanthacoley/10-olympians-reveal-their-lucky-charms/

180 Attributions: https://www.nbcnews.com/news/olympics/what-are-twisties-simone-biles-explains-gymnastics-struggle-tokyo-olympics-n1275460

181 By KEBman - Own work, Public Domain | https://commons.wikimedia.org/w/index.php?curid=744133

182 This image was created by Tyler Vigen on 7April2016 and generously released under the Creative Commons Attrition License. This is an example of data produced by data dredging. | http://www.tylervigen.com/spurious-correlations

9.7 The Minimalist Approach to Data Preservation and Presentation

Now you are ready and primed to write your report to correspond and communicate clearly. Unlike most commercial labs, I routinely give my clients all of my data. I reason that if they don't like my interpretation, they can show the data to someone else. What they cannot do is give the part to someone else; the other experts will never see what I saw because, most of the time, the part is no longer in the condition in which I received it.

Over the years, much of my work has been related to manufacturing troubleshooting or concerns about routine quality issues at manufacturing companies. This has been my primary business frame of reference.

In addition, a portion of my work was related to non-conforming parts that had slipped through the cracks of a quality system and arrived at my client's customer's factory. Sometimes when that happens, a factory staff member gets very upset with my client and demands a formal inquiry. Other times – especially when the individuals involved had been working together for a long time – it was enough to know that their supplier was taking action to solve the problem. Just knowing that they had hired an outside expert often provided breathing room for everyone involved so I could perform my work; I could ask for additional documentation search and review, or plant trials. By providing expert outside data and a brief explanation of the findings, as well as suggestions for minimizing the chances of recurrence, companies which had received "bad parts" were often satisfied.

As time moved on, we were able to deliver my information to all the interested parties during a conference call. And then, suddenly, most companies had access to video conferencing. That made it easy for me to highlight the features of interest while speaking. I could show and explain images or other data outputs. Pointing out the specific features that depict a particular phenomenon is much easier to do with a moving cursor. Of course, my clients had to trust me to speak directly to their customers.

Sometimes, the issues that I investigated were of great concern. During these report-outs, the parties involved wanted or needed a formal report.

However, even if a formal report is required, when people are willing to participate in a video conference, they usually gain a more complete understanding of what work was performed, why it was deemed important to perform, and what was revealed. It is difficult to create a written document that is as customized to each involved individual and what they need to understand within a given project.

As a materials engineer leading the failure analysis (or at least the physical cause part of the failure analysis), I am often the only one involved who has a materials engineering background. Thus, it is helpful to give all interested parties the chance to ask questions in a way that is comfortable for them. I can be as technical as each member of my audience needs. This increases the chances that the stakeholders gain the understanding they require to move forward.

The modern global manufacturing environment can create language barriers, but despite those issues, video conferencing offers an important option for communication of the data, its interpretation, and a flexible way to explain the relationships between different aspects of the findings.

GOING DEEPER – DISINFORMATION & UNQUESTIONED BELIEF

The United States of America has been struggling with a flood of disinformation that two weeks ago (as I write this) resulted in an insurrection at the national capitol in Washington, DC. As far as I am concerned, a root cause of this event is the almost complete ignorance of epistemology. Many Americans are taught to have blind faith in the existence of events that are nominally physically impossible in the material world. This importance of unquestioned belief is promoted by many religious leaders. Belief of this type, **based on the word of authorities, is inherently different from belief based on personal observation subject to rational, creative, and systems thinking, followed by calibration of our confidence**, such as promoted throughout this book. Given the human condition, faith and belief are necessary as is the recognition and honoring of our spiritual natures. **But we must learn to discern the difference between factual and metaphorical evidence.**

One of the frequent speakers at a church I used to attend gave two sermons on the beneficial effects on human consciousness of the "poton belt". Sadly, when I acted on my compulsion to inform him that "photons" were different from "protons" and that there was no such thing as a "poton". He resisted. When I said that the "photon belt", as it is more commonly known, is a metaphorical truth about humanity having an opportunity to learn to treat each other better, he became very hostile. This is an individual claiming to have been trained in a scientific field. Presumably such an individual had to take a chemistry course at some point.

Constructing competence in a technical field is one of the best ways to acquire a foundation of knowledge based on physical reality. Without some minimum number of people who bother to build such a solid personal knowledge foundation and a membership in the accompanying "club of consensual reality", humanity's future looks dark. Too bad this individual did not benefit broadly from his technical education.

When the political leadership of a human organization fails to conform to a previously established "consensual reality", long term social stability is put at risk. Social stability reinforces the status quo. Yet, putting social stability at risk is not always an undesirable situation. Some changes would be desirable for the majority of humanity. A great example is Dr. Martin Luther King, Jr. (a Christian preacher and a great civil rights leader). He said, "… the arc of the moral universe is long but it bends toward justice."[183]

The Prophet Zarathushtra promoted the idea that truth will eventually win. According to this teaching, it is a simple fact that falsehood is based on things that do not actually exist. When we rely on things that do not exist, eventually, we run into obstacles.

(more next page)

183 The full sentence by Dr. Martin Luther King Jr is, "We shall overcome because the arc of the moral universe is long but it bends toward justice." This was taken from the Remaining Awake Through a Great Revolution speech given at the National Cathedral on March 31, 1968.

On the other hand, when enough people believe something for long enough, it is sometimes possible for it to manifest in some manner. That is why breaking out of false beliefs is so difficult. We see what we expect to see. For most people, seeing is believing; and sadly, believing is knowing. Perhaps these two paragraphs may be understood as representing another pair of great truths. In cultivating curiosity about the natural world, and the perceptual filters that we use to experience it, we may create a structure that lets us understand the difference between a physical situation and a metaphorical picture.

For many years, humanity was saved the trouble of having to sort this out, because humans have a tendency to keep their knowledge in silos (or files). When in a Christian church, for example, the equation 3 = 1 is a legitimate fact (God The Father, Jesus The Son, and The Holy Spirit are One). But most people realize that their elementary school math teachers would not give them a gold star for this equation, even if that teacher had been a nun at a Catholic school.

As science has increasingly been used to enhance military might and, also, has unintentionally resulted in the poisoning of the environment, many non-scientists have become hostile to the consensual reality built on the work of the Enlightenment in the 17th and 18th centuries CE. We modern humans have our work cut out for ourselves. An important step in the correct direction is constructing competence by using our mental capabilities to serve our fellow humans.

I find my investigation work to be more efficient when I can share the data or findings early and often over the course of the investigation.

SOME THOUGHTS!

Sharing my investigation data and findings early and often allows me to gather feedback. If something I am telling the client seems wrong, there's time to address it. That said; many laboratories routinely do not provide data until the investigation is complete. They do this to allow for internal peer review at the lab before the data is released.

When work is related to litigation, it's a good idea to keep quiet until you have had a chance to get peer review from a confidential, appropriate source. This could be sensitive during an internal investigation of a catastrophic failure. But for routine work, it's usually helpful to have a second opinion. When the people performing the investigation are not experienced enough to have legitimate confidence in their data, it is best to wait to distribute the results until a thorough review has been performed.

Whenever you (the analyst) think your client may take action based on incomplete data, it's a good idea to keep quiet until you have had a chance to get peer review from a confidential, appropriate source.

What about sharing photos? It's impossible to capture a view of every feature from every angle. Videos can help. Speaking while capturing the video images aids the memory. Sharing these narrated videos may become problematic for some forensic evaluations though. It is generally required to keep our thoughts to ourselves in work subject to overview by courts.

In order to effectively prepare for a meeting, and to facilitate keeping my clients updated during larger investigations that may take a week, a month, or longer, I sometimes make the image files available as I complete each step of the project.

Figure 9.6: *Windows Explorer format record of file folders for a typical failure analysis project*

EOS and TG5 are my regular Cannon and Olympus digital cameras. OM and SEM stand for Optical Metallography and Scanning Electron Microscopy respectively.

I log the project after receiving formal (written) assurance by the client company that they will pay for the work (usually a purchase order). During the logging, I photograph the items submitted. That's when I might email or post a folder with the overview images. More often, however, I wait to send that folder until after I have accomplished at least a visual exam. I might even wait until I completed an optical microscope exam. Then I usually have something helpful to say in the text accompanying the photographic data. "Helpful" in this case means that there is something that the client did not already know.

I can also confirm the planned details of the test protocol. I may provide documentation of the marked-up parts to show how I plan to cut them (in order to get the required test coupons). So, I might send two files at once: Overviews and Optical Microscopy. Especially for smaller parts, it is often the optical microscopy that helps me decide what other testing needs to be performed and what test locations will be most informative.

- I try to name the folders so that they will end up in an order that makes sense and follows the flow of the work.
- Obviously, the folder names may change as the project moves along.
- But having a systematic method to name the files and organize the data helps avoid wasting valuable project time.

Figure 9.6 shows a screen capture from Microsoft Windows 7 File Explorer, documenting the complete set of folders that applies to a single project.

Perhaps it seems unusual to start this discussion of data evaluation with a minimalist approach to client communication. But everything discussed here is going to be part of any more detailed documentation scheme. There are already many other publications that take a linear, formulaic approach that assumes every report will resemble the one for the collapse of the New York City World Trade Towers on September 11, 2001 or the Union Carbide Bhopal disaster of December 2, 1984. Most people reading an introductory failure analysis book are not routinely doing such high-profile work. For smaller projects, one of the most important parts of the scope negotiation (this is detailed in Chapter 7) is what type of report will be required. Maybe no formal report is necessary, only a verbal communication. Verbal communication is also often the desired manner of transmitting the results for a preliminary investigation related to a lawsuit. If a potential client, who is an attorney, does not find your conclusions helpful, they usually will not want to have a written document in the file for the opposing side to see.

If all parties agree to a minimalist approach, people dealing with similar future problems will thank you for having a simplified system to ensure that the required data will be preserved. They will really thank you for ensuring that people down the road will be able to find out what testing was performed and what was learned. This is true even if the minimalist documentation approach doesn't capture all of the whys.

From the very beginning of the project, we must know what level and type of documentation will be required.

IMPORTANT TIP: Saving a series of photographs taken in a way that each image contains a feature to communicate its position within the Subject of the investigation is an efficient way to keep the relevant evidence. Usually, this means a series of photos that gradually zooms in to a feature of interest. Then, if there are other features of interest, zoom back out and start zooming in to the new area. This advice probably cannot be repeated often enough. We have already covered this ground at the end of Chapter 3, about the Visual Inspection and helping people follow photographic records of various visual inspections.

Good reports provide enough evidence to retroactively link the relevant features back to a particular macroscopic position on the assembly or component. Usually that means keeping the orientation of the view fairly constant from image to image as you zoom in to finer and finer details.

The biggest difference between good and useless reports is that the good ones leave enough evidence so that the microscopic features can be retroactively linked back to a particular macroscopic position on the assembly or component.

My metallograph provides a mirror image. This is sometimes a bit of a headache when preparing my reports. I did not want to spend the extra $7000.00 in 2016 money to have the images match my other microscopes.

Figure 9.6 documents a complete set of folders for a single project. The reader may note the first folder is named "*00 For Report*". In this folder, I include a small selection of images that capture the main points of the investigation. This will generally include an overview to show what the part looks like. If it is the scene of a large component that was part of a major industrial accident, for example, there may be a few images required to capture the overall scene and then the component. Arrows are helpful to show the component of interest in its place in the general environment.

- For a typical industrial investigation of a durability part failure, a failure of a component during assembly, or a field return with no associated consequential major property damage and no human injuries, there would probably be 1-4 images from each of the individual tests.
- If the project included optical microscopy, scanning electron microscopy, EDS microchemical analysis, micro-indentation hardness testing, tensile testing, bulk composition analysis, and optical microscopy of cross-sections, there might be a dozen images in this folder.
- Even if there is no formal report, it can be helpful to prepare this type of summary folder as a guide to what was done and what was found.

Each person will follow their own personal preferences or relevant company guidelines when writing a report. I find that starting out with the preparation of this selected photo-set helps me write any required report in an efficient manner.

- In a formal report, each of these images would have a figure number and caption and be included in the "00 For Report" folder.
- In addition, any freestanding formal reports from other labs, such as the mechanical testing and composition analysis provider(s), would be included in this "00 For Report" folder.
- For formal reports, I often include, within the report itself, a screen capture of the subfolders (as shown in Figure 9.6).
- For forensic work, I often include the number of individual files within each folder in parenthesis after each sub-folder name.

While we are on the subject of file names, it is helpful to think about the individual file names as well as the folder names very early in the project. Within each folder, I give each file a name that will force software programs to automatically display the photos in the correct order. Nominally that means in the order in which they were captured. Keeping track of the order of the image capture means it will be easier to follow the trail of the investigation. It will be easier to see which image is a detail of another image. This method is especially important when a project takes a month or a year (or more) to complete. In fact, I'd say this is critical.

It is also critical if you work in a company that performs similar work on related types of components or failures. These projects and investigative flows can be difficult to keep track of and nearly impossible to recall if you are using only your memory. One set of cleavage facets blends into another in the analyst's memory banks.

To me, this small amount of foresight as well as an almost religious adherence to the use of systematic file naming protocols is far, far preferable to having to make up slideshows (for example, PowerPoint) for each project, even if that process is automated. The system promoted here automatically makes a slide show easy to create. And even if the project's protocol requires preparation of a lengthy, detailed formal report, this file naming system will help you prepare it in an efficient and quick manner.

9.8 How to Approach Report Preparation

Let us look at the common sections of a failure analysis investigation report. Most formal reports have:

- an executive summary that briefly documents the background and usually includes consequences,
- the scope,
- the findings,
- the conclusions, and
- the results;
- recommendations may also be included if they were requested.

Photographs will generally be most helpful to convey background information and findings. Remember that a picture is worth a thousand words[184] and a story is worth a thousand data points.[185] When the format allows, it is helpful to have some photos incorporated into the introductory portions of the report.

My personal approach to completing failure analysis project documentation is explained below in eleven steps.

> **Basic Skills Allowing Us to Competently Obtain, Evaluate and Communicate Our Data so it Becomes Useful Information Include:**
>
> **Reason, logic, epistemology, knowing how-to-*look-at* and knowing what-to-*look-for*, along with the knowledge of how to use the instrumentation that facilitates our observations.**

9.8.1 First Step – List and Consider Self-Consistency of Key Points

The First Step is the time to mentally review the key points of the investigation. Pay particular attention to whether there are anomalous data points that detract from confidence in the most likely candidate scenario(s).

The best time to do this is ***many times along the project journey***. Waiting until you are writing the report to determine what happened will make a clear description of what happened more difficult and time consuming, at best. The best time to catch a problem with data is as near to the time of its creation as possible.

---

184 This old proverb is attributed to Fred R. Barnard in the 1920s. He first used the phrase to teach how using illustrations and photographic images is extremely effective in advertising.

185 This is an Aliya proverb!

One of the book reviewers suggested attaching a numerical probability estimate to each conclusion in the report. For example, if lots of data is analyzed and no anomalies are found, the conclusion could be "almost certain". Other lower values of confidence could be described as "most likely", then "likely", and lower on the scale could be "possible", then "unlikely", and finally "impossible".

I thought this suggestion had merit, but I would never personally encourage people to state they are "almost certain". Professionally, I am more comfortable using the term "highly confident". Once I told an arbitration board, "I had never found such self-consistent data sets in my entire career. Every single test I ran confirmed the previous data as well as the 'most likely' scenarios based on background data collection."

It is because of so many years in this field that my common attitude is of exercising caution when communicating extremely high confidence. This caution is based on time-and-again discovering hidden information that may or may not have surfaced. In fact, there is usually too much that remains permanently hidden in the past. It's difficult to do our job without relying on hearsay, at least to some degree.

> *Decades ago, at a conference seminar that I attended for expert witnesses, I was warned, "For forensic work, never give numerical values to confidence intervals." It was good advice.*
>
> *However, before attending this seminar, I told a client the work was essentially, "Easy peasy." And then I couldn't do it. I learned my lesson! That's why I went to the seminar!*

A better approach to take when trying to convey your calibrated confidence is by stating that:

- "Finding A is consistent with Findings B, C, and D, etcetera."
- "All of the findings are self-consistent except Finding E."
- Then you can go on to explain whether you think there is a practical way to move beyond the epistemological discrepancy.

9.8.2 Second Step – In Case of No Anomalies; Explanations Build Confidence

STEP 2 occurs when there are no anomalies and there is a likely, confidence-inspiring explanation available for the string of events within the scenario you've been given to investigate. STEP 2 occurs when the findings from the investigation are consistent with knowledge about the timeframe from design through manufacturing through assembly and/or service. STEP 2 includes that confidence-inspiring explanation for the damage/failure event. If this is the case, then start selecting photos, images, and other key data to highlight your chain of thought. If there is an analytical team, determine the best way to communicate the joint or various explanations.

9.8.3 Third Step – Selection of Key Data to be Presented When Anomalies Exist

STEP 3 occurs when there **are** anomalies or missing data. Any anomalous findings must be addressed (or at least acknowledged) in the project record. Only then may the report preparation process move forward with the selection of key data to be presented.

Until that happy moment arrives, you will be working within an activity loop that returns to STEP 1.

9.8.4 Fourth Step – Begin Writing

During STEP 4 you begin writing. The opening section of the document is often a narrative covering the key background points and the description of the damage/failure event.

Early acknowledgment of the scope of the project is paramount in the document. This protects the integrity of the project team. Someone who is knowledgeable about the type of situation will be able to formulate an opinion as to whether the resources allocated to the project were adequate.

> *Describing the scope of the project is necessary self-protection for the investigative team. If you didn't evaluate xyz, because xyz was not funded as part of the project, and the technical leads had requested such funding, it's important to note that fact. You don't have to blame anyone. "XYZ was beyond*

the scope of the project" is a tactful and truthful way to convey the fact.

REMINDER: Converse and Compose Clearly. The point here is to remember that word choice and sentence structure will be different in verbal and written communications.

- People naturally speak in incomplete sentences and that is acceptable and desirable in verbal communication.
- However, incomplete sentences are generally frowned upon during written communications. The exception to this, perhaps, is to emphasize a point. Do it only occasionally.
- In many cultures, people naturally interrupt each other during conversation. However, in business settings, people will generally speak more sequentially to avoid chaos and enhance understanding.

Generally, if you determine how to present the project in writing, it will be easy to present a spoken version (unless you have fear of public speaking). As you become more competent, you will naturally be asked to present your findings verbally. If you do have discomfort associated with public speaking, as is said to be the case for most humans, you may wish to join a Toastmasters Club[186] or another program teaching or facilitating skill development in verbal communications.

9.8.5 Fifth Step – Present Key Findings

During STEP 5 you present a bullet point list of key findings. This is the next part of the document. In a slideshow type document (i.e., PowerPoint), the relevant key photo or tabulated selection from another report may accompany each finding on its own page. An example of this is the bulk composition analysis.

In other, more formal, types of documentation, the images and tables may follow in the results section.

HERE IS AN EXAMPLE. If a hardened steel part has been found to have hydrogen embrittlement, a macro-photo of the crack surface and an SEM image documenting the classical intergranular micro-scale crack features would be appropriate as well as their position with respect to the overview. This would give the reader an immediate clue that the writer is knowledgeable and that the conclusions are based on fact. Hardness data would also be appropriate here since that is a key contributor to hydrogen embrittlement in steels.

9.8.6 Sixth Step – Bulleted Conclusions

Conclusions may be effectively presented as a bullet list. STEP 6 is the place in the communication to address any specific questions that were asked at the project's initiation.

When conclusions follow the findings, they have a higher chance of flowing in a manner resembling articulation of self-evident truth, rather than an arbitrary revelation from a Magic Eight Ball®.[187] *A very scientific colleague just informed me that he has an app on his phone that simulates the old Eight Ball toy. I am sure he uses it responsibly.*

When relevant data is presented appropriately, we appear to be expert, trustworthy, and professional. Clients do not feel we are asking them to have blind faith in our conclusions. Well-presented information also allows stakeholders to follow along with our thought processes. Some people might not understand every detail that we considered, accepted or rejected in the same way a technical specialist would understand it; but if they can follow our thinking process and it makes

[186] Go to the following URL for more information: https://www.toastmasters.org

[187] A Magic Eight Ball is a toy sold by Mattel, Inc. The original Magic 8 Ball toy was made to look like its billiard namesake, but larger. It will return a random answer to a yes or no question. It is marketed as an entertainment device.

sense to them, they will more readily trust our findings.

9.8.7 Seventh Step – Present Recommendations

Prior to STEP 7 we have laid the factual groundwork. Now, if recommendations were requested as part of the project scope, recommendations may be presented. This is the Seventh Step - to provide meaning and put the recommendations in a context; conveniently place justification for your recommendations. But don't overcrowd them. Let the recommendations stand out clearly.

9.8.8 Eight Step – Results

Results are usually the last major section of the document. This is STEP 8. If I am writing a formal report for a major failure, this section will generally have the relevant details of each test in its own sub-section.

For those who have followed along as you collected data, this may appear repetition. However, as more data crowds the project folder, you can provide different ways to display it. You may highlight the epistemological accord (or discord) with other data.

The Results section is often the best place to discuss any issues related to accuracy or precision of numerical values. It is often the place to present related details of conformance to specifications.

Discussion of the significance of the data may be included in the Results section as well.

> *For example, does this set of hardness data strengthen the theory that the failure was due to hydrogen embrittlement? Or does it weaken it?*

Some companies include this in a separate discussion section, but I think it is helpful to have it when the data are presented. Otherwise, the data are just data. A knowledgeable person discussing the data helps transform it into information. Information is easier to remember than raw data. Information from one segment of the Results section will naturally help transform the next segment from data to information as well. Discussing the data in individual sections lets less-knowledgeable people learn by providing "instruction" within the context of facts directly in front of their eyes.

If two possible explanations exist for one set of data, bringing both forward in the Results section will facilitate a deeper understanding of the analytical process that led to the conclusions stated earlier in the report.

If some work was performed at another lab, this section provides the name and location of the outside service.

9.8.9 Ninth Step – Discussion

Sometimes the discussion section is included in the Results section. If it is not included there, STEP 9 is a good place to put it.

Sometimes additional discussion of a particular finding or conclusion is warranted, even if most of the discussion is included with the Results.

If there has been a lively debate over the course of the investigation about a particular data point or failure theory, it might be helpful to acknowledge it here in the report's discussion section. This may smooth the acceptance on the part of the person or party whose theory was "shot down".

9.8.10 Tenth Step – Appendix

STEP 10 includes the entire collection of file folders and forms the appendix. This was highlighted in Figure 9.6.

9.8.11 Eleventh Step – Cover Letter for Additional Issues

Once all of these segments of the report are complete, there may be a reason to prepare a cover letter. In STEP 11, you can address additional issues of interest to management that are confidential. There may be aspects of the investigation that management may not want to share with other parties. All parties may have a right to the investigation data and conclusions drawn by the investigators and that is all included in the other steps. This section is for management's eyes only.

9.8.12 The Entire Project Report Communication Process

Figure 9.7 (below) shows a flow chart of the project communication process. The arrows on the individual sub-tasks all originate from the same major task step (each sand colored oval has its own arrow). This indicates that these individual sections may be written in any convenient order.

Likewise, the individual result sections may be created in any convenient order and then edited to flow together into one unified section when the report is being completed.

Fiction writers call this writing technique "helicoptering". You "fly in" to a given scene, write it, and then "fly out", stitching the pieces into a unit at the end, or at earlier convenient times.

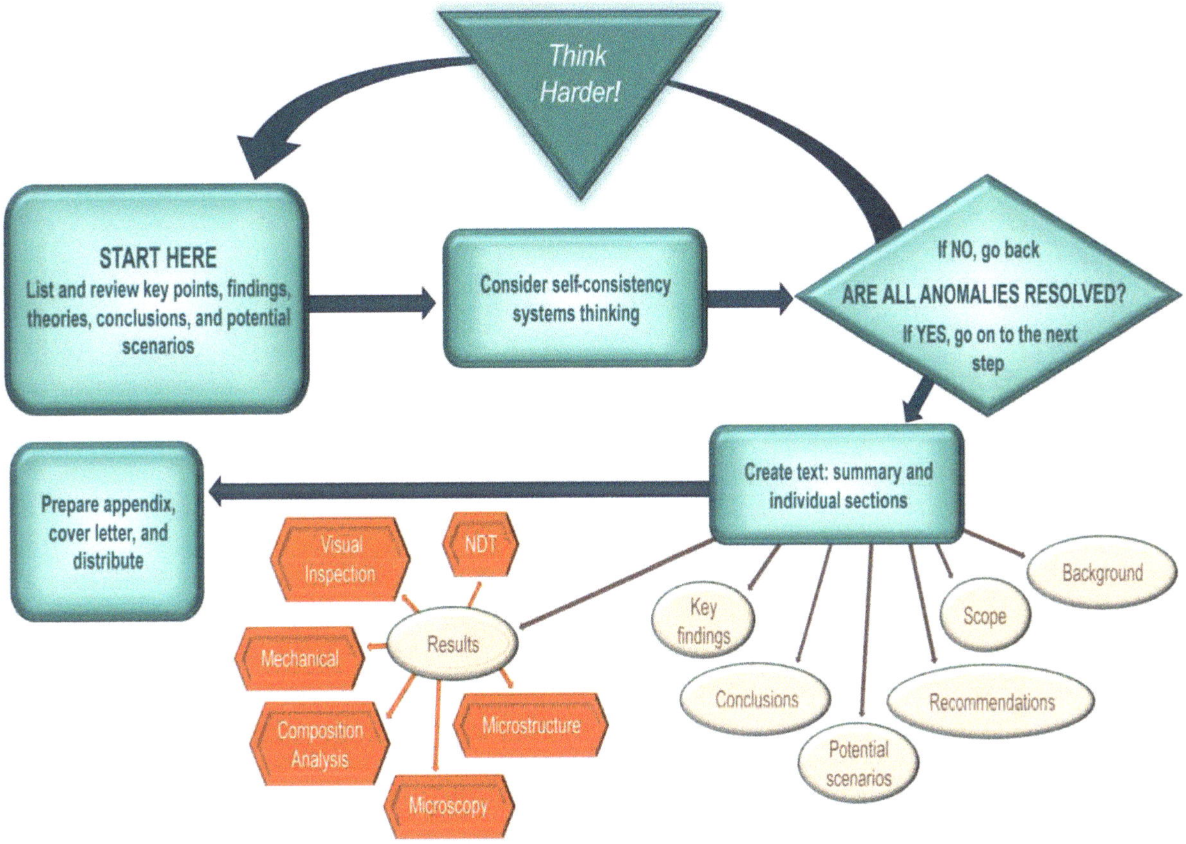

Figure 9.7: *Flow Chart applicable to many types of report formats for failure investigations*

9.9 Results

In Chapter 6, we reviewed the details of how selecting specimens for failure analysis differs from other metallurgical work. As we work to complete our communication aids, we have one more chance to contemplate the challenges associated with interpretation of the results of what some would consider simple lab work.

Evaluation of composition and mechanical test data obtained in support of failure analysis work is inherently different from the testing performed for quality assurance or basic materials science research. Intermediate between the two, we might consider composition and mechanical testing for product development.

What are the factors that make failure analysis related data interpretation more challenging than standard quality assurance work? Let us explore the characteristics that are most likely to change as a result of manufacturing processes and service conditions: composition, mechanical properties and microstructure. Our focus will be on these topics not yet covered as communication topics in this chapter. Note the orange hexagons in Figure 9.7. Visual inspection, NDT and Microscopy were covered in previous chapters.

9.9.1 Composition Analysis Data Evaluation Concerns

Alloy determination of metallic materials is usually confined to true chemical analysis techniques, especially for the American steel standards developed by the American Iron and Steel Institute. European and Asian steel specifications are more likely to have integrated composition and mechanical property requirements. Aluminum standards generally have a two-part naming system in the USA that separately calls out the composition and then a suffix that specifies the process condition (see Chapter 4) and associated mechanical properties. Many design engineers do not realize that they must specify the process condition suffix along with the alloy designation. Overlooking this requirement can be a cause for failure. The American Society for Testing and Materials has a lot of product form specifications that call out both composition and mechanical (and sometimes other) performance characteristics. ASTM and other trade groups also have detailed requirements for dimensional consistency.

Material identification of polymers is much more complicated, and requires methods that are truly looking at both chemical characteristics (molecular bonds), and what are essentially physical (phase change or temperature related) or mechanical property measurements, often just to identify the basic type of resin.

For both families of materials, deviations from specified ranges may be a result of improper processing (viscosity or Melt Flow Index for polymers, and surface carbon on steel parts, for example) or someone using the wrong material to make the part in the first place. When trying to identify polymer resins, the presence of small amounts of contamination in polymers of the same color and general family (nylon, or polypropylene) may be very difficult to demonstrate with FTIR alone. Differential Scanning Calorimetry (DSC) may be more likely in some cases to find trace contaminants, especially in semi-crystalline polymers. See Chapter 6 for information about FTIR and DSC.

Deviations from specification may also be a result of changes during service, particularly corrosion or environmental damage. It is important to attempt to distinguish any deviations that are a natural result of service in harsh conditions from those resulting from improper manufacturing. If recurrence prevention is a key goal of the investigation, we need to be able to determine if the part was sub-par at the time it came into being, or whether it degraded over time. This may be difficult in certain circumstances.

Both mechanical properties and composition are subject to degradation or change during service, as we discussed in the Damage Categorization section of this book. We will discuss this again soon, as it applies to data interpretation, when we get to mechanical properties.

Exposure to UV light can break down some polymeric materials. This might result in an undesirably high Melt Flow Index. Though I am unaware of any published data on this, another potential source of undesirably high Melt Flow Index, due to molecular breakdown, is the etching process used to prepare ABS resins for electroplating. Presumably the etching is only removing the little styrene spheroids, but it's possible that other damage could happen.

Be sure to communicate any concerns regarding composition analysis data and its evaluation.

9.9.2 Mechanical Property Evaluation

In the section on Specimen Selection in Chapter 6, we discussed issues of concern in metallic components, related to variation in properties that result from differential heating and cooling rates, strain gradients from cold forming operations, and changes due to material that has been machined away. We also discussed the issue of separate test bars being used to certify strength of cast, forged, and heat-treated materials. When we are evaluating numerical data, once again we must take care. Test coupons cut from parts as manufactured have all of the issues we just listed. But parts that have served in harsh environments have additional changes that are expected to accompany the conditions they were made for. The Results section is a good place to remind the stakeholders of *the difficulties inherent in determining whether a used component was made to its specifications.* One more issue that surfaces is relevant when an exact manufacture date is not available. If there were multiple versions of a particular design, it can be challenging to determine which engineering print applied at the time of manufacture.

EXERCISE 9-B: Compare-Contrast - test results, specification requirements, and certification documentation.

If you are working on a failure analysis project and have obtained composition or mechanical property data, then compare-contrast the specification requirements, the certification documentation (if available), and the test results from the Subject of the investigation, as well as any Reference Specimens. It's helpful to prepare a table to show the values side-by-side. This may be a useful document to include in the report or in the appendix.

EXERCISE 9-C: Connect to Create – review the tabulated data.

Review the data that you just tabulated. Consider the size, shape and extraction position from within the part for the two or three data sets you compared.

- Do the variations make sense?
- Is the strength and elongation lower for the Subject Part than the specifications require?
- How do these values compare with a Reference Part that may have been tested alongside the Subject?
- Is the strength higher and the elongation lower due to cold working in the part, compared to the data for flat sheet metal?
- Are both strength and elongation lower for plastic parts as a result of expected resin degradation that happens during normal processing?

With plastic materials, the issues related to extracting specimens from parts can be even more challenging. There are strain gradients, especially with injection molded parts. Extrusions have more uniform elongation of the molecules than individually molded components.

9.9.3 Microstructure

If your failure analysis project included microstructure work, you may be comparing the specimen from the Subject of the investigation to images from the literature, trying to find examples of the same (or similar) grade of material, processed in the same (or similar) manner. Alternatively, if the scope negotiations allowed the inclusion of Reference Specimens – especially if you have known good parts – then your work will be more straightforward.

It's likely when comparing microstructures that we will have a difficult time deciding whether visible differences are significant. Of course, if there is a quantitative specification for grain size or decarburization depth or another specific feature, it would not hurt to quantify that feature.

However, as we have previously noted, the part doesn't know the specification. Many specifications are arbitrary to some degree. It is rare that a slightly-out-of-specification condition is the entire cause of a failure.

So, what should you put into the microstructure section of your report? It is helpful to pair images of the Subject and Reference Part if they were obtained:

- in the same condition (as-polished or etched),
- at the same magnification, and

- at the same position within the component.

It is often helpful to include a pair of images of the surface layers near the damage area. For larger parts, it is also helpful to include another pair of images in the core. This helps the reader note what looks different. Most humans are comfortable acknowledging the results of the expert's basic pattern recognition activities. This is true even if the person has little technical background and won't be able to grasp the significance of the differences without your help.

9.10 Development of the Main Narrative of the Report

This chapter started with a very cursory discussion of the report preparation. Let us go back to Figure 9.7 (the flow chart) and take a more detailed look at preparing the primary sections. We continue along our path of avoiding general admonishments without also providing detailed examples of every step of the data evaluation and communication process.

Usually, a formal report will include an introduction, which includes a background paragraph. Why was the project undertaken? What happened? Were there consequential or cascading damages in addition to the first "failure"?

Next, outline the agreed-upon scope of work. In projects with ethical issues, it will be necessary to state what was recommended but not performed. Tactfully indicate the agreed-to-tasks. The scope will include a brief list of what tests were performed and which specimens were evaluated. It might also include a description of documents that were reviewed.

After the scope description, I usually transition to a numbered list of Findings. In Figure 9.7, the flow chart calls it, "Key Findings". Here is an example. "The Subject Specimen had enlarged grains, which are often associated with high temperature exposure in service." All key findings include a few words about why the reader should care. Remember, not all stakeholders are technically inclined.

Below are some more examples:

KEY FINDING EXAMPLE A: *The Subject leaf spring was found to have a thick layer of complete decarburization, which is often associated with a significantly reduced fatigue life. The Reference Part from last year's production did not have a damaging decarburized layer.*

KEY FINDING EXAMPLE B: *The Subject nylon injection molded part was found to have a knit line at a location shown by a finite element analysis to have a high stress during installation. Components with inauspiciously located knit line discontinuities will display lower strength than sound parts.*

KEY FINDING EXAMPLE C: *The analyst who performed a replication microstructure analysis on a high-pressure steam pipe fifteen years ago did not recognize that the microstructure degradation was already advanced.*

KEY FINDING EXAMPLE D: *The stainless steel alloy used to make the collapsed frame was found to have been consistent with the specified alloy. A review of the material specification showed no strength requirements.*

KEY FINDING EXAMPLE E: *The weld throat on the Subject Part was found to be 10% larger than the specified code minimum requirement. This is desirable.*

All of these findings set the stage for *potential recommendations.* Despite understanding the value of establishing conclusions before formulating recommendations, sometimes a Failure Analyst cannot resist the lure of jumping to a solution. This methodology gives you the skills to renounce this urge. For example, in the case of a structural collapse due to a fatigue crack, the absence of a strength requirement provides an opportunity for recommending a different grade of material that has strength requirements. It also opens the door to suggesting that a civil or mechanical engineer perform an analysis to determine what strength level is required.

EXERCISE 9-D: If you have the technical background required, take some time now to suggest relevant, actionable recommendations for each of the unrelated Key Findings in Examples A-E.

Conclusions might cover such issues as whether any relevant codes were adhered to, or not. Conclusions might include a diagnosis of a complex damage mechanism. Conclusions often attempt to identify timing of the damage that led to the failure. Although basic factual findings may allow recommendations, conclusions allow assignment of blame. As I contemplate the above, I realize that there is no hard boundary between a factual finding and a conclusion. This paragraph is an invitation to contemplate dichotomous and continuous variables. Be sure to consider their relationship to gleaning insights.

FOOD FOR THOUGHT: Conclusions are consequences of the summation of the findings.

CONCLUSIONS EXAMPLE A: *My clients in the metal stamping and deep drawing industries often want to know if a discontinuity was present in the raw material; and if not, was it present when the part was making its way through the stations of the stamping press line. Or, did the discontinuity happen as a result of heat-treating? Or later, during installation into a larger assembly?*

Identifying the time frame of the evolution of a discontinuity (whether during raw material production, later manufacturing processes, normal service, or maintenance) is often a key conclusion. This key conclusion provides direct insight into whose "fault" the problem is.

CONCLUSIONS EXAMPLE B: *A knit line in an injection molded thermoplastic part is usually either the "fault" of the tool designer or the injection molder. Running the process at the wrong temperature can cause even a properly designed tool to create flow discontinuities.*

If there are multiple broken parts within an assembly, determining which one set the train of events in motion is a *key task*. Usually, evidence of fatigue in one component identifies the earliest damage. See Figures 9.8A-C for examples of how to determine the sequence of multiple fractures in a complicated assembly.

.

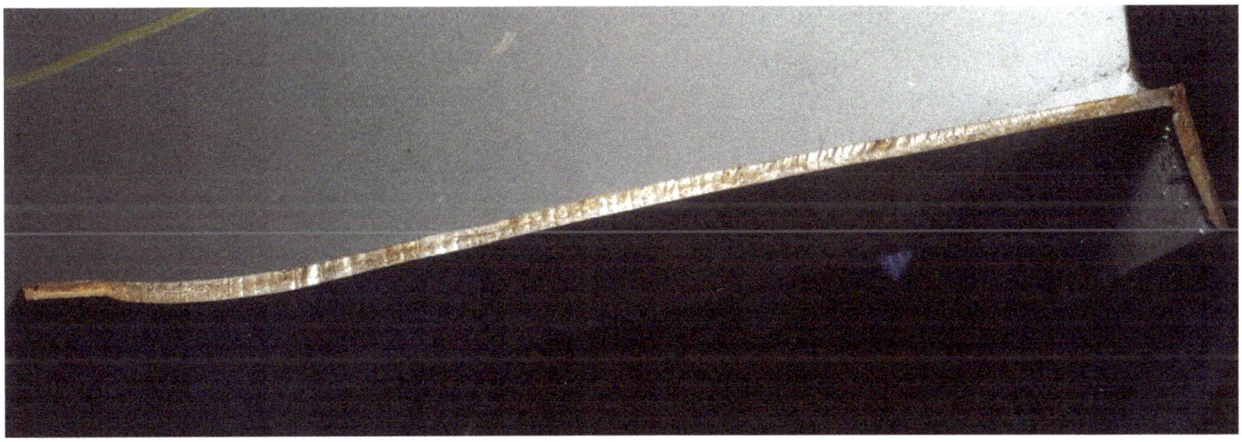

Figure 9.8A: *This sheet metal weldment has obvious beach marks along most of its length. There is a shear lip at the far left. The crack started on the right side.*

Figures 9.8B and 9.8C (below) show two other components that appear to be simple overload (monotonic) fractures. Likely they happened after the final separation at the shear lip shown in 9.8A.

Figure 9.8B: *This cast iron pillow block held a bearing assembly. It has two chips missing. This is consequential damage from the broken sheet metal component of 9.8A finally having torn itself apart. The broken sheet metal component was a subassembly on rotating equipment.*

We are just now arriving at the part of the report that relates to revealing the potential scenarios that could have resulted in the failure or damage event.

Figure 9.8C: *Another (partially rusted) monotonic crack surface in another area of the same grey iron casting (bottom edge of photo) is shown in 9.8B. Also note the visibly stretched fastener threads (red arrows). These are both most likely consequential damage resulting from the fatigue crack in 9.8A (previous page).*

While the flow chart for report preparation includes scenario development as the first step, all of the other steps we have just described, including analysis of composition, determination of mechanical properties and microstructure, are more straightforward than speculating on a damage sequence or scenario, which requires a higher level of cognitive skills. This is the essence of the communication task, which describes what happened.

Coming up with a potential scenario that is consistent with the data requires a strong familiarity with the data.

In some cases, the scenario will be mentioned earlier in the formal report, but I think it is more effective to state the clear basic facts before proposing a complicated scenario. In this way, the reader will be able to follow the logic, rather than being asked to have faith.

Once you are an experienced Failure Analyst and are involved in a straightforward investigation, it's fine to write all of the primary sections (background, scope, findings, conclusions, recommendations, and scenario) prior to writing the results sections. However, beginner or Failure Analysts dealing with very complex projects do well to write the results sections first. This is especially true when the project's interrelations between data sets are not obvious. This allows the writer to get to know the data better and have time for the information to percolate on the brain's back burner (involve the subconscious creative processes). When I let this happen, I find connections and anomalies that might not have been obvious to me earlier.

9.11 Assigning Fault or Blame

In the work I do, it is rare that I am asked to assign fault or blame. I do some work for insurance companies, in particular for damage to pleasure boats. Again, my task is to determine a likely and self-consistent damage scenario or sequence. The insurance company uses my data and explanation of the data to determine fault. The same is true when I perform litigation work. I determine a likely and self-consistent damage scenario or sequence and the courts use the data to assign blame. That is fine with me.

When I first starting doing the pleasure boat work, I thought it was amusing that when I determined someone made a mistake (even a big mistake), the insurance company often paid the claim. If the boat was not maintained properly, the insurance company generally did not pay. I had a sad case of a rubber boot that covered a cable. The maintenance team at the marina thought it looked ok and did not replace the $3^{\underline{00}}$ part during a big overhaul. Sometime later, there was a leak and significant damage occurred to the boat. The boot was involved. The damage cost hundreds of dollars. See Figures 9.9A-C.

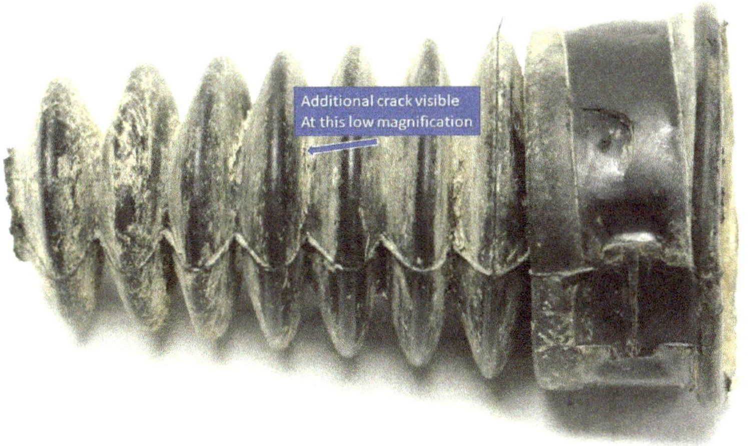

***Figure 9.9A**: Overview of rubber boot*

The tip at left is broken off. You can see an additional crack in a similar accordion root position if you zoom into the point of the blue arrow.

***Figure 9.9B**: The main crack is at 45 degrees to the "X" legs. Additional cracks form at least three of the four legs of an "X" pattern. A fragment of rubber has been "liberated" from a position below the far right arrow. This indicates the potential of reversed torsional loading. It is clear that this rubber is completely embrittled.*

Figure 9.9C: *Note the broken cable boot and the newly purchased cable boot used as a reference. The new cable boot has a smooth and relatively reflective surface finish. The broken one looks rough and dull.*

Which would you prefer? Even if the tip was not missing, which would you prefer? As a customer, if I knew nothing about cable boots for pleasure boats and saw both of these products on a shelf being offered for purchase, I'd take the one labeled "New" in this image.

The owner of the boat was not a rich man. He had to pay for repairs out of his own pocket. There are a lot of people who love boats, being on and around the water. That is a part of why such a large percentage of humans live along coastlines. Emotional preferences are acting in all aspects of our lives.

9.12 Defending a Report

Despite all of our thoughtful rational, creative and systematic attempts to help others understand what happened to cause a failure, the need to assign responsibility (if not outright blame) means that, at times, our work will invite criticism. Numerous colleagues over the years have stopped doing this work because they found it too stressful. I continue to enjoy the work overall, although there have been times in my career when I turned down individual jobs that I thought would interfere with my long-term peace of mind.

Even if we don't find that a particular party "did wrong", the folks who are "under suspicion" during the investigation may hold our suspicion against us. We are required to consider all possibilities and some people resent that type of professionalism.

> EXAMPLE ONE: *Years ago, the head engineer for a fan company told me "Nothing personal, but I hope I never have to talk to another metallurgist for the rest of my life!"*
>
> *Well, how could I respond, other than with silence? I did not bother to inform him that maybe if he talked to a metallurgist while he selected the material for his equipment, he would not have to have the painful conversation later.*
>
> *In this particular example, nobody was injured. There were product manufacturing losses, however. I was not informed about the cost of the downtime that the equipment manufacturer may have been asked to reimburse.*
>
> EXAMPLE TWO: *One of my manufacturing clients had a "deal" with a customer. They would pay for an independent analysis if there were more than a predetermined number of warranty failures of a specific sub-assembly. If I determined it was the manufacturer's fault, they would pay their customer an amount based on a formula. If I determined that the parts were being made properly, but their customer was causing the damage (perhaps by improper assembly), then their customer had to cover the cost of the damage.*

The point that I am trying to make here is perhaps obvious. There will always be someone who is threatened by having to take responsibility for their mistake. They may (or may not) try to "poke holes" in the beautiful theory that you, the Failure Analyst, have created. In fact, we might as well get used to assuming that people will try to poke holes in our theories. That is why it is so important to do our best to ensure that we obtain credible data, and explain it clearly.

> *I learned this lesson early in my career, during a short stint in the quality department at a manufacturing company. I was responsible for assigning blame for mistakes made during a certain step of the process. Soon, I realized that whoever I pointed at, someone was going to be unhappy.*

Sometimes it may not be obvious to us just who will be unhappy with our findings. There are many layers of hidden suppliers, sub-suppliers, contractors, etcetera. Many people and companies may have performed some aspect of the manufacturing or maintenance on the Subject of our investigation. So be prepared.

9.13 Report Preparation Approach Wrap-Up

The basic approach outlined here is intended to be flexible. Since I almost always provide my clients with all of the data, sometimes the formal report will be a short (one or two page) text-only document (economy-scope type project). If it is to be legitimately considered a failure analysis, as opposed to damage characterization, or "putzing around in the lab", it must include at least a basic discussion of the self-consistency of the different data sets. Key findings and conclusions must also be included.

A report for a large investigation of a major disaster would potentially include additional sections relevant to the event. Long investigations of big disasters generally require agreement from all affected parties before any intentional evidence destruction is permitted. Getting agreement may be a long, drawn-out process. Then, more documentation of every detail of every step of the investigation is required.

Plan your time well because documentation is time consuming. Photographing a component to show the location where the tensile test coupon will be removed takes time. Photographing the machining process takes time. Was a dry band saw used, or a wet saw with coolant? If someone wants to be able to determine whether the deposit on the component was associated with the service life or the failure analysis itself, photos of the machining in process will allow that to be determined.

I place most of these failure analysis process images in the Appendix, not the formal body of the report.

THE FINAL WORDS

Dear Reader,

Thank you for sticking with me. If you skipped some parts of the book, I hope you will come back to review them later when you have more experience in this field.

We've taken a long journey together. We questioned how we see, how we know, and how we reach conclusions. We've reviewed the significance of every major step of a failure analysis. We also explored how to select individual specimens for each test. We've worked together to learn how to negotiate a project and write reports. We've traversed many suggestions for explaining what you do for people who lack a strong technical background but must understand your work.

***

This is not the book I wrote. Much about how the book was edited will make it better for most people. What I feel has been lost is the friendly tone of sitting around a table at lunch during a technical conference. The tone is now more that of a lecturer. Maybe compared to all other textbooks, this book still seems like a chat between friends. Nevertheless, something has been gained and something lost.

While both my colleagues in my writing group and the publisher insisted that the authoritative tone was necessary in order for people to respect the content, the friendly tone I had used to write the manuscript was intended to show my own struggles with self-confidence and how I gradually overcame them. The methods described in this book are HOW I gained the confidence that I now have. I feel (yes, I feel!) that my personal journey of confidence improvement is not depicted in this final version of the book in the way that I had intended. My depiction of my journey from total worrywart to being self-confident enough to write this book was not explicit, and it has been lost to some degree. So, I want to explicitly take this moment to call everyone's attention to the work of some social scientists who have studied the relationship between self-confidence and competence. There is an article in the Annals of Medicine that looks at this issue related to nurses leaving the field.[188] It saddens me when people I think are competent stop doing failure analysis or other engineering work because they are too worried about making a mistake. There probably are not enough failure analysts for anyone to bother with such a study of us! The bottom line in many studies is that often it is the more competent people who lack confidence. I wrote this book because I hope to help others overcome their similar confidence problems in a legitimate and useful way. Overall, after almost 38 years of doing failure analysis, I still think it is a good way to earn a living.

Ok. On to the next topic!

***

I have provided, in several places, a new way for materials and mechanical engineers to talk about mechanical behavior.

Unlike my mentor in the field, Dr. William T. Becker, I had the chance to write my failure analysis book. Bill left the community his major contribution via the *Metals Handbook, Volume 11, 2002* article on Fracture Appearance, as well as his editing work for the entire volume. In the Fracture Appearance article, he explained his dual track method of classifying ductile and brittle fracture. But as far as I know, he never came out and explained to anyone why it was so significant.

So, I will not let this moment pass – in honor of Dr. Becker and the time I took to grok[189] his breakthrough. If there's a single hard-technical legacy of this book, I want experts, technical leaders and mechanical engineers evaluating

---

188 https://www.ncbi.nlm.nih.gov/pmc/articles/PMC8843132/

189 Grok is a word used in the book, *Stranger in a Strange Land* by Robert A. Heinlein, 1961, G. P. Putnam's Sons | ISBN : 978-0441790340 | 438 pages | Grok means to fully comprehend something in its entirety.

broken structures to adopt the method, explained in Chapter 8, of using fracture appearance to determine the actual loading geometry.

I also encourage you to adopt and leverage my 7-Step Method for Constructing Competence – a useful technique that will serve you and others in finding important verifiable facts in your work, as well as the limits of human knowledge. Now is the time to commit the Seventeen Cs to memory.

CONSTRUCTING COMPETENCE

Cultivate Curiosity

Confirm Context

Compare-Contrast

Connect-Create

Contemplate Complexity

Calibrate Confidence

Converse & Compose Clearly

If you work as a failure analyst, you will need many other references and resources; but this book will lay a foundation for constructing competence in failure analysis - and any other intellectual activity you pursue.

> *My curiosity about identifying mushrooms has led me to notice more details in everything I see. Cultivating curiosity and developing individuality are perhaps the two characteristics that make us most human.*

Genius has been shown to be only very weakly linked to IQ. But it is linked to curiosity. If you follow a path to understand something that nobody else has understood, perhaps even that nobody else CAN ever truly understand, you will gain untold and untellable riches. The seven steps to constructing your competence described in this book will serve you well as you practice them. My commitment to developing and practicing the method has given me insights that have let me feel like I am leaving a valuable legacy. The sooner you start your journey with these techniques, the sooner they will bear fruit for you.

Failure analysis is a great way to challenge yourself to keep learning. Just yesterday morning, I had a new insight about the reason that it can be so hard to distinguish between causes and effects. This is a big part of Chapter 10 of David Levy's book that has been referred to here many times. But even if it is self-evident in a way, gaining a new handle on a new meta-thought is exciting to me. In this case, I heard for the xth time that XYZ was a SYMPTOM of ABC, not its cause. And I suddenly realized that when something happens where we don't see the cause until after we look for it "because" the symptoms are undeniable, it does make it tough to see the primary path by which we generally recognize cause and effect. The primary path is timing. In the mundane world, something has to happen first if it is even potentially be considered the cause of something else. Think of lightning and thunder. If we see the effect (symptom) first, most of us will classify it as the cause!

***

The world is a crazy place. We need competent people who are willing to do the work to improve their competence, their clarity, and courage. Humans need this; all that we can get. The more competent we are, hopefully, the less need we will have to express our greed and selfishness. Life is hard. Thinking clearly might not automatically make it better, but I have found that I do have more peace about all of my personal failures as a result of using these techniques, and I have extracted satisfaction in helping others understand their materials and parts behaviors – essentially seeing reality more for what it is than what I or my client wants it to be. Taking well-considered risks and leveraging mistakes are the best ways to learn.

Good luck, skill, and practice in constructing your competence!

Debbie Aliya

February 20, 2024

LIST OF FIGURES

LIST OF TABLES

INDEX

A

B

C

E

G

H

I

K

N

O

Q

R

S

T

U

V

W

X

Y

Z

www.ingramcontent.com/pod-product-compliance
Ingram Content Group UK Ltd.
Pitfield, Milton Keynes, MK11 3LW, UK
UKHW060225061026
15625UKWH00008B/499

9 781938 28233